Microstructure of Dairy Products

Microstructure of Dairy Products

First Edition

Edited by

Mamdouh Mahmoud Abdel-Rahman El-Bakry
Universitat Autònoma of Barcelona
Barcelona, Spain

Antoni Sanchez
Universitat Autònoma of Barcelona
Barcelona, Spain

Bhavbhuti M. Mehta
Anand Agricultural University
Gujarat, India

WILEY Blackwell

This edition first published 2018

Registered Office(s)
John Wiley & Sons, Inc., 111 River Street, Hoboken, NJ 07030, USA
John Wiley & Sons Ltd, The Atrium, Southern Gate, Chichester, West Sussex, PO19 8SQ, UK

Editorial Office
The Atrium, Southern Gate, Chichester, West Sussex, PO19 8SQ, UK

For details of our global editorial offices, customer services, and more information about Wiley products visit us at www.wiley.com.

Wiley also publishes its books in a variety of electronic formats and by print-on-demand. Some content that appears in standard print versions of this book may not be available in other formats.

Library of Congress Cataloging-in-Publication Data has been applied for:
9781118964224

Cover Design: Wiley
Cover Image: © Biophoto Associates/Science Source Images

Set in 10/12pt WarnockPro by SPi Global, Chennai, India
Printed in Singapore by C.O.S. Printers Pte Ltd
10 9 8 7 6 5 4 3 2 1

Contents

List of Contributors

M. Auty
Food Chemistry & Technology Department
Teagasc Food Research Centre
Teagasc, Moorepark
Co. Cork, Ireland

J. Chandrapala
School of Science
RMIT University
Bundoora, Victoria
Australia

R.W. Hartel
University of Wisconsin-Madison
Madison, WI, USA

J. Hazekamp
Unilever R&D
Colworth Science Park
Sharnbrook, Bedfordshire
United Kingdom

I. Hernando
Departamento de Tecnología de Alimentos
Universitat Politècnica de València
Valencia, Spain

G. Impoco
CoRFiLaC
Ragusa, Italy

E. Llorca
Departamento de Tecnología de Alimentos
Universitat Politècnica de València
Valencia, Spain

A.G. Marangoni
Food Science Building
University of Guelph
Guelph, Ont. Canada

Bhavbhuti M. Mehta
Dairy Chemistry Department
SMC College of Dairy Science
Anand Agricultural University
Anand, Gujarat, India

M.N. Oliveira
Department of Biochemical and Pharmaceutical Technology
University of São Paulo, Brazil

S. Otles
Department of Food Engineering
Faculty of Engineering
Ege University
Izmir, Turkey

V. Ozyurt
Graduate School of Natural and Applied Sciences
Food Engineering Branch
Ege University
Izmir, Turkey

P.H.P. Prasanna
Department of Animal & Food Sciences
Faculty of Agriculture
Rajarata University of Sri Lanka
Anuradhapura, Sri Lanka

A. Quiles
Departamento de Tecnología de Alimentos
Universitat Politècnica de València
Valencia, Spain

P.R. Ramel
Food Science Building
University of Guelph
Guelph, Ont. Canada

C.S. Ranadheera
Advanced Food Systems Research Unit
College of Health & Biomedicine
Victoria University
Werribee Campus
Werribee, Victoria, Australia

I.T. Smykov
All-Russian Scientific Research Institute for Butter and Cheese Making
Uglich, Russia

M.H. Tunick
Center for Food & Hospitality Management
Drexel University
Philadelphia
PA, USA

S.R. VanWees
University of Wisconsin-Madison
Madison, WI, USA

J.K. Vidanarachchi
Department of Animal Science
Faculty of Agriculture
University of Peradeniya
Peradeniya, Sri Lanka

Preface

The idea of the edited book *Microstructure of Dairy Products* has evolved from the fact that it is difficult to find recent books related to and focusing on the microstructure analyses of dairy products in addition to their detailed microstructural aspects. Microstructure has been studied for several decades; however, the few recent specialized books that discuss the microstructure of food matrices have no focus on the microstructure of dairy products. *Microstructure of Dairy Products* is considered as a timely multi-author text, with contributors from the USA, Europe, Canada, Australia, Brazil, India and other parts of the world.

Microstructure significantly affects all end-product processing characteristics. These are functional properties, which include mainly textural and rheological parameters, and flavor or sensory properties. Therefore, it is essential to understand fully the microstructure of these food products. *Microstructure of Dairy Products* provides a full description and updates of the most recent developments of microscopy techniques used and analyses of the microstructure of dairy products. The book may help in areas related to microstructural analyses, and to studies on the relationship between microstructure, functionality and flavor. Currently, there is a need for systematic microstructure studies of most dairy products, in order to provide an insight into the understanding of the different phenomena that relate to functional and biochemical changes during processing and subsequent storage.

The topics in *Microstructure of Dairy Products* are of great interest to everyone involved in the manufacture of dairy products, through to dairy consultants and scientists who are involved in product development and troubleshooting. The topic has been extensively researched, and the result of this widespread interest is that many articles on the structure of different dairy products have been published in scientific journals, targeted at very specific groups of scientists. This book enables an easy knowledge transfer of a comprehensive and global overview to the reader. For the sake of clarity, there are various chapters dealing separately with different dairy products. For example, a researcher might be working/interested only in dairy fat products rather than fermented milks.

The book consists of 14 chapters, spanning about 400 pages. There are two main parts:

1) Overview of microscopy techniques used, where different types of the following techniques are discussed in detail: light microscopy, electron microscopy and emerging microscopy techniques (Chapters 1–4). This is followed by a chapter on the quantitative analyses of micrographs and the software used in microstructural analyses (Chapter 5).

2) Microstructure of various dairy foods (Chapters 6–12). This part is divided into sections related to the microstructure of milk, cheeses, yoghurts, fat products, ice cream and frozen dairy desserts, dairy powders and selected traditional Indian dairy products. In addition, Chapter 13 reviews the localization of microorganisms within the microstructure of various dairy products.

The last chapter discusses the challenges and future trends of the microstructure of dairy products (Chapter 14).

This book is aimed at the following potential audiences.

Microstructure of Dairy Products, which serves as an essential stand-alone source, is primarily recommended in academia as well as the food industry and especially dairy science and technology. In addition, physicists and microscopists can benefit from this book as it provides updated information on the description, uses and applications of microscopy techniques in the microstructural analyses of dairy products. Other audiences include graduate students and researchers in the field of pure sciences such as biology, physics and chemistry. It is always useful to have an updated knowledge about microscopy techniques and their benefits and challenges, and accordingly the future research that is required.

It is hoped that this text will become an important component of the book series in the field of dairy microstructure.

Mamdouh Mahmoud Abdel-Rahman El-Bakry, Antoni Sanchez,
and Bhavbhuti M. Mehta
Editors

1

Microscopy Techniques for Dairy Products – An Introduction

Mark A.E. Auty

Food Chemistry and Technology Department, Teagasc Food Research Centre, Teagasc, Moorepark, Co. Cork, Ireland

1.1 Introduction

The textural properties of a particular dairy product are strongly influenced by the three-dimensional arrangement of its structural elements and their interactions (Heertje, 1993). To fully understand the behavior of dairy products, it is therefore not enough to know the chemical composition and bulk physical properties but how they interact and affect the spatial arrangement or organization of the food constituents at the nano- and micro-length scales. Food microstructure studies therefore provide a link between physico-chemical properties, process behavior and organoleptic qualities of a particular dairy product (Figure 1.1). Linking microscopy with rheological and sensory techniques in particular is necessary for a fuller understanding of food behavior, requiring a multivariate approach to experimental design. This chapter gives a brief overview of the different types of microscopy used to study dairy foods with a focus on confocal microscopy as this is arguably the most useful single technique of benefit to both researchers and food industry technologists.

1.1.1 Brief History and Background

In the seventeenth century, Antonie van Leeuwenhoek, using a high magnification hand lens, first viewed fat droplets in milk (Leeuwenhoek, 1674). Despite this early start, microscopy of dairy products, and food in general, remained unexplored, with little published literature until well after the development of electron microscopy techniques in the 1940s. As food manufacturers began using microscopes in the 1950s and 1960s, it became apparent that the structural arrangement of food components strongly influenced food processing and quality. Most of the early food-related electron microscopy work was performed on dairy products, mainly yoghurt and cheese (for reviews see Brooker, 1979; Kalab, 1979a, b, c, 1981, 1993; Holcomb, 1991; Schmidt and Bucheim, 1992). Despite the enormous influence of light microscopy on medical research at the end of the last century and the improvement in optic materials and design, conventional light microscopy of food products remained largely neglected although Lewis (1978)

Microstructure of Dairy Products, First Edition.
Edited by Mamdouh Mahmoud Abdel-Rahman El-Bakry, Antoni Sanchez, and Bhavbhuti M. Mehta.

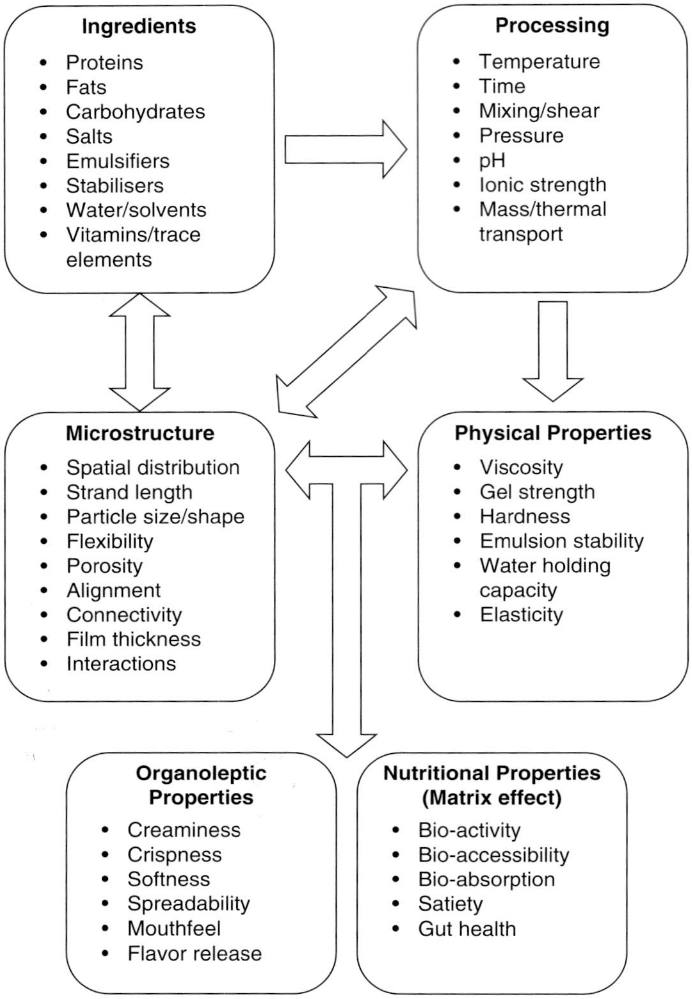

Figure 1.1 Diagram showing inter-relationships between microstructure and functionality of dairy products.

and Flint (1994) published selected methods for light microscopic examination of a range of food ingredients and products, including milk powders and dairy spreads. Contrast between the component of interest and surrounding food material may be achieved by optical techniques, chemical staining, or a combination of both (Flint, 1994). Despite this, optical microscopy of dairy products remained largely neglected until the development of commercial confocal microscopes in the 1990s. In the past 20 years, there has been considerable research interest in food microstructure as a key to understanding structure-function relationships. A wide range of microscopy techniques is now available for the study of food microstructure, with more being developed (for a review see Morris and Groves, 2013). These techniques are frequently employed to study dairy products such as cheese (El-Bakry and Sheehan, 2014). The food researcher now has a large toolbox of techniques, the choice of which depends on the particular application. However, a correlative approach employing various microscopy techniques is required to provide a fuller understanding of complex multiphase nano- and microstructures (Aguilera and Stanley, 1990; Lewis, 1993). This approach has led to the development of hybrid microscopes such as the RISE (WITech, Ulm, Germany) system which combines a scanning electron microscope with Raman confocal, focused ion beam and even atomic force microscopes, permitting examination of the same sample area by different microscopy techniques.

Many of the common techniques used to study food microstructure have been adapted from specimen preparation procedures for biological tissue. However, there are particular problems associated with the preparation of food products for microscopic examination that the researcher should be aware of. Many foods have high levels of moisture, fat or sugar and preserving the original microstructure of such materials may be difficult, particularly for electron microscopic studies that may require low moisture, conductive specimens. Dried ingredients, such as spray dried powders, crystalline sugars, starches etc. with a moderately small particle size (<100 μm) may be examined in their natural state and require little sample preparation. Highly refractile or opaque solid and semi-solid food materials however, need to be rendered thin enough to transmit light and generally this is achieved either by compression or sectioning. Soft materials may be compressed or smeared across a microscope slide. Solid food materials and cellular tissues may be chemically fixed, dehydrated then embedded in paraffin wax or plastic resin prior to sectioning by microtomy. Alternatively, frozen sections, approximately 5–20 μm thick, may be cut in a cryostat. Sectioned material can then be observed using any of the optical or chemical contrast techniques described below. Powdered ingredients, for example spray-dried milk powder particles, should be mounted in a clear, immiscible liquid such as sunflower oil that should be viscous enough to restrict Brownian movement of the particles.

The main microscopy techniques used to study dairy products are listed in Table 1.1.

Table 1.1 Main microscopy techniques used in food microscopy.

Technique	Radiation type	Incident radiation wavelength	Approximate resolution	Application
Stereo-microscopy	Photons	400–750 nm	5 μm	Overview of microstructure
Light microscopy	Photons	400–750 nm	200 nm	Ingredient localization
Fluorescence microscopy	Photons	350–400 nm (u.v.) 400–750 nm (visible)	200 nm	Ingredient localization
Confocal microscopy	Photons	350–400 nm (u.v.) 400–750 nm (visible)	200 nm	Ingredient localization, 3D information
SEM	Electrons	0.001–0.01 nm	4 nm	Large depth of field-simulated 3D view
TEM	Electrons	0.01–0.01 nm	1 nm	Fine structural detail, macromolecular interactions
AFM	N/A (physical cantilever)	N/A		Surface topology, nano-mechanical behavior

1.2 Conventional Optical Microscopy Techniques

1.2.1 Conventional Light Microscopy – Optical Contrast

1.2.1.1 Bright Field

Bright field illumination employs an axial cone of light from the condenser, which is transmitted through the specimen and is commonly based on Koehler illumination. This technique is useful if there is inherent contrast in the specimen, for example in highly colored food products, otherwise stains or dyes may be used to impart color contrast to the specimen (see below).

1.2.1.2 Polarized Light

A polarized light microscope consists of two polarizing plates arranged perpendicularly, one below the condenser (the polarizer) and a second above the objective (the analyzer). If the sample is isotropic, incident polarized light is not rotated and no light is transmitted. If the polarized light passes through an anisotropic substance, such as a lactose crystal, part of the light is rotated and passes through analyser and appears bright (birefringence). Examples of polarized light microscopy include the study of microcrystalline inclusions in cheese (Brooker, 1979) and lactose crystallization in spray dried milk (Saito, 1985; Maher *et al.*, 2014). Employing "partially uncrossed polars" allows visualization of non-birefringent material while retaining the polarizing effect as shown in Figure 1.2.

1.2.1.3 Phase Contrast

This technique has traditionally been used to study transparent biological material including eukaryotic cells and bacteria. An annular phase ring within the condenser below the sample retards the phase of light by ¼λ. Diffracted and non-diffracted light passing through the sample is recombined using a similar phase ring in the objective. Contrast is obtained due to difference in refractive index between the sample and its surroundings. This technique, in conjunction with Sudan Black staining, has been used

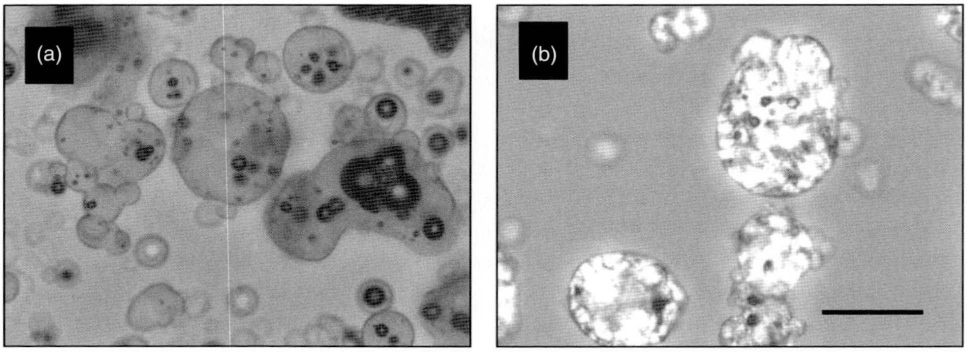

Figure 1.2 Spray dried skim milk powder particles (a) fresh powder; (b)powder stored at 55% relative humidity for 24 h. Polarized light micrographs taken using partially uncrossed polarizing filters, reveal extensive birefringent lactose crystals (b) while allowing visualization of particle shape and occluded air bubbles (dark circles). Scale bar = 100 μm.

to view fat droplets in mayonnaise, which would otherwise be transparent in bright field (Tung and Jones, 1981).

1.2.1.4 Differential Interference Contrast

This technique requires the addition of fairly expensive optical elements to a basic light microscope setup but gives excellent results and is generally preferred to phase contrast for studying cells, bacteria and other transparent objects. A polarizer and prism is located above and below the specimen, analogous to phase contrast. Differences in refractive index are visualized in relief. This technique is particularly useful for studying fat droplets in milk in addition to phase separation and depletion flocculation in dairy-based emulsions (Hibberd *et al.*, 1997).

1.2.1.5 Fluorescence

Fluorescence is the process by which light in the shorter wavelength regions of the spectrum is absorbed by an object and almost immediately re-emitted as light of longer wavelengths. Some materials, for example chlorophyll, are intrinsically auto-fluorescent and appear self-luminous when viewed under the appropriate wavelength of light. However, most dairy ingredients exhibit little or no autofluorescence and fluorescent dyes are added to the specimen to increase contrast and fluorescence intensity. Most modern fluorescence microscopes employ epi-fluorescence, where a monochromatic beam of light is used to illuminate the specimen; the emitted fluorescence, either auto-fluorescence or fluorophore-induced, is detected via the same optical path using appropriate emission filters. For the study of dairy products however, epi-fluorescence microscopy has largely been superseded by confocal scanning laser microscopy (see section 1.2.2 below).

1.2.2 Chemical Contrast Techniques in Light Microscopy

Although optical contrast techniques may be useful for determining phase differences, crystalline properties or structural orientation, they give little information on the chemical nature of specific food components. Dyes are defined as colored substances that impart color to either manufactured products or particular plant/animal tissues, whereas stains may be uncolored. Various histological and histochemical methods, originally developed for animal and plant tissues, have been applied to food materials (Lewis, 1978; Flint, 1990). Identification of the main food constituents, such as protein and fat may be performed using relatively simple staining protocols. Flint (1990) successfully identified proteins, fats, starches and food gums in a wide variety of foodstuffs using Toluidine Blue to stain proteins and gums, iodine/potassium iodide or iodine vapor to stain starches and Oil Red O to stain lipids. The stain is added to the prepared food sample as a dilute solution prior to bright field light microscopic examination. King (1958) first used transmitted fluorescence microscopy in conjunction with fluorescent stains to study milk, curd and cheese. For cheese, frozen sections were stained with either Basic Fuschin or Acridine Orange to stain protein and Phosphine to visualize fat. Green *et al.* (1980) used Sudan Black and Carbol Fuschin to stain lipid and protein, respectively, in frozen sections of Cheddar cheese and samples were examined under bright field illumination. The microflora, fat and protein components of Cheddar, Mozzarella, Camembert, Brie, Blue and processed cheeses were studied by epi-fluorescence in conjunction with fluorescent staining (Yiu, 1985). Chemically fixed samples were sectioned either frozen or embedded in glycol methacrylate, then stained with Acridine Orange and Nile Blue A to stain protein and fat, respectively. Shimmin (1982) used a mixture of Acridine Orange and Phosphine to stain fat in cheese prior to visualization by incident light fluorescence microscopy. The chemical basis for many staining reactions is complex and interpretation of images often relies on empirical knowledge (Aguilera and Stanley, 1990).

1.3 Confocal Scanning Laser Microscopy

1.3.1 Confocal Principle

There are three basic confocal microscope designs: (i) Spinning (Nipkow) disk, also called tandem scanning; (ii) stage scanning; and (iii) beam scanning. The spinning disk-type can use conventional white light and have very fast acquisition rates. Recent disk-scanning instruments such as the Yokogawa-type have greatly improved laser though-put and give real-time imaging of biological specimens in real color. Beam (or point) scanning instruments, although relatively slower and more complicated (and therefore expensive), rely on detection of fluorescent light using laser excitation. Multiple lasers and separate emission detection channels facilitate specific labeling of different food components simultaneously. The stage scanning instrument is rarely used today and nearly all published food research studies use the point or beam scanning design.

Confocal scanning laser microscopy (CSLM) is a form of epi-fluorescence optical microscopy. The key feature of confocal imaging is that both the illumination and detection systems are focused simultaneously on a single volume element in the specimen, which is elegantly achieved by positioning a pinhole close to the detection source (Minsky, 1957). Conventional light microscopy is described as "wide filed" illumination where the volume of sample above and below the plane of focus is uniformly and simultaneously illuminated. This requires thin, relatively transparent, samples but often results in out-of-focus blur that reduces resolution and specimen contrast. CSLM employs a diffraction-limited spot which is detected by means of a small aperture (pinhole) placed in front of the emitted light detector, greatly reducing out-of-focus information (Brakenhoff *et al.*, 1988; Wright *et al.*, 1993). The illuminated spot is then scanned across the specimen (Figure 1.3).

CSLM employs lasers with defined emission wavelengths, typically in the range 488–647 nm, but may also include blue 405 nm or ultra-violet lasers. Modern instruments may be fitted with 4 or more separate confocal channels and acousto-optical tuneable filters can optimally excite a wide range of fluorochromes. In addition, acousto-optical tuneable filters and beam splitters may be used to replace conventional epi-fluorescence filters, greatly reducing emission spectral overlaps and improve bleed-through/crossover where multiple dyes are employed or autofluorescence is an issue.

Incident photons from the laser beam are absorbed by fluorophores within the sample and emitted at a longer wavelength via a dichroic mirror to a photon detector such as a photomultiplier tube or CCD. The signal is amplified and converted into pixels. The beam is scanned to produce a two-dimensional image. A z-stepping motor on the

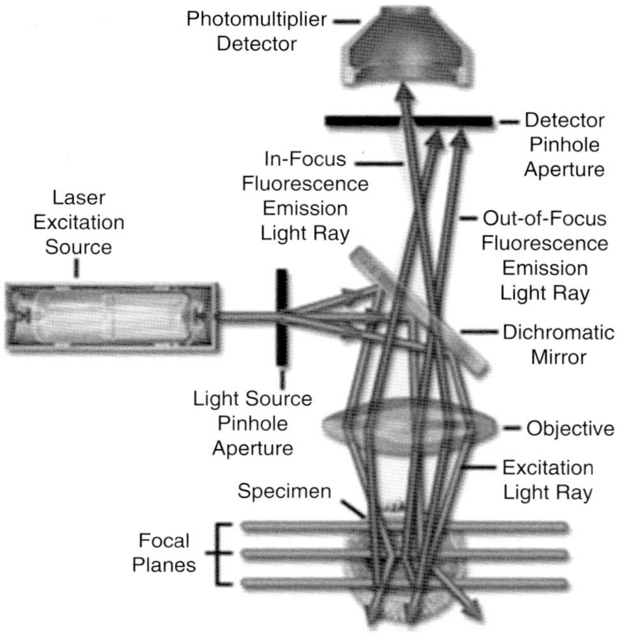

Figure 1.3 Schematic of confocal scanning laser microscope. (*See color plate section for the color representation of this figure.*)

sample stage facilitates consecutive x-y planes of focus through the z-plane to produce three-dimensional data, which can then be recombined and rendered into a three-dimensional image projection. The resolution of a confocal microscope is governed by the same optical limitations affecting conventional light microscopy with some subtle differences. The theoretical resolution of an optical microscope is limited by diffraction but by reducing out of focus blur, confocal microscopes under ideal conditions can approach the theoretical resolutions predicted. In practice, x-y resolution of 200–250 nm and z resolution of ~750 nm (depending on wavelength) can easily be achieved in food materials using a high quality objective lens. The z-thickness of the optical sections can be increased or decreased by opening or closing the pinhole aperture, respectively. For most modern instruments, the pinhole diameter will be automatically set to optimize resolution depending on the objective magnification and numerical aperture (N.A.), typically equivalent to the diameter of one Airy disc which approximates to a minimum axial resolution of ~730 nm with a high quality lens such as a ×63, 1.4 N.A. oil-immersion apochromatic objective.

The advantages of CSLM over conventional light microscopy thus include:

1) Three-dimensional imaging of bulk samples by "optical" sectioning and digital reconstruction.
2) Sub-surface imaging minimizes sample microstructure disturbance.
3) Slightly improved resolution over conventional optical microscopy.
4) Sensitive simultaneous detection of two or more fluorochrome probes.

5) Dynamic processes can be studied under controlled environmental conditions using appropriate sample stages and fast acquisition rates.

Disadvantages include:

1) Food components usually require labeling with a suitable fluorochrome which may involve solvents and/or multiple processing steps that could adversely affect the sample. This is particularly true for polysaccharides.
2) The sample surface has to be flat to ensure an even illumination and emission signal across the field of view.
3) Diffraction-limited lateral resolution is ~200 nm at best, making it difficult or impossible to properly resolve colloidal food systems such as casein micelles, or interfacial films such as the milk fat globule membrane.

The choice of laser excitation wavelength should match as closely as possible to the fluorochrome excitation wavelength. Laser diode arrangements containing semiconductor materials have recently been developed for obtaining laser excitation wavelengths in the range 300–460 nm; gas lasers such as Argon ion or mixed gas laser were originally used to for wavelengths in the range 450–633 nm, Laser technology has advanced rapidly and newer CSLMs will generally incorporate solid state lasers for enhanced stability. In addition, white light, continuous wave or tunable lasers can be employed to ensure precise matching of excitation wavelengths to the fluorophores of choice.

Despite these disadvantages, CSLM is now routinely used by food researchers and industry both for characterizing the microstructure of dairy products and for use as a troubleshooting tool (Vodovotz, *et al.*, 1996; Loren *et al.*, 2007; Tamime *et al.*, 2011). The ability to clearly visualize internal microstructure of a food product gives unique insight into its true three-dimensional arrangement and allows for differential labeling of specific components such as fats or proteins. Furthermore, samples can be imaged under ambient or temperature-controlled conditions in real time, unlike other microscopy techniques such as electron microscopy which require the sample to be held under a vacuum, made relatively conductive and/or extremely thin.

1.3.2 Identifying Dairy Primary Components in CSLM: Labeling Strategies

Three general approaches or strategies may be used for labeling dairy components with fluorochromes: (a) *generic;* (b) *specific; or* (c) *covalent pre-labeling* or occasionally a mixture of these (see for example Li *et al.*, 2016). Generic labeling is the most convenient, enabling fats and protein to be rapidly distinguished. Specific labeling will involve multiple steps and may not be possible for very shear-sensitive foods such as whipped cream ort yoghurts. Specific labeling may involve immunolabeling of specific proteins or lectin labeling of exo-polysachharides while covalent pre-labeling of specific biopolymers is also possible where binding is via functional groups (Patonay *et al.*, 2004).

1.3.2.1 Generic Labeling

Fluorescent dye molecules may interact non–covalently to the dairy component via ionic, electrostatic or hydrophobic interactions. Protein and fat distribution in most dairy products may be easily visualized, using either a single fluorochrome such as

Nile Blue or preferably a dual labeling approach. Acid Fuchsin, Bodipy® 665/676 and DM-NERF has been used to label proteins, lipids and whey, respectively, in model dairy-based gels (Herbert *et al.* 1999). Auty *et al.* (2001a) has developed this dual labeling approach for a range of dairy products including dairy spreads, various cheeses and milk powders using a single application of a dye mixture, for example Nile Red and Fast Green FCF. This approach has now been used by many researchers and examples of various dairy products as shown by generic labeling in the CSLM are shown in Figure 1.4.

Generic labeling with fluorescent probes relies on passive diffusion of dye molecules to the target site and minimizes sample disturbance. Nile Red, an oxazone derivative of Nile Blue, is a lipophilic probe that fluoresces when in contact with a liquid lipid phase.

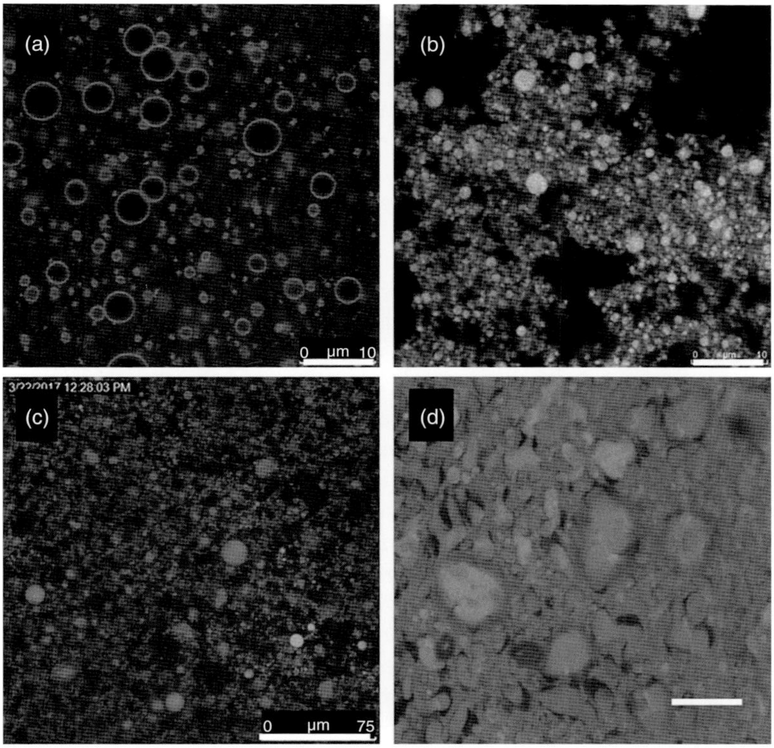

Figure 1.4 Confocal scanning laser micrographs of various dairy products labeled with fluorochromes. b, c, d and f are dual labeled with Nile Red and Fast Green FCF to show fat (green) and protein (red), respectively. (a) Whey protein-stabilized emulsion labeled with Rhodamine B (pseudo-colored green) to visualize protein at the droplet interface, scale bar = 10 µm. (b) Full-fat dairy yoghurt showing protein network and small fat droplets in addition to pores (dark regions), scale bar = 10 µm. (c) Mayonnaise showing fat droplets and discrete aggregated protein particles, scale bar = 75 µm. (d) Cheddar cheese showing irregular shaped fat pools and continuous protein matrix, scale bar = 25 µm. (e) Butter labeled with Nile Red (greyscale image) showing characteristic rounded butterfat crystals (dark grey, arrowed), water droplets (dark circles) and continuous fat phase (bright), scale bar = 5 µm. (f) Fat-filled whey protein enriched dairy powder showing fat droplets entrapped in continuous protein matrix, scale bar = 5 µm. (*See color plate section for the color representation of this figure.*)

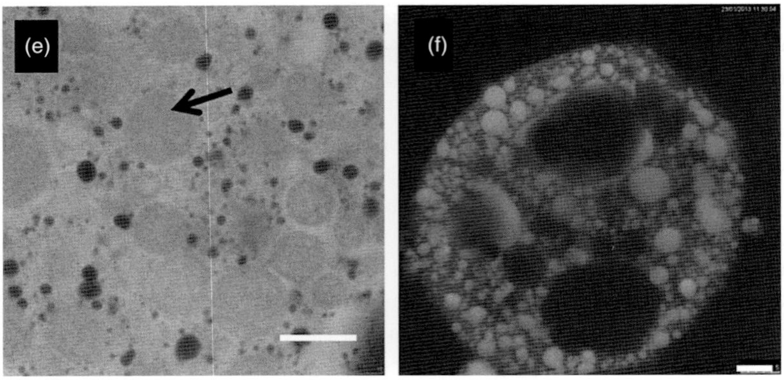

Figure 1.4 (Continued)

A relatively new new fat dye, VO03-01136 (Dyomic, Jena, Germany) was successfully used to label fats in a mixed food biopolymer gel (Heilig *et al.*, 2009). For generic labeling of proteins there are several fluorochromes available such as Rhodamine B, Fast Green FCF, or Nile Blue (when excited at 633 nm). These dyes are typically used in aqueous solutions. In some dairy spreads, solid dye crystals can be directly sprinkled onto the product surface to facilitate solubilization and migration of the dye into the sample over time (Brooker, 1995). For polysaccharides *in situ* labeling is more difficult although fluorescein isothiocyanate (FITC), Rhodamine B and Safranin O have been used for non-covalent labeling of starch (van de Velde *et al.*, 2002). When preparing the sample for CSLM, the sample must have flat surface to ensure even illumination and prevent shading. Laser penetration depends on the sample and ranges from ~100 μm for transparent gels to <15 μm for optically dense, refractile material such as very hard cheeses. For solid or self-supporting viscoelastic materials for example Cheddar or Emmental cheese, hand cut sections may be used. Frozen sections can be cut using a microtome but this may introduce freezing artifacts and disrupt the microstructure. For low-moisture ingredients such as milk powders, small quantities of powder are suspended in the liquid dye formulation, which must be optimized to prevent dissolution of the powder whilst facilitating diffusion of dye molecules to the site of interest (Auty *et al.*, 1999). Two advantages of this single-step approach are speed, with results being obtained within five minutes, and that it does not require multiple treatments, such as sequential labeling, washings etc., which may remove or distort unstable food constituents. The principal disadvantage is non-specificity.

1.3.2.2 Specific Labeling

For specific labeling, there are two main approaches: (1) immunolabeling of proteins or localization of polysaccharides with lectins and (2) covalent labeling. Immunolabeling is more commonly used for proteins and requires a specific antibody to target the protein of interest and methodology is usually adapted from its extensive use in cell biology (Tsien *et al.*, 2006). Immunolabeling depends heavily on the specificity of the primary antibody to the target epitope. In particular, the effect of heat denaturation on the immunoreactivity of the target protein should be characterized and ideally antibodies raised against the denatured protein, rather than native protein, as appropriate.

Two-step immunolabeling, using a fluorescently labeled secondary antibody to bind to the primary antibody, is generally recommended to increase the signal-to-noise ratio. This approach can also be used to localize bacteria within food systems (Auty *et al.*, 2005a).

Lectins are naturally occurring proteins and glycoproteins that bind to specific carbohydrate monomers or residues present in a biopolymer (Brooks *et al.*, 1997). A commonly used lectin for carbohydrate recognition is Concanavalin A (ConA, Loris *et al.*, 1998). In the presence of selected metal ions, ConA has the ability to bind with the D-glucose and D-mannose residues at the non-reducing terminus of polysaccharides and glycoproteins. Arltoft *et al.* (2007) used enzyme-linked immunosorbent assay (ELISA) techniques to investigate the affinity of ConA with the sugar residues in different food polysaccharides and observed that ConA can bind with galactomannans, xanthans and bacterial exopolysaccharides. In dairy microscopy, this approach has been used to localize bacterial exopolysaccharides in dairy yoghurt gels and Feta cheese (Arltoft *et al.*, 2007; Hassan *et al.*, 2002; Hassan *et al.*, 2003). The specificity of ConA may also allow localization of Konjac glucomannan in a mixed biopolymer system using CLSM techniques. Arltoft *et al.* (2007) screened two lectins: Con A and wheat germ agglutinin (conjugated with Alexa Fluor 488 dyes) and one polyclonal anti-carageenan antibody, JIM7, as probes for wide range of polysaccharides in foods. Despite the cross-reactivity of some of these probes, the authors were able to demonstrate good localization of pectin and carrageenan in dairy products. For cell wall components including cellulose, glucans and glycoproteins, a range of monoclonal antibodies and carbohydrate binding molecules are now available (Hervé *et al.*, 2011).

1.3.2.3 Covalent Labeling

Polysaccharide localization by covalent labeling usually involves heating the polysaccharide and fluorescent probe at high temperature followed by alcohol precipitation of the polysaccharide–fluorescent probe conjugate (Belder and Granath, 1973). There is a concern that the harsh chemical labeling process involving solvents, high temperatures and pH adjustment, may adversely modify the functionality of the target molecule (van de Velde *et al.*, 2003; Abhyankar *et al.*, 2011a). Garnier *et al.*, 1998 showed that there may be some depolymerization of the polysaccharide molecules. Despite this drawback, covalent labeling techniques have been successfully used to localize polysaccharides including Konjac glucomannan to study phase behavior in a mixed biopolymer system containing whey protein (Abhyankar *et al.*, 2011a). Covalent conjugation of Konjac with FITC led to a shift in the absorbance spectrum peak of FITC to a lower wavelength and a decrease in the average molecular weight distribution of Konjac. Furthermore, covalently labeled Konjac showed reduced apparent viscosity compared to unlabeled Konjac. Van de Velde *et al.* (2003) were able to discriminate between starch, polysaccharides and various covalently-labeled proteins in mixtures containing starch and gelatine. Kett *et al.* (2013) were able to localize covalently labeled milk protein fractions in biopolymer mixtures containing modified starch, suggesting that this technique could be very useful for tracking specific proteins in heterogeneous food models. Li *et al.* (2016) covalently labeled β-casein using 5-(and 6)-carboxytetramethylrhodamine, succinimidyl ester (NHS-Rhodamine) prior to mixing with whey protein and sunflower oil in conjunction with Nile Blue generic labeling to characterize the lipid-aqueous phase interface of oil droplets. There appeared to be little effect of covalent labeling on the physico-chemical properties of the emulsions. A major disadvantage of covalent labeling is that the component must be pre-labeled in a purified form making it unsuitable for characterizing actual dairy products, being limited to model systems.

1.3.3 Some Applications of Confocal Microscopy to Dairy Products and Ingredients

1.3.3.1 Spreads

Dairy products have been studied since confocal instruments were commercially available (Heertje *et al.*, 1987; Brooker, 1991; Blonk and van Aalst, 1993). The non-invasive nature of confocal microscopy makes it a useful technique for analysing shear-sensitive samples such as dairy fat spreads and mayonnaises (Blonk and van Aalst, 1993, Langton *et al.*, 1999). Heertje *et al.* (1987) developed a plunger-like sampling tube to extract a sample of spread for CSLM with minimal sample disturbance. The spread sample was then pushed out of the tube for cutting prior to labeling with Nile Blue. Brooker (1995) applied solid crystals of fluorochrome such as Nile Red or Nile Blue to the spread surface to facilitate slow solubilization and diffusion of the stain into the product. Peltier-based heating/cooling stages can be used to characterize fat crystal networks to simulate refrigeration temperatures (4–10°C). The microstructure of low fat spreads containing sodium caseinate has been studied by CSLM (Clegg *et al.*, 1996) where the fat phase was stained with 0.1%, w/v, Nile Red in polyethylene glycol. One beneficial effect of labeling with fat dyes in CSLM is that fat crystals can be seen in negative contrast within the stained liquid fat phase. This can be used to distinguish between characteristic rounded clusters of butterfat produced by the churning process and needle-like crystal networks produced via scrape surface heat exchanger process.

1.3.3.2 Emulsions and Foams

CSLM is also very useful for characterizing interfaces in shear-sensitive food systems such as food foams. An example of this is whipped cream, where dual labeling reveals that the air bubble interface is stabilized by a mixture of partially coalesced fat globules and protein aggregates. Fat droplet and protein aggregate sizes were measured from CSLM images of mayonnaises stained with Nile Blue (Langton *et al.*, 1999). The displacement of sodium caseinate, covalently labeled with FITC, by monoacylglycerols, was observed and quantified by measuring emitted fluorescence intensity using CSLM to study emulsifier displacement at the oil: water interface (Heertje *et al.*, 1990, 1996).

1.3.3.3 Fermented Milks

Fermented milks such as yoghurts have been studied using CSLM which can be used to characterize the aggregated protein network as well as fat droplet distribution and starter bacteria. Hassan *et al.* (1995a and 1995b) first demonstrated the use of reflected light confocal microscopy to visualize protein aggregation in directly acidified and fermented milks. There authors also applied the pH-sensitive probe, CL-NERF, to visualize pH gradients around encapsulated bacteria. The effect of ropy yoghurt cultures on the structure of yoghurt gels was studied by CSLM using Rhodamine B staining and image analysis to quantify pore size and protein aggregates. The effect of high shear milk processing on sensory and rheological properties of yoghurts has been studied by CSLM using the dual labeling technique (Ciron *et al.*, 2012).

1.3.3.4 Cheese

Cheeses generally comprise a viscoelastic protein-continuous phase with dispersed fat globules and may be described as fat-filled gels. The microstructure of rennet curd made from non-fat dried milk has been studied by CSLM where the protein phase was visualized by reflectance and the fat was labeled with Nile Red. (Hassan and Frank, 1997). CSLM imaging of fat and protein phases is relatively straightforward using a dual-labeling approach (Auty, 2001a). Several cheese varieties have been studied by CSLM employing dyes added to the cheese surface; these include Gouda (Heertje *et al.*, 1987; Blonk and van Aalst, 1993), Cheddar (Brooker, 1991; Everett *et al.*, 1995; Gunasekaran and Ding, 1999; Guinee *et al.*, 1999; Guinee *et al.*, 2000a and b; Everettt and Olson, 2003) Feta (Hassan *et al.*, 2002), Mozzarella (Guinee *et al.*, 2002; Rowney *et al.*, 2003a and b) and processed cheese (Sutheerawattananonda *et al.*, 1997). Different cheese varieties have very different microstructures. Processed cheeses have discrete, spherical fat globules in a homogeneous protein continuum; Cheddar (and Gouda) type cheeses have a protein continuous phase with curd junctions and irregular, fairly large fat globules or pools; stretched curd cheeses such as Mozzarella have aligned protein fibers with interstitial fat while cream cheeses contain small fat droplets clustered within an aggregated protein matrix (Auty *et al.*, 2001a). Several CSLM studies at various key stages in cheese manufacture have been published. Studies using CLSM have examined the fat globule structure in cheese (Gunasekaran and Ding, 1999) milk gelation and cheese melting (Auty *et al.*, 1999), permeability of rennet casein gels (Zhong *et al.*, 2004), the effect of the pasta filata process on fat globule coalescence in Mozzarella cheese (Rowney *et al.*, 2003b) and Cheddar cheese (Everett and Olson, 2003; Everett *et al.*, 1995), localization and viability assessment of probiotic bacterial cells (Auty *et al.*, 2001b) and starter cells in cheese (Hannon *et al.*, 2006), location of exopolysaccharides (EPS) in cheese (Hassan *et al.*, 2002), and correlation with sensory data of acid milk gels (Pereira *et al.*, 2006). O'Reilly *et al.* (2002) studied the effect of high pressure treatment on Mozzarella cheese. CSLM proved a useful technique to show the increased hydration and swelling of the protein phase. Three-dimensional reconstructions from CSLM image stacks confirmed the more swollen nature of the protein phase in the 1-day-old HP-treated sample compared to the control at 1 day. More recently, Ong *et al.* (2011) used dual labeling of the fat and protein phases to study the effect of processing on microstructural changes at gel formation, curd and cheddaring stages, and comparing CSLM results with cryo-scanning electron micrographs.

1.3.3.5 Dairy Powders

Confocal microscopy was first described by McKenna (1997) to localize fat and phospholipid in whole milk powder. To visualize the fat distribution, milk powder particles were immersed in various mounting media containing Nile blue. Glycerol prevented dissolution of powder particles and enabled localization of fat globules. Lecithin, prelabeled with BODIPY 3806, was visualized by CSLM and was located on the surfaces of instantized milk powder agglomerates. By careful formulation of staining mixtures, it is possible to simultaneously visualize both the fat and protein phases of spray dried milk powders in situ (Maher *et al.*, 2014). An alternative to post-labeling powders is to pre-label one of the phases prior to spray drying. Although impractical for large scale pilot or commercial dryers, this approach has been used to spray dry microcapsules containing fish oils pre-labeled with Nile Red on a laboratory scale using a bench-top dryer. (Drusch and Berg, 2008). Recently, the effect of protein content and homogenization on emulsion droplet size in spray dried infant milk formula was studied using in situ powder labeling (McCarthy *et al.*, 2012).

1.3.3.6 Milk Protein Gel Systems

Bremer *et al.* (1993) used CSLM to demonstrate the fractal nature of particulate casein gels. Sodium caseinate was labeled with fluorescein isothiocyanate (FITC) or Rhodamine B prior to acidification by glucono-delta-lactone. Schorsch *et al.* (2001) pre-labeled casein-whey mixtures with Rhodamine B prior to the study of gelation by CSLM. Herbert *et al.* (1999) employed multiple labeling using acid Fuschin, Bodipy 665/676 and DM-NERF to label milk proteins, lipid and the aqueous phase, respectively, in acid-coagulated raw milk prior to CSLM imaging. Post-labeling of milk protein gels has also been performed: β-lactoglobulin gels have been imaged by CSLM following immersion of heat-induced gels in 0.001 wt% FITC (Hagiwara *et al.*, 1997), and Olsson *et al.* (2002) used 0.01% Texas Red to visualize the protein phase of β-lactoglobulin/amylopectin gels by CSLM.

1.3.3.7 Dynamic CSLM Techniques

Dairy manufacture is not a steady-state process and many foods undergo several transformations during manufacture, storage and consumption. These processes often include mixing, heating, pH adjustment as well as complex biochemical transformations such as fermentation. CSLM can be used to effectively model these transformations to give insight into how food structures are formed and deformed. Environmental effects such as temperature can be monitored using temperature-controlled heating and cooling stages that can be purchased from specialist suppliers such as Linkam Scientific (Surrey, UK). Heuer *et al.* (2007) used a pressure stage to study bubble size distribution and coalescence by direct CSLM imaging. In order to understand how the different food processing stages affect the microstructure and ultimately behavior, of a food product, confocal images taken at key process steps can be useful and combining CSLM with rheological measurement is a common research approach (see for example Manski *et al.*, 2007). These "snapshots" cannot however show what is happening during a fast-moving process such as acid gelation or during breakdown of a product under shear and dynamic imaging techniques are needed. This has led to the development of new hybrid configurations combining shear/oscillatory rheometry and biaxial compression with a confocal microscope to facilitate real-time imaging during small and large deformation measurement (Nicolas *et al.*, 2003). An early example of dynamic CSLM was the direct monitoring of acid gelation of milk using time lapse CSLM imaging (Auty *et al.*, 1999). Using this technique, it was possible to follow the particle movement, aggregation and subsequent network formation of casein micelles during real-time acidification by glucono-delta-lactone at a controlled temperature. The progress of gelation could be followed using low amplitude oscillatory rheometry and images mapped to the storage modulus. CSLM analysis of skim milk clearly showed the progression from initiation of aggregation of casein micelles at pH 5.58, network

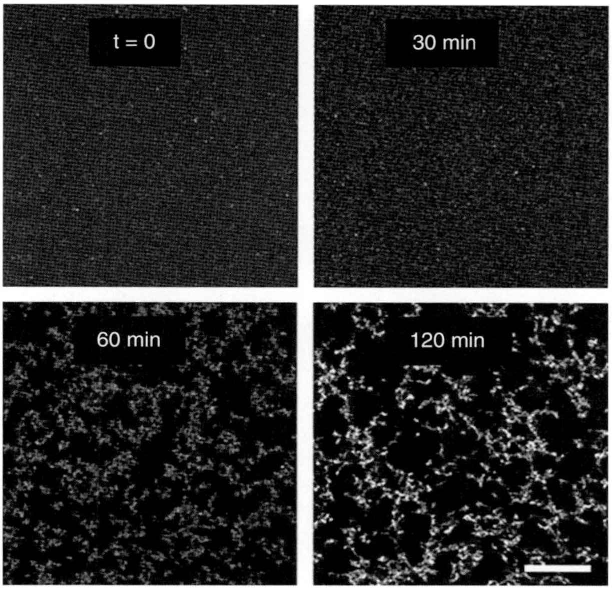

Figure 1.5 Confocal scanning laser micrographs taken from time lapse series of skim milk acidified with glucono-delta lactone, labeled with Nile Blue. Scale bar = 25 μm.

formation and the reduction in protein mobility at pH 5.48 and increase in storage modulus followed by consolidation of the three-dimensional gel network. Time-lapse animation of the CSLM image sequence enabled dynamic visualization of the gelation process (Figure 1.5).

A second example demonstrated fracture behavior of filled gels. In addition to structure formation described above, it should be remembered that food is designed to break down in the mouth and the fracture behavior of food has been little studied using CSLM. Microtensile stages are now commercially available can be very useful for studying deformation of solid food materials at the microscopic scale (Brink *et al.*, 2007). Uniaxial compression and fracture analysis of food composites and their relationship with sensory perception has been studied by van den Berg *et al.* (2007, 2008) Tensile testing using fluorescent particles embedded in zein films has also been monitored dynamically with CSLM (Emmambux and Stading, 2007). Abhyankar *et al.* (2011b) employed a standard failure analysis technique called notch propagation to monitor the fracture of filled whey protein gels. A microtensile stage (Deben Ltd., Bury St Edmonds, UK) was fitted to an upright confocal microscope and the progress of a fracture through a heat-induced whey protein gel filled with sunflower oil droplets monitored in real time both visually and through a force transducer (2N load cell). The effect of the gel type (coarse particulate versus fine stranded) on the fracture properties and release of fat was studied. Time-lapse animations graphically show how the two different gels fracture behavior and the mobility of the fat phase (Figure 1.6). This type of study highlights the importance of microstructure on the fat release properties of food and how new reduced-fat products can be designed with defined fat release and sensory properties.

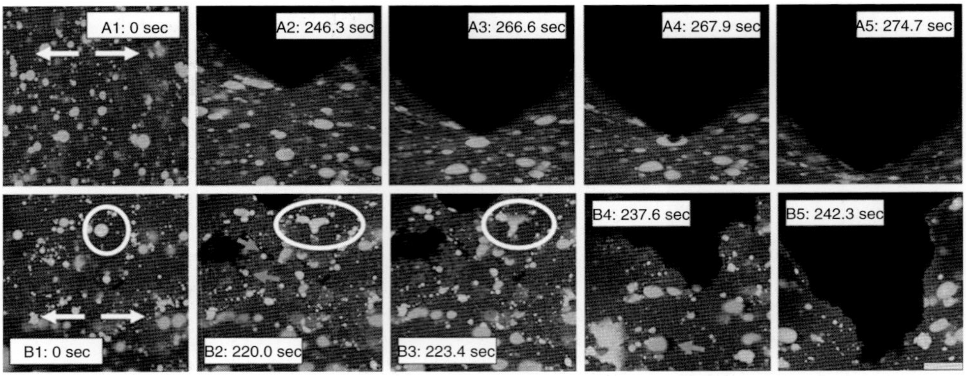

Figure 1.6 Confocal scanning laser micrographs of fat-filled whey protein gels monitored using a microtensile stage. Protein is pseudocolored red, fat pseudocolored yellow/green. A1–A5 gelled at pH 7.0 and is fine stranded, B1–B5 is gelled at pH 5.4 and is particulate. Notch propagation is vertical and fracture properties measured in extension. Note: deformation and movement of fat in the particulate gel (circle) while the fine stranded gel fractures through fat droplets. Adapted from Abhyankar (PhD thesis). Scale bar is 25 μm. (*See color plate section for the color representation of this figure.*)

1.4 Electron Microscopy (EM) Techniques

Electron microscopes basically consist of an electron gun encased in a vacuum. Accelerated electron beams have a much shorter wavelength and consequently greatly increased resolution compared to light radiation. The electron beam is focussed with electromagnets and an image is produced either by passing the beam through a thin section of material as in transmission electron microscopy (TEM), or by electrons impinging on the surface of a bulk sample and emitting further, secondary electrons as in scanning electron microscopy (SEM). Electron microscopy has been used extensively in the study of milk and milk products (for a review see Heertje, 1993). For many years, chemical fixation and dehydration protocols were necessary to preserve biological specimens from the harsh environment of electron microscopes. Consequently, interpretation of EM images requires a thorough understanding of the effects of sample processing on the integrity of microstructural elements and the possible generation of artifacts (Schmidt, 1982; Kalab, 1984). The recent introduction of cryo-stage technology for both TEM and SEM has greatly reduced the number of sample preparation steps and consequently reducing preparation time. These procedures now allow products to be viewed frozen and close to their native state (Schmidt and Buchheim, 1992; Heertje and Pâques, 1995). However, care is still needed to minimize ice crystal formation which can interfere with correct image interpretation.

1.4.1 Transmission Electron Microscopy

Transmission electron microscopy involves passing a narrow beam of electrons through a thin specimen at accelerating voltages in the range 40–120 kV. The sample may be prepared either as a negatively stained dispersion or in the form of a thin section or a metallic replica. Negative staining is a relatively rapid technique where the sample is immersed in a solution of a heavy metal salt such as uranyl acetate. When dried, the sample appears translucent but enables observation of internal or surface structures. This technique is useful for dilute protein dispersions and has been used to study casein micelles and their association with whey proteins (Creamer *et al.*, 1978).

For resin-embedded thin sections, sample preparation can be extensive and usually involves chemical fixation in glutaradehyde and osmium tetroxide, solvent dehydration and embedding in a polymer-based resin such as Spurrs or LR White. Ultrathin sections ~90–150 nm thick are cut using a glass or diamond knife with an ultramicrotome. The sections are post-stained to increase contrast and carbon coated to increase resistance to beam damage. Thin sectioning has been applied to milk (Henstra and Schmidt, 1970) yoghurt (Kalab *et al.*, 1983) and various cheese types including Cheddar, Mozzarella, Gouda, cream cheese and processed cheese (*Green et al.*, 1980; Rayan *et al.*, 1980; Kalab, 1977, 1993). Image analysis has been applied to thin sections of acidified milks to measure pore size (Figure 1.7a, Auty *et al.*, 2005b) and also yoghurt (Skriver *et al.*, 1997), whey protein gels (Langton and Hermansson, 1996) and Mozzarella cheese (Cooke *et al.*, 1995).

A further TEM technique is replica shadowing, usually in combination with freeze fracturing or etching. A small sample of the product is rapidly frozen and fractured to expose internal features. The fracture face may then be warmed under vacuum to allow sublimation of water molecules from the fracture surface to reveal ultrastructural details, Carbon or a heavy metal is then evaporated at an oblique angle to the sample, which is supported on a thin grid. The evaporated coating thus matches the contour features of the sample. This technique has been used extensively to study interfacial features such as milk fat globule membranes (for review, see Buchheim, 1986).

More specific histochemical or immunolocalization procedures to identify particular food proteins have been developed for electron microscopy techniques (Kalab *et al.*, 1995; Armbruster and Desai, 1993; Armbruster *et al.*, 1995). Localization of specific proteins or polysaccharides may be achieved using immunogold labeling in conjunction with TEM. Small gold particles of a known diameter (from 1 to 25 nm) are conjugated

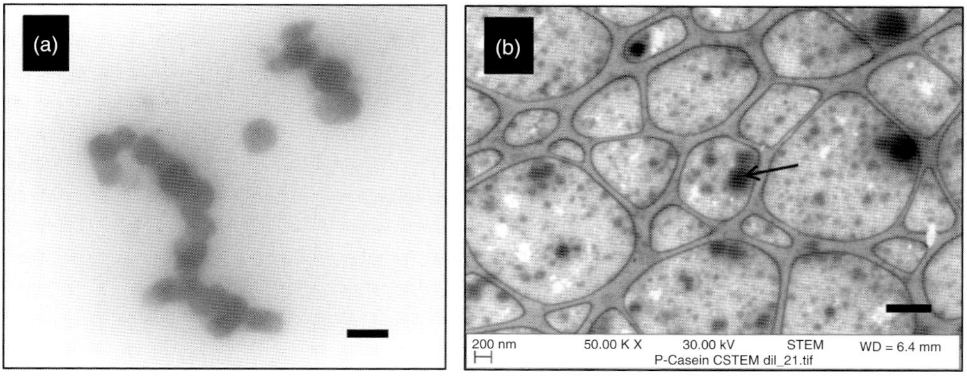

Figure 1.7 Transmission electron micrographs. (a) Resin embedded section of acidified milk showing chain of acidified micellar casein; (b) Cryo-scanning transmission electron micrograph of frozen hydrated skim milk supported on a lacey carbon film showing electron-dense rounded particles, presumed casein micelles (arrow). Scale bar = 500 nm.

to a ligand or receptor that may be an antibody (for proteins) or a lectin (for polysaccharides). β-Lactoglobulin, casein and bovine whey protein have been localized in reduced fat cheese using this approach (Armbruster and Desai, 1993).

1.4.2 Scanning Electron Microscopy

In scanning electron microscopy (SEM), secondary electrons emitted by the sample provide topographic information with a high depth of field. Typical accelerating voltages used are in the range 1–20 kV, depending on the electron source. The high depth of field gives a simulated three-dimensional view of samples. Typical sample preparation includes glutaraldehyde fixation and alcohol dehydration followed by critical point or freeze-drying. To prevent charging of the sample surface in the electron beam, a thin coating of gold or platinum is applied under vacuum. Many dairy products have been studied using this technique; these include milk gels, various cheese types including cream cheese, Cheddar, Mozzarella and Gouda (for reviews, see Kalab, 1981, 1993; Schmidt and Buchheim, 1992). Conventional SEM has been extensively used to characterize the surface morphology of a range of powdered food ingredients including dairy powders such as whole milk powder, skim milk powder, caseinates, whey concentrates and isolates, milk concentrates, infant milk formulae and many other dried ingredients derived from milk (Figure 1.8).

Modern SEM instruments may be fitted with field emission electron sources which allow for much lower voltages (0.1–5 kV) thereby reducing charging effects. However, for powders with high fat contents, cryo-SEM may still be preferable to minimize physical distortion of liquid pools of fat due to the high vacuum (Maher *et al.*, 2014).

1.4.3 Other EM Techniques

1.4.3.1 X-ray Microanalysis

Localization of specific elements in food products is possible using energy dispersive X-ray microanalysis, in conjunction with TEM and, more commonly, with SEM. This

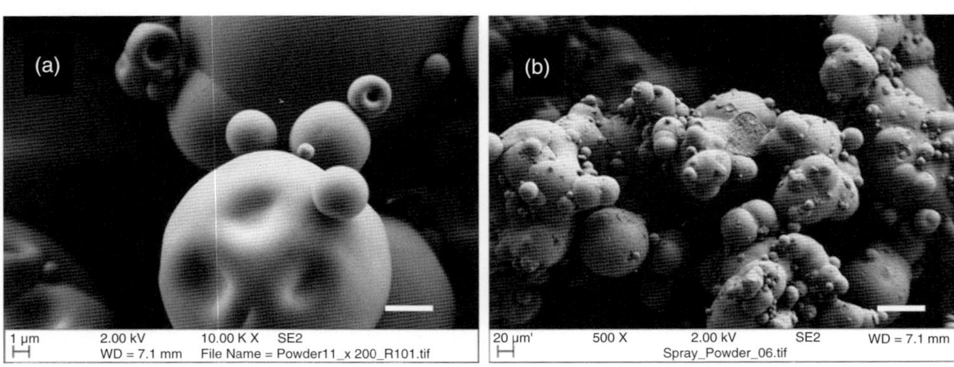

Figure 1.8 Scanning electron micrographs of spray dried milk derived powders. (a) Milk protein concentrate primary particles showing smooth surfaces and some fine particles, scale bar = 2 μm. (b) Infant milk formula spray dried powder, showing agglomeration of primary particles and fines into large particles. Scale bar = 50 μm. Images taken at 2 kV accelerating voltage.

technique has been used to map the distributions of calcium, potassium and phosphorus in chocolate (Brooker, 1990).

1.4.3.2 Cryo-electron Microscopy

Cryo-SEM is particularly useful for dairy products and involves rapid freezing of small samples, fracturing under vacuum, and then transferring to the SEM chamber onto a special cryo-stage, where frozen hydrated bulk samples can be directly imaged under the electron beam. Cryo-SEM has been successfully used to study dairy spreads, mayonnaise and ice cream (Sargent, 1988), although freezing-induced artifacts can occur (Schmidt, 1982; Kalab, 1984). Freeze fracturing allows visualization of internal structures, such as air and fat distribution in whipped cream (Schmidt and van Hooydonk, 1980; Brooker, 1985). The applications of cryo-SEM have increased in recent years, in conjunction with techniques such as CLSM, for the study of dairy food microstructure and microbial populations (Hassan *et al.*, 2003a and 2003b; Romeih *et al.*, 2012; Martinovic *et al.*, 2013; Ong *et al.*, 2013). Costa *et al.* (2013) used cryo-SEM to study exopolysaccharide producing bacteria in Cheddar cheese showing that exopolysaccharide producers were concentrated at the fat-protein interface. Examples of cryo-SEM for whipped cream, Cheddar cheese and yoghurt are shown in Figure 1.9. Care should be taken to reduce ice crystal damage particularly for high moisture products such as yoghurt, since ice crystals will form resulting in a honeycomb-like appearance (Figure 1.9d).

Cryo-TEM techniques, such as thin film freezing, have been used to study casein micelles close to their native state (van Bruggen *et al.*, 1986). This technique involves small (<10 μm) samples being rapidly frozen to vitrify water, thus preventing ice crystal damage. Samples are observed at −80 to −150°C. Gee *et al.* (2010) developed a variant of this technique to visualize milk proteins by using a transmitted electron detector in a field emission SEM fitted with a cryo-stage (Figure 1.7b). For larger samples, cryoprotectants are required and the samples are cryo-sectioned (Schmidt and Buchheim, 1992).

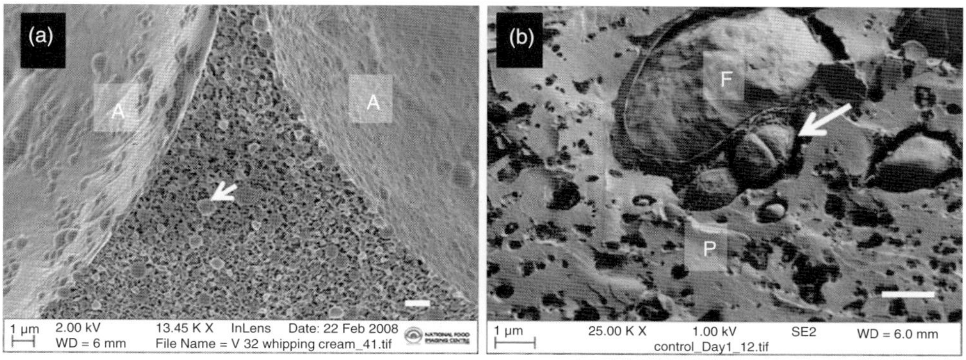

Figure 1.9 Cryo-scanning electron micrographs of various dairy products images at −125°C and prepared by plunge freezing into melting nitrogen, fracturing (a, b and d) and moderate sublimation. (a) Whipped cream showing the interface between two air bubbles (A) that contains fat droplets (arrow) and milk proteins dispersed in the aqueous phase. (b) Cheddar cheese showing fat globule (F), (P) and streptococci-like bacteria at the fat/protein interface (arrow). (c) Full-fat yoghurt prepared by thin film freezing (not fractured) showing starter bacteria (arrows) and fat droplet partially embedded in the protein matrix. (d) Full-fat yoghurt cryo-fractured fracture showing large pores caused by ice crystal damage during the freezing process (arrows). Scale bar = 1 μm.

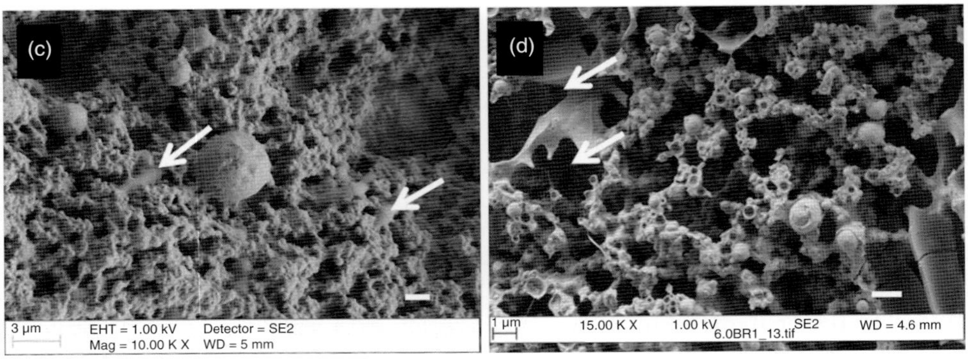

Figure 1.9 (Continued)

1.4.3.3 Environmental and Variable Pressure SEM

One of the main problems with conventional EM techniques is the damage to non-conductive specimens caused by the high-energy electrons used to visualize the sample. Biological materials, being relatively non-conductive, are especially susceptible to beam damage and require coating with a conductive layer of carbon or heavy metal. In addition, the electron beam requires a vacuum to prevent dissipation of electrons in the column. Variable pressure (sometimes termed low vacuum mode) or true "Environmental" scanning electron microscopes overcome both specimen-charging effects due to the electron beam, and also the need for dehydration by allowing a gas of 1–20 Torr within the imaging chamber. The gas may be inert or water vapor, consequently samples may be observed at saturated water vapor pressure in the case of true ESEMs (Danilatos, 1991). Variable pressure SEM employs a weak vacuum that still allows direct observation of dairy samples with moderate moisture content such as cheese.

1.5 Emerging Microscopy Techniques

There is a need to develop new imaging tools that are relevant to the metastable complexity of dairy products. These new tools ideally will require the following features:

- **Speed of acquisition:** many dairy products are highly metastable or liquids.
- **Chemical mapping:** particularly for complex organics such as proteins.
- **High resolution:** spanning the nano-micro length scales.
- **Controlled environment:** for example temperature.

There are several exciting new developments in microscopy and imaging that can address most of these issues and can be further exploited by dairy researchers. Some of these are discussed below.

1.5.1 Atomic Force Microscopy

Scanning probe microscopies are a group of techniques originally designed to characterize surfaces at the atomic scale. Atomic force microscopy (AFM) is one such technique and, allows imaging of soft composite materials and liquids at molecular scales with a resolution of <1 nm. AFM is increasingly beginning to be applied to food materials including dairy products. A cantilever with a needle-like probe tip is scanned close to or touching the sample surface and deflection monitored by a laser and photodiode and converted to topographical and/or nano-mechanical information. Atomic force microscopy has been used to study food materials including polysaccharides, starches and cereal proteins (Miles *et al.*, 1991, Thomson *et al.*, 1992). There have been few reported studies of dairy applications, although there is great potential for probing the surfaces of dried food ingredients such as spray dried powders (Figure 1.10b). Lent *et al.* (1998) successfully measured β-lactoglobulin molecules (~7 nm diameter) in whey protein films using non-contact AFM. Recently Oboroceanu *et al.* (2010) used AFM in liquid mode to visualize whey proteins such as β-lactoglobulin fibrils in water (Figure 1.10a). Scanning near field optical microscopy (SNOM) is a similar technique that enables fluorescence or infrared spectroscopic imaging of

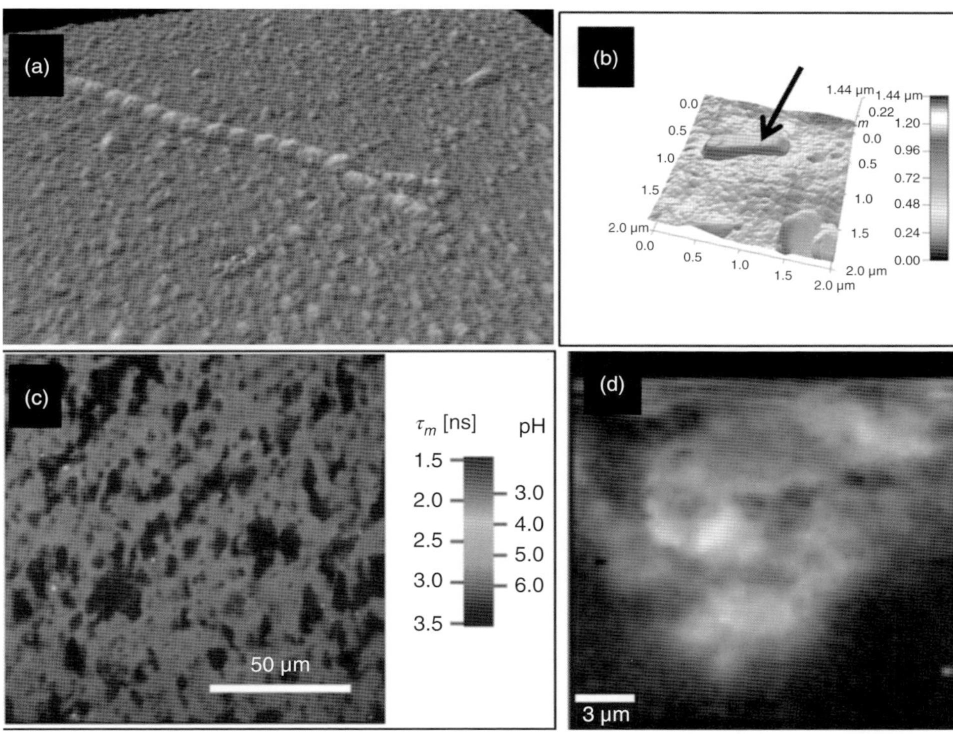

Figure 1.10 Some examples of advanced/emerging imaging techniques. (a) Atomic force 3D micrograph (height image) of β-lactoglobulin fibrils imaged in water; fibrils are ~3 nm thick. (b) Atomic force microscope image of surface of spray dried skim milk powder stored at 55% relative humidity, showing lactose crystallizing at the surface (arrow). (c) Fluorescence lifetime imaging (FLIM) of Cheddar cheese labeled with Oregon Green 488 nm, showing heterogeneous pH distribution (adapted from Burdikova *et al.*, 2015b). (d) Confocal Raman microscope z-depth image of Cheddar cheese from a discolored region showing presence of lycopene (red). (*See color plate section for the color representation of this figure.*)

biological structures at sub-wavelength resolution based on near-field evanescence (Betzig and Chichester, 1993).

1.5.2 Advanced Fluorescence Microscopy Techniques

Fluorescence recovery after photo-bleaching (FRAP), in combination with CSLM, can be used to measure diffusion coefficients by timing the rate fluorescent molecules diffuse or percolate into a pre-defined photo-bleached region of a fluorescently labeled sample. Floury *et al.* (2012) used FRAP to study diffusion kinetics in a model cheese and this technique could be applied to other gel systems such as yoghurts or low-moisture dairy products. Burdikova *et al.* (2015a) has demonstrated the potential of advanced optical microscopic techniques including fluorescence lifetime imaging (FLIM), confocal Raman spectroscopy (see for example Figure 1.10 c and d) in addition to non-linear microscopy (two-photon excitation and second harmonic generation microscopy), to study localized physicochemical properties of cheese matrices. This study showed that second harmonic generation microscopy imaging facilitates visualization of calcium lactate crystals and two-photon excitation fluorescent lifetime microscopic techniques enabled more precise determination and location of fats and phospholipids within the cheese matrix. This research group has also successfully FLIM to map the pH distribution in cheese using the fluorochrome Oregon Green 488® (Burdikova *et al.*, 2015b, Figure 1.0c). It is hoped that these advanced techniques will be further developed and applied to study the complex metabolic processes in fermented products such as yoghurts and cheeses.

1.5.3 Confocal Raman Microscopy

The development of confocal Raman microscopy will become a key tool in the microstructural characterization of dairy products (Pudney *et al.*, 2004). Raman imaging is a powerful technique for generating detailed chemical images based on a sample's Raman spectrum. A complete spectrum is acquired at each pixel of the image, which is then processed to generate false color images based on material composition, phase, crystallinity and strain. This allows for 3-dimensional molecular mapping of a material with a spatial resolution of < 1 μm. Since there is only one Raman photon for every $\sim 10^6$ fluorescence photons, care is needed to avoid natural fluorescence which can swamp the weak Raman signal. For spectral mapping, acquisition and processing times are also much slower than those of traditional confocal microscopes and can take several minutes or hours rather than seconds. Recent developments in technology have reduced the time to acquire a full spectrum down to just under a millisecond. Despite these limitations, there have been promising early studies that demonstrate the power of the technique. These include characterization of the milk fat globule (Gallier *et al.*, 2011) and identifying the cause of a "pinking" discoloration in natural hard cheeses (Quigley *et al.*, 2016, Figure 1.10d), where confocal Raman microscopy successfully identified the microbially metabolized pigment lycopane as the source.

1.5.4 X-ray Nano/Microtomography

X-ray microtomography is a non-destructive analysis technique used to visualize and characterize objects in three dimensions. It is the process of imaging an object from many directions using penetrating radiation such as X-rays, then using a computer to determine the interior structure of that object from these projected images. It has been used to probe the structure of dairy products such as cream cheeses and yoghurts (Laverse *et al.*, 2011a and 2011b). Chawanji *et al.* (2012) used X-ray micro tomography (XMT) to visualize the microstructure of loose-packed and compacted samples of spray-dried skim milk powder (SMP) and whole milk powder (WMP) and to quantify the proportion of both interstitial and occluded air voids in each sample. Individual beam scans can be analysed and reconstructed for true volumetric morphological analysis. Disadvantages include the lack of atomic number contrast in many food materials and the relatively slow acquisition speed (minutes or hours). These studies showed moderately high resolution images (~10 μm) but newer high resolution instruments are now available that can image samples at the nanometer scale. New phase contrast imaging and faster data processing rates will further improve XMT making it a very powerful technique for direct imaging of liquid and solid dairy products in their bulk state and at sub-micron resolution in full 3D.

1.5.5 Super-Resolution Microscopy

Super-resolution techniques can increase optical resolution beyond the diffraction limit of light and into the nano-scale. There are now many approaches to achieving this, but they fall into two main groups: (1) Deterministic super-resolution, such as STED, GSD, RESOLFT and SSIM; and (2) Stochastic super-resolution, for SPDM, SPDMphymod, PALM, FPALM, STORM and dSTORM. To date there have been very few, if any, reported dairy applications, so there is enormous potential to exploit these techniques in characterizing interfacial proteins and colloidal dairy systems such as milk which typically contains casein micelles ranging from 50–300 nm in diameter and cannot be resolved using conventional CSLM.

1.6 Image Analysis

The continuing increase in computing power combined with advances in unbiased, statistically rigorous stereological procedures should greatly improve the reliability of quantitative data from microscopic images. Both image analysis and stereological techniques are increasingly used in combination in food structure studies. Langton and Hermansson (1996) and Auty *et al.* (2005b) used a combination of image analysis and stereology techniques to study particle and pore sizes of whey protein gels. van Dalen (2002) used image analysis with shape unfolding algorithms to quantify the water droplet size distribution of fat spreads. Fenoul *et al.* (2008) applied image analysis of CSLM images to cream cheese. It is anticipated that these measurement techniques, in combination with newer image deconvolution techniques, will in future enable accurate quantification of three-dimensional microstructures by CSLM. A typical basic approach, for example when measuring droplet size, would be to threshold or segment the image to a binary image, separate touching objects then measure equivalent diameter. It is very important when quantifying image data to ensure consistent sampling and imaging conditions and to ensure that the software is measuring the object of interest and the aim is always to understand the true microstructure of the sample, rather than generating large datasets. For more information on image analysis of food materials the reader should consult the book *Image Analysis of Food Microstructure* by John C. Russ (Russ, 2004).

1.7 Conclusions

This chapter has summarized the main techniques that have been used to characterize the structure of dairy products. Microscopy offers a powerful set of tools to help understand the complex relationships between structure and function in dairy products. There are now a wide variety of microscopy techniques to choose from and new techniques develop continuously. It is therefore important to know the strengths and weaknesses of each microscopy technique for a particular application. Ideally, the researcher should use a correlative approach to ensure a thorough understanding of their particular dairy product's structure throughout the molecular, nano, micro and macro-length scales.

References

Abhyankar, A.R., Mulvihill, D.M., Chaurin, V. and Auty, M.A.E. (2011a). Techniques for localization of Konjac glucomannan in model milk protein–polysaccharide mixed systems: physico-chemical and microscopic investigations. Food Chemistry 129, 1362–1368.

Abhyankar, A.R., Mulvihill, D.M. and Auty, M.A.E. (2011b). Combined microscopic and dynamic rheological methods for studying the structural breakdown properties of whey protein gels and emulsion filled gels. Food Hydrocolloids 25, 275–282.

Aguilera, J.M. and Stanley, D.W. (1990). Microstructural Principles of Food Processing and Engineering, Elsevier Science Publishers, Oxford, p. 47.

Arltoft, D., Madsen, F. and Ipsen, R. (2007). Screening of probes for specific localization of polysaccharides. Food Hydrocolloids 21, 1062–1071.

Armbruster, B.L. and Desai, N. (1993). Identification of proteins and complex carbohydrates in some commercial low-fat dairy products by means of immunolocalization techniques. Food Structure 12, 289–299.

Armbruster, B.L., Chastain, S. and Desai, N. (1995). Immunolocalisation and microstructure of milk proteins and fat mimetics in reduced fat cheese. In: Chemistry of Structure-Function Relationships in Cheese. E.L. Malin and M.H. Tunick, Eds. Plenum Press, New York, pp. 277–294.

Auty, M.A.E., Fenelon, M.A., Guinee, T.P., Mullins, C., Mulvihill, D.M. (1999). Dynamic Confocal Scanning Laser Microscopy Methods for Studying Milk Protein Gelation and Cheese Melting. Scanning 21, 299–304.

Auty, M.A.E., Twomey, M., Guinee, T.P. and Mulvihill, D.M. (2001a). Development and applications of confocal scanning laser microscopy methods for studying the distribution of fat and protein in selected food products. Journal of Dairy Research 68, 417–427.

Auty, M.A.E., Gardiner, G., McBrearty, S., O'Sullivan, E., Mulvihill, D.M., Collins, J.K., Fitzgerald, G.F., Stanton, C. and Ross, R.P. (2001b). Direct in situ viability assessment of bacteria in probiotic products using viability staining in conjunction with confocal scanning laser microscopy. Applied and Environmental Microbiology 67, 420–425.

Auty, M.A.E., Duffy, G., O'Beirne, D., McGovern, A., Gleeson, E. and Jordan, K. (2005a). In situ localization of Escherichia coli O157:H7 in food by confocal scanning laser microscopy. Journal of Food Protection 68, 482–486.

Auty, M.A.E., O'Kennedy, B.T., Allan-Wojtas, P. and Mulvihill, D.M. (2005b). The application of microscopy and rheology to study the effect of milk salt concentration on the structure of acidified micellar casein systems. Food Hydrocolloids 19, 101–109.

Belder, A. N. D., and Granath, K. (1973). Preparation and Properties of Fluorescein-labeled dextrans. Carbohydrate Research 30, 375–378.

van den Berg, L., van Vliet, T., van der Linden, E., van Boekel, M.A.J.S. and van de Velde, F. (2007). Breakdown properties and sensory perception of whey proteins/polysaccharide mixed gels as a function of microstructure. Food Hydrocolloids 21, 961–976.

van den Berg, L., Klok, H.J., van Vliet, T., van der Linden, E., van Boekel, M.A.J.S. and van de Velde, F. (2008). Quantification of a 3D structural evolution of food composites under large deformations using microrheology. Food Hydrocolloids 22, 1574–1583.

Betzig, E. and Chichester, R.J. (1993). Single molecules observed by near-field scanning optical microscopy. Science 262, 1422–1425.

Blonk, J.C.G. and van Aalst, H. (1993). Confocal scanning light microscopy in food research. Food Research International 26, 297–311.

Brakenhoff, G.J., van der Voort, H.T.M., van Spronsen, E.A. and Nanninga, N. (1988). 3-Dimensional imaging of biological structures by high resolution confocal scanning laser microscopy. Scanning Microscopy 2, 33–40.

Bremer, L.G.B., Bijsterbosch, B.H., Schrijvers, R., van Vliet, T. and Walstra, P. (1993). Formation, properties and fractal nature of particle gels. Advances in Colloid and Interface Science 46, 117–128.

Brink, J., Langton, M., Stading, M. and Hermansson, A-M. (2007). Simultaneous analysis of the structural and mechanical changes during large deformation of whey protein isolate/gelatin gels at the macro and micro levels. Food Hydrocolloids 21, 409–419.

Brooker, B.E. (1979). Milk and its products. In: Food Microscopy. J.G. Vaughan, Ed. Academic Press, London, pp. 273–331.

Brooker, B.E. (1985). Observations on the air-serum interface of milk foams. Food Microstructure 4, 289–296.

Brooker, B.E. (1990). Identification and characterization of cocoa solids and milk proteins in chocolate using X-ray microanalysis. Food Structure 9, 9–21.

Brooker, B.E. 1991. The study of food systems using confocal laser scanning microscopy. Microscopy and Analysis 27, 13–15.

Brooker, B.E. (1995). Imaging food systems by confocal scanning laser microscopy. In: New Physico-chemical Techniques for the Characterization of Complex Food Systems. Dickinson E., Ed. Blackie Academic and Professional, London, pp. 53–68.

Brooks, S.A., Leathem, A.J.C. and Schumacher, U. 1997. Lectin Histochemistry. BIOS Scientific Publishers Limited, Oxford.

van Bruggen, E.F.J., Booy, F.P., van Breemen, J.F.L., Brink, J., Keegstra, W. and Schmidt, D.G. (1986). Cryo-EM of various biomolecular assemblies. In; Proceedings of the 11th International Congress on Electron Microscopy, Kyoto.

Buchheim, W. (1986). Membranes of milk fat globules: ultrastructural, biochemical and technological aspects. Kieler Milchwirt. Forsch 38, 227–246.

Burdikova, Z., Svindrych, Z., Hickey, C., Wilkinson, M.G., Auty, M.A.E., Samek, O., Bernatova S., Krzyzanek, V., Periasamy, A. and Sheehan, J.J. (2015a). Application of advanced light microscopic techniques to gain deeper insights into cheese matrix physico-chemistry, Dairy Science and Technology 95: 687–700.

Burdikova Z, Svindrych Z, Pala J Hickey, C., Wilkinson, M.G., Panek, J., Auty, M.A.E., Periasamy, A. (2015b). Measurement of pH micro-heterogeneity in natural cheese matrices by fluorescence lifetime imaging. Frontiers in Microbiology, 6: 183. https://dx.doi.org/10.3389%2Ffmicb.2015.00183.

Chawanji, A.S., Baldwin, A.J., Brisson, G. and Webster, E. (2012). Use of X-ray micro tomography to study the microstructure of loose-packed and compacted milk powders. Journal of Microscopy, 248, 49–57.

Ciron, C.I.E., Kelly, A.L. and Auty, M.A.E. (2012). Modifying the microstructure of low-fat yoghurt by microfluidization of milk under different pressures to enhance rheological and sensory properties. Food Chemistry, 130, 510–519.

Clegg, S.M., Moore, A.K. and Jones, S.A. 1996. Low-fat margarine spreads as affected by aqueous phase hydrocolloids. Journal of Food Science 61, 1073–1079.

Cooke, P.H., Tunick, M.H., Malin, E.L., Smith, P.W. and Holsinger, V.H. (1995). Electron-density patterns in low-fat mozzarella cheeses during refrigerated storage. In: Chemistry of Structure-Function Relationships in Cheese. Malin, E.L. and Tunick, M.H., Eds. Plenum Press, New York, pp. 311–320.

Costa, N.E., Wang, L., Auty, M.A.E., Hannon, J.E., McSweeney, P.L.H. and Beresford, T.P. (2013). Rheological, microscopic and primary characterization of the exopolysaccharide produced by *L. lactis* subsp. *cremoris* DPC6532. Dairy Science and Technology, 92: 219-235.

Creamer, L.K., Berry, G.P. and Matheson, A.R. (1978). The effect of pH on protein aggregation in skim milk. New Zealand Dairy Science and Technology 13, 9–15.

van Dalen, G. (2002). Determination of the water droplet size distribution of fat spreads using confocal scanning laser microscopy. Journal of Microscopy 208, 116–133.

Danilatos, G.D. (1991). Review and outline of environmental SEM at present. Journal of Microscopy 162, 391–402.

Drusch, S. and Berg, S. (2008). Extractable oil in microcapsules prepared by spray-drying: Localization, determination and impact on oxidative stability. Food Chemistry 109, 17–24.

El-Bakry, M. and Sheehan, J.J. (2014). Analysing cheese microstructure: A review of recent developments, Journal of Food Engineering 125, 84–96.

Emmambux, M.N. and Stading, M. (2007). In situ tensile deformation of zein films with plasticizers and filler materials. Food Hydrocolloids 21, 1245–1255.

Everett, D.W., Ding, K., Olson, N.F. and Gunasekaran, S. (1995). Applications of confocal microscopy to fat globule structure in cheese. In: Chemistry of Structure-Function Relationships in Cheese. Malin, E.L. and Tunick, M.H., Eds. Plenum Press, New York, pp. 321–330.

Everett, D.W. and Olson, N.F. (2003). Free oil and rheology of Cheddar cheese containing fat globules stabilised with different proteins. Journal of Dairy Science 86, 755–763.

Fenoul, F., Le Denmat, M., Hamdi, F., Cuvelier, G. and Michon, C. (2008). Technical note: Confocal scanning laser microscopy and quantitative image analysis: Application to cream cheese microstructure investigation. Journal of Dairy Science 91, 1325–1333.

Flint, FO. (1990). Micro-technique for the identification of food hydrocolloids. Analyst 115, 61–63.

Flint, O. (1994). Food Microscopy. Bios Scientific Publishers Ltd., Oxford.

Floury, J., Madec, M-N., Waharte, F., Jeanson, S. and Lortal, S. (2012). First assessment of diffusion coefficients in model cheese by fluorescence recovery after photobleaching (FRAP). Food Chemistry 133, 551–556.

Gallier, S., Gordon, K.C., Jimenez-Flores, R. and Everett, D.W. (2011). Composition of bovine milk fat globules by confocal Raman microscopy. International Dairy Journal 21, 402–412.

Garnier, C., Bourriot, S. and Doublier, J-L. (1998). The use of confocal laser scanning microscopy in studying mixed biopolymer systems. In: Gums and Stabilisers in the Food Industry 9; 9th conference. Royal Society of Chemistry, Cambridge, pp. 247–256.

Green, M.L., Turvey, L.A. and Hobbs, D.G. (1980). Development of structure and texture in cheddar cheese. Journal of Dairy Research 48, 343–355.

Guinee, T.P., Auty, M.A.E. and Mullins, C. (1999). Observations on the microstructure and heat-induced changes in the viscoelasticity of commercial cheeses. Australian Journal of Dairy Technology 54, 84–89.

Guinee, T.P., Auty, M.A.E. and Fenelon, M.A. (2000a). The effect of fat content on the rheology, microstructure and heat-induced functional characteristics of Cheddar cheese. International Dairy Journal 10, 277–288.

Guinee, T.P., Auty, M.A.E., Mullins, C., Corcoran, M.O. and Mulholland, E.O. (2000b). Preliminary observations on the effects of fat content and degree of emulsification on the structure-functional relationship of Cheddar cheese. Journal of Texture Studies 31, 645–663.

Guinee, T.P., Feeney, E.P., Auty, M.A.E. and Fox, P.F. (2002). Effect of pH and calcium concentration on some textural and functional properties of Mozzarella cheese. Journal of Dairy Science 85, 1–15.

Gunasekaran, S. and Ding, K. (1999). Three-dimensional characteristics of fat globules in Cheddar cheese. Journal of Dairy Science 82, 1890–1896.

Hagiwara, T., Kumagai, H., Matsunaga, T. and Nakamura, K. (1997). Analysis of aggregate structure in food protein gels with the concept of fractal. Bioscience, Biotechnology, Biochemistry 61, 1663–1667.

Hannon, J.A., Lopez, C., Madec, M.N. and Lortal, S. (2006). Altering Renneting pH Changes Microstructure, Cell Distribution, and Lysis of Lactococcus lactis AM2 in Cheese Made from Ultrafiltered Milk. Journal of Dairy Science 89, 812–823.

Hassan, A.N., Frank, J.F., Farmer, M.A., Schmidt, K.A. and Shalabi, S.I. (1995a). Observation of encapsulated lactic acid bacteria using confocal scanning laser microscopy. Journal of Dairy Science 78, 2624–2628.

Hassan, A.N., Frank, J.F., Farmer, M.A., Schmidt, K.A. and Shalabi, S.I. (1995b). Formation of yoghurt microstructure and three-dimensional visualization as determined by confocal scanning laser microscopy. Journal of Dairy Science 78, 2629–2636.

Hassan, A.N. and Frank, J.F. (1997). Modification of microstructure and texture of rennet curd by using a capsule-forming non-ropy lactic culture. Journal of Dairy Research 64, 115–121.

Hassan, A.N., Frank, J.F. and Qvist, K.B. (2002). Direct observation of bacterial exopolysaccharides in dairy products using confocal scanning laser microscopy. Journal of Dairy Science 85, 1705–1708.

Hassan, A.N., Ipsen, R., Janzen, T. and Qvist, K.B. (2003a). Microstructure and rheology of yoghurt made with cultures differing only in their ability to produce exopolysaccharides. Journal of Dairy Science 86, 1632–1638.

Hassan, A.N., Frank, J.F., and Elsoda, M. (2003b). Observation of bacterial exopolysaccharide in dairy products using cryo-scanning electron microscopy. International Dairy Journal 13, 755–762.

Heertje, I., van der Vlist, P., Blonk, J.C.G., Hendrickx, H.A.C. and Brackenhof, G.J. (1987). Confocal scanning laser microscopy in food research: some observations. Food Microstructure 6, 115–120.

Heertje, I., Nederlof, J., Hendrickx, H.A.C.M. and Lucassen-Reynders, E. (1990). The observation of the displacement of emulsifiers by confocal scanning laser microscopy. Food Structure 9, 305–316.

Heertje, I. (1993). Structure and Function in Food Products – A Review. Food Structure 12, 343–364.

Heertje, I. and Pâques, M. (1995). Advances in Electron Microscopy. In: New Physico-chemical Techniques for the Characterization of Complex Food Systems. Dickenson, E., Ed. Blackie Academic and Professional, London, pp. 1–52.

Heertje, I., van Aalst, H., Blonk, J.C.G., Nederlof, D.J. and Lucassesn-Reynders, E.H. (1996). Observations on emulsifiers at the interface between oil and water by confocal scanning light microscopy. Lebensmittel Wissenschaft und Technologie 29, 217–226.

Heilig, A., Goeggerle A. and Joerg, H. (2009). Multiphase visualization of fat containing beta-lactoglobulin-kappa-carrageenan gels by confocal scanning laser microscopy, using a novel dye, V03-01136, for fat staining. Lwt-Food Science and Technology 42, 646–653.

Henstra, S. and Schmidt, D.G. (1970). Ultrathin sections from milk using the microcapsule method. Naturewissenschaften 57, 247–248.

Herbert, S., Bouchet, B., Riaublanc, A., Dufour, E. and Gallant, D.J. (1999). Multiple fluorescence labeling of proteins, lipids and whey in dairy products using confocal microscopy. Lait 79, 567–575.

Hervé, C., Marcus, S.E. and Knox, J.P. (2011). Monoclonal antibodies, carbohydrate-binding modules, and the detection of polysaccharides in plant cell walls. In: The Plant Cell Wall: Methods and Protocols, Methods in Molecular Biology, vol. 715. Popper Z.A., Ed. Springer/Humana Press, pp. 103–113.

Heuer, A., Cox, A.R., Singleton, S., Barigou. M. and Ginkel, M.-v. (2007). Visualization of foam microstructure when subject to pressure change. Colloids and Surfaces A: Physicochemical and Engineering Aspects 311, 112–23.

Hibberd, D., Holmes, A, Farood, M., Fillery-Travis, A., Robins, M and Challis, R. (1997). Ultrasonic characterization of flocculation in oil-in-water emulsions. In: Food Colloids: Proteins, Lipids and Polysaccharides. Dickinson, E. and Bergenståhl, B., Eds. Royal Society of Chemistry, Cambridge, pp. 137–149.

Holcomb, D.N. (1991). Structure and rheology of dairy products. A compilation of references with subject and author indexes. Food Structure 10, 45–108.

Kalab, M. (1977). Milk gel structure VI. Cheese texture and microstructure. Milchwissenschaft 32, 449–458.

Kalab, M. (1979a). Scanning electron microscopy of dairy products: an overview. Scanning Electron Microscopy III, 261–272.

Kalab, M. (1979b). Microstructure of dairy foods. 1. Milk products based on protein, Journal of Dairy Science 62, 1352–1364.

Kalab, M. (1979c). Microstructure of dairy foods. 2. Milk products based on fat, Journal of Dairy Science 68, 3234–3248.

Kalab, M. (1981). Scanning electron microscopy of dairy products: a review of techniques. Scanning Electron Microscopy III, 453–472.

Kalab, M. (1984). Artefacts in conventional scanning electron microscopy of some milk products. Food Microstructure 3, 95–111.

Kalab, M. (1993). Practical aspects of electron microscopy in dairy research, Food Structure 12, 95–114.

Kalab, M., Allan-Wojtas, P. and Phipps-Todd, E. (1983). Development of microstructure in set-style nonfat yoghurt – a review. Food Microstructure 2, 51–66.

Kalab, M., Allan-Wojtas, P. and Miller, S.S. (1995). Microscopy and other imaging techniques in food structure analysis. Trends in Food Science and Technology 6, 177–186.

Kett, A.P., Chaurin, V., Fitzsimons, S.M., Morris, E.R., O'Mahony, J.A. and Fenelon, M.A. (2013). Influence of milk proteins on the pasting behavior and microstructural characteristics of waxy maize starch. Food Hydrocolloids, 30, 661–671.

King, N. (1958). Observations by fluorescence microscopy of casein in milk, curd and cheese. Journal of Dairy Research 25, 312–319.

Langton, M. and Hermansson, A-H. (1996). Image analysis of particulate whey protein gels. Food Hydrocolloids 10, 179–191.

Langton, M., Jordansson, Elvy., Altskär, A., Sørensen, C. and Hermansson, A-H. (1999). Microstructure and image analysis of mayonnaises. Food Hydrocolloids 13, 113–125.

Laverse, J., Mastromatteo, M., Frisullo, P. and Del Nobile, M.A. (2011a). X-ray microtomography to study the microstructure of cream cheese-type products. Journal of Dairy Science 94, 43–50.

Laverse, J., Mastromatteo, M., Frisullo, P., Albenzio, M., Gammariello, D. and Del Nobile, M.A. (2011b). Fat microstructure of yoghurt as assessed by X-ray microtomography. Journal of Dairy Science 94, 668–675.

van Leeuwenhoek, A. (1674). Microscopical observations concerning blood, milk, bones, the brain, spittle, and cuticula. Philosophical Transactions 9, 121–128.

Lent, L.E., Vanasupa, L.S. and Tong, P.S. (1998). Whey protein edible film structures determined by atomic force microscope. Journal of Food Science 63, 824–827.

Lewis, D.F. (1978). Manual of Microscopical Methods. Leatherhead Food R.A., Surrey, UK.

Lewis, D.F. (1993). Development of the food microscopist. Food Structure 12, 275–283.

Li, M., Auty, M.A.E., O'Mahony, J.A., Kelly, A. and Brodkorb, A. (2016). Covalent labeling of β-casein and its effect on the microstructure and physicochemical properties of a mixed casein and whey protein stabilised emulsions. Food Hydrocolloids 61, 505–513.

Loren, N., Langton, M. and Hermansson, A.M. (2007). Confocal fluorescence microscopy (CLSM) for food structure characterization. In: Understanding and Controlling the Microstructure of Complex Foods. McClements D.J., Ed., pp. 232–260.

Loris, R., Hamelryck, T., Bouckaert, J. and Wyns, L. 1998. Legume lectin structure. Biochimica et Biophysica Acta – Protein Structure and Molecular Enzymology, 1383, 9–36.

Maher, P.G., Auty, M.A.E., Roos, Y.H., Zychowski, L.M. and Fenelon, M.A. (2014). Microstructure and lactose crystallization properties in spray-dried nanoemulsions. Food Structure 3, 1–11.

Manski, J.M., van der Goot, A.J. and Boom, R.M. (2007). Influence of shear during enzymitic gelation of caseinate-water and caseinate-water-fat systems. Journal of Food Engineering 79, 706–17.

Martinovic, A., Moe, K.M., Romeih, E., Aideh, B., Vogensen, F.K., Østlie, H. and Skeie, S. (2013). Growth of adjunct *Lactobacillus casei* in Cheddar cheese differing in milk fat globule membrane components. International Dairy Journal 31, 70–82.

McCarthy, N.A., Gee, V.L., Hickey, D.K., Kelly, A.L., O'Mahony, J.A., Fenelon, M.A. (2012). Effect of protein content on the physical stability and microstructure of a model infant formula. International Dairy Journal, doi: 10.1016/j.idairyj.2012.10.004.

McKenna, A.B. (1997). Examination of whole milk powder by confocal scanning laser microscopy. Journal of Dairy Research 64, 423–432.

Miles, M.J., Carr, H.J., McMaster, T.J., L'Anson, K.J., Belton, P.S., Morris, V.J. Field, J.M. Shewry, P.R. and Tatham, A.S. (1991). Scanning tunnelling microscopy of wheat seed storage protein reveals details of an unusual supersecondary structure, Proceedings of the National Academy of Science USA 88, 68–71.

Minsky, M. (1957). U.S. Patent #3013467, Microscopy Apparatus.

Morris, V. and Groves, K. (2013). Food Microstructure: Microscopy, Measurement and Modelling. Morris, V. and Groves K., Eds. Woodhead Publishing, Cambridge, 438 pp.

Nicolas, Y., Paques, M., van den Ende, D., Dhont, J.K.G., van Polanen, R.C., Knaebel, A., *et al.* (2003). Microrheology: new methods to approach the functional properties of food. Food Hydrocolloids, 17, 907–913.

Oboroceanu, D., Wang, L., Brodkorb, A., Magner, E. and Auty, M.A.E. (2010). Characterization of β-lactoglobulin fibrillar assembly using atomic force microscopy, polyacrylamide gel electrophoresis and in situ Fourier transform infrared spectroscopy. Journal of Agricultural and Food Chemistry, 58, 3667–3673.

Olsson, C., Langton, M., Hermansson, A-M. (2002). Microstructures of β-lactoglobulin/amylopectin gels on different length scales and their significance for rheological properties. Food Hydrocolloids 16, 111–126.

Ong, L., Dagastine, R.R., Auty, M.A.E., Kentish, S.E. and Gras, S.L. (2011). Influence of coagulation temperature on the microstructure and composition of full fat Cheddar cheese. Dairy Science and Technology 91, 739–758.

Ong, L., Dagastine, R., Kentish, S. and Gras, S. (2013). Microstructure and composition of full fat Cheddar cheese made with ultrafiltered milk retentate. Foods 2, 310–331.

O'Reilly, C.E., Murphy, P.M., Kelly, A.L., Guinee, T.P., Auty, M.A.E. and Beresford, T.P. (2002). The effect of high-pressure treatment on the functional and rheological properties of Mozzarella cheese. Innovative Food Science and Emerging Technologies 3, 3–9.

Patonay, G., Salon, J., Sowell, J. and Strekowski, L. (2004). Noncovalent labeling of biomolecules with red and near-infrared dyes. Molecules 9, 40–49.

Pereira, R., Matia-Merino, L. Jones, V. and Singh, H. (2006). Influence of fat on the perceived texture of set acid milk gels: a sensory perspective. Food Hydrocolloids 20, 305–313.

Pudney, P.D.A., Hancewicz, T.M., Cunningham, D.G and Brown, M.C. (2004). Quantifying the microstructures of soft solid materials by confocal Raman spectroscopy. Vibrational Spectroscopy 34, 123–135.

Rayan, A.A., Kalab, M. and Ernstrom, C.A. (1980). Microstructure and rheology of process cheese. Scanning Electron Microscopy III, 635–643.

Romeih, E.A., Moe, K.M., and Skeie, S. (2012). The influence of fat globule membrane material on the microstructure of low-fat Cheddar cheese. International Dairy Journal 26, 66–72.

Quigley L., O'Sullivan D.J., Daly D., O'Sullivan O., Burdikova Z., Vana R., Beresford T.P., Ross R.P., Fitzgerald G.F., McSweeney P.L.H., Giblin L., Sheehan J.J. and Cotter P.D. (2016). Thermus and the pink discoloration defect in cheese. mSystems 1(3):e00023-16. doi:10.1128/mSystems.00023-16.

Rowney, M.K., Roupas, P., Hickey, M.W. and Everett, D.W. (2003a). The effect of compression, stretching and cooking temperature on free oil formation in Mozzarella curd. Journal of Dairy Science 86, 449–456.

Rowney, M.K., Roupas, P., Hickey, M.W. and Everett, D.W. (2003b). The effect of homogenization and milk fat fractions on the functionality of Mozzarella cheese. Journal of Dairy Science 86, 712–718.

Russ, J. (2004). Image Analysis of Food Microstructure. CRC Press, 384 pp.

Saito, Z. (1985). Particle structure in spray-dried whole milk and in instant skim milk powder as related to lactose crystallization. Food Microstructure 4, 333–340.

Sargent, J.A. (1988). The application of cold stage scanning electron microscopy to food research. Food Microstructure 7, 123–135.

Schmidt, D.G. (1982). Electron microscopy of milk and milk products: problems and possibilities. Food Microstructure 1, 151–165.

Schmidt, D.G. and van Hooydonk, A.C.M. (1980). A scanning electron microscopical investigation of the whipping of cream. Scanning Electron Microscopy III, 653–658.

Schmidt, D.G. and Buchheim, W. (1992). The application of electron microscopy in dairy research. Journal of Microscopy 167, 105–121.

Schorsch, C., Wilkins, D.K., Jones, M.G. and Norton, I.T. (2001). Gelation of casein-whey mixtures: effects of heating whey proteins alone or in the presence of casein micelles. Journal of Dairy Research 68, 471–481.

Shimmin, P.D. (1982). Observations of fat distribution in cheese by incident light fluorescence microscopy. Australian Journal of Dairy Technology 37, 33–34.

Skriver, A., Hansen, M.B. and Qvist, B. (1997). Image analysis applied to electron micrographs of stirred yoghurt. Journal of Dairy Research 64, 135–143.

Sutheerawattananonda, M., Fulcher, R.G., Martin, F.B. and Bastian, E.D. (1997). Fluorescence image analysis of process cheese manufactured with trisodium citrate and sodium chloride. Journal of Dairy Science 80, 620–627.

Tamime, A.Y., Muir, D.D., Wszolek, M., Domagala, J., Metzger, L., Kneifel, W., Durrschmid, K., Domig, K.J., Hill, A., Smith, S., Guinee, T. and Auty, M. (2011). Quality control in processed cheese manufacture. In: Processed Cheese and Analogues. Tamime, A.Y., Ed. Blackwell Publishing, Oxford, pp. 245–332.

Thomson, N.H., Miles, M.J., Tatham, A.S. and Shewry, P.R. (1992). Molecular images of cereal proteins by STM. Ultramicroscopy 42, 1204–1213.

Tsien, R.Y., Ernst, L. and Waggoner, A. (2006). Fluorophores for confocal microscopy: photophysics and photochemistry. In: Handbook of Biological Confocal Microscopy, 3rd edn. Pawley J.B., Ed. Springer Science+Business Ledia, New York. pp. 338–348.

Tung, M.A. and Jones, L.J. (1981). Microstructure of mayonnaise and salad dressing. Scanning Electron Microscopy 3, 523–530.

Van de Velde, F., van Riel, J and Tromp, R.H. (2002). Visualization of starch granule morphologies using confocal scanning laser microscopy (CSLM). Journal of the Science of Food and Agriculture 82, 1528–1536.

Van de Velde, F., Weinbreck, F., Edelman, M.W., van der Linden, E. And Tromp, R.H. (2003). Visualization of biopolymer mixtures using confocal scanning laser microscopy

(CSLM) and covalent labeling techniques. Colloids and Surfaces B: Biointerfaces 31, 159–168.

Vodovotz, E., Vittadini, E., Coupland, J., McClements, D.J. and Chinachoti, P. (1996). Bridging the gap: the use of confocal microscopy in food research. Food Technology 74–82.

Wright, S.J., Centonze, V.E., Stricker, S.A., DeVries, P.J., Paddock, S.W. and Schatten, G. (1993). Introduction to confocal microscopy and three-dimensional reconstruction. Methods in Cell Biology 38, 1–45.

Yiu, S.H. (1985). A fluorescence microscopic study of cheese. Food Microstructure 4, 99–106.

Zhong, Q., Daubert, C.R. and Velev, O.D. (2004). Cooling effects on a model rennet casein gel system: part II. Permeability and microscopy. Langmuir 20, 7406–7411.

2

Light Microscopy and CSLM Techniques, Principles and Applications

Johan Hazekamp

Unilever R&D, Colworth Science Park, Sharnbrook, Bedfordshire, MK44 1LQ, UK

2.1 Introduction

2.1.1 The History of Microscopy

The increase in resolving power of optically aided observations was initiated by the first practical telescope patented by Hans Lippershey in Middelburg, The Netherlands in 1608. Based on this system, Galileo Galilei in the following years produced telescopes capable of up to ×30 magnification and published his initial astronomical observations in March 1610 in the *Starry Messenger*.

The first well-documented micrographs were, literally, from the hand of Anthonie van Leeuwenhoek (1632–1723). Although in their times, another Dutchman Jan Swammerdam (1637–1680) and the Englishman Robert Hooke (1635–1703) had already used compound microscopes with an objective and ocular, the resolving power of their microscopes was by far exceeded by the resolution which Van Leeuwenhoek obtained with his strong lens. Hooke's compound microscope magnified up to ×30 whereas the magnification of the simple Van Leeuwenhoek design reached up to ×480.

Recently, as reported in *Nature News* (Ball, 2011), the work of Brian Ford showed that the imaging capabilities of the Van Leeuwenhoek microscope are comparable to the modern light microscope.

Van Leeuwenhoek used his microscopes to study many different biological specimens and published his findings in letters addressed to the Royal Society in London. In 1678 Hooke was assigned by the Society to verify Van Leeuwenhoek's findings in which he succeeded. Van Leeuwenhoek's main discoveries were the Infusoria (Protista) in 1674, bacteria in 1676, spermatozoa in 1677, but also the banded pattern of muscular fibers, in 1682. Robert Hooke published his book *Micrographia* in 1665 in which he described insects, sponges, shells and bird feathers, establishing his reputation in the history of biology.

Throughout the 18th century, technical innovations (e.g. the combination of two glass types in the lenses solving the problem of spherical aberration) improved the microscopes, leading to microscopy becoming popular in science. In light microscopy,

Microstructure of Dairy Products, First Edition.
Edited by Mamdouh Mahmoud Abdel-Rahman El-Bakry, Antoni Sanchez, and Bhavbhuti M. Mehta.

the final improvement leading to the modern wide field microscope came from Ernst Abbe in 1872, describing the physical laws about the collection of light by an objective and being able to maximize this collection of light by using water and oil immersion lenses. Abbe improved the resolution of Van Leeuwenhoek's achievements by 10-fold, and not far from the modern wide field optical instruments in which Abbe's concepts are still employed.

The success of the light microscope as it is, despite all modern developments, lies in the fact that the cost of maintenance and investment for the instrument are relatively low. Besides the cost aspects, other advantages lie in the fact that the image returned, by using visible light rather than a monochromatic source, is full-color and hence enabling selective and colorful staining of different phases and components. Last but not least, the specimen can be observed in a hydrated state and under ambient conditions, providing options to perform microstructural observations *in situ* rather than after a sample preparation procedure as required for electron microscopy (Armbruster *et al.*, 1982; Kellenberger, 1991). The disadvantages of the technique may come from a limited focal depth, although auto-focus and through-focus routines are used; the diffraction-limited spatial resolution; and difficulties in eliminating optical aberrations.

The modern light microscope is fully automated and self-aligning and the embedding of sensitive digital cameras in the user interface complete the light microscope as a ubiquitous and versatile tool in structural and dynamic imaging of foods in general.

2.1.2 Evolution of Confocal Microscopy

The invention of the confocal microscope is usually attributed to Marvin Minsky, who produced a working instrument in 1955. The development of the confocal approach was largely driven by the desire to image biological events as they occur in living tissue (*in vivo*), as Minsky had the goal of imaging neural networks in unstained preparations of living brains (Minsky, 1988). The confocal principle enhances the resolution of collected light by eliminating out-of-focus interference. To date, the principle of confocal imaging advanced by Minsky and patented in 1957 is employed in all modern confocal microscopes.

Minsky, in his original configuration generated a spot illumination by placing a pinhole in front of a point source of light which was focused by an objective lens at the desired focal plane. Light passing through was then focused by a second objective lens at a second confocal pinhole which had the same focal point as the first pinhole. The second pinhole prevents light originating from above or below the plane of focus in the specimen from reaching the detector. This use of spatial filtering to eliminate out-of-focus light, in specimens that are thicker than the plane of focus, is the essence of the confocal approach by returning information from the focal plane only.

To build a confocal image, the focused spot illumination must be scanned across the specimen and in Minsky's original instrument a vibrating stage was used to move the specimen under a stationary beam. An advantage of this approach is the elimination of lens defects as the scanning beam is held stationary on the optical axis of the microscope, where spherical aberration effects are minimal. In biological samples, however, movement of the specimen stage may cause wobbling and distortion, resulting in a loss of resolution in the image. In Minsky's original design, a real image was not formed, but instead the output from the photomultiplier detector was translated into an image on the screen of an oscilloscope that had no provision for recording. In 1955 the technology to fully demonstrate the potential of the confocal approach, especially for imaging biological structures, was not yet available.

The growth and refinement of the confocal approach to the modern state of art is a direct result of the advancements gradually becoming available in modern technology (Pawley, 2006). Among these are stable multi-wavelength lasers for improved point light sources, improved dichroic mirrors, sensitive low-noise photo-detectors, fast computers with modern image processing capabilities, sophisticated image analysis software packages, and high-resolution video displays and image printers.

The modern confocal scanning laser microscope (CSLM) uses scanning mirrors to scan the sample over the x- and y-axis and, by changing the focal plane, scans the sample in the z-axis as well. This provides the possibility for studying the sample in 3D, albeit with some restrictions, e.g. in the case of highly scattering components that hinder the passage of the returning signals.

The resolution of optical instruments, besides aberrations of the optical system and diffraction of photons in the sample volume, is primarily dependent on the wavelength of the light source used. For light microscopy, the light source is a visible light having a wavelength spectrum between 400–750 nm. By physical laws as proven by the equations of Ernest Abbe (1872), it is impossible to resolve structural details below 200 nm when using visible light for illumination. Modern developments, using the near field scattering phenomena (SNOM, scanning near field optical microscopy) rather than the classical far field image, or structured illumination microscopy (SSIM) applying patterned illumination combined with extensive post-exposure analysis as well as the photo-activated localization microscopy (PALM/STORM) increase the resolving power of the optical microscope down to the single molecule level. With the introduction of STED (stimulated emission depletion microscopy) high resolution 3D data from the distant field were reported down to a 10–30 nm resolution limit (Hell and Wichmann, 1994).

2.1.3 Food Microscopy

The significance of microscopy to food science was probably first described by the work of Louis Pasteur (1822–1895). Initially contributing his work to health and medicine, Pasteur gained insight on the relationship between food spoilage and bacteria. In the same period the research of Arthur Hill Hassal, describing methods for preparation and microscopic studies of foods led in 1860 to the first Parliament legislation against adulteration of foods in the United Kingdom. In late 19th and 20th century microscopy, the work of Greenish and Collin (1904), Clayton (1929) and Winton and Winton (1939) show the increasing significance of using microscopy to study food microstructure, processing and safety. For comprehensive and practical descriptions of modern techniques the work of Brooker (1993), Aguilera and Stanley (1999) and Olga Flint (1993) cover the preparation and microstructural evaluation of basic foodstuffs as well as complex full formulations.

In the scientific area of food microstructure, structure–function relations in processed foods are correlated as a function of processing and varying composition, using a growing range of imaging and correlative analytical tools that return structural data ranging from centimeters (wide field optical microscopy, CSLM, Raman microscopy, MRI and μCT) to nanometers (SEM, TEM, XRD). Optical microscopy typically uses a variety of selective stains, chromatic or fluorescent, to identify and discriminate between different phases and components while the same instrument in polarization or phase contrast mode returns information about the physical state of molecules and components (Morris and Groves, 2013).

Both wide field microscopy and confocal microscopy provide data about the presence and localization of specific components, either statically in a carefully preserved and prepared microscopic specimen, or dynamically during processing using in line fast imaging or time lapse duration experiments (Flint, 1993). The confocal microscope, to some extent, is non-invasive and capable of recording volumes that can be visualized after reconstruction of a series of optical slices through a sample volume (Vaughan, 1979; Pawley, 2006; Tanke, 2006).

2.1.4 Wide Field Microscopy

The incidence of light with a microscopic specimen results in a series of events contributing to image formation as listed in Table 2.1. When, in normal bright field modes, the sample does not return a sufficient contrast by absorption or reflection, the addition of specific chromatic staining and/or specific labeling may enhance the absorption of light resulting in a colored contrast image (Table 2.2). These methods are widely applied for bulk specimens of dairy products, as well as on specimens that are chemically stabilized and embedded in resin to facilitate thin sectioning for transmitted light microscopy.

Foods and, more specifically, dispersions and emulsified processed or non-processed materials, quite frequently cannot readily be mixed with chemicals to enhance an imaging contrast, as changes in pH and osmolarity may lead to conformational changes in the colloid chemistry of the sample that do not related to the product properties. Microscopic characterization, in these cases, is restricted to the unstained and native appearance of the specimen using scattering and refractive interaction of the incident light with in the volume (Table 2.1).

The main advantage of light microscopy, as mentioned earlier, lies in the fact that it enables the food material to be studied in its hydrated state facilitating the additions of chromatic stains to overcome contrast issues discussed previously. The addition of chromatic stains greatly facilitates phase identification as well as component identification and, combined with the physical contrast enhancing mechanisms, result in a versatile toolbox. Furthermore, the chromatic nature of the stains often result in a fluorescence that bridges the application of fluorescence microscopy and confocal scanning laser microscopy. The chromatic stains as listed in Table 2.2 represent some commonly used and versatile stains, but the list is probably not complete and many different protocols described for biomedical histology apply. In general, these chromatic stains are considered mutagenic or carcinogenic and require special protocols compliant to local regulations.

When light waves travel through a medium other than a vacuum, interactions with the medium cause the amplitude and phase of the light wave to change in a manner dependent on properties of the medium. These changes in refraction are visible as changes in amplitude (brightness) arising from the scattering and the wavelength-dependent absorption of light may give rise to colors in the image obtained. Phase changes however, as image recording devices such as digital cameras, photographic equipment and the human eye are only sensitive to amplitude variations, and without special arrangements are invisible. Nevertheless these phase changes often carry

Table 2.1 Imaging modes of the optical microscope.

Absorption, transmission, reflection, staining	Bright field imaging (BF)
Scattering and diffraction	Dark field (DF)
Refraction and polarization	Polarization microscopy (POL), DIC
Phase change	Phase contrast microscopy (PhC)
(auto) Fluorescence, staining	Fluorescence microscopy (FL)

important information and when used to enhance the contrast, boundaries and interfaces can be visualized. Especially in food material, these may be rather colorless or even translucent; imaging of the unstained microstructure may rely on recording differences in the refracting behavior of the different phases. The typical imaging mechanisms are described as polarized light (POL), phase contrast (Ph) and differential interference contrast (DIC).

Polarized light microscopy relies on the phenomenon that the direction of the incident light can change upon interaction with crystalline phases in the specimen. In polarized light mode, the specimen is illuminated with linearly, cross or circular polarized light by placing a so-called “polarizer” in the optical path. A second filter set, the “analyzer”, positioned after the specimen and at a 90° rotated orientation, is used to exclude all transmitted and un-refracted polarized light and only the information generated by interactions with the specimen, typically after undergoing a change in polarization, will pass the second filter set. In most set-ups the filter sets can be rotated independently, enabling image mixing with the bright field information so that specific staining of phases can be combined with birefringent information.

Table 2.2 Commonly used chromatic stains for the identification of food components.

Starch	Iodine	Black/Blue
Native starch	Polarization	Birefringent
Polysaccharide gums	Toluidine Blue (acidified)	Shades of pink
	Iodine	Dark blue
Oil and lipids	Oil Red O	Red
	Sudan Black B	Black/blue
	Sudan Black IV	Red/orange
	Nile Red	Red/orange
Fat	Polarization	Birefringent
	Oils stains	Negative stained
Proteins	Potassium iodide	Yellow
	Toluidine Blue	Blue/purple/lilac shades
Fungi, yeast	Tryphan Blue	Blue
Yeast	DIC	Phase contract
	Calcofluor White	White (fluorescence)
Bacteria	Toluidine Blue	Blue

With the polarized set-up, information can be obtained about crystallinity (see Figure 2.3), absorption color and optical path boundaries between crystalline structures of differing refractive indices. In addition, the contrast-enhancing technique exploits the optical properties specific to anisotropy and thereby reveals detailed information concerning the structure and composition of materials.

Phase contrast microscopy (Frits Zernike, Nobel prize 1953) is a technique that converts phase shifts in light passing through a transparent specimen to brightness changes in the image by separating the illuminating background light from the light scattered by the specimen and to manipulate these differently. Phase shifts themselves are invisible, but become visible when shown as brightness variations. In a phase contrast microscope, the background light is phase shifted −90° by passing it through a phase shift ring and thereby eliminating the phase difference between the background and the scattered light leading to an increased intensity difference between foreground and background. To further increase contrast, the background is dimmed by a set of grey-filter rings. Although some of the scattered light will be phase shifted, the background light is affected to a much greater extent which creates the phase contrast effect.

The amendments to a bright field microscope being transferred into a phase contrast set-up are relatively low-cost and most microscopes are equipped with a phase plate and filter set. Disadvantages of the phase contrast imaging lie in the fact that it may suffer from out-of-focus information that is visualized as phase halos around the objects, thereby complicating the overall interpretation of the image.

Differential interference contrast (DIC) is a variation of the phase contrast technique displaying an enhanced contrast by using a set of prisms to cast a shadow illumination on the specimen. It is a significant addition in terms of cost that is justified by an enhanced resolution over standard phase contrast by using the full numerical aperture of the objective. DIC returns a high edge contrast in the image and an improved optical sectioning, not suffering from out-of-plane information. This improved optical section can be used to collect extended focus, imaged to create depth in the image.

A disadvantage of DIC illumination may be found in the fact that the contrast around the edges may vary upon orientation and is almost non-existent towards the middle of objects reducing options for image segmentation and analysis.

2.1.5 Confocal Scanning Laser Microscopy (CSLM)

The added value of CSLM over conventional light microscopy lies in the improved resolution of using a monochromatic (laser) light source and the very thin focal plane (0.5–1 μm) by applying a confocal pinhole that rejects the out-of-focus light (Figure 2.1). The confocal design of the optical system enables the recording of 3D images of the sample that are built by scanning a series of thin optical sections at increasing depth in the sample (Heertje *et al.*, 1987; Auty *et al.*, 1999). The main advantage of CSLM in food research however lies in the non-invasive imaging of bulk material which is significantly less destructive compared to the sample preparation (sectioning, dilution, etc.) required for conventional light microscopy (Loren *et al.*, 2007; Fenoul *et al.*, 2008). In addition, the interaction of fluorescent light with free and unbound fluorescent molecules may return data about diffusion and pore size in structured liquids using FRAP (fluorescence recovery after photo bleaching) or FCS (fluorescence correlation spectroscopy) methodologies.

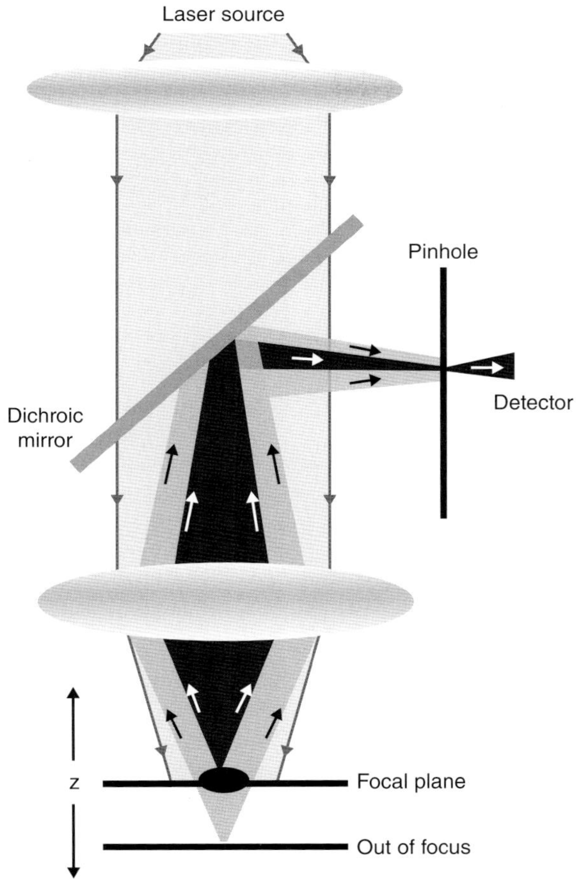

Figure 2.1 The principle of confocal imaging. The out-of-focus frequencies in the returning signal are blocked out by the pinhole enhancing the resolution of the final image.

In confocal scanning laser microscopy (CSLM), the use of fluorescent dyes is applied to identify different phases in food products; the most commonly used fluorochromes and their applications are listed in Table 2.3. The fluorescent dyes introduce contrast and differentiate between various components in a specimen, based on chemical properties like hydrophobicity, covalent binding (e.g. aqueous vs. oil, protein and starch). The fluorescent molecules are excited with laser light of a specific wavelength and emit fluorescence, which is detected using a photon detection system. The detected light is represented in a chosen, hence artificial, color (Minsky, 1988; Pawley, 2006). The modern CSLM instrument uses multiple laser lines to excite and discriminate between fluorescent molecules, sequential scanning techniques to eliminate cross-talk between detected signals. The laser lines cover the visible light spectrum as well as the UV and near infra-red, enabling the use of multiple stains simultaneously in one specimen to differentiate and localize different phases.

Fluorescence correlation spectroscopy (FCS) is one of the many different modes of high-resolution, spatial and temporal analysis of extremely low concentrated

Table 2.3 Fluorescent stains used for phase staining of food material.

Fluorochrome	Excitation	Emission	Component	Working dilution
Nile Red	488	500–600	Oil	0.01% in Glycol/glycerol/water (50/45/5)
Old Nile Blue (Nile Red present)	488 633	500–600 640–750	Oil Proteins (also others)	1% w/v aqueous
Acridine Orange	488 561	469–555 568–646 654–723	Carbohydrates Proteins	0.5% aqueous
Rhodamine B	561	700	Protein	0.05% aqueous
Congo Red	561	565–650	Cellulose (fibers)	0.5% aqueous
FITC	495	519		
Safranin O	495	587		
Calcofluor White	350	450–500	Cellulose, chitin	0.1% aqueous
Eosin	524	545		
Sudan Red (1–4) Ponceau R, Scarlet R	450–590		Fat, lipids, triglycerides	
Propidium iodine	536	623		

bio-molecules. In contrast to other fluorescence techniques, the parameter of primary interest is not the emission intensity itself, but rather spontaneous intensity fluctuations caused by the movement of fluorescent bodies through the focal spot of the objective. In general, all physical parameters that give rise to fluctuations in the fluorescence signal are accessible by FCS. It is, for example, straightforward to determine local average concentrations of fluorescently labeled biomolecules in nanomolar concentrations. This can translate to mobility coefficients or characteristic rate-constants of inter- or intramolecular reactions as there are diffusion coefficients, hydrodynamic radii, kinetic chemical reaction rates, singlet-triplet dynamics, mean squared displacements (MSD) of molecules or particles as a function of time (Digman *et al.*, 2005).

Combining the FCS technique with confocal detection, the incoming laser light is strongly focused by a high numerical aperture objective (ideally NA > 0.9) to a diffraction limited spot exciting only the few fluorophores within the illuminated region. A pinhole is introduced in the image plane which blocks all out of focus light in order to limit the detection volume also in axial direction. Most FCS measurements are performed on fluorescently labeled bio-molecules diffusing in an aqueous buffer solution and typically any sufficiently dilute and small fluorescent probe can potentially be used.

Raster image correlation spectroscopy (RICS) is the variation on FCS, performing the experiment in multiple spatial points in parallel, using the scanning beam of the laser scanning confocal microscope and returning a map of diffusion coefficients of the specimen under the beam. In principle RICS is the image transformation that converts an intensity image into a correlation image of the same XY dimensions. Where FCS is most suitable for fast moving particles/molecules and for determining MSDs, RICS can be applied to larger structures as beads, vesicles, aggregates and to slow processes.

Fluorescence recovery after photobleaching (FRAP) denotes an optical technique capable of quantifying the diffusion of a molecularly thin film containing fluorescently labeled probes (Blonk *et al.*, 1993; Braeckmans *et al.*, 2003). In FRAP, at a localized spot in the specimen the fluorescence is quenched by a high intensity laser beam, after which the recovery of fluorescence is a measure of the diffusion of the fluorochrome (or fluorescent probes with defined size) in the matrix.

2.2 Sample Preparation and Specific Staining and Labeling

The range of products in dairy science consists of soft solids, foams, emulsions and dispersions, all highly hydrated and structured by a delicate balance of chemical interactions between the components (proteins, biopolymers, emulsifiers and fats). For microstructural studies using optical instruments, a range of sample preparation techniques are available aiming to retain the original structural integrity and composition of the sample and should be aligned to the sample composition and research questions to be answered (Figure 2.2). In all cases, a sub-volume of the material is studied on the microscope stage and should be a true representation of the bulk material.

Ideally, but only applicable to a very limited number of specimens, the unstained sample is imaged, employing a non-invasive procedure that resolves structural parameters as well as phase differences. In most cases however, a simple sub-volume of the sample is applied to a microscope glass and imaged using transmission or reflective illumination. The non-invasive requirement then only applies for thin films made from dispersions and fine aggregates that are not disturbed by the sampling technique. Typically,

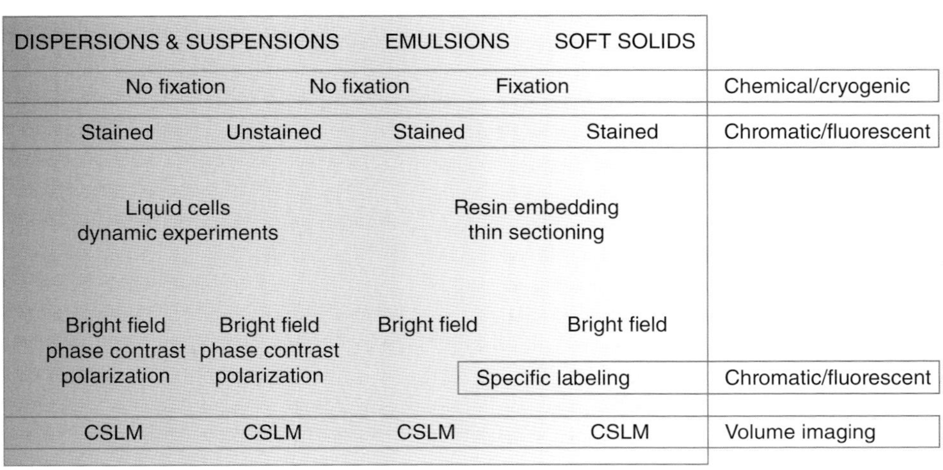

Figure 2.2 Schematic representation of choices and options to make during sample preparation and imaging. The dairy product consists of a range of hydrated colloid systems and although the difference in viscosity of both extremes obviously dictate a different experimental approach, the emulsions dependent on texture and viscosity may be studied in a hydrated and dynamic state as a fixed and embedded specimen for localization studies.

contrast enhancements from polarization, phase contrast and DIC are used to discriminate between boundaries and interface changes, an example of which is depicted in Figure 2.3, showing the added value of polarized light illumination on the identification of crystalline material. The images were taken of a disintegrated frozen dessert (Goff, 2002; Cook and Hartell, 2010) and at low temperature using a Linkam cooling stage (see Figure 2.7). The unstained image (Figure 2.3a) shows the aggregates of protein, fat and ice crystals that are identified by the polarized light image (Figure 2.3b).

For imaging food emulsions that are stable at ambient or refrigerated conditions, the different phases can be stained and identified using a range of chromatic or fluorescent stains. A straightforward mixing usually facilitates the infusion and homogeneous distribution of stains provided that the shear of the process does not alter the microstructure, e.g. shear-induced coalescence, aggregation, phase separation etc.

For structured emulsions (spreads, butter, shortenings) or soft hydrated solids (e.g. cheese products), shear applied to the sample during infusion of the stains may disrupt the microstructural properties of the material. For these materials it is good practice to rely on diffusion of the stains rather than mixing. To achieve this, typically a filter material saturated with a staining solution is used to facilitate a diffusion of stains into the sample rather than immersing the specimen into a stain solution with the risk of dissolving the specimen in the bulk liquid. The diffusion of stains in a bulk specimen is then ideal for imaging the specimen using a confocal imaging system as opposed to transmission imaging systems that require a thin specimen. Although a pre-stained specimen for bright field imaging could be compressed into a thin film between coverslips, this procedure disintegrates the microstructure.

From an imaging perspective, the structured emulsions and soft solids are preferably imaged using a confocal and non-invasive illumination resulting in a minimal structural distortion. The examples given in Figure 2.4 shows the microstructure of a liquid margarine, a fresh cheese spread and an ice cream emulsion imaged using (old) Nile Blue as fluorescent stain diffused into the specimen over 24 h at 4°C. The images show the presence of protein (red channel) present in the water phase and a continuous oil phase (green channel) and clearly illustrate the different nature of the material. The oil

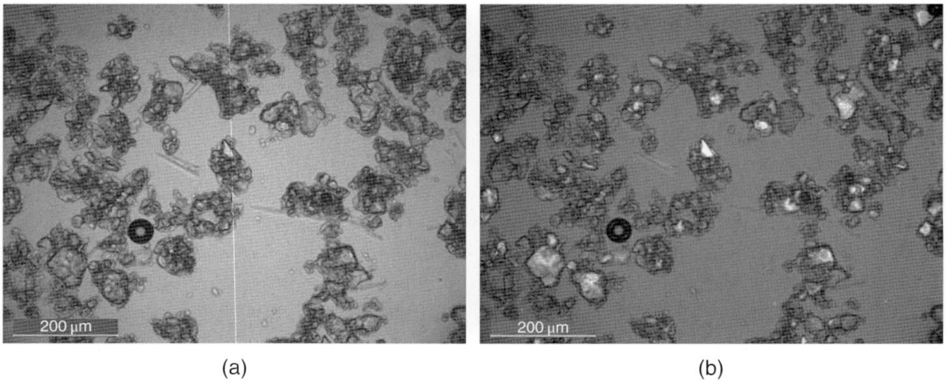

Figure 2.3 Images of ice crystals isolated from ice cream. The image depicted in (a) shows the presence of disintegrated ice cream consisting of clusters of various size. The polarized light image (b) identifies the presence, size and shape of the ice crystals. (*See color plate section for the color representation of this figure.*)

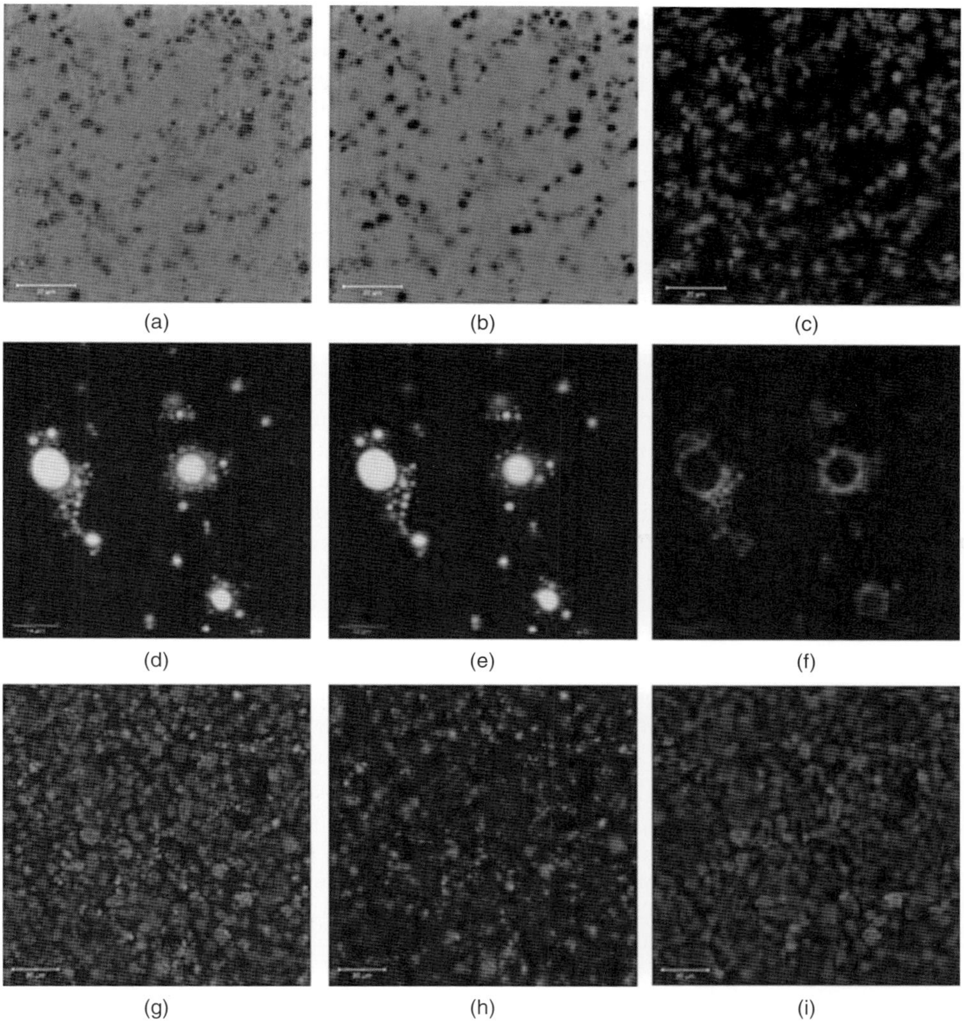

Figure 2.4 CSLM images of different emulsions stained with old Nile Blue, co-localizing protein and fat contents by assigning the protein contents of the emulsion to a red channel and the oil phase to a green channel. The images depicted in (a–c) show a liquid margarine as an example of a water in oil emulsion whereas the images shown in (d–f) show an ice cream emulsion at high magnification, as an example of an oil in water emulsion clearly showing the stabilization of the fat droplets by protein material. The fresh cheese sample as an example of a more complex system, depicted in the images (g–i) shows the co-localization, the homogenization and the distribution of the two phases. (*See color plate section for the color representation of this figure.*)

continuous w/o emulsion shows the droplet size of the unstained water phase and the presence of protein material added to the aqueous phase (Figure 2.4a). The water continuous o/w emulsion, an ice cream pre-emulsion, depicts the stabilization of oil droplets by aggregates of protein inducing stability to the dispersion. The fresh cheese visualized in the images Figure 2.4(g–i) shows the level homogenization and aggregation of protein and oil into a viscous emulsion.

In addition to the specific staining described, the application of bio-recognition protocols for specifically labeling components in the product cannot be carried out on the hydrated and unfixed material. Typically the products are fixed and embedded in resin of wax that retain structure and original composition. To retain this structural integrity from soft hydrated solids requires a carefully designed fixation, embedding and sectioning protocol to preserve the original microstructure and chemical composition of the material. The subsampling should be representative for the overall product and as a first step, the fixative ideally does not interfere with chemical processes in the specimen that may lead to structural alterations such as cross-linking and/or extraction. After fixation, the specimen is stained en-bloc, dehydrated and infiltrated followed by final curing of the embedding material. The cured sample block is usually stable for lengths of time and can be used for thin sectioning, post-embedding localization studies and en-bloc imaging.

The example given in Figure 2.5 shows the microstructure of frozen dessert, fixed at low temperature and resin embedded for light microscopy. The image depicts the presence of the protein after Rhodamine staining and shown as red in the image; the stabilizing carbohydrates stained using Calcofluor White, assigned to a cyan color. The procedure used, applies for ice cream and frozen desserts and involves a sub-sampling at low temperature followed by freeze substitution in organic solvents and in the presence of fixatives and fluorescent stains using an automated freeze substitution system (e.g. Leica microsystems). The specimen is infiltrated using a low viscosity resin, cured and further processed at room temperature. The low temperature fixation stabilizes the aqueous phase and the components present, the ice crystals dissolve completely and are seen as resin filled voids. Similarly, the air phase present in some frozen desserts will be filled with the resin.

2.2.1 Specific Labeling

Although displaying a reasonable specificity, the chromatic and fluorescent stains as described earlier, label the phases rather than specific molecules. To discriminate

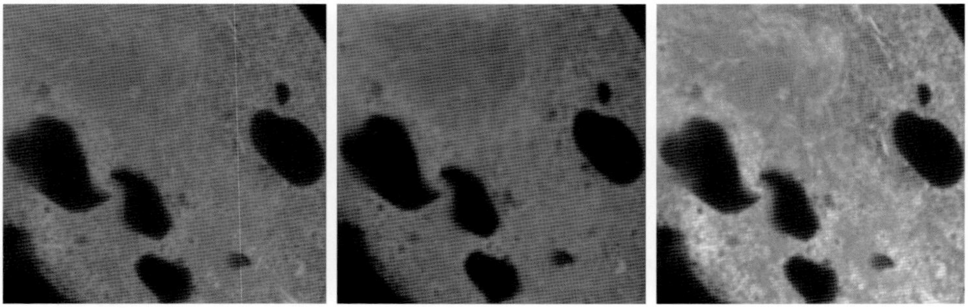

(a) (b) (c)

Figure 2.5 Fluorescence imaging of a frozen aerated dessert substituted at low temperature in the presence of fluorescent stains and imaged as a resin embedded block (a). The protein phase was stained using Rhodamine and assigned to a red color (b); Calcofluor White was used to stain carbohydrate stabilizers, depicted in cyan (c). The fat phase was extracted during the freeze substitution procedure; air and ice show as voids. (*See color plate section for the color representation of this figure.*)

between different proteins or subclasses of proteins, the use of highly specific markers is required. These can have an immunological mechanism of action, antibodies raised against a well-defined protein or sequence of amino acids or a biochemical affinity as in lectin molecules binding specific sugars or carbohydrate binding modules for identification of carbohydrate polymers.

Specific labeling relies on the availability of specific probes and, although selected for a highly specific binding to the targeted molecules, these biological probes display a general sticking and a specific binding that needs cancelling out by applying blocking and pre-treatment of the section with similar molecules. Similarly, irregularities in the sample surface causing capillary action, as there are creases and crevasses as a result of sectioning, and they may collect specific primary probes that can be mistaken as a false positive signal, requiring intensive washing steps and optimizing of the antibody concentration applied.

In Figure 2.6, the immunofluorescent staining of molecules present at the air–water interface of a foam using a combination of fluorescent imaging and transmitted light is shown as an example of specific labeling in a low viscosity liquid. This foam was isolated from frozen material and labeled in suspension using a two-step procedure as depicted in the schematic representation (Figure 2.6b). The labeled air bubble suspension was imaged using a combination of fluorescent and transmitted imaging. Figure 2.6b shows a two-step procedure generally used as a universal labeling procedure involving a first step in which a specific antibody against a domain is applied relying on bio-recognition action. The detection is done in a second step, applying secondary antibodies specifically targeted against the primary antibody, covalently bound to a detection system. These can either be fluorescently labeled secondary antibodies for CSLM, or gold labeled secondary antibodies for TEM. Although reported successfully on labeling bulk hydrated systems using diffusion of probes (Johnson and Spencer, 2010), labeling is usually carried out on thin sections not dependent on diffusion and access of the matrix.

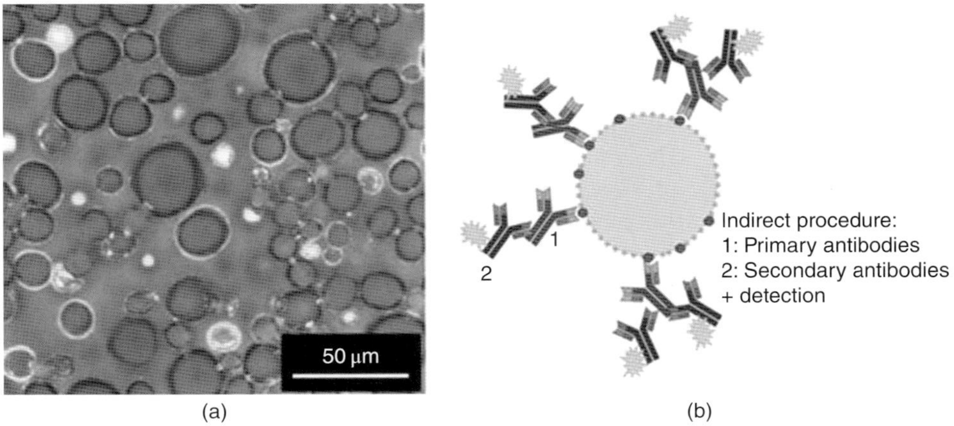

Figure 2.6 Immunofluorescent staining of molecules present at the air–water interface of a foam using a combination of fluorescent imaging and transmitted light (a) and (b) a schematic representation of a two-step universal immunolabeling in which a highly specific primary antibody binds to its targeted epitope, itself targeted by a secondary antibody-marker complex that is visualized by the imaging system. Similar mechanisms apply for lectin and CBM labeling procedures. *(See color plate section for the color representation of this figure.)*

The success of specific labeling for bio-medicine is inevitably reflected in the vast number of applications and publications using the methodology since the mid-1990s, but in food science the number or reports has also increased. However it should be taken into account that bio-recognition using immunoglobulins relies on the binding to specific conformations of the targeted proteins. In processed foods these conformations may have been altered as a function of food processing, heat treatment and acidification, and often immunoglobulins raised against the native conformation do not recognize the processed state of the protein.

2.2.2 Dynamic Imaging

Food products and the production of food is subject to dynamics during aging, ripening and processing that all build to the specific attributes of a product. Imaging the microstructure using a light microscope that does not require a vacuum chamber to operate opens opportunities for imaging sample dynamics as a function of processing and aging. For this purpose, a wide range of specially designed stages enabling temperature-control and shearing stages are available that are universally applied to most available upright, and some inverted microscopes. Using an inverted microscope design provides opportunities of introducing reaction vessels and flow cells to the instrument.

Temperature controlled stages are available, covering temperature ranges from liquid nitrogen temperature up to 600°C and divided into different ranges depending on the application required. Typically these systems require a non-contact objective at a long working distance as both cooling and heating may affect an immersion liquid. This has a negative effect on the resolution of the instrument as the light path will be affected by difference in the refraction between air and the continuous phase of the sample. Nevertheless, for the resolution required to study dynamic processes in food material, a 20× and 40× long working distance objective should adequately image the microstructure, or at least phase changes as a result of processing.

Cooling is achieved either by actively cooling the stage using thermoelectric cooling (Peltier effect) or a flow of cold nitrogen gas or liquid nitrogen through the cell. The specimen chamber of the cooling stage is usually flushed with a dry nitrogen gas flow to prevent condensation of the viewing windows and microscopic glasses. An example is given in Figure 2.7, depicting a population of ice crystals isolated from a frozen dessert using the methodology as described by Yuennan *et al.* (2014).

Figure 2.7a shows the typical set-up for dynamic cooling experiments, here the development of ice crystallization as a function of temperature dynamics. The wide field illumination (Figure 2.7b) is applied in transmission mode whilst temperature cycling the specimen using a Linkam cooling device.

The application of heat as a parameter requires a sealed specimen compartment to prevent dehydration or evaporation of the sample which is designed to endure pressure and high temperatures of the experiment. The application of digital cameras and time lapse software facilitates the recording of structural and phase changes over extended periods of time.

2.2.3 Future Perspectives

The modern light microscope has the benefits of increasing levels of automation and is operated via a user interface controlling alignment, focusing, and stage scanning if

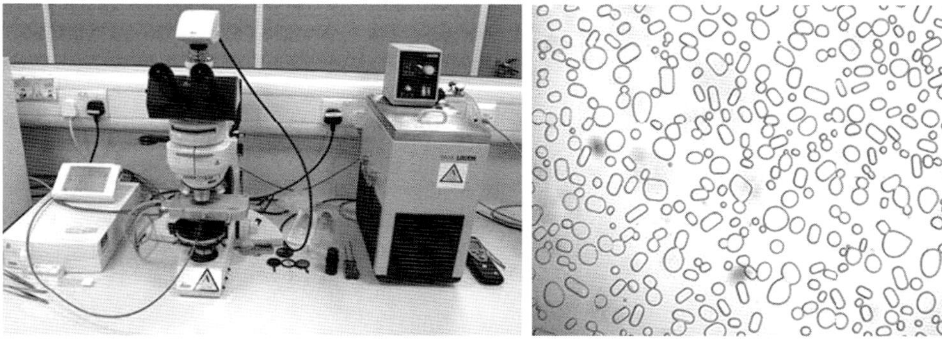

(a) (b)

Figure 2.7 Cold stage microscopy (a) and the bright field image (b) obtained depicting ice crystals isolated from a frozen dessert and imaged at −20°C using a Linkam cooling stage. Image width 0.9 mm, 10× LD objective.

required. Objectives are usually of high quality and optimized to generate the best resolution and least aberration for numerous applications, including long working distance and water objectives to image with the objective immersed in a liquid. Image analysis can be an integral tool of the user interface as is the generation of reporting documents, all contributing to the versatile use of the microscope. As the physics of light propagation and scattering cannot be changed, the real advances are made by improving signal-to-noise ratios in the detection, constantly improving the quality of the digital cameras.

Advances in confocal laser scanning microscopy are represented in the super resolution techniques acknowledged by the Noble Prize in Chemistry (2014) for pushing resolution limits to 20 nm, by applying stimulated emission depletion techniques (STED) as described by Hell and Wichmann in 1994. Similarly, the developments in SSIM (structured illumination microscopy), SNOM (scanning near-field microscopy), and PALM/STORM (photo activated localization microscopies) significantly increase the resolution of fluorescent optical microscopy.

Developmental biology studies report that the use of light sheet imaging, capable of following embryonic development over impressive lengths of time result in 3-dimensional imaging and localization studies in living tissues and embryos. The processes involved in food science perhaps may not require such lengthy time lapse imaging, but it may certainly benefit from the non-invasive character of the imaging.

With the recently launched Zeiss 880, the resolution of CSLM imaging was reportedly reduced to 140 nm (X,Y) and 400 (Z) by applying Airyscan technology. The Airyscan principle is based on reducing the minimal pinhole size as a result of improved photon detection using the 32 channel array detector collecting all the light passing through a small airy disc.

References

Armbruster, B.L., Carlemalm, E., Chiovetti, R., Garavito, R.M., Hobot, J.A., Kellenberger, E. and Villiger, W. (1982). Specimen preparation for electron microscopy using low temperature embedding resins. *Journal of Microscopy* **126**, 77–85.

Aguilera, J.M. and Stanley, D.W. (eds) (1999). Microstructural Principles of Food Processing and Engineering. Aspen Publishers Inc, New York.

Auty, M.A.E., Fenelon, M.A., Guinee, T.P., Mullins, C. and Mulvihill, D.M. (1999). Dynamic confocal scanning laser microscopy methods for studying milk protein gelation and cheese melting. *Scanning* **21**, 299–304.

Ball, Ph. (2011). *Nature News. Nature*, doi:10.1038/news.2011.116.

Blonk, J.C.G., Don., A., Aalst, H. van and Birmingham, J. (1993). Fluorescence photo-bleaching recovery in the confocal scanning light microscope. *Journal of Microscopy* **169**, 363–374.

Bradbury, S. and Bracegirdle, B (1998). Introduction to light microscopy. Bios Scientific Publishers in association with the RMS, Abingdon.

Braeckmans, K., Peeters, L., Sanders, N.N., De Smedt, S.C. and Demeester, J. (2003). Three-dimensional fluorescence recovery after photo-bleaching with the confocal scanning laser microscope. *Biophysical Journal* **85**, 2240–2252.

Brooker, B.E. (1991). The study of food materials under confocal microscopy. *Microscopy and Analysis* **27**, 13–15.

Cavalier, A., Spehner, D. and Humbel, B.M. (2009). The Handbook of Cryo Preparation Methods for Electron Microscopy. CRC Press, Boca Raton, FL.

Cook, K.L.K. and Hartel, R.W. (2010). Mechanisms of ice crystallization in ice cream production. *Comprehensive Reviews in Food Science and Food Safety* **9**, 213–222.

Digman, M.A., Brown, C.M., Sengupta, P., Wiseman, P.W., Horwitz, A.R. and Gratton, E. (2005). Measuring fast dynamics in solutions and cells with a laser scanning microscope. *Biophysical Journal* **98**, 1317–1327.

Fenoul, F., Le Demat, M., Hamdi, F., Cuvelier, M. and Michon, C. (2008). Technical Note: confocal scanning laser microscopy and quantitative image analysis: application to cream cheese microstructure investigation. *Journal of Dairy Science* **91**, 1325–1333.

Flint, O. (1993). Food Microscopy: A Manual of Practical Methods Using Optical Microscopy. *RMS Handbook 3.* BIOS Scientific, Abingdon.

Goff, H.D. (2002). Formation and stabilization of structure in ice cream and related products. *Current Opinions in Colloid Interface Science* **7**, 432–437.

Heertje, I., van der Vlist, P., Blonk, J.C.G., Hendrickx, H.A.C. and Brackenhof, G.J. (1987). Confocal scanning laser microscopy in food research: some observations. *Food Microstructure* **6**, 115–120.

Hell, S.W. and Wichmann, J. (1994). *Breaking the diffraction resolution limit by stimulated emission: stimulated-emission-depletion fluorescence microscopy. Optics Letters* 19.

Johnson, I. and Spencer M.T.Z. (eds) (2010). *The Molecular Probes Handbook; A Guide to Fluorescent Probes and Labeling Techniques.* Molecular Probes, Waltham, MA.

Kellenberger, E. (1991). The potential of cryofixation and freeze substitution: observations and theoretical considerations. *Journal of Microscopy* **161**, 183–203.

Loren, N., Langton, M. and Hermansson, A.M. (2007). Confocal fluorescence microscopy (CLSM) for food structure characterization. In: Understanding and Controlling the Microstructure of Complex Foods. D.J. McClements, ed. Elsevier, Amsterdam, pp. 232–260.

Minsky, M. (1988). Memoir on inventing the confocal scanning microscope. *Scanning* **10**, 128–138.

Morris, V.J. and Groves, K. (eds) (2013). Food Microstructures, Microscopy Measurements and Modelling. Woodhead Publishing, Cambridge.

Pawley, J.B. (ed.) (2006). Handbook of Biological Confocal Microscopy. Springer Publishing, New York.

Tanke, H.J. (2006). *Fluorescence Microscopy. Wiley Online Library*, doi: 10.1038/ npg.els.0005780.

Vaughan, J.G. (ed.) (1979). Food Microscopy. Academic Press, Waltham, MA.

Yuennan, P., Sajjaanantakul, T. and Goff, H.D. (2014). Effect of okra cell wall and polysaccharide on physical properties and stability of ice cream. *Journal of Food Science* **79**, 1522–1527.

3

Electron Microscopy Techniques

Semih Otles[1] *and Vasfiye Hazal Ozyurt*[2]

[1] *Department of Food Engineering, Faculty of Engineering, Ege University, 35100 Bornova, Izmir, Turkey*
[2] *Graduate School of Natural and Applied Sciences, Food Engineering Branch, Ege University, 35100 Bornova, Izmir, Turkey*

3.1 Introduction

An electron microscope (EM) is a microscope which uses a beam of accelerated electrons as a source of illumination. EMs have a higher resolving power than light microscopes because the wavelength of an electron is approximately 100,000 times shorter than that of visible light photons and therefore can discern the structure of tiny objects. EMs can be used to research the ultrastructure of a great number of biological and inorganic samples including microorganisms, cells, large molecules, biopsy samples, metals, and crystals. Moreover, EMs are prefered for quality control and conformation analysis and produce electron micrographs using specialized digital cameras and frame grabbers to capture the image (Fazaeli *et al.*, 2012). Recently, developed electron microscopy techniques have changed the perceptions related to both food structures and the types of information which can be now expected to glean regarding structure (Wilson, 1991). Different electron microscopy techniques have been employed to evaluate the structure of food such as transmission electron microscopy (TEM), scanning electron microscopy (SEM) (Kalab, 1979), cryo-SEM, cryo-TEM, and environmental scanning electron microscopy (ESEM) (Flint, 1994).

This paper is reviewed to understand exactly the structure of dairy products using some examples including microscopic studies and to illustrate the potential of these techniques to reveal detailed and specific information.

3.2 Types of EM

3.2.1 Scanning Electron Microscopy (SEM)

The scanning electron microscopy technique is used to form the image by scanning a focused electron beam across the sample. The first electrons penetrate the solid sample and are wrapped by a wide range of elastic scattering processes. As long as the electrons collide, various signals are generated. Therefore, the sample surface is analyzed or an

Microstructure of Dairy Products, First Edition.
Edited by Mamdouh Mahmoud Abdel-Rahman El-Bakry, Antoni Sanchez, and Bhavbhuti M. Mehta.

image is formed. The energies of a few tens of eV, high-energy electrons back-scattered from the primary beam and characteristic X-rays are secondary electrons (Bogner *et al.*, 2007).

SEM has been explained as a combination of light microscopy and transmission electron microscopy (Aguilera and Stanley, 1999).

SEM has many advantages such as simple sample preparation, a great deal of magnification, high depth of field and the fact that the image is a representation of electronic data allowing for image analysis and quantification. However, it has some disadvantages such as examining insulating samples and the impossibility of examining hydrated samples without altering their state in some way (either drying or freezing). Therefore, these pretreatments may introduce artifacts (Harker *et al.*, 2006; James and Fonseca, 2006).

SEM uses an electron beam that scans the surface of the sample to generate secondary electrons. Then a three-dimensional (3D) image is formed. SEM gives intermediate resolution (3–4 nm) and exaggeration (20–100,000×) (Pérez-Munuera *et al.*, 2008).

3.2.2 Transmission Electron Microscopy (TEM)

Transmission electron microscopes (TEM) are electron optical instruments like light microscopes. However, the sample is illuminated by electron beam, not light. Since gas molecules are turned over the electrons, a vacuum is needed. An electron gun is at the top of the microscope column; in addition a system of electromagnetic lenses focuses the electron beam on the sample. In TEM, the image is formed by the interactions between electrons and the sample. The resolution is in proportion with the acceleration voltage of the electrons. While the voltage increases, the resolution is higher because of the short wavelength of the electrons. However, increasing acceleration voltage results in poorer contrast since scattering of electrons is decreased in the higher velocity. TEM has a capability of voltages from 40 to 120 kV and microscopes in the range of 200–400 kV. Moreover, when colloidal systems are used, voltages between 80 and 200 kV are usually employed (Kuntsche *et al.*, 2011; Klang *et al.*, 2012).

TEM is used when the samples are very thin and a two-dimensional (2D) image is produced. TEM gives an image with high resolution (0.2–1 nm) and exaggeration (200–300,000×) (Pérez-Munuera *et al.*, 2008).

3.2.3 Cryo-SEM

Cryo scanning electron microscopy (cryo-SEM) is a direct method of analyzing the ice crystals of frozen samples without resolving. Moreover, the sample is coated with metal for enhanced conductance of the electrons (Preetz *et al.*, 2010). The frozen sample used in cryo-electron microscopy can be immediately transferred to a low temperature stage within the microscope and viewed directly (Belkoura *et al.*, 2004). In cryo-SEM, rapid physical detection is provided, and the risk of introducing artifacts occurring in chemical detection, structural collapse or shrinkage is refrained. Thus, in this technique, triphasic structure of a sample is preserved with the distribution of solid, liquid and gas in a state closer to its natural state (James and Fonseca, 2006). However, this technique is considerably expensive like SEM, and it is one of the biggest disadvantages.

3.2.4 Cryo-TEM

Cryo-TEM is used to analyze directly the colloids in the glazed frozen-hydrated state and thus information about the internal structure of colloidal system is obtained (Kuntsche *et al.*, 2011). The sample preparation step is by step; the formulation microstructure is displayed in its original state and a clear differentiation between nano-sized droplets and other structures can be obtained (Klang *et al.*, 2012). When it is compared to conventional TEM techniques, this techniques represents major improvement, because aqueous compounds of the system will evaporate rapidly under vacuum within an electron microscope in conventional TEM, during the analysis of hydrated systems (Klang *et al.*, 2012).

3.2.5 Environmental Scanning Electron Microscopy (ESEM)

Environmental SEM has the gas in the sample chamber and thus samples do not need to high vacuum; a low vacuum is supplied by a special design of electron optics column that permits differential pumping and the column is divided into different pressure zones separated by pressure limiting apertures (Danilatos, 1993). ESEM has two main roles. One of these roles is electronic and the gas acts as an electrical charge conductor avoiding sample charging and facilitates signal detection. Moreover, the second one is thermodynamic and the gas is a conditioning medium, preventing evaporation of liquids from a sample (Thiel and Toth, 2005).

3.3 Sample Preparation for EMs

3.3.1 Scanning Electron Microscopy

1–3 mm^3 cubes of the samples are treated with 2.5% glutaraldehyde (first detection) and 2% osmium tetroxide (second detection) and dehydrated in a series of 10%, 20%, 40%, 60%, 80%, and 100% ethanol (ethanol/water, v/v), for 20 min in each (Echlin, 2009; Goldstein *et al.*, 1992). Then, they are flushed in acetone and dehydrated with CO_2 by the critical point method (7.58 MPa bar, 31.5°C). The last step is that the samples are coated with gold and observed in a SEM at 10–15 kV (Ayache *et al.*, 2010; Hernando *et al.*, 2010).

3.3.2 Transmission Electron Microscopy

1–3 mm^3 cubes of samples are mixed with firstly 2.5% glutaraldehyde and then with 2% osmium tetroxide like SEM, and dehydrated with 30%, 50%, and 70% ethanol for 10 min (Hayat, 1986). Then, formed samples are engrained in a resin and the resulting blocks are downscaled with an ultramicrotome. Then, the obtained parts (≅100 Å) are collected on copper grills and stained with 4% lead citrate to allow their observation in a transmission microscope at 100 kV (Hernando *et al.*, 2010).

3.3.3 Cryo-Scanning Electron Microscopy

Samples are immersed in slush nitrogen (–210°C) and transferred to a cold stage linked to a SEM operating at a temperature below –130°C (Echlin, 2009). Samples are cryofractured at −180°C and corroded at −80°C to −90°C. The observations in the microscope are carried out at 10–15 kV (Hernando *et al.*, 2010).

3.4 Dairy Microstructure

The milk proteins occur as soluble proteins, such as lactalbumin and lactoglobulin, and casein micelles. The structure of the casein micelle is about 100 nm in diameter and therefore, it is spherical. The casein micelles have smaller units, called sub-micelles, which primarily contain various fractions of caseins and colloidal calcium phosphates. In addition to protein, fat globules are also dispersed in the milk (Lluch *et al.*, 2003).

3.5 Electron Microscopy for the Dairy Product

The large part of dairy food is structured from casein micelles and whey proteins and to understand the structure of dairy food requires the observation of casein micelles and their interactions between each other using electron microscopy (EM).

Verheul and Roefs (1998) investigated the microstructure of heat-induced whey protein isolate (WPI) gels. These gels were made at near-neutral pH at 68.5°C, over 20 h. The formed WPI gels were identified as particulate and obtained by SEM. This observation was confirmed by a strong increase in gel permeability. Gel permeability revealed that gel structure did not change much after gel formation.

Martin *et al.* (2006) described a new method to examine thoroughly the structure of casein micelles and their interactions with κ-carrageenan and λ-carrageenan using scanning electron microscopy (SEM). The result of this study showed that immobilization of casein micelles at pH 6.7 to a gold substrate by covalent binding resulted in a homogeneous distribution of isolated, spherical casein micelles on the substrate. Therefore, the images of the casein micelles were provided in detail. The obtained electron micrographs revealed that casein micelle–carrageenan mixtures formed larger structures with increasing carrageenan concentration. When λ-carrageenan and κ-carrageenan were compared, larger and more aggregates were found with λ-carrageenan at similar carrageenan concentrations (Figure 3.1).

Ong *et al.* (2011) used cryo-scanning electron microscopy (cryo-SEM) to observe changes in the microstructure of milk, rennet-induced gel and curd during the manufacture of Cheddar cheese. The results showed that sample preservation for cyro-SEM did not alter the microstructure of the sample, but artifacts such as the formation of ice crystals could be observed in some samples. In this situation, the formed ice crystals caused alteration of the sample shape. However, such artifacts are minimized by changing the etching time, thickness of gold coating, and accelerating voltage and type of detector used for cryo-SEM observation. According to this study, the mixture of casein micelles and fat globules were best observed in the gel and the cooked curd. In the cyro-SEM, the first sample preparation step was freezing of sample for 30 min, and then the frozen sample was coated with a mixture of gold and palladium alloy approximately 6 nm thick at 140°C. It was then observed using a backscattered electron detector at 15 kV (Figure 3.2).

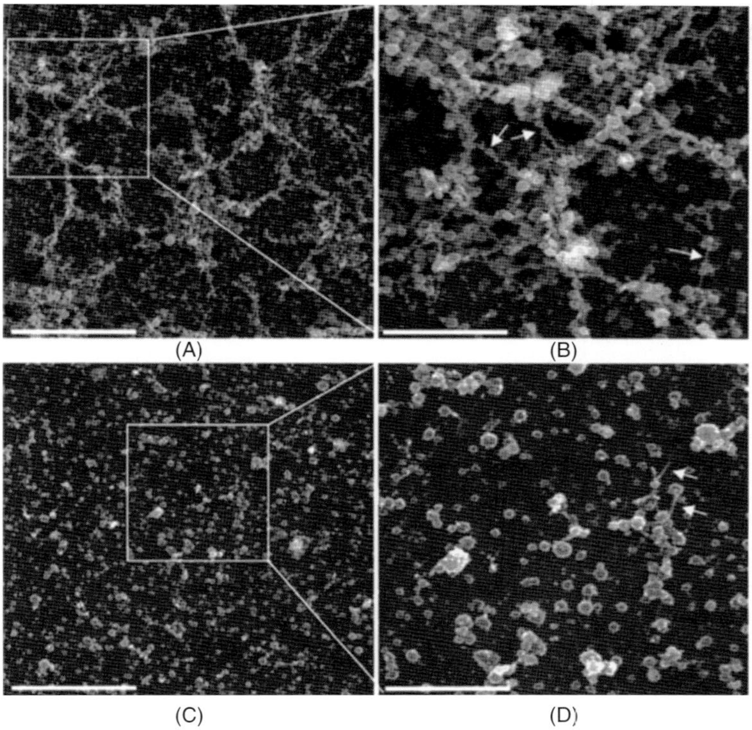

Figure 3.1 Casein micelles containing 0.03% κ-carrageenan (A, B) or 0.03% λ-carrageenan (C, D). For A, C, scale bar is 3000 nm and for B, D 1200 nm (Martin *et al.*, 2006).

Hassan *et al.* (2003) visualized the microstructure of two types of cheese (Karish and Feta) and milk fermented with different ropy and non-ropy strains of lactic acid bacteria using cryo-SEM. Milk fat and exopolysaccharide (EPS) were observed in pores within the protein network when cyro-SEM was used. Differences were observed in the microstructure of EPS between moderately ropy and highly ropy strains (Figure 3.3).

Caldwell *et al.* (1992) aimed to examine the influence of processing parameters, incorporation of polysaccharide stabilizers, freezing rates and storage times and temperatures on the structure of ice cream using SEM. Quenching and batch scraped-surface freezing used for the production of ice-cream were compared (Figure 3.4). Figure 3.5 showed the comparison of the different polysaccharide stabilizers and Figure 3.6 indicated the effect of storage conditions. According to Figure 3.6a, freezing was applied with a liquid nitrogen quench at −40°C and in Figure 3.6b, a plate freezer was used at −40°C. In Figure 3.6c, the temperature of the freezing room was −25°C.

Bottazzi and Bianchi (1980) applied SEM to examine the micro-organisms on a kefir granule. The results showed that kefir granules included mainly bacteria and a few yeasts in the exterior of the granules, and a few bacteria were found in the center of granule (Figure 3.7).

McManus *et al.* (1993) used to a technique for the preparation of milk products for scanning electron microscopy, resulting in high-resolution images. In this study,

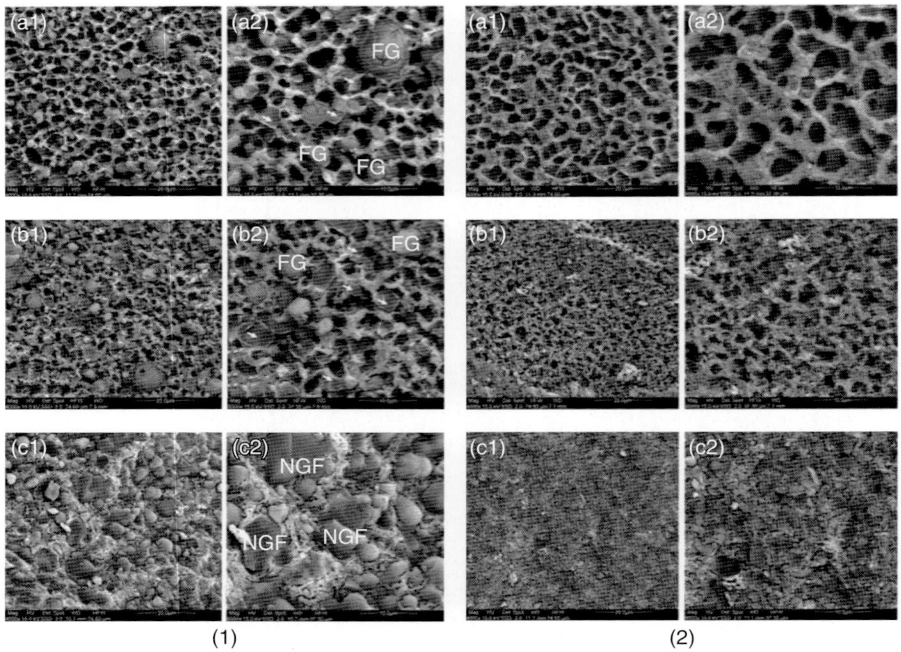

Figure 3.2 Cryo-scanning electron microscopy images of (a) the gel; (b) the cooked curd; and (c) cheese prepared from unhomogenized milk (1) and homogenized milk (2). The scale bars are 20 mm (a1, b1, c1) and 10 mm (a2, b2, c2) in length (Ong *et al.*, 2011).

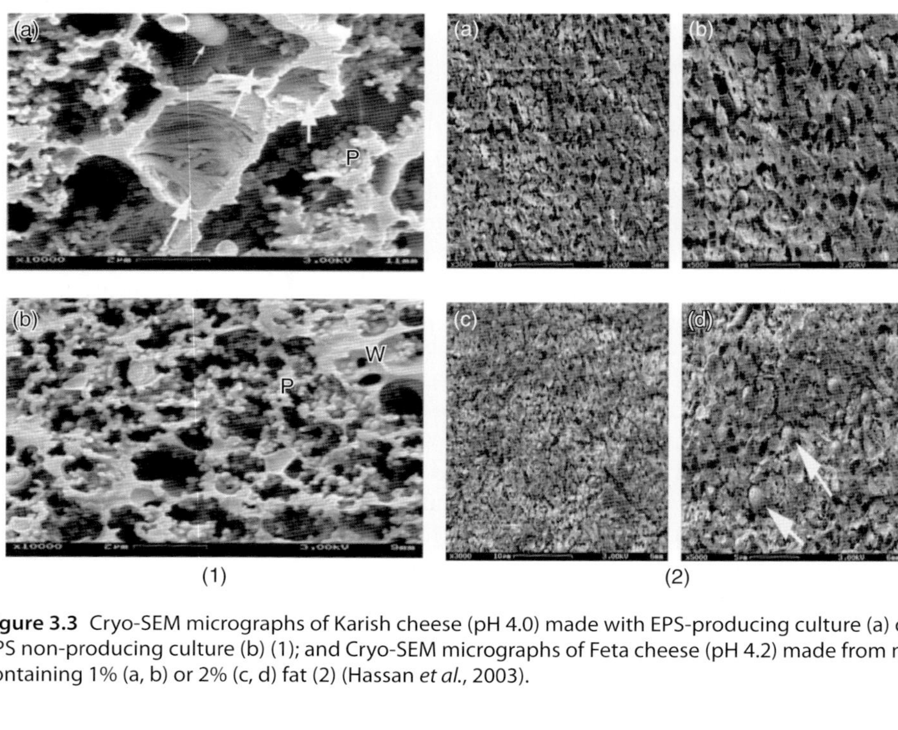

Figure 3.3 Cryo-SEM micrographs of Karish cheese (pH 4.0) made with EPS-producing culture (a) or EPS non-producing culture (b) (1); and Cryo-SEM micrographs of Feta cheese (pH 4.2) made from milk containing 1% (a, b) or 2% (c, d) fat (2) (Hassan *et al.*, 2003).

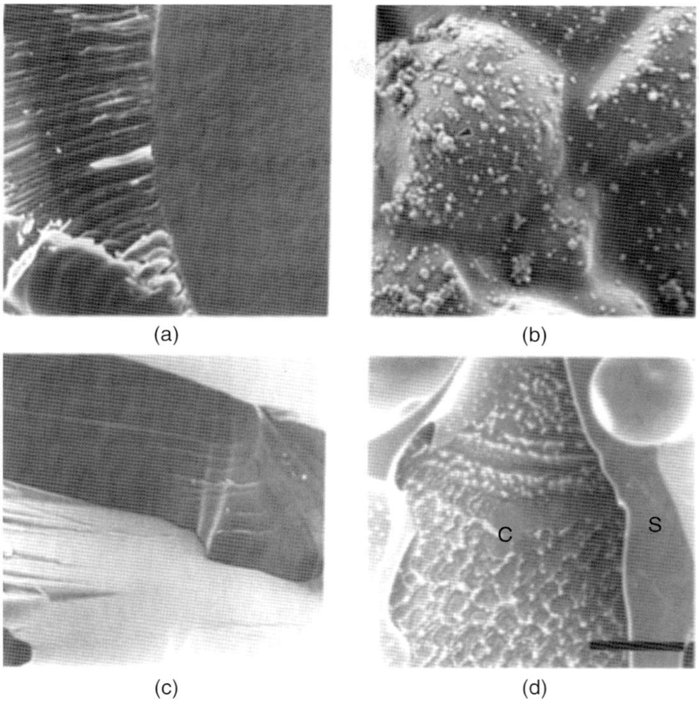

Figure 3.4 Liquid nitrogen quenching and batch scraped-surface freezing. (a) 20% sucrose, quenched; (b) 20% sucrose, batch-frozen; (c) 40% sucrose, quenched; and (d) 40% sucrose, batch-frozen, showing ice crystal socket (c) and serum phase (S) (Caldwell *et al.*, 1992).

samples were prepared with 2% glutaraldehyde, cryoprotected in 70% ethanol, and cryofractured. Using this procedure, hard cheeses, cream cheese and yogurt were observed and recorded.

Kalab and Harwalkar (1972) studied the structure of many foods such as renneted milk gels, whey gels, tofu, coagulated egg white, and gelatin gels using scanning electron microscopy. Renneted milk gels contained a high proportion of long fibers, and casein micelles were fused together. Whey gels contained particles of various sizes, and the occurrence of fiber-like material was low. Tofu and coagulated egg white had similar structures consisting of many corpuscular fibers.

Marchesseau *et al.* (1997) investigated the influence of pH on biochemical characteristics during the production of cheese (Figure 3.8).

Karlsson *et al.* (2007) used transmission electron microscopy using three different preparation methods to investigate the microstructure of casein micelles in ultrafiltrated (UF) skim milk concentrate at pH 6.5 and 5.8. The results of this study revealed that the micellar core appeared rougher and differences in the micellar core microstructure due to changing pH could be observed when this preparation method was used (Figure 3.9).

Calderon-Miranda *et al.* (1999) investigated using transmission electron microscopy (TEM) the morphological changes on *Listeria innocua* as a result of PEF treatment in skimmed milk containing nisin. The combination of pulsed electric fields and nisin exhibit an additive effect in the morphological damage observed on *L. innocua*. Pore formation was observed on *L. innocua* for an electric field intensity of 40 kV/cm. The

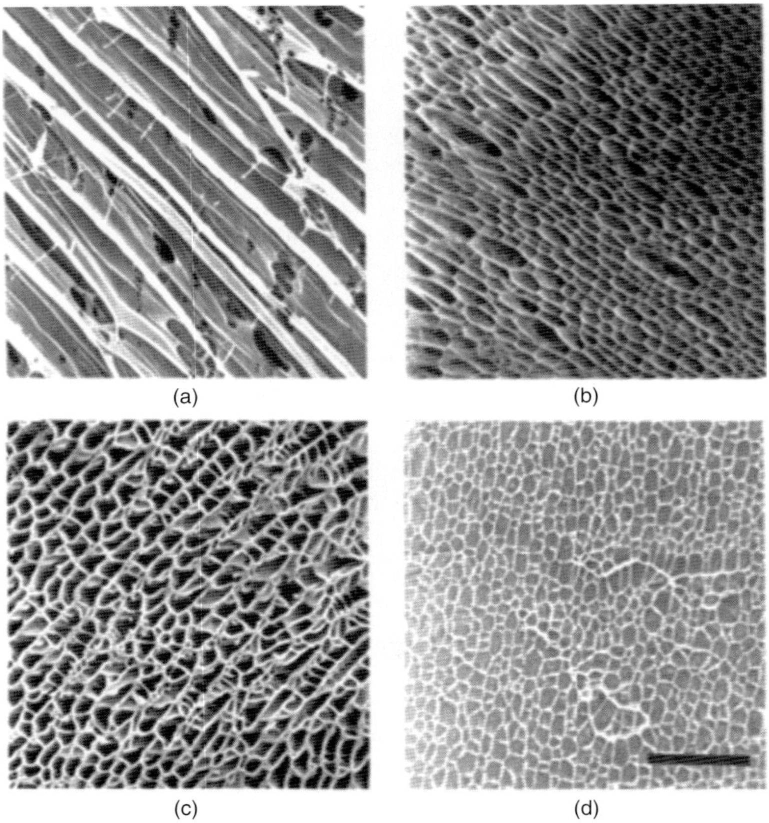

Figure 3.5 Quench-freezing of 0.2% polysaccharide solutions. (a) Xanthan; (b) locust bean gum; (c) carrageenan; and (d) guar gum (Caldwell *et al.*, 1992).

inactivation of *L. innocua* was a consequence of rupture of the cell membrane and loss of cell membrane functionality (Figure 3.10).

Knudsen and Skibsted (2010) applied high hydrostatic pressure to milk in order to modify the casein micelles. The casein micelle structure in untreated and pressure-treated skimmed milk were examined by using cryo-transmission electron microscopy (cryo-TEM). The electron micrographs indicated that the substructures were similar for pressure-modified casein micelles and casein micelles in untreated milk (Figure 3.11).

Cameron *et al.* (2008) researched the effect of ultrasonication on the inactivation of dairy microbes such as *Escherichia coli, Saccharomyces cerevisiae and Lactobacillus acidophilus* and showed that effect with transmission electron microscopy (Figure 3.12).

Marchin *et al.* (2007) used cryo-TEM to provide fine structural details on casein micelles structure (Figure 3.13). Differential centrifugation was applied to fractionate casein micelles and these micelles were observed using cryo-TEM.

Tunick *et al.* (2002) compared the structure of control cheeses made from untreated milk and that of Mozzarella cheeses prepared from microfluidized milk. The effects on the casein particles and fat globules in the cheese were investigated by TEM after 1 day and 6 weeks of storage at 4°C (Figure 3.14).

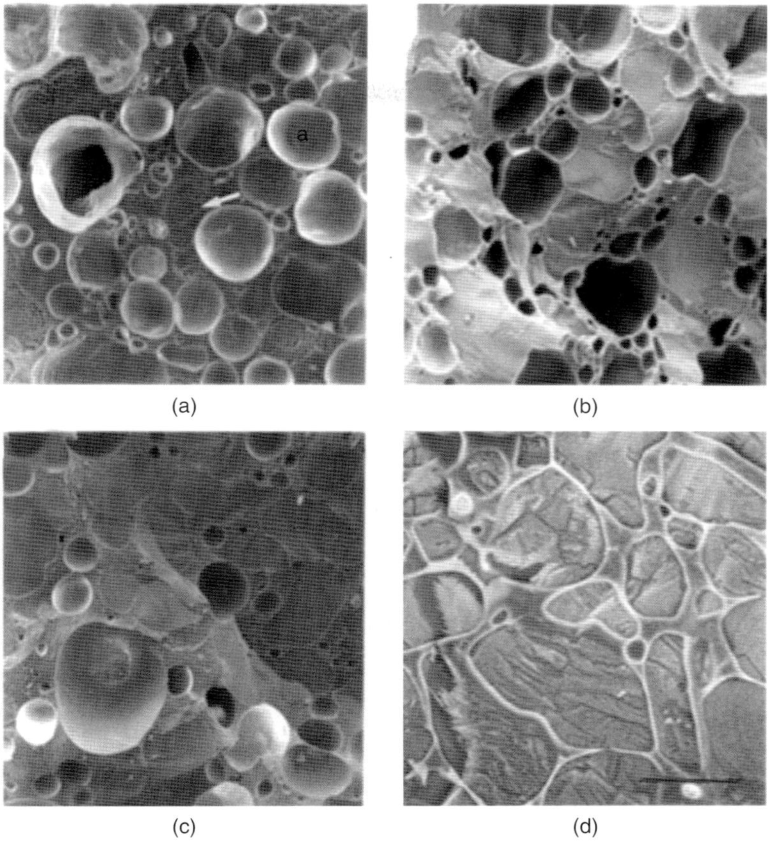

Figure 3.6 The influence of freezing rate on the microstructure of ice cream. Freezing method: (a, b, c) continuous scraped-surface freezing; (d) batch-scraped-surface freezing (Caldwell *et al.*, 1992).

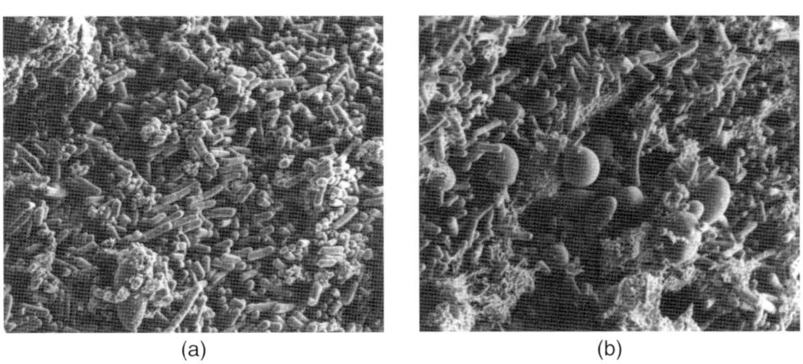

Figure 3.7 Rod-shaped bacteria in the exterior layer (a); rod-shaped bacteria and yeasts in the median part (b); and yeast at the center (c) of a kefir granule (Bottazzi and Bianchi, 1980).

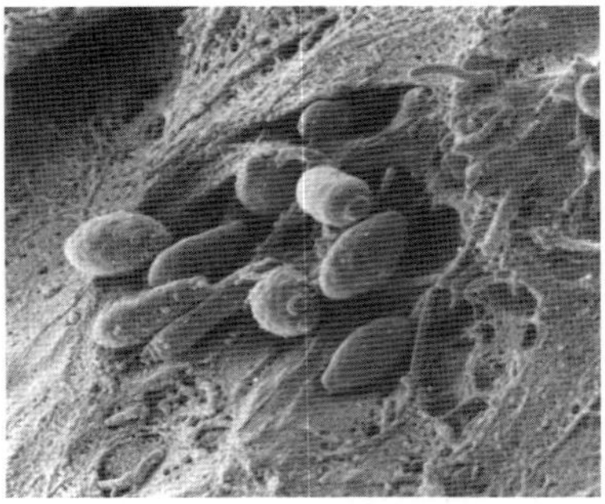

(c)

Figure 3.7 (Continued)

Figure 3.8 Scanning electron micrographs of processed cheese cooked at 115°C for 1 minute, stirred at 95°C for 8 minutes, and made at pH 5.2 (a), 5.7 (b), 6.1 (c), or 6.7 (d) Marchesseau *et al.* (1997).

3.6 Summary

Milk and other dairy foods have been known as important foods since time immemorial. Remains of cheese found in Egyptian tombs have been proof. However, the importance of dairy products varied geographically. Dairy products were especially consumed by the public throughout Europe; in Greece and Rome cheese was immensely popular.

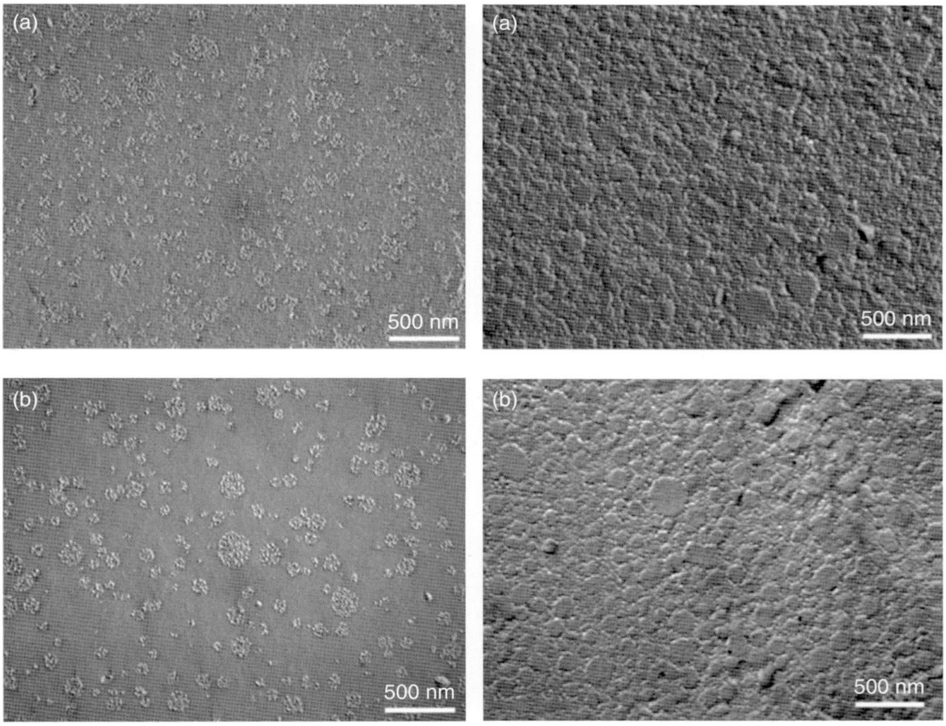

(1) (2)

Figure 3.9 (1) The casein micelles from ultrafiltrated skimmed milk of pH 6.5 (a) and 5.8 (b) observed in TEM. The casein micelles were mixed by glycerol, fixed in Freon 22 at −160°C and freeze-fractured at −120°C. (2) The casein micelles from ultrafiltrated skimmed milk of pH 6.5 (a) and 5.8 (b) observed in TEM. The undiluted ultrafiltrated skimmed milk was quickly frozen at 180°C in a propane jet and freeze-fractured at 120°C (Karlsson *et al.* (2007).

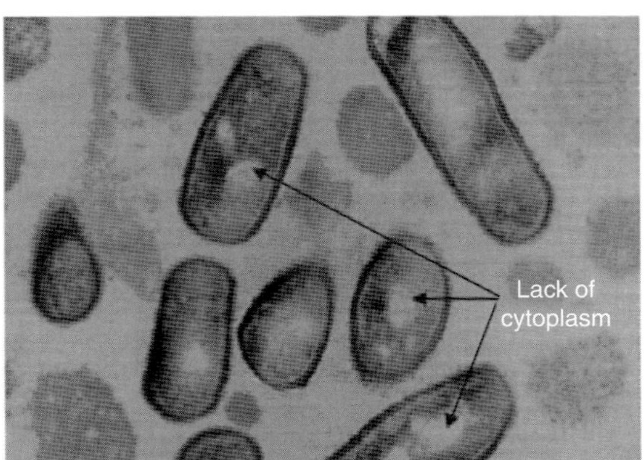

a (bar = 200 nm)

Figure 3.10 *L. innocua* cells applied to PEF with electric field intensity of 30 kV/cm and nisin (Calderon-Miranda *et al.*, 1999).

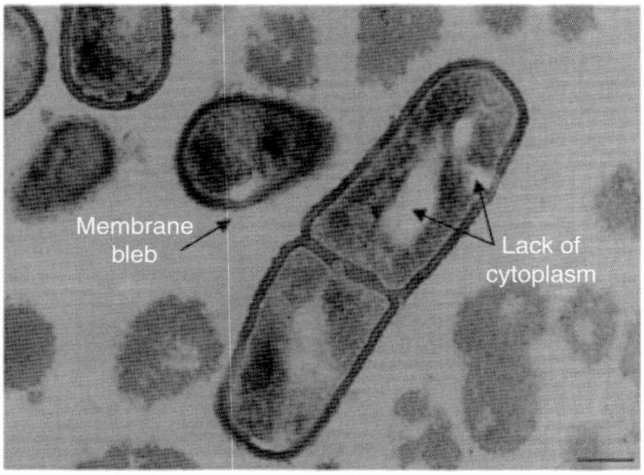

b (bar = 200 nm)

Figure 3.10 (Continued)

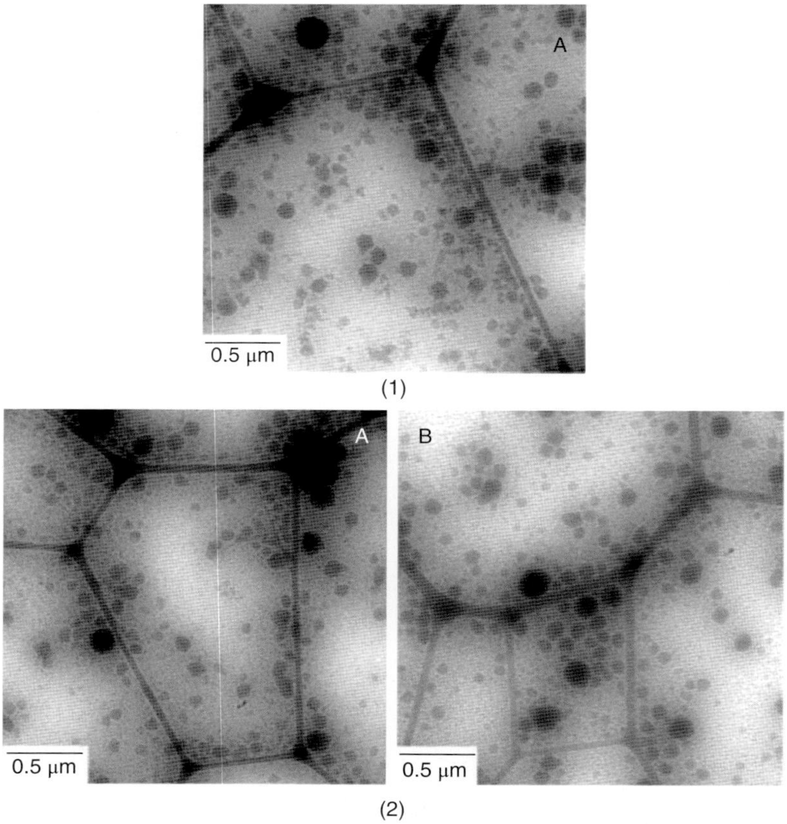

Figure 3.11 (1) Cryo-transmission electron micrographs of casein micelles in untreated skimmed milk. (2) Cryo-transmission electron micrographs of casein micelles in skimmed milk subjected to high pressure treatment at 150 MPa (A), 200 MPa (B) (Knudsen and Skibsted, 2010).

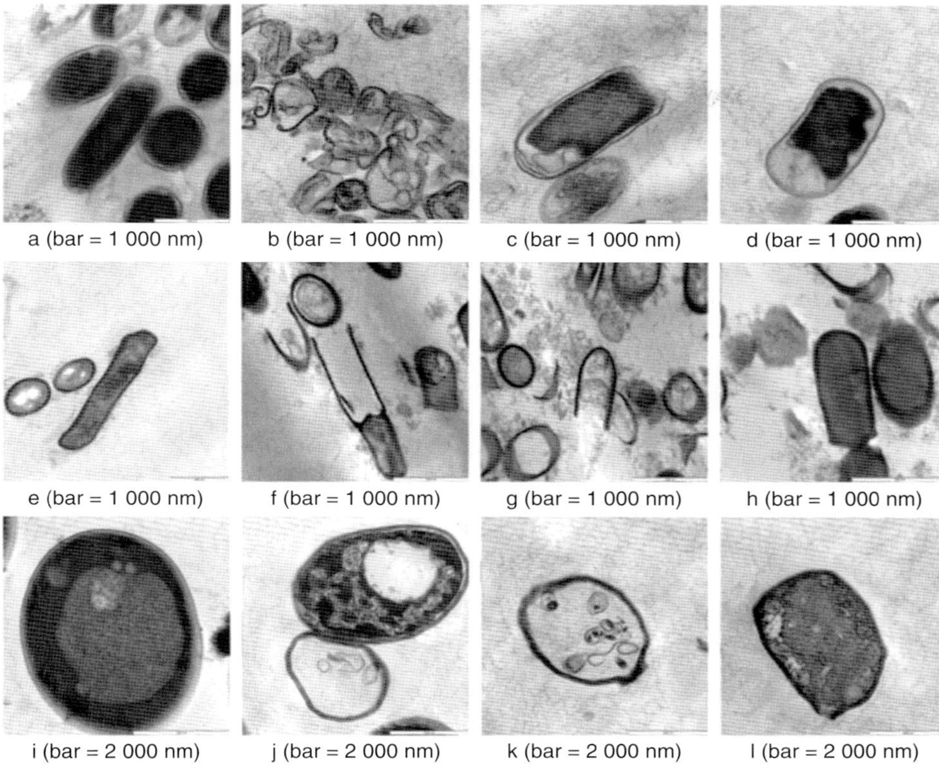

Figure 3.12 Transmission electron micrographs of *E. coli* [untreated (a) and ultrasonicated (b–d)], *L. acidophilus* [untreated (e) and ultrasonicated (f–h)] and *S. cerevisiae* [untreated (i) and ultrasonicated (j–l)] cells (Cameron *et al.*, 2008).

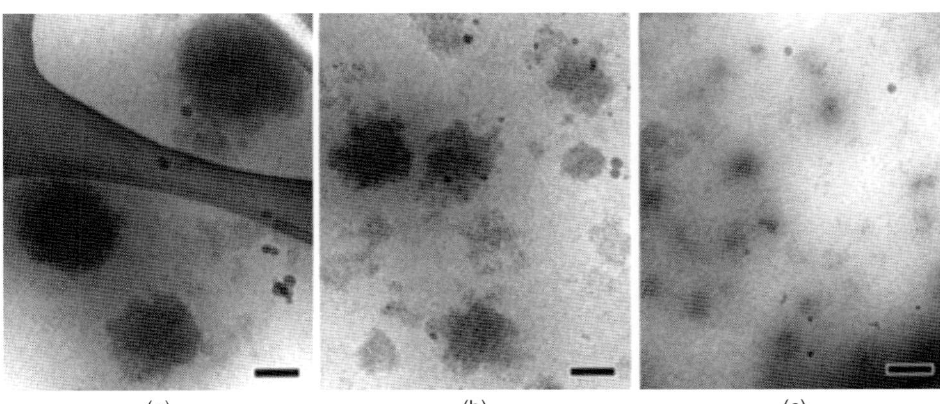

Figure 3.13 Cryo-TEM images of casein micelles from fractions (a) 1, (b) 4, and (c) 6.

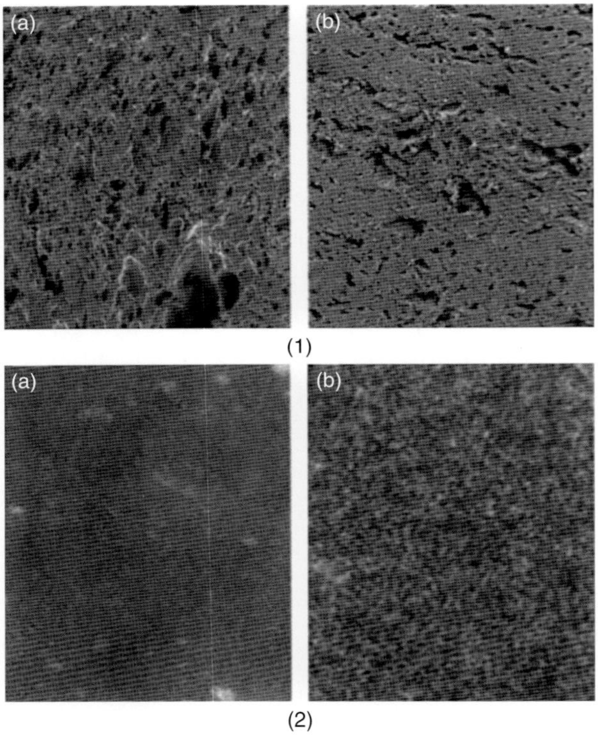

Figure 3.14 (1) SEM of Mozzarella cheeses. (a) Low-fat cheese. (b) Low-fat cheese made from milk microfluidized at 10°C and 34 MPa. (2) (a) Low-fat Mozzarella cheese after 1 day of refrigerated storage. (b) Same, after 6 weeks of refrigerated storage (Tunick *et al.*, 2002).

However, the consumption of fresh milk and butter (except cheese) were popular in northern Europe and Asia.

Nowadays, determination of the structure of dairy products has been an important topic because dairy products may be divided into two groups based on their composition, one consisting mainly of milk proteins, like casein and whey protein, and the other consisting mainly of fat. However, during the applied processes, maturation or storage, this structure changes and these changes can be observed by electron microscopy. The development of electron microscopy (EM) has allowed food structure to be studied in greater detail. In this chapter, the different electron microscopy techniques applied to dairy products are discussed and compared.

References

Aguilera, J.M., Stanley, D.W. 1999. Examining Food Microstructure. Microstructural Principles of Food Processing and Engineering, 2nd edn. Aspen Publishers, Gaithersburg, MD, pp. 1–70.

Ayache, J., Beaunier, L., Boumendil, J., Ehret, G., Laub, D. 2010. Sample preparation handbook for transmission electron microscopy techniques. Springer, New York, pp. 1–363.

Belkoura, L., Stubenrauch, C., Strey, R. 2004. Freeze fracture direct imaging: a new freeze fracture method for specimen preparation in cryo-transmission electron microscopy. Langmuir, 20, 4391–4399.

Bogner, A., Jouneau, P.H., Thollet, G., Basset, D., Gauthier, C. 2007. A history of scanning electron microscopy developments: Towards "wet-STEM" imaging, Micron, 38, 390–401.

Bottazzi, V., Bianchi, F. 1980. A note on scanning electron microscopy of microorganisms associated with the kefir granule. Journal of Applied Bacteriology, 48, 265–268.

Calderon-Miranda, M.L., Barbosa-Canovas, G.V., Swanson, B.G. 1999. Transmission electron microscopy of *Listeria innocua* treated by pulsed electric fields and nisin in skimmed milk. International Journal of Food Microbiology, 51, 31–38.

Caldwell, K.B., Golf, H.D., Stanley, D.W. 1992. A low temperature scanning electron microscopy study of ice cream. II. Influence of selected ingredients and processes. Food Structure, 11, 192, 11–23.

Cameron, M., McMaster, L.D., Britz, T.J. 2008. Electron microscopic analysis of dairy microbes inactivated by ultrasound. Ultrasonics Sonochemistry, 15, 960–964.

Danilatos, G.D. 1993. Introduction to the ESEM instrument. Microscopy Research Technique, 25, 354–361.

Echlin, P. 2009. Handbook of Sample Preparation for Scanning Electron Microscopy and X-Ray Microanalysis. Springer, New York, pp. 1–329.

Fazaeli, M., Tahmasebi, M., Emam Djomeh, Z. 2012. Characterization of food texture: application of microscopic technology. In: Current Microscopy Contributions to Advances in Science and Technology. A. Méndez-Vilas (ed.) Formatex, Spain.

Flint, O., 1994. Using Optical Microscopy. In: Food Microscopy: A Manual of Practical Methods. Acribia, S.A. (ed.) Bio Scientific Publishers Limited, Zaragoza, Chapter 4.

Goldstein, J.I., Newbury, D.E., Echlin, P., Joy, D. C., Romig, A.D., Lyman, C.E., Fiori, C., Lifshin, E. 1992. Scanning Electron Microscopy and X-Ray Microanalysis. Plenum Press, New York, pp. 1–829.

Harker, F.R., White, A., Gunson, F.A., Hallett, I.C., De-Silva, H.N. 2006. Instrumental measurement of apple texture: a comparison of the single-edge notched bend test and the penetrometer. Post-Harvest Biology and Technology, 39, 185–192.

Hassan, A.N., Frank, J.F., Elsoda, M. 2003. Observation of bacterial exopolysaccharide in dairy products using cryo-scanning electron microscopy. International Dairy Journal, 13, 755–762.

Hayat, M.A. 1986. Basic techniques for Transmission Electron Microscopy. Academic Press, London, pp. 1–415.

Hernando, I., Pérez-Munuera, I., Quiles, A., Lluch, M.A. 2010. Microstructure, in Handbook of Dairy Food Analysis. Toldrá F. and Nollet L. (eds.) CRC Press, Boca Raton, FL, Chapter 13.

James, B.J., Fonseca, C.A. 2006. Texture studies and compression behavior of apple flesh. International Journal of Modern Physics, 20, 3993–3998.

Kalab, M. 1979. Scanning electron microscopy of dairy products: An overview. Scanning Electron Microscopy, III, 261–272.

Kalab, M., Harwalkar, V.R. 1972. Milk gel structure. I. Application of scanning electron microscopy to milk and other food gels. Journal of Dairy Science, 56, 7.

Karlsson, A.O., Ipsen, R., Ardo, Y. 2007. Observations of casein micelles in skim milk concentrate by transmission electron microscopy. LWT, 40, 1102–1107.

Klang, V., Matsko, N.B., Valenta, C., Hofer, F. 2012. Electron microscopy of nanoemulsions: An essential tool for characterisation and stability assessment. Micron, 43, 85–103.

Knudsen, J.C., Skibsted, L.H. 2010. High pressure effects on the structure of casein micelles in milk as studied by cryo-transmission electron microscopy. Food Chemistry, 119, 202–208.

Kuntsche, J., Horst, J.C., Bunjes, H. 2011. Cryogenic transmission electron microscopy (cryo-TEM) for studying the morphology of colloidal drug delivery systems. International Journal of Pharmaceutics, 417, 120–137.

Lluch, M.A., Hernando, I., Pérez-Munuera, I. 2003. Lipids in food structures. In: Chemical and Functional Properties of Food Lipids. Sikorski, Z.E. (ed.) CRC Press, Boca Raton, FL, Chapter 2.

Marchesseau, S., Gastaldi, E., Lagaude, A., Cuq, J.L. 1997. Influence of pH on protein interactions and microstructure of process cheese. Journal of Dairy Science, 80, 1483–1489.

Marchin, S., Putaux, J.L., Pignon, F., Léonil, J. 2007. Effects of the environmental factors on the casein micelle structure studied by cryo transmission electron microscopy and small-angle x-ray scattering/ultrasmall-angle x-ray scattering. Journal of Chemical Physics, 126, 045101.

Martin, A.H., Goff, H.D., Smith, A., Dalgleish, D.G. 2006. Immobilization of casein micelles for probing their structure and interactions with polysaccharides using scanning electron microscopy (SEM). Food Hydrocolloids, 20, 817–824.

McManus, W.R., McMahon, D.J., Oberg, C.J. 1993. High-resolution scanning electron microscopy of milk products: a new sample preparation procedure. Food Structure, 12, 475–482.

Ong, L., Dagastine, R.R., Kentish, S.E., Gras, S.L. 2011. Microstructure of milk gel and cheese curd observed using cryo scanning electron microscopy and confocal microscopy. LWT – Food Science and Technology, 44, 1291–1302.

Pérez-Munuera, I., Larrea, V., Quiles, A., Lluch, M.A. 2008. Microstructure in muscle foods. In: Handbook of Muscle Foods Analysis. Toldrá, F. and Nollet, L. (eds.) CRC Press, Boca Raton, FL, Chapter 19.

Preetz, C., Hauser, A., Hause, G., Kramer, A., Mader, K. 2010. Application of atomic force microscopy and ultrasonic resonator technology on nanoscale: distinction of nanoemulsions from nanocapsules. European Journal of Pharmaceutical Science, 39, 141–151.

Thiel, B.L, Toth, M. 2005. Secondary electron contrast in low-vacuum/ environmental scanning electron microscopy of dielectrics. Journal of Applied Physics, 97, 1–18.

Tunick, M.H., Hekken, D.L.V., Cooke, P. H., Malin, E.L. 2002. Transmission electron microscopy of mozzarella cheeses made from microfluidized milk. Journal of Agricultural Food Chemistry, 50, 99–103.

Verheul, M., Roefs, S.E.E.M. 1998. Structure of whey protein gels, studied by permeability, scanning electron microscopy and rheology. Food Hydrocolloids, 12, 17–24.

Wilson, J. 1991. Microscopical methods for examining frozen foods. In: Bald, W.B. Food Freezing: Today and Tomorrow. London, Springer, pp. 97–112.

4

Emerging Techniques for Microstructural Analysis

I. Hernando, E. Llorca, and A. Quiles

Departamento de Tecnología de Alimentos, Edificio 3F, Universitat Politècnica de València, Camino de Vera s/n, 46022, Valencia, Spain

4.1 Introduction

When we think of microscopes, generally the first ones that come to mind are light microscopes and electronic microscopes. They are powerful tools that can generate magnified images of the samples by focusing electromagnetic radiation, such as photons and electrons, on their surface. Light and electron microscopes can easily generate two-dimensional images of a sample's surface, but they do not readily supply the vertical dimensions of the sample, nor can they achieve atomic resolution levels in unstained samples, or detect the structure of samples in their natural environment.

Other technologies do exist. They are generally used in other disciplines such as physics, materials science, chemistry, electrical engineering, biology and biotechnology, and are less well-known in the area of food science, although they can also be used in this field. This chapter is intended to give information on microscopy techniques such as Scanning Probe Microscopy, X-ray Computed Tomography, Scattering methods, Raman and FTIR spectroscopy and Magnetic Resonance Imaging, which can help us to describe and study changes or events that may occur in the sphere of food science.

4.2 Scanning Probe Microscopy

In 1981, applying tunneling effect theory, Gerd Binnig and Heinrich Rohrer invented Scanning Tunneling Microscope (STM). This opened the door to Scanning Probe Microscopy (SPM). As a result of their invention, in 1986 they won the Nobel physics prize. The way that SPM works is based on the tunneling effect, a quantum effect that occurs at distances below a thousand-millionth part of a meter (a nanometer, 1 nm). Controlling this type of phenomenon allows us to study the topography of surfaces at atom level. From the point of view of classic mechanics, an electron with lower energy than the potential barrier cannot pass through it. According to quantum mechanics, however, electrons are not defined by a precise position but by a probability cloud. This

Microstructure of Dairy Products, First Edition.
Edited by Mamdouh Mahmoud Abdel-Rahman El-Bakry, Antoni Sanchez, and Bhavbhuti M. Mehta.

means that in some systems, the probability cloud extends to the other side of a potential energy barrier. As a result, the electron can cross the barrier and generate electrical intensity. This type of intensity is called tunneling intensity and is a control parameter that makes it possible to plot the topography of a surface. Binnig and Rohrer demonstrated that by controlling the vibrations of an instrument very similar to Young's Topographiner, it was possible to monitor the electron tunneling current between a sharp probe and a sample. Since then, SPM technology has been expanding and can be used to measure chemical, electrical, magnetic, etc. forces (Firtel and Beveridge, 1995).

SPM is a method that makes it possible to plot the surface of a sample by placing a tip or probe close to the surface and measuring the interaction between the two. When the probe is moved over the surface, it generates a map of the interaction and, therefore, a picture of the sample being studied. This technique can be used with different microscopes to measure different types of interaction between the probe and the sample. For instance, STM (Scanning Tunneling Microscope) measures the electric current between the tip and the sample when a difference in potential between them is applied. AFM (Atomic Force Microscope) measures the interaction forces when the tip and the sample are close to each other. MFM (Magnetic Force Microscope) is a variation on AFM that measures magnetic interaction. SNOM (Scanning Near-Field Optical Microscope) measures the evanescent light reflected or transmitted by the sample. SECM (Scanning Electrochemical Microscope), an electrochemical version of scanning probe microscopy, measures the local electrochemical behavior of liquid/solid, liquid/gas and liquid/liquid interfaces.

Because the interaction parameters diminish with distance, with these microscopes the distance between the probe and the sample has to be one nanometer or less. As a result, these microscopy techniques are very sensitive to small vibrations and noises. However, they provide very high resolution, even to the point where the atoms in the surface being observed become visible.

Since the invention of the first microscope that could provide atomic resolution, SPM has rapidly developed into a high resolution technique with applications in surface physics, material science, chemistry, electrical engineering, and other disciplines such as biology and biotechnology (You and Lowe, 1996). Soon after the first STM made its appearance, an image of a biological specimen was achieved (Baró *et al.*, 1985). This showed that it was possible to form images of small specimens, in the order of nanometers, at atmospheric pressure and without staining. The rapid expansion of SPM studies in biology/biotechnology has stemmed from the fact that SPM techniques offer several unique advantages: first, they require little or no sample preparation, with native biomolecules usually being imaged directly; second, they can provide a three-dimensional reconstruction of the sample surface in real space at ultra-high resolution; third, they are less destructive than other techniques (e.g. electron microscopy) commonly employed in biology; and fourth, they can operate in several environments, including air, liquid and vacuum (You and Lowe, 1996). The operation of SPM in an aqueous solution makes it possible not only to image biomolecules in their physiological environments without denaturation, but also to study biologically important processes at liquid/solid interfaces (You and Lowe, 1996).

Out of all the probe microscopes (STM, AFM, SNOM, SECM), the ones with most references in the biology/biotechnology area are STM and AFM, owing to their high resolution. STM and AFM are potentially the most powerful techniques for probing biomolecular structures (Hoh *et al.*, 1991). The number of recent papers reporting STM studies on biomolecules is substantially smaller than for AFM, probably because of difficulties in obtaining reproducible and interpretable images from uncoated biomolecules. Scanning tunneling microscopy is suitable for analyzing conductive and semi-conductive surfaces (Firtel and Beveridge, 1995). Consequently, biomacromolecules are often first prepared by coating the surface with a conductive film of metal (e.g. platinum-iridiumcarbon). However, the cost of the image stability provided by the metal coating is a loss of lateral resolution due to metal graininess. Considerable advances have been made in the production of fine-grained metal films suitable for STM applications (Amrein and Gross, 1992; Cheon *et al.*, 1993; Wepf *et al.*, 1991). The remarkable success of AFM is due to the fact that this technique can be used for imaging non-conducting biomaterials (Firtel and Beveridge, 1995). The growing interest in AFM rather than STM for obtaining images of biomaterials is due to the capacity of AFM to deliver stable images of many types of surface (Gould *et al.*, 1990).

4.2.1 Scanning Tunneling Microscope (STM)

The task of scanning tunneling microscopy is to calculate the current flowing from the tip to the sample surface or vice versa depending on the voltage applied. In a single-particle picture, this can be viewed as a scattering process where e.g. an incident electron from the interior of the metal tip scatters from the barrier and has a certain probability of penetrating into the sample surface (Doyen *et al.*, 1990). STM offers the possibility of direct, real space determination of surface structure. The result is essentially a contour map of the surface (Tersoff and Hamann, 1983). For an STM, a good resolution is considered to be 0.1 nm lateral resolution and 0.01 nm depth resolution. STM can be used in an ultra-high vacuum, in air, in liquid or gas environments, and at temperatures ranging from near zero Kelvin to a few hundred degrees Celsius. It requires: (i) an ultra-thin conducting tip, made from tungsten, platinum-iridium or gold, for example; (ii) excellent vibration control; and (iii) sophisticated electronics. Ideally, the tip of the probe is a single atom. When the conducting tip is placed at a distance of approximately 1 nm, the tunneling effect causes the electrons of the sample to jump to the tip, or vice versa, depending on the polarity of the voltage. For the tunneling effect to occur, both the sample and the tip must be conductors or semiconductors. This is a limitation, as it means that STM cannot create images of insulating materials. The measurement parameter is the intensity of the tunneling current. This depends on: (i) the distance between the surface and the tip; (ii) the bias (voltage difference) between the tip and the sample; and (iii) the local density of states (LDOS) of the sample. STM can be a difficult technique to apply, since it requires very clean and stable surfaces, very sharp tips, excellent vibration control and sophisticated electronics, but it is a very sensitive one that can obtain images with resolutions of less than 0.1 nm.

To study a sample with STM, it must be scanned with a very fine conducting tip placed on the surface of the sample at a constant tunneling current. The microscope can operate in two different modes: constant height or constant current. In the constant height mode, the tip moves horizontally over the sample and the current changes according to the topography and electronic properties of the sample. In the constant current mode, the feedback loop adjusts the tip height to keep the tunneling current constant at all the points measured. The constant height method has the advantage of being faster as the scanner does not need to be moved up and down. The downside is that it only provides useful information on relatively smooth surfaces. The constant current mode can measure irregular surfaces with great precision, but it takes longer.

4.2.2 Atomic Force Microscope (AFM)

An AFM is a very high resolution (nanometer level) microscope that can also measure forces at nano Newton scale (Meyer, 1992).It is made up of a cantilever, usually between 100 and 200 micro in length, with a very fine tip or probe at one end that scans the surface of the sample (Figure 4.1). By monitoring the motion of the tip across the surface, a three dimensional image of the surface is constructed. When the tip approaches the surface of the sample the cantilever is deflected by the forces operating between the tip and the sample (mechanical contact, Van der Waals, electrostatic, magnetic and chemical bonding).The deflection of the cantilever obeys Hooke's law and there are several ways of measuring it. The most common method is to use a laser and photodiodes, or piezo resistive cantilevers (deformation changes their electrical resistance).The cantilever is usually made of silicon and has a tip at one end with a curvature radius measured in nanometers. The tips are usually formed either by chemical vapor deposition of the chosen material or, more usually, by anisotropic etching of silicon. AFM scanners are made of piezoelectric materials that expand or contract depending on whether the voltage is applied in one direction or the other. AFM includes various modes of operation that can be divided into static (contact mode) and dynamic (non-contact mode) (Figure 4.1). In contact mode, the outline of the surface is measured

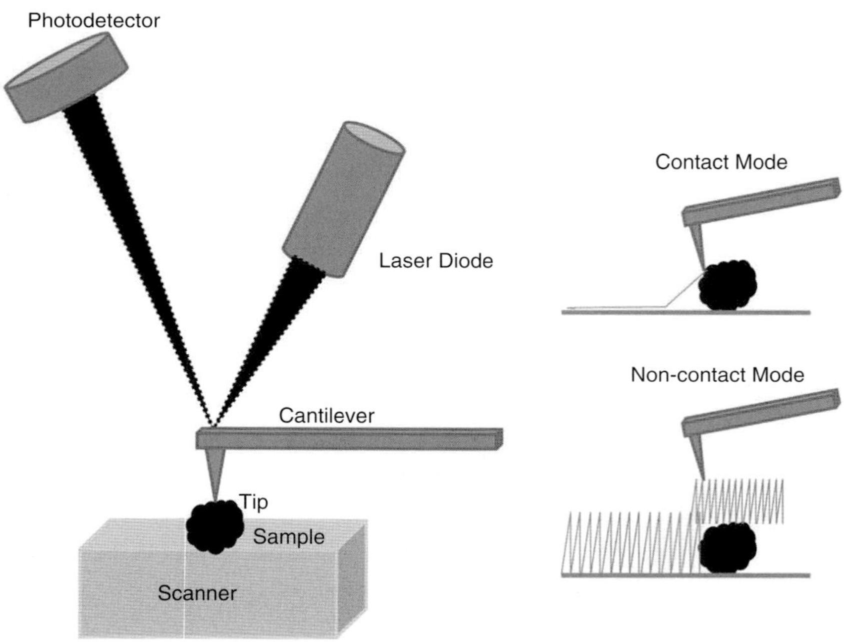

Figure 4.1 Schematic diagram showing the modes of action of an atomic force microscope.

directly through the deflection of the cantilever. An optical device is normally used for this measurement: a laser beam strikes the surface of the cantilever and is reflected onto two position-sensitive photodiodes. Owing to long-range attractive forces (generally Van der Waals), the tip may dig into the material. For this reason, the tip is normally placed in contact with the surface, where repulsive forces act, which is why this is generally known as the contact mode. In this mode, the deflection remains constant: when a variation in the fixed deflection value is detected, a current is applied to return it to the original value, and it is the measurement of this current that provides information about the surface. The disadvantage of this mode is that it can damage the sample surface and the AFM tip. In the non-contact mode, the cantilever is made to oscillate near its resonance frequency. In this mode the information is obtained through the changes in amplitude, phase or frequency of oscillation as a result of forces interacting with the surface. There is no contact between the tip and the sample at any moment and the tip is not worn down. Other modes of operation include the tapping mode, which is halfway between the dynamic and static modes, and force spectroscopy, which is based on direct measurement of the interaction forces between the tip and the sample in relation to the distance between them.

As already commented, the main advantage of AFM over STM is that AFM is able to work with insulating materials as well as with conductors, but its resolution can be lower. The advantages of AFM over SEM (scanning electron microscope) are that AFM has higher resolution, can obtain 3-D images and can operate under ambient conditions and without applying any previous treatment to the sample, making it possible to study living organisms, whereas SEM only works under special conditions (e.g. in a vacuum).

4.2.3 Applications of the Main Probe Microscopes

In the area of food science, STM and AFM have been used to study food-related systems, ranging from relatively large structures such as starch granules to the organization of secondary structures in proteins and the interaction of proteins. STM and AFM, one scanning the tunneling current and the other the atomic repulsive force between sample and probe, can give high-quality surface topographies of proteins. These have been difficult to obtain by more conventional methods such as transmission electron microscopy (TEM) (Arakawa *et al.*, 1992).

STM and AFM have been used to investigate some food proteins (Arakawa *et al.*, 1992; Mao *et al.*, 2013; Miles *et al.*, 1991; Tatham *et al.*, 1999), and Wang *et al.* (2016) demonstrated that the structure, morphology, and dynamics of a protein molecule can be examined under physiological conditions by STM.

The interfaces in food systems can be very complex, as they contain various proteins, different surfactant like molecules, and polysaccharides. AFM can be used to understand such systems by probing simpler model interfacial structures (Burgess *et al.*, 1999; Ducker and Grant, 1996; Manne *et al.*, 1994; Morris, 2004).

AFM is an important tool that has aided the understanding of different aspects of dairy products. For instance, Qi and Onwulata (2011) investigated the effect of extrusion and moisture content on whey protein structure to gain further insight on the relationship between extrusion conditions and protein structure and Burgain *et al.* (2014, 2015) studied the interactions between milk proteins (micellar casein, native and denatured whey proteins) and *Lactobacillus rhamnosus*. Burgain *et al.* (2013) showed that AFM enabled identification of specific interactions between bacteria (*Lactobacillus rhamnosus*) and whey proteins, in contrast to the observed non-specific interactions with micellar casein. In addition, Gebhardt *et al.* (2006) studied the reversible and irreversible states of pressure dissociated casein micelles.

Using AFM, Handojo *et al.* (2009) determined the thickness of a thin film of milk left on a drinking glass after a washing protocol. This thickness could be used to estimate the strength of adhesion and, by extension, the difficulty of removing milk contamination from drinking glasses. James *et al.* (2003) characterized the surface of unused microfiltration and ultrafiltration membranes, the fouled layer on the surface of membranes used for milk filtration, and the internal fouling within the used membranes.

An overview of recent advances concerning the organization of lipids in the milk fat globule membrane has been provided by López *et al.* (2015), while Murrieta-Pazos *et al.* (2011) studied the microstructure of skim and whole milk powers.

Picart *et al.* (2002) used AFM to study the inactivation by pulsed electric fields (PEF) of *Listeria innocua* in dairy fluids (in this case, suspended in skim milk). The 3D images revealed collapsed and/or segmented bacteria.

On occasion, combining different techniques makes it possible to reach deeper levels of knowledge. Doak *et al.* (2008) showed an efficient sample preparation method to facilitate coupled atomic force microscopy and confocal laser scanning microscopy (AFM–CLSM), providing a novel means of concurrent high resolution structural and fluorescence imaging. They demonstrated that AFM–CLSM is a powerful tool for the life sciences that has the potential to provide novel insights to further our understanding of cellular structure–function relationships. Andreeva *et al.* (2009) applied scanning near-field optical microscopy and atomic force microscopy and showed that the AFM/SNOM technique is of interest not only for obtaining versatile information about the object of study, but also for improving the resolution of SNOM topographical measurements in combination with AFM.

4.3 X-Ray Tomography

X-rays are a type of electromagnetic radiation with a shorter wavelength than visible light. The direction of an X-ray and the distance it travels through the sample depend on the energy of the beam and the composition of the sample. X-rays can interact with materials in different ways (Barigou and Douaire, 2013): (i) transmission, when the rays pass through the sample without affecting it; (ii) absorption, when the energy of the X-rays is transferred to the sample; and (iii) scattering, when the photons are deflected in different directions. In X-ray microscopy, the image is obtained through the transmitted light from the beam, projected on a 2-D plane.

X-ray tomography, like high-resolution X-ray tomography, micro-computed tomography (micro-CT or μCT), and similar terms use X-rays to create cross-sections of a physical object that can be used to recreate a virtual model (3-D model) without destroying the original object. High-resolution X-ray tomography (X-ray micro-CT) is a non-destructive technique and can provide information about the internal structure and properties of food. The prefix micro (prefix μ) is used to indicate that the pixel sizes of the cross-sections are in the range. With μCT, the internal structure of most food products can be viewed by measuring the different attenuations of X-rays for the materials in the product. The level of transmission of these rays depends mainly on the mass density and mass absorption coefficient of the material (Herremans *et al.*, 2013; Verboven *et al.*, 2008). Because absorption is different in gas and water, gas-filled spaces can be distinguished from the matrix material (Duquenne *et al.*, 2016).

X-ray μCT has several advantages over other methods, including the ability to image low-moisture materials. It uses the differences in X-ray attenuation arising, principally, from differences in density within the specimen. A series of 2-dimensional X-ray images are obtained as a sample is rotated. The series of slices, covering the entire sample, can be rendered into a 3-D image that can either be presented as a whole or as virtual slices of the sample at different depths and in different directions. Manipulation and analyses of μCT data using special software also allows reconstruction of cross-sections at depth increments as low as 15 μm, and along any desired orientation of the plane of cut. A series of noninvasive μCT slices of the same sample in any direction can provide much more information than just one scanning electron or optical microscopy image (Laverse *et al.*, 2011a).

This technique has been widely used in medicine, but not so much in food technology, and still less in dairy products. In this field, in recent years, the microstructure of cream cheeses has been successfully analyzed using μCT (Laverse *et al.*, 2011b).

Laverse *et al.* (2011a, 2011b) also demonstrated that X-ray microtomography could be a useful tool for quantifying fat deposits and fat distribution in emulsion-based food products such as yogurts and cheese. O'Sullivan *et al.* (2016) used this technique to study defective gas formation in cheese: X-ray CT provided an accurate, non-invasive, overall image, not only of eye formation but also of eye size and distribution and overall void percentage (Figure 4.2). This method also makes it possible to establish relationships between the biochemical characteristics of the cheese and the physical manifestation of the eyes.

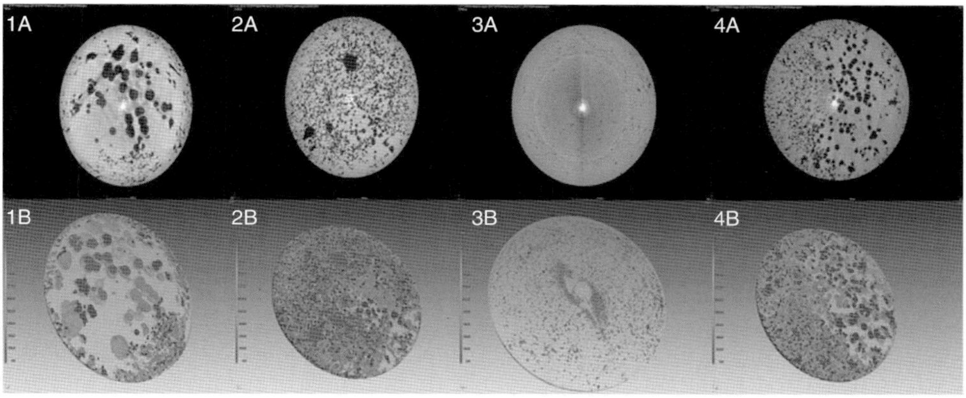

Figure 4.2 Eye formation in Swiss-type cheeses as determined by X-ray computed tomography. Colors in the blue spectrum represent voids of 0 to 6000 mm^3, those in green represent 9000 to 21,000 mm^3 and those in red represent 24,000 to 30,000 mm^3 (O'Sullivan *et al.*, 2016). (*See color plate section for the color representation of this figure.*)

4.4 Small-Angle-Scattering (SAS) Methods: SAXS and SANS

Scattering methods, using X-ray, light and neutron sources are important techniques for studying the structure of biological macromolecules. These three types of radiation cover a large range of magnification, from 1 nanometer to 1 micrometer. The scattering of visible light arises through differences in polarizability, that of X-rays from differences in the electronic structure of the atom, while for neutron scattering it depends on the nuclear structure of the atom (Lopez-Rubio and Gilbert, 2009). Small angle scattering (SAS) is the collective name given to the techniques based on analyzing the near-zero angle scattering of a beam of light by the material it passes through. These techniques are called small angle neutron (SANS), small angle X-ray (SAXS) and small angle light (SALS, or just LS) scattering, depending on the source of radiation. In each of these techniques, radiation is elastically scattered by a sample and the resulting scattering pattern is analyzed to provide information about the size, shape and orientation of some component of the sample. The type of sample that can be studied by SAS, the sample environment that can be applied, the actual scales of length that can be probed, and the information that can ultimately be obtained, all depend on the nature of the radiation employed. For example, LS cannot be used to study optically opaque samples and SAXS cannot (easily) be employed to study thick samples or samples requiring complex containers, whilst SANS (and SAXS) probe different length scales to LS (King, 1999), giving access to length scales smaller than 10 nanometers that are not accessible to light scattering (Nicolai, 2007).

Using small angle X-ray and neutron radiation scattering makes it possible to reach scales as small as 1 nm, as in the case of SAXS and SANS. This section discusses the basic use and some applications of SAXS and SANS in dairy products.

4.4.1 Small-Angle X-Ray Scattering (SAXS)

SAXS is a high resolution characterization technique that can resolve features in the range between 1 and 100 nm. An important benefit of using this technique is that biological samples can potentially be studied in their natural environment (Allec *et al.*, 2015). There is no need the sample to be in a crystallized form as in the case of X-ray diffraction.

In the SAXS method, the sample is irradiated with a beam of monochromatic X-rays. Very small angle scattering of these X-rays provides information on the size or size distribution of the particles and on their shape and internal structure. This method is used for particles in the 0.5 to 50 nm size range in materials such as liquid glass, polymer films, microemulsions, proteins, etc. There are important differences between a conventional diffractometer and the equipment required for this type of analysis. It requires a very fine beam so that the X-rays can be intercepted without blocking the scattered intensity. In addition, a greater distance between the sample and the detector allows the scattered beam to be separated from the incident beam and reduces the background. There also has to be a vacuum from the source of the X-rays to the detector. In addition, a PSD (position sensitive device) detector is used. The main technical challenges of SAXS are separating the strong primary X-ray beam from the weak scattering information near to the primary beam, and increasing the intensity of the scattered X-rays.

4.4.2 Small-Angle Neutron Scattering (SANS)

SANS is the most popular neutron technique for the study of food matrices. This technique allows the structure of biological macromolecules to be investigated in solution, under native conditions identical to those used in biochemical studies, and does not depend on the availability of crystals (May, 2002). Its applications extend from elucidating the quaternary structure of a protein to investigating the conformation of a polysaccharide chain or the lamellar structure of granular starch (Lopez-Rubio and Gilbert, 2009). It is complementary to SAXS, and both techniques are often jointly used when studying biological materials. SANS and SAXS techniques are unique tools compared to other time-resolved methods such as fluorescence, UV-vis, or even light scattering, even if the resolution time is somewhat longer. X-rays are more sensitive to heavy atoms, and thus to inorganic species (mineral, metal, etc.), whereas neutrons allow more specific following of rearrangements in organic assemblies (Grillo, 2009). The advantages of SANS over SAXS are its sensitivity to light elements, the possibility of isotope labeling, and the strong scattering by magnetic moments.

4.4.3 Applications of Small-Angle-Scattering Methods

Amongst food proteins, milk caseins have been widely studied using SANS and SAXS. Aschi *et al.* (2007) used SANS to elucidate the structure of unfolded beta-casein at various concentrations of salt, while De Kruif *et al.* (2012) described the internal structure of casein micelles by calculating the SANS and SAXS spectrum. Other models describing the casein micelle substructure had previously been built from SAXS and SANS data by Holt *et al.* (2003). In recent years, a new model for fitting SAXS data on milk casein micelle structure has been proposed (Ingham *et al.*, 2016). Not only has the internal structure of casein been investigated, but the nanoscale structure of milk fat crystal networks has also been described using SAXS (Ramel *et al.*, 2016).

Regarding milk and dairy products, the effect of ultrafiltration and diafiltration on the distribution of the calcium phosphate clusters of the casein micelle in ultrafiltered fresh milk has been investigated using SANS by Alexander *et al.* (2011). Darnay *et al.* (2015) proposed the SANS method for characterizing the fermentation process in low fat yogurts prepared with transglutaminase. Another dairy product that has been investigated using SAXS is milk protein concentrate (Mata *et al.*, 2011). Moreover, the thermal and structural behavior of butter has been studied by traditional X-ray diffraction techniques, using SAXS to describe the structural dynamics of milk fat fractions in crystallization (Huck-Iriart *et al.*, 2011).

4.5 Vibrational Spectroscopies (Fourier Transform Infrared-FTIR and Raman Microscopy)

Vibrational spectroscopies such as Fourier transform infrared (FTIR) and Raman microscopy, combined with microscopy techniques, can be used to examine the chemical composition and structure of foods. These spectrometry methods are based on the interaction of the sample with light, which can be excited, leading to the molecules entering vibrational transition states (Wellner, 2013).

FTIR spectroscopy is based on studying the interaction between electromagnetic radiation in the infrared region of the spectrum and the material to which it is applied. In other words, it measures light absorption by the material over a range of wavelengths of approximately 600–4000 cm^{-1}, which corresponds to fundamental molecular vibration modes. FTIR spectroscopy is a method that makes it possible to study the vibration and rotation of the molecules in the infrared region of the electromagnetic spectrum.

Unlike FTIR, which involves infrared absorption, Raman is a scattering technique. Raman spectroscopy is based on Raman scattering, in other words, on inelastic scattering of monochromatic light in the visible, near infrared or near ultraviolet range, generally from a laser (typically 785 nm). It is effective in the 250–2900 cm^{-1} range. The laser beam interacts with phonons (quasiparticle or quantized vibrational mode) or other excitations in the system, causing the energy of the laser photons to shift up or down. The shift in energy gives information on the vibrational modes in the system. FTIR spectroscopy provides information that complements that of Raman spectroscopy.

4.5.1 Fourier Transform Infrared (FTIR) Spectroscopy

The nature of the interaction with the material differs according to the energy range of the radiation being used. When the molecules absorb infrared radiation, they can change their vibrational and rotational energy state. In the case of gaseous samples, it is only possible to observe the transitions between rotational states that require very little energy. In the case of solid and liquid samples, the only changes taken into account are those between vibrational energy states, which make it possible to characterize the main functional groups in the molecular structure of the compound.

The basic principle of molecular spectrometry is that molecules absorb energy from light at specific wavelengths, known as their resonance or vibration frequencies. A small sample is placed in an infrared cell and subjected to a source of infrared light, ramped through wavelengths from 4000 cm^{-1} to 600 cm^{-1}.The intensity of the light that passes through the sample is measured at each wavelength, enabling the quantity of light absorbed by the sample to be calculated as the difference in intensity before and after passing through the sample cell. This is known as the infrared spectrum of the sample.

Fourier transform infrared spectroscopy (FTIR) is a less intuitive way of obtaining information. Instead of allowing only one wavelength at a time to pass through the detector, this technique uses a beam that contains many different wavelengths at the same time, and measures the total intensity of the beam. The beam is then modified to obtain a different combination of wavelengths, giving a second data point. This process is repeated many times. All the data are then processed by a computer to measure how much light is detected at each wavelength.

Specifically, a certain configuration of mirrors placed between the light source and the detector allows some wavelengths to pass and blocks others (through wave interference). For each new data point, the beam is altered by moving one of the mirrors. This changes the set of wavelengths that can pass through and interact with the sample.

The core of an FTIR system is the interferometer (Subramanian and Rodriguez-Saona, 2009; Wellner, 2013). A conventional interferometer works by splitting the IR light from a spectrum source into two beams and reflecting one from a fixed mirror and the other from a moving mirror (Figure 4.3). The moving mirror introduces a retardation that

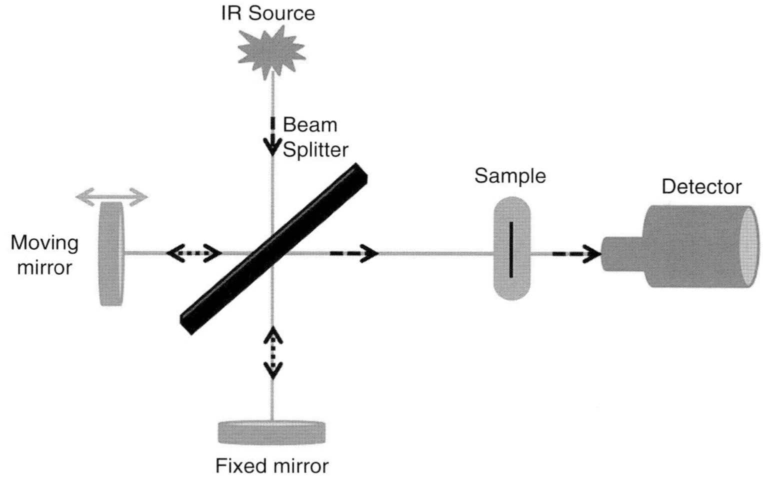

Figure 4.3 Schematic diagram of an interferometer.

generates a mixture of constructive and destructive interference when the beams are recombined. Measuring this signal in relation to the position of the mirror produces an interferogram that can be reconstructed as an IR spectrum by a mathematical algorithm called the Fourier transform. While the interferometer manages the IR light required for FTIR spectrum measurements, in order to perform chemical identification procedures it is essential to place a trustworthy standard sample in the FTIR system. When a sample is placed in contact with the sensor, a small amount of infrared light enters the sample and is absorbed at certain frequencies that are characteristic of that material. These characteristic frequencies can be recorded without any sample preparation or destruction.

The main difference between FTIR microscopy and a normal confocal microscope is that the former must not contain any glass lenses, because glass would absorb all the IR light. Therefore the optical elements are gold- or aluminum-coated mirrors and some IR-transparent windows. Focusing the beam on the sample is achieved by Cassegrain reflector-style condensers and collectors, and both transmission and reflection modes are available. These mirror optics extend the usable frequency range into the IR region. The IR light passes through several openings in the interferometer, where it is modulated by a moving mirror. The modulated light beam is then directed towards the microscope optics, which focus it on the sample. The transmitted or reflected IR light is collected and beamed to the IR detector, which converts the intensity of the light into an electrical signal. The intensity measured, which depends on the position of the mirror, is converted into a simple beam spectrum by the Fourier transform.

Infrared spectroscopy is a powerful tool for determining the constituent concentrations and qualitative characteristics of dairy products. Many examples and applications show that the technique is accurate and fast and can be used for process control (Fagan *et al.*, 2009).

Fourier transform infrared (FTIR) spectroscopy is widely used to determine milk composition (Wang *et al.*, 2016). FTIR data have shown nor bixin-induced conformational changes in whey proteins and caseins (Zhang and Zhong, 2013). Ferraro *et al.* (2015) studied the interaction of polyphenol with milk proteins by FTIR.

This technique has been used to characterize changes in the conformation of proteins in dairy products (Boye *et al.*, 1996), and to test differences in the protein conformation of milk protein concentrate powder and determine whether such differences could be associated with nitrogen solubility (Kher *et al.*, 2007). The solubility of commercial milk protein concentrates and milk protein isolates has also been studied by this technique (Sikand *et al.*, 2011). FTIR spectroscopy has been used to determine the concentration of proteins in raw milk (Etzion *et al.*, 2004), the structure of β-lactoglobulin (Boye *et al.*, 1996), thermal denaturation of whey proteins in milk (Parris *et al.*, 1991) and changes in the protein secondary structure of various proteins during drying (Izutsu *et al.*, 2006; Mauerer and Lee, 2006). Hewavitharana and Brakel (1997) used FTIR to study milk casein and Ioannou *et al.* (2015) used it to study whey proteins. The FTIR spectra of casein and whey hydrolysates have been shown to be capable of predicting several physical parameters such as bitterness, solubility and the emulsifying and foaming properties of hydrolysates (Van der Ven *et al.*, 2002). Nicolaou *et al.* (2011) demonstrated that when combined with chemometrics, FTIR provides a powerful approach for bacterial enumeration in milk and in cocultures without the need for their prior separation.

4.5.2 Raman Spectroscopy

Raman spectroscopy (RS) is a high-resolution photon technique that provides structural information on any material, facilitating its identification. It provides information related to the vibrational motions of the molecules to be used in prediction or identification of their chemical structures (Nonaka *et al.*, 1993). In this method, the analysis is based on beaming monochromatic light over the sample. Most of the scattered light is of the same frequency as the incident light (Rayleigh scattering), but part of the light is scattered inelastically and undergoes slight changes in frequency that depend on the material under study rather than on the frequency of the light striking it.

When the photons of the incident light beam hit the sample, most of them pass through it but a small fraction are scattered (around 1 scattered photon for every 1011 incident photons). In this case, the incident photon briefly excites the molecule to a higher vibrational or rotational state of energy that is not allowed, so rapidly drops to an allowed energy state, emitting a photon. The frequency at which it is emitted depends on the jump in energy level made by the molecule. If the result of the photon-molecule interaction is a photon that is scattered at the same frequency as the incident photon, the collision is said to be elastic, as there is no change in the energy state of either the photon or the molecule. This gives rise to Rayleigh scattering. If the result of the photon–molecule interaction is a photon that is scattered at a different frequency from the incident photon, the collision is said to be inelastic (energy is transferred between the photon and the molecule). In this case, one of two effects can occur: (i) if the scattered photon is at a lower frequency than the incident one, energy is transferred from the photon to the molecule, which, after jumping to an energy level that is not allowed, drops to an allowed one that is higher than its initial level, producing what is known as Stokes Raman scattering; or (ii) if the scattered photon is at a higher frequency than the incident one, energy is transferred from the molecule to the photon, which means that the molecule was not in its ground vibrational state but in a higher-energy one before the photon struck it, and after the collision it drops to its ground state. This is called Anti-Stokes Raman scattering. The Raman spectrum covers all these phenomena (Figure 4.4a). Each material has a set of frequencies that are characteristic of its structure.

Raman microscopy is carried out with conventional light microscopes, which have high-power lasers and Raman detectors attached, either directly or via optical fibers. The simple position and focus are selected in normal white light, then the system is switched over to the Raman mode. A range of diode-pumped solid state lasers and tunable diode lasers in the wavelength range of 400–1100 nm can be used for Raman excitation. The laser light is passed through the microscope onto the sample, a lens is used to collect the light, and a filter is used to block the wavelength of the incident light. Longer wavelengths (Raman scattering) are transmitted to the detector (Wellner, 2013). A diagram of Raman instrumentation is provided in Figure 4.4b.

The heterogeneous nature of foods demands that studies at the microstructural level be able to localize specific microregions and retrieve valuable physicochemical information in situ, with minimal intrusion and sample preparation (Celedón and Aguilera, 2002). RS may become a valuable complementary tool in food materials science, to interpret the mechanical, thermal and rheological behavior of complex biomaterials based on microstructural information (Aguilera and Lillford, 1997). RS is an effective tool for investigating the conformational transitions in proteins by detecting changes in the secondary structure (Nonaka *et al.*, 1993), and has been used to examine protein–lipid interactions in milk (Howell *et al.*, 2001).

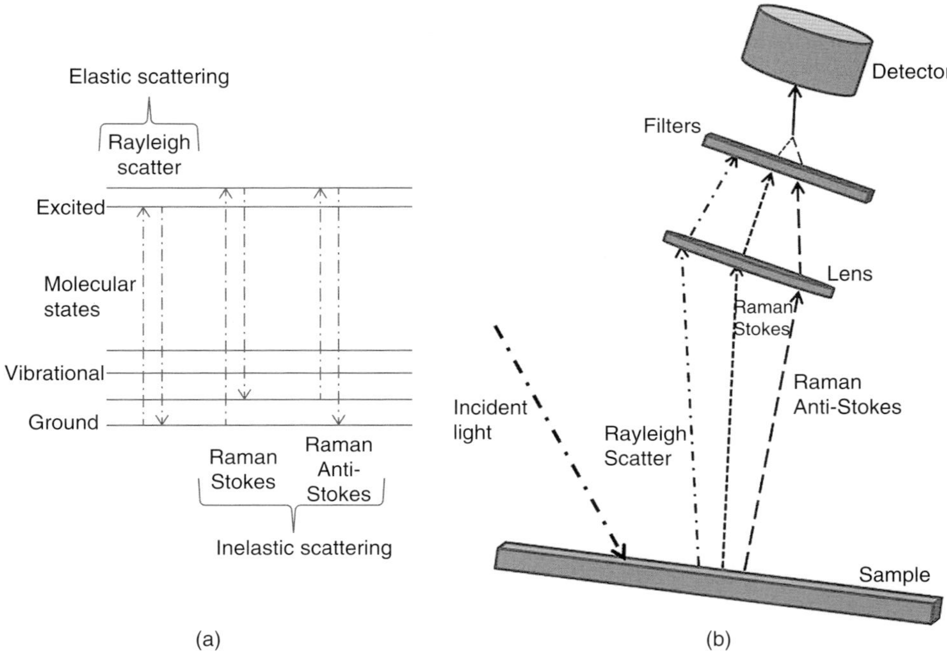

Figure 4.4 (a) Schematic diagram of the energy levels involved in Raman spectroscopy; (b) diagram of Raman instrumentation.

RS and FTIR are the perfect complement for a wide variety of analyses in foods. FTIR spectroscopy is sensitive to hetero-nuclear functional group vibrations and polar bonds, especially OH stretching in water. Raman, on the other hand, is sensitive to homo-nuclear molecular bonds. RS is also a non-destructive method requiring minimal sample preparation that can be considered to be complementary to FTIR (Ferraro *et al.*, 2003). Both are powerful metabolic fingerprinting methods as they reflect the phenotype of a sample accurately, including changes to its metabolism (Ellis and Goodacre, 2006; Goodacre *et al.*, 2004). FTIR and RS have indicated structural conformations in α-lactalbumin during hydrolysis and nanotube growth, destabilized the native tertiary structure of α-lactalbumin, and enabled observation of some conformational changes in protein secondary structure (Tarhan *et al.*, 2017).

4.6 Magnetic Resonance: NMR and MRI

Nuclear Magnetic Resonance (NMR) is a non-destructive spectroscopy technique, based on the magnetic properties of materials, which can be applied to any material in a liquid or solid state that has nuclei with nuclear spins. Hydrogen and carbon are the most important atoms in biomolecules, and therefore in foods, and can be observed easily with not very large quantities of the sample. As well as H and C atoms, other atoms can also be observed, always provided there is a sufficient abundance of a magnetically active isotope (one that has magnetic moment).In the case of H, the most abundant isotope is 1H, and is the one observed. For C, the isotope observed is 13C, as the most abundant, 12C, is magnetically inactive. Other magnetically active isotopes that are present in biomolecules and can be observed are 15N and 31P. Magnetically active nuclei have spin, like electrons. When a sample is briefly irradiated by an intense pulse of radiation, the nuclei in an α spin or low-energy state are flipped into the higher-energy spin β state. When the excitation field is turned off, the protons tend to return to their lower-energy state. This "relaxation" or return to a state in which their spin is parallel to the static magnetic field produces a small amount of radiation that is detected as an NMR signal. Typically, two different relaxation time constants are observed: T1 or longitudinal or spin-lattice relaxation time, associated with the component of the field parallel to the static magnetic field applied, and T2 or transverse or spin-spin relaxation time, which is shorter and is associated with the component of the field perpendicular to the external field applied.

From the 1970s onwards, several applications of NMR for food science were developed, but they were limited to determining water and fat content and measuring the solid fat index. In the 1990s, new applications such as droplet size measurement were proposed. An important advance came with the proposal to use NMR relaxation time not only for determining water and fat contents, but also for investigating food structure (Mariette, 2009). NMR relaxation time makes it possible not only to determine the composition of foods, but also to elucidate their structure. Particularly in the Magnetic Resonance Imaging (MRI) technique, NMR relaxation time permits quantification of the image contrast, therefore enabling visualized monitoring of structural changes in foods during processing and storage (Marcone *et al.*, 2013). MRI is a non-destructive and non-invasive technique that can be used to obtain two-dimensional and even three-dimensional images of food products, even if these are opaque systems. The signal of each voxel (the smallest processable unit of a three-dimensional matrix) depends on the physical properties of the sample, such as proton density, relaxation times, temperature, diffusion, flow, and local differences in magnetic susceptibility. The information within the images can be manipulated and used for spatially resolved measurements of concentration, structure, temperature, velocity, and diffusivity (Mariette *et al.*, 2012).

Nowadays, therefore, NMR analysis for food applications is mainly based on the relaxation of NMR active nuclei (1H and 13C being the most used for food) in a magnetic field and a pulsed radio frequency irradiation (Marcone *et al.*, 2013). Proton NMR spectroscopy is used for compositional analysis (water, fat, protein and amino acids), as many foods are proton-rich, and MRI is successfully used to monitor moisture transportation in food systems(Ramos-Cabrer *et al.*, 2006), but both techniques are also used for structural analysis.

The dairy product most often studied using NMR/RMI is cheese. In this area, NMR has been successfully applied to studying the diffusion of fat (Callaghan *et al.*, 1983) and water mobility in cheese (Kuo *et al.*, 2001). Tellier *et al.* (1993) used transverse relaxation time data to study the microscopic evolution of water during milk coagulation and during syneresis of the curd. Mapping fat and moisture distribution in cheese blocks using an MRI technique showed that moisture distribution was less uniform than fat distribution (Ruan *et al.*, 1998). Fat and water distribution during heating was also studied in Mozzarella and Cheddar cheese by Vogt *et al.* (2015) (Figure 4.5). Mulas *et al.* (2013, 2016) developed MRI protocols to discriminate between cheeses made from raw or heat-treated milk, and to assess the ripening stage of Grana Padano cheese (Figure 4.5). More information is available in the interesting review on cheese structure by Everett and Auty (2008).

Regarding other dairy products, a method for rapid remote determination of casein concentration in unopened packages of dairy products such as whole milk and whey has been developed by Alekseev and Khripov (2015). Other systems have also been studied by means of MRI and relaxation spectrum analysis, like following the swelling of whey protein-based gels when placed in aqueous solutions (Oztop *et al.*, 2010). A nice

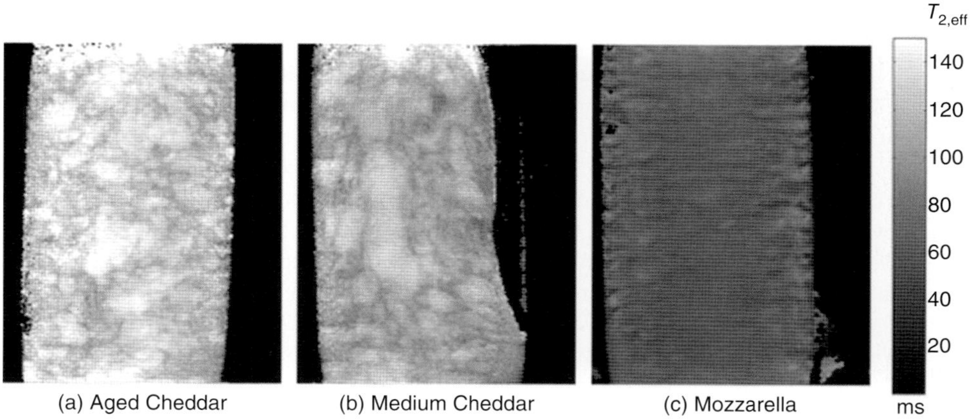

Figure 4.5 T2 maps of cheese after heating. The lighter areas indicate longer relaxation times (fat) and darker areas indicate shorter relaxation time (water) (Vogt *et al.*, 2015).

review by Tellier *et al.* (1993) and others by Mariette (2008a, 2008b) have summarized NMR relaxation applications in dairy products such as cheese or ice cream.

4.7 Conclusions

The use of the techniques described in this chapter for the investigation of food microstructure is an emerging field. Although generally used in other disciplines such as materials, physics, biology or medicine, they can complement the traditional microscopies – light and electron microscopies – to better characterize different food structures and their changes during processing. Moreover, due to continuous technical developments, these techniques are nowadays powerful and easy to perform.

Some of them, such as AFM or X-ray tomography, are non-invasive techniques and allow the non-destructive visualization of the sample, avoiding artifacts due, for example, to the sample fixation or the sample dehydration associated with the preparation of the specimen for light and electron microscopies. Other emerging techniques such as NMR provide information on both structure and composition. Although the application of these techniques is helpful for dairy food research, their use is not widely extended. Researchers on dairy products should take into account the complementary information that emerging technologies provide on food structure.

References

Aguilera, J. and Lillford, P. (1997). Microstructural and imaging analyses as related to food engineering. *Food Engineering* 2000.

Alekseev, G.V. and Khripov, A.A. (2015). Method of rapid remote control of casein concentration in dairy products in unopened packages. *Journal of Food Process Engineering, 38*(1), 11–18.

Alexander, M., Nieh, M.P., Ferrer, M.A. and Corredig, M. (2011). Changes in the calcium cluster distribution of ultrafiltered and diafiltered fresh skim milk as observed by small angle neutron scattering. *Journal of Dairy Research, 78*(3), 349–356.

Allec, N., Choi, M., Yesupriya, N., Szychowski, B., White, M.R., Kann, M.G. and Badano, A. (2015). Small-angle X-ray scattering method to characterize molecular interactions: Proof of concept. *Scientific Reports, 5*.

Amrein, M. and Gross, H. (1992). Scanning tunneling microscopy of biological macromolecular structures coated with a conducting film. *Scanning Microscopy, 6*(2).

Andreeva, N., Ferrinit, G., Bassi, D., Cappa, F., Cocconcelli, P.S., Prato, S. and Parmigiani, F. (2009). Preliminary results of combined scanning near-field optical microscopy and atomic force microscopy applied to a model biological system: *Clostridium tyrobutyricum* spores. *Bollettino Di Geofisica Teorica Ed Applicata, 50*(4), 369–377.

Arakawa, H., Umemura, K. and Atsushikai, K. (1992). Proteins images obtained by STM, AFM and TEM.pdf. *Nature, 358*, 171–173.

Aschi, A., Gharbi, A., Daoud, M., Douillard, R. and Calmettes, P. (2007). Study of structure and scaling behavior of chemically unfolded β-casein by means of small-angle neutron scattering. *Polymer International, 56*, 606–612.

Barigou, M. and Douaire, M. (2013). X-ray micro-computed tomography for resolving food microstructures. In: V.J. Morris and K. Groves (eds.) *Food Microstructures: Microscopy, Measurement and Modelling*. Woodhead Publishing, Cambridge, pp. 246–267.

Baró, A.M., Miranda, R., Alamán, J., García, N., Binnig, G., Rohrer, H. and Carrascosa, J.L. (1985). Determination of surface topography of biological specimens at high resolution by scanning tunnelling microscopy. *Nature, 315*(6016).

Boye, J.I., Ismail, A.A. and Alli, I. (1996). Effects of physicochemical factors on the secondary structure of β-lactoglobulin. *Journal of Dairy Research, 63*(1).

Burgain, J., Gaiani, C., Francius, G., Revol-Junelles, A.M., Cailliez-Grimal, C., Lebeer, S. and Scher, J. (2013). In vitro interactions between probiotic bacteria and milk proteins probed by atomic force microscopy. *Colloids and Surfaces B: Biointerfaces, 104*, 153–162.

Burgain, J., Scher, J., Lebeer, S., Vanderleyden, J., Cailliez-Grimal, C., Corgneau, M. and Gaiani, C. (2014). Significance of bacterial surface molecules interactions with milk proteins to enhance microencapsulation of Lactobacillus rhamnosus GG. *Food Hydrocolloids, 41*, 60–70.

Burgain, J., Scher, J., Lebeer, S., Vanderleyden, J., Corgneau, M., Guerin, J. and Gaiani, C. (2015). Impacts of pH-mediated EPS structure on probiotic bacterial pili-whey proteins interactions. *Colloids and Surfaces B: Biointerfaces, 134*, 332–338.

Burgess, I., Jeffrey, C. A., Cai, X., Szymanski, G., Galus, Z. and Lipkowski, J. (1999). Direct visualization of the potential-controlled transformation of hemimicellar aggregates of dodecyl sulfate into a condensed monolayer at the Au(111) electrode surface. *Langmuir, 15*(8), 2607–2616.

Callaghan, P.T., Jolley, K.W. and Humphrey, R.S. (1983). Diffusion of fat and water in cheese as studied by pulsed field gradient nuclear magnetic resonance. *Journal of Colloid And Interface Science, 93*(2).

Celedón, A. and Aguilera, J.M. (2002). Applications of Microprobe Raman Spectroscopy in Food Science. *Food ScienceTechnology International, 8*(2), 101–108.

Cheon, G.W., Guckenberger, R., Fuchs, K.H., Gross, H. and Baumeister, W. (1993). The structure of the surface layer of Methanoplanus lim{i}cola obtaines by a combined electron microscopy and scanning tuneling microscopy approach. *Journal of Structural Biology, 111*, 125–134.

Darnay, L., Len, A., Koncz, Á., Friedrich, L. and Rosta, L. (2015). Small angle neutron scattering study of nanostructural changes in microbial transglutaminase-treated low-fat yogurt during fermentation. *Food Science and Biotechnology, 24*(6), 2125–2128.

De Kruif, C.G., Huppertz, T., Urban, V.S. and Petukhov, A.V. (2012). Casein micelles and their internal structure. *Advances in Colloid and Interface Science, 171–172*, 36–52.

Doak, S.H., Rogers, D., Jones, B., Francis, L., Conlan, R.S. and Wright, C. (2008). High-resolution imaging using a novel atomic force microscope and confocal laser scanning microscope hybrid instrument: Essential sample preparation aspects. *Histochemistry and Cell Biology, 130*(5), 909–916.

Doyen, G., Koetter, E., Vigneron, J.P. and Scheffler, M. (1990). Theory of scanning tunneling microscopy. *Applied Physics A – Materials Science and Processing, 51*(4), 281–288.

Ducker, W.A. and Grant, L.M. (1996). Effect of substrate hydrophobicity on surfactant surface-aggregate geometry. *Journal of Physical Chemistry, 100*(28), 11507–11511.

Duquenne, B., Vergauwen, B., Capdepon, C., Boone, M.A., De Schryver, T., Van Hoorebeke, L. and De Block, J. (2016). Stabilising frozen dairy mousses by low molecular weight gelatin peptides. *Food Hydrocolloids, 60*, 317–323.

Ellis, D.I. and Goodacre, R. (2006). Metabolic fingerprinting in disease diagnosis: biomedical applications of infrared and Raman spectroscopy. *The Analyst, 131*(8), 875.

Etzion, Y., Linker, R., Cogan, U. and Shmulevich, I. (2004). Determination of protein concentration in raw milk by mid-infrared fourier transform infrared/attenuated total reflectance spectroscopy. *Journal of Dairy Science, 87*(9), 2779–2788.

Everett, D.W. and Auty, M.A.E. (2008). Cheese structure and current methods of analysis. *International Dairy Journal, 18*(7), 759–773.

Fagan, C.C., O'Donnell, C.P., Rudzik, L. and Wüst, E. (2009). *Milk and Dairy Products. Infrared Spectroscopy for Food Quality Analysis and Control.* Elsevier, Oxford.

Ferraro, J.R., Nakamoto, K. and Brown, C.W. (2003). *Introductory Raman Spectroscopy*, Second Edition. Academic Press, Amsterdam.

Ferraro, V., Madureira, A.R., Sarmento, B., Gomes, A. and Pintado, M.E. (2015). Study of the interactions between rosmarinic acid and bovine milk whey protein α-lactalbumin, β-lactoglobulin and lactoferrin. *Food Research International, 77*, 450–459.

Firtel, M. and Beveridge, T.J. (1995). Scanning probe microscopy in microbiology [Review]. *Micron, 26*(4), 347–362.

Gebhardt, R., Doster, W., Friedrich, J. and Kulozik, U. (2006). Size distribution of pressure-decomposed casein micelles studied by dynamic light scattering and AFM. *European Biophysics Journal, 35*(6), 503–509.

Goodacre, R., Vaidyanathan, S., Dunn, W.B., Harrigan, G.G. and Kell, D.B. (2004). Metabolomics by numbers: acquiring and understanding global metabolite data. *Trends in Biotechnology, 22*(5), 245–252.

Gould, S.A.C., Drake, B., Prater, C.B., Weisenhorn, A.L., Manne, S., Hansma, H.G. and Hansma, P.K. (1990). From atoms to integrated circuit chips, blood cells and bacteria with the atomic force microscope. *Journal of Vacuum Science and Technology A:* Vacuum, Surfaces and Films, 8(1).

Grillo, I. (2009). Applications of stopped-flow in SAXS and SANS. *Current Opinion in Colloid and Interface Science, 14*(6), 402–408.

Handojo, A., Zhai, Y., Frankel, G. and Pascall, M.A. (2009). Measurement of adhesion strengths between various milk products on glass surfaces using contact angle measurement and atomic force microscopy. *Journal of Food Engineering, 92*(3), 305–311.

Herremans, E., Verboven, P., Bongaers, E., Estrade, P., Verlinden, B.E., Wevers, M. and Nicolai, B.M. (2013). Characterisation of "Braeburn" browning disorder by means of X-ray micro-CT. *Postharvest Biology and Technology, 75*, 114–124.

Hewavitharana, A.K. and van Brakel, B. (1997). Fourier Transform Infrared Spectrometric method for the rapid determination of casein in raw milk. *The Analyst, 122*(7), 701–704.

Hoh, J.H., Lal, R., John, S.A., Revel, J.-P. and Arnsdorf, M.F. (1991). Atomic force microscopy and dissection of gap junctions. *Science, 253*(5026).

Holt, C., de Kruif, C., Tuinier, R. and Timmins, P. (2003). Substructure of bovine casein micelles by small-angle X-ray and neutron scattering. *Colloids and Surfaces A: Physicochemical and Engineering Aspects, 213*(2), 275–284.

Howell, N.K., Herman, H. and Li-Chan, E.C.Y. (2001). Elucidation of protein–lipid interactions in a lysozyme–corn oil system by Fourier Transform Raman Spectroscopy, *49, 1529–1533.*

Huck-Iriart, C., Morales-Mendoza, N.J., Candal, R.J. and Herrera, M.L. (2011). Applications of X-ray scattering in edible lipid systems. In: *X-Ray Scattering*. Elsevier, Oxford.

Ingham, B., Smialowska, A., Erlangga, G.D., Matia-Merino, L., Kirby, N.M., Wang, C. and Carr, A.J. (2016). Revisiting the interpretation of casein micelle SAXS data. *Soft Matter, 12*(33), 6937–6953.

Ioannou, J.C., Donald, A.M. and Tromp, R.H. (2015). Characterising the secondary structure changes occurring in high density systems of BLG dissolved in aqueous pH 3 buffer. *Food Hydrocolloids, 46*, 216–225.

Izutsu, K., Fujimaki, Y., Kuwabara, A., Hiyama, Y., Yomota, C. and Aoyagi, N. (2006). Near-infrared analysis of protein secondary structure in aqueous solutions and freeze-dried solids. *Journal of Pharmaceutical Sciences, 95*(4), 781–789.

James, B.J., Jing, Y. and Chen, X.D. (2003). Membrane fouling during filtration of milk – A microstructural study. *Journal of Food Engineering, 60*(4), 431–437.

Kher, A., Udabage, P., McKinnon, I., McNaughton, D. and Augustin, M.A. (2007). FTIR investigation of spray-dried milk protein concentrate powders. *Vibrational Spectroscopy, 44*(2), 375–381.

King, S.M. (1999). Small angle neutron scattering. In: *Modern Techniques for Polymer Characterisation*. Pethrick, R.A. and Dawkins, J.V. (eds). New York: John Wiley & Sons, pp. 171–232.

Kuo, M.-I., Gunasekaran, S., Johnson, M. and Chen, C. (2001). Nuclear magnetic resonance study of water mobility in pasta filata and non-pasta filata Mozzarella. *Journal of Dairy Science, 84*(9).

Laverse, J., Mastromatteo, M., Frisullo, P., Albenzio, M., Gammariello, D. and Del Nobile, M.A. (2011a). Fat microstructure of yogurt as assessed by x-ray microtomography. *Journal of Dairy Science, 94*(2), 668–675.

Laverse, J., Mastromatteo, M., Frisullo, P. and Del Nobile, M.A. (2011b). X-ray microtomography to study the microstructure of cream cheese-type products. *Journal of Food Engineering*, 43–50.

Lopez-Rubio, A. and Gilbert, E.P. (2009). Neutron scattering: a natural tool for food science and technology research. *Trends in Food Science and Technology, 20*, 576–586.

Lopez, C., Cauty, C. and Guyomarc'h, F. (2015). Organization of lipids in milks, infant milk formulas and various dairy products: role of technological processes and potential impacts. *Dairy Science and Technology, 95*(6), 863–893.

Manne, S., Cleveland, J.P., Gaub, H.E., Stucky, G.D. and Hansma, P.K. (1994). Direct visuallization of surfactant hemimicelles by force microscopy of the electrical double layer. *Langmuir, 10*(10), 4409–4413.

Mao, X., Guo, Y., Luo, Y., Niu, L., Liu, L., Ma, X. and Wang, H. (2013). Sequence effects on peptide assembly characteristics observed by using scanning tunneling microscopy. *Journal of the American Chemical Society, 8*(Scheme 1), 4–10.

Marcone, M.F., Wang, S., Albabish, W., Nie, S., Somnarain, D. and Hill, A. (2013). Diverse food-based applications of nuclear magnetic resonance (NMR) technology. *Food Research International, 51*(2), 729–747.

Mariette, F. (2008a). NMR imaging of dairy products. In: *Modern Magnetic Resonance*. G.A. Webb (ed.) Springer, New York, pp.1801–1806.

Mariette, F. (2008b). NMR relaxation of dairy products. In: *Modern Magnetic Resonance*. G.A. Webb (ed.) Springer, New York, pp. 1697–1701.

Mariette, F. (2009). Investigations of food colloids by NMR and MRI. *Current Opinion in Colloid and Interface Science, 14*(3), 203–211.

Mariette, F., Collewet, G., Davenel, A., Lucas, T. and Musse, M. (2012). Quantitative MRI in Food Science and Food Engineering. *eMagRes, (1)*, 1–8.

Mata, J.P., Udabage, P. and Gilbert, E.P. (2011). Structure of casein micelles in milk protein concentrate powders via small angle X-ray scattering. *Soft Matter, 7*(8), 3837–3843.

Mauerer, A. and Lee, G. (2006). Changes in the amide I FT-IR bands of poly-l-lysine on spray-drying from α-helix, β-sheet or random coil conformations. *European Journal of Pharmaceutics and Biopharmaceutics, 62*(2), 131–142.

May, R.P. (2002). Biological applications of small-angle neutron scattering. In: *Neutrons, X-Rays and Light: Scattering Methods Applied to Soft Condensed Matter.* P. Lindner and Th. Zemb (eds). Elsevier Science B.V., Oxford, pp. 463–480.

Meyer, E. (1992). Atomic force microscopy. *Progress in Surface Science, 41*, 3–49.

Miles, M.J., Carr, H.J., McMaster, T.C., I'Anson, K.J., Belton, P.S., Morris, V.J. and Tatham, A.S. (1991). Scanning tunneling microscopy of a wheat seed storage protein reveals details of an unusual supersecondary structure. *Proceedings of the National Academy of Sciences of the United States of America, 88*(1), 68–71.

Morris, V. (2004). Probing molecular interactions in foods. *Trends in Food Science and Technology, 15*(6), 291–297.

Mulas, G., Anedda, R., Longo, D.L., Roggio, T. and Uzzau, S. (2016). An MRI method for monitoring the ripening of Grana Padano cheese. *International Dairy Journal, 52*, 19–25.

Mulas, G., Roggio, T., Uzzau, S. and Anedda, R. (2013). A new magnetic resonance imaging approach for discriminating Sardinian sheep milk cheese made from heat-treated or raw milk. *Journal of Dairy Science, 96*(12), 7393–7403.

Murrieta-Pazos, I., Gaiani, C., Galet, L., Cuq, B., Desobry, S. and Scher, J. (2011). Comparative study of particle structure evolution during water sorption: Skim and whole milk powders. *Colloids and Surfaces B: Biointerfaces, 87*(1), 1–10.

Nicolai, T. (2007). Food characterization using scattering methods. In: *Understanding and Controlling the Microstructure of Complex Foods.* J. McClements (ed.) CRC Press Bocaraton, FL, pp. 288–310.

Nicolaou, N., Xu, Y. and Goodacre, R. (2011). Fourier transform infrared and raman spectroscopies for the rapid detection, enumeration and growth interaction of the Bacteria *Staphylococcus aureus* and *Lactococcus lactis* ssp. *cremoris* in milk. *Analytical Chemistry, 83*(14), 5681–5687.

Nonaka, M., Li-Chan, E. and Nakai, S. (1993). Raman spectroscopic study of thermally induced gelation of whey proteins. *Journal of Agricultural and Food Chemistry, 41*(8).

O'Sullivan, D., McSweeney, P.L.H., Cotter, P.D., Giblin, L. and Sheehan, J.J. (2016). Compromised *Lactobacillus helveticus* starter activity in the presence of facultative heterofermentative *Lactobacillus casei* DPC6987 results in atypical eye formation in Swiss-type cheese. *Journal of Dairy Science, 99*(4), 2625–2640.

Oztop, M.H., Rosenberg, M., Rosenberg, Y., McCarthy, K.L. and McCarthy, M.J. (2010). Magnetic resonance imaging (MRI) and relaxation spectrum analysis as methods to investigate swelling in whey protein gels. *Journal of Food Science, 75*(8), 508–515.

Parris, N., Purcell, J.M. and Ptashkin, S.M. (1991). Thermal denaturation of whey proteins in skim milk. *Journal of Agricultural and Food Chemistry, 39*(12).

Picart, L., Dumay, E. and Cheftel, J.C. (2002). Inactivation of Listeria innocua in dairy fluids by pulsed electric fields: Influence of electric parameters and food composition. *Innovative Food Science and Emerging Technologies*, *3*(4), 357–369.

Qi, P.X. and Onwulata, C.I. (2011). Physical properties, molecular structures and protein quality of texturized whey protein isolate: Effect of extrusion temperature. *Journal of Agricultural and Food Chemistry*, *59*(9), 4668–4675.

Ramel, P.R., Peyronel, F. and Marangoni, A.G. (2016). Characterization of the nanoscale structure of milk fat. *Food Chemistry*, *203*, 224–230.

Ramos-Cabrer, P., Van Duynhoven, J.P.M., Timmer, H. and Nicolay, K. (2006). Monitoring of moisture redistribution in multicomponent food systems by use of magnetic resonance imaging. *Journal of Agricultural and Food Chemistry*, *54*(3).

Ruan, R., Chang, K., Chen, P.L., Fulcher, R.G. and Bastian, E.D. (1998). A magnetic resonance imaging technique for quantitative mapping of moisture and fat in a cheese block. *Journal of Dairy Science*, *81*(1).

Sikand, V., Tong, P.S., Roy, S., Rodriguez-Saona, L.E. and Murray, B.A. (2011). Solubility of commercial milk protein concentrates and milk protein isolates. *Journal of Dairy Science*, *94*(12), 6194–6202.

Subramanian, A. and Rodriguez-Saona, L. (2009). Fourier Transform Infrared (FTIR) Spectroscopy. In *Infrared Spectroscopy for Food Quality Analysis and Control*.

Tarhan, Ö., Tarhan, E. and Harsa, E. (2017). Investigation of the structure of alpha-lactalbumin protein nanotubes using optical spectroscopy. (in press).

Tatham, A.S., Thomson, N.H., McMaster, T.J., Humphris, A.D.D., Miles, M.J. and Shewry, P.R. (1999). *Scanning Probe Microscopy Studies of Cereal Seed Storage Protein Structures*, *21*, 293–298.

Tellier, C., Mariette, F., Guillement, J.P. and Marchal, P. (1993). Evolution of water proton nuclear magnetic relaxation during milk coagulation and syneresis: Structural implications. *Journal of Agricultural and Food Chemistry*, *41*(12), 2259–2266.

Tersoff, J. and Hamann, D.R. (1983). Theory and application for the scanning tunneling microscope. *Physical Review Letters*, *50*(25), 1998.

Van der Ven, C., Muresan, S., Gruppen, H., de Bont, D.B.A., Merck, K.B. and Voragen, A. G.J. (2002). FTIR Spectra of whey and casein hydrolysates in relation to their functional properties. *Journal of Agricultural Food Chemistry*, *50*, 6943–6950.

Verboven, P., Kerckhofs, G., Mebatsion, H.K., Ho, Q. T., Temst, K., Wevers, M. and Nicolaï, B.M. (2008). Three-dimensional gas exchange pathways in pome fruit characterized by synchrotron x-ray computed tomography. *Plant Physiology*, *147*(2), 518–527.

Vogt, S.J., Smith, J.R., Seymour, J.D., Carr, A.J., Golding, M.D. and Codd, S.L. (2015). Assessment of the changes in the structure and component mobility of Mozzarella and Cheddar cheese during heating. *Journal of Food Engineering*, *150*, 35–43.

Wang, Q., Hulzebosch, A. and Bovenhuis, H. (2016). Genetic and environmental variation in bovine milk infrared spectra. *Journal of Dairy Science*, *99*(8), 6793–6803.

Wellner, N. (2013). Fourier transform infrared (FTIR) and Raman microscopy: principles and applications to food microstructures. In: *Food Microstructures: Microscopy, Measurement and and Modelling*. Morris, V.J. and Groves, K. (eds). Woodhead Publishing, Cambridge, pp. 163–191.

Wepf, R., Amrein, M., Bürkli, U. and Gross, H. (1991). Platinum/iridium/carbon: a high-resolution shadowing material for TEM, STM and SEM of biological macromolecular structures. *Journal of Microscopy, 163*(1).

You, H.X. and Lowe, C.R. (1996). Progress in the application of scanning probe microscopy to biology. *Current Opinion in Biotechnology, 7*, 78–84.

Zhang, Y. and Zhong, Q. (2013). Probing the binding between norbixin and dairy proteins by spectroscopy methods. *Food Chemistry, 139*, 611–616.

5

Quantitative Image Analysis in Microscopy

Gaetano Impoco

CoRFiLaC, S.P. 25 Km 5, Ragusa Mare, I-97100, Ragusa, Italy

5.1 Aim and Scope

Image analysis has been used in conjunction with microscopy since its early development stages, a few decades ago. Googling around, it is easy to stumble upon old-fashioned books, articles, reviews, and relatively sophisticated image analysis software packages, which are not worth mentioning here. Digital image analysis has become ubiquitous since the advent of digital sensors in microscopy.

Visual observation was early recognized, by influential thinkers, as being biased by subjective evaluation of image content [4]. However, for some time the food research community employed image analysis mainly for qualitative inspection. Although microscopists have long been advised that microscopy is not just a matter of beautiful pictures [40], the potential of quantitative analysis for food microscopy has only recently been fully appreciated, thanks also to some fundamental books [4, 33].

Enthusiasm, however, has somehow elevated image analysis to a reference tool to prove one's thesis on data, without bothering for a serious assessment of results. In fact, image analysis has become the latest trend in food microscopy. Every respectable paper is somehow compelled to show some shiny image analysis tricks. Even the traditional paper structure has been modified to make room for a brand-new "image analysis" section. Wonderful. However, under the surface of image analysis exercises, results and even its very usefulness for some published work are often questionable. Ramírez and colleagues [30] point out that most scientific papers lack a careful description of image processing techniques, as well as the composition and extent of the image set used in experiments. They conclude that "images have become a sui generis type of data in scientific research". That is, images do not undergo the same rigorous experimentation standards as other data. We completely agree with their argument. Let us add that, even when authors provide a careful description of image analysis experiments, there is no scientific rigor in accepting image analysis results without any validation.

Being that these practices are so widespread in the food microscopists' community, we think that there is still much to learn about image analysis in order to enable

Microstructure of Dairy Products, First Edition.
Edited by Mamdouh Mahmoud Abdel-Rahman El-Bakry, Antoni Sanchez, and Bhavbhuti M. Mehta.

microscopists to put its use on solid scientific grounds. Here, we hope to give a little contribution in this direction.

This chapter is thus intended to lift some of the fog lying on image analysis in food microscopy. Rather than providing ready-made answers, we try to give readers the means to find solutions for themselves. The following pages are organized to cover, each in its own section, three fundamental issues in image analysis. First, we address the problem of choosing software packages for your application. We try to assemble typical needs in food microscopy analysis, and to find matching software in a selection of packages. At this point most people are bewildered by the amount of seemingly similar functions with funny names, in their software. Thus, we lead readers through the world of image analysis, taking advantage of a few applications of image analysis to real problems. Typical solutions and workflows are briefly discussed, to serve as a guide. Finally, we devote the last part of this chapter, large indeed, to show how image analysis should, and how it should not be used, in an attempt make images find their place among standard scientific data. We conclude this last section trying to sketch a sort of generic protocol, a workflow model to serve microscopists as a checklist to conduct sound image analysis experiments.

Although stereology is of great importance in several contexts, even a basic introduction would require a chapter on its own. Here, we entirely avoid this topic, referring the interested reader to the clear introduction in ref [33].

5.2 Image Analysis Software

There is a plethora of software packages to process and analyse digital images, both commercial and in the public domain. Making one's way through this crowded land may not be easy for the beginner, but also for experienced image analysis users. Every product has different capabilities, offers a different set of functions, is designed following its own working philosophy, exposes a different graphical interface, provides compatibility with other software packages and different ways of collecting and outputting data.

For some of these characteristics, choosing among the possibilities offered by different software packages is a matter of taste. That is certainly true for graphical interfaces, since some users might find one interface friendly and others do not. To some extent, it is also true for working philosophy. The way a program is built influences productiveness, both due to the scientific application at hand, and to user's taste. For example, some programs leave users (often too much) free to act on images with single atomic actions. Others work as wizards, strictly guiding and directing one's actions, in a step-by-step fashion. Some users feel comfortable when guided by wizards, others feel compelled, unable to develop new and personal ways of using the program. Conversely, the former often feel like being left alone by menu-based programs, whereas the latter feel like being in more control of their actions and of the program. Either case, is a matter of productiveness and taste.

On the other hand, properties such as capabilities and functions are required by the application. Functional analysis might require temporal tracking; cell counting might require advanced segmentation algorithms, and so on. Different software packages have different sets of capabilities and functions, so that some might not be suitable for one's application. Moreover, if for example two packages can do image segmentation, one could provide only simple tools whereas the other provides more sophisticated algorithms. If your application can do well, e.g., with simple thresholding, both packages will do the job. Conversely, if you need more advanced methods, there is no choice but the latter package.

In order to get out of this entangled forest, microscopists should clearly define what their requirements are, and what functions and capabilities they need for their image analysis experiment. Only then they will be ready to bravely engage the harsh world of image analysis software. Compiling a check-list of your needs may turn out to be a good starting point for your quest. Without any intention to be exhaustive, let us discuss some commonly required capabilities in microscopy.

Image processing is one of the most fundamental steps in image analysis. Most packages supply at least a minimum set of image processing functions, such as noise suppression, thresholding, and morphological filtering. Some provide more advanced processing tools, such as color segmentation, image arithmetic, and selection tools. Microscopists should carefully look into the image processing toolbox before choosing a software package. All needed functions must be available and give sufficiently accurate results for one's needs. Nonetheless, there is no reason to choose the application supplying the largest set of more sophisticated processing functions. Complex programs are usually more expensive, more burdensome for your computer, and often more difficult to learn, due to the bunch of functions provided. Moreover, complex packages may lack other important capabilities, being as they are intended for uses other than microscopy. Before deciding, then, let us move on.

Image alignment is an important processing function, helping the user to align and fuse images to obtain larger pictures (image mosaics). From a logical standpoint, this is just a processing function that could be included among other processing tools. However, due to its importance for some applications, some people prefer to use different software packages for processing and alignment.
One reason for that choice is that they might need only basic processing tools, but advanced and very specialized tools for alignment.

Image deconvolution follows the same destiny as alignment. That is, it is a processing function but one which is of paramount importance for some applications. There exist advanced software tools to deconvolve very specialized kinds of images, some providing no additional processing functions. Again, different programs may be employed for basic processing and for deconvolution.

Quantitative analysis is usually only supplied in packages that are specifically designed for the sciences. This is one point where programs for science and for image enhancement differ. Being designed for photo retouching, the latter do not usually provide any tools for quantitative measurement of image features. Although they might provide astonishing processing capabilities, they are thus of little use for the sciences.

3D-viewing and inspection is an important function in some applications. Nonetheless, it is not provided by all image processing programs, mainly because it requires complex rendering software on top of which the viewer is to be programmed. The simplest packages, but also some of the more advanced ones, simply do not cope with the 3D world. Without going into details, the reader should know that the majority of image processing software packages only supply a very limited support for truly 3D processing.

Automating command sequences is an important capability if multi-step processing is to be run on large image sets. Macro programming or batch processing are provided by most programs. Batch processing is simply a means of recording one's actions, in order to repeat them on a bunch of images. On the other hand, macros are simple programs written in simplified programming languages supplied by, and specialized for, a particular image processing program. Macros are more flexible and allow more complex behavior to be defined by the user, but are more complex to learn, requiring basic programming skills.

Temporal analysis of marked image features is of paramount importance in applications, such as functional analysis. Tracking and analysis of time series require specialized image processing and analysis tools, often involving some form of interaction with microscope drivers. A few software packages are equipped with such a capability, and usually very specialized ones. We know of a few applications of temporal analysis to food image analysis, so this topic is a little out of our main interest in this chapter.

Interaction with third-party software is a feature that does not directly affect image analysis itself. It is provided by every image processing program by reading and writing several image formats, such as PNG, JPEG, TIFF, and so on. However, this is not the kind of interaction we intend here. Rather, we refer to the possibility to include or simply use some functions of other software packages inside one program, or to make some functions available to other programs. 3rd-party support is a useful capability in order to allow your program to easily share data with other programs, such as processed images, spreadsheets, and so on. Some programs can even import and integrate functions from popular third-party software packages. For example, Matlab [38], and its free counterpart Octave [26], have simple and powerful programming environments capable of using and incorporating external programs, provided that parameters can be passed to these programs using the command line (e.g., the program Convert released with the ImageMagick library [16]). Conversely, ImageJ [35] is an example of a program exposing functions that can be used in other programs, by means of its Java library. Clearly, interaction with third-party software gives more exibility to an image processing package.

Expandability is another step towards exibility. In its simplest form, a program can be extended by adding plugins providing new functions. If plugins can be produced by independent programmers, the program can be easily adapted for scopes not intended by its designers. In more general terms, a software package can be extended by adding new software as plugins, libraries, and so on. Open-source software can even be modified and adapted to new uses and demands. Macros can be somehow regarded as software expansions.

Workflow is strictly defined by some programs. The user is guided, on a step-by-step basis, through pre-defined stages of data analysis. Most work flow-oriented software packages are designed for specific applications, such as cell analysis. Such programs often interact with the microscope drivers, allowing the user to acquire image and guiding her from the very first stages of image acquisition and processing. Processed images can be analyzed using built-in tools and the gathered data can be plotted, stored in files or databases, or sent to third-party spreadsheet or presentation programs. Although these programs ease the burden of analysis, they define a precise workflow that some people might find uncomfortable or unsuitable for their needs.

Supported operating systems and platforms. Some programs only come with versions for some platforms. This might not seem a limitation at all, as long as one uses the right operating system. However, a program that has versions for all the major operating systems allows easier sharing of results, procedures, macros, plugins, and so on. Do not underestimate this point.

Supported image formats. Virtually all image processing software packages support the major formats, such as PNG, TIFF, and JPEG. Some programs also have proprietary formats. However, there are some formats specifically designed for particular applications that may not be supported by common image processing programs. Image format converters sometimes fill this gap, at the expense of a loss of information (usually, meta-information not related to image pixels). We are not interested in further discussing very peculiar or specific image formats. Each of the packages we discuss below supports all of the formats most readers will ever know about.

Commercial or free software is not much of a Shakespearean dilemma. A few image analysis packages are beyond most budget limits. However, no doubt it is not worth paying for a program if we make a very limited use of its full capabilities. Free alternatives exist for most commercial programs, even for some high-end ones. On the other hand, commercial software tends to surpass free software on smart, well-polished graphical interfaces, and on user assistance (although in this respect, there are remarkable exceptions). As we pointed out above, for some people these features may be relevant. Notice that free software is not synonym to open-source (see [12] for more information).

Now that we have listed a few important distinctive properties for image analysis software products, let us have a brief tour to some of them, with reference to Table 5.1. Here, the columns account for properties, such as image processing capabilities, image alignment, image deconvolution, quantitative analysis, 3D viewing, built-in macro language, possibility to connect with third-party software, expandability by means of extra packages, whether its philosophy is workflow-oriented or not, and whether it is commercial or publicly available. A star indicates that a capability is only partially implemented by the software, or that it is implemented as a (possibly commercial or third-party) expansion package. Apart from temporal analysis and image formats, of which we are not interested to give more details here, supported operating systems are not included in the table. It is assumed that all the three major platforms (i.e., Windows, Mac OS, and GNU/Linux) are supported. We explicitly report below when this is not the case.

Here, we list a few well-known as well as less popular packages. Our selections by no means intends to be exhaustive. It is not even a selection of best software whatsoever. It is just useful to make our discussion more concrete. Let us start with Photoshop [2], a must-have for graphical designers but of little use for scientists. For some reason, some people began to use it for image analysis even if it has no analysis capabilities, with the exception of some basic form of pixel counting. We believe that those people, not knowing of any image analysis software, and being contented with simple thresholding and pixel-wise analysis (trying to equate pixel counting to porosity), turned to Photoshop

Table 5.1 A selection of image processing programs, together with their ability to meet some common requirements in microscopy. The columns account, respectively, for the presence of: a minimal set of image processing functions, image alignment, image deconvolution, quantitative analysis, 3D viewing, built-in macro language; the possibility to connect with third-party software, expandability by means of plugins or extra packages; whether its philosophy is workflow-oriented (see the text) or not, and whether it is commercial or publicly available. A star indicates that a capability is only partially implemented by the software, or that it is implemented as a (possibly commercial or third-party) expansion package.

Package	Processing	Alignment	Deconvolution	Analysis	3D	Macro	3rd-Party	Expandability	Workflow	Commercial
Cell Profiler	✓			✓		✓	✓			
ImageJ	✓	*	*	✓	✓	✓	✓	✓		
Imaris®	✓	*	*	✓	✓		*	✓		✓
Leica LAS	✓	✓		✓	✓	✓	*	*	✓	✓
Matlab®	✓	*	*	✓	✓	✓	✓	✓		✓
MetaMorph®	✓	✓	✓	✓	✓	✓			✓	✓
Octave	✓	*	*	✓	✓	✓	✓	✓		
Photoshop®	✓					✓				✓
Olympus Stream	✓	✓		✓	✓	✓	*	*	✓	✓

without pondering if it was an appropriate tool for their research. In fact, Photoshop is one of the most powerful and complete programs available for photo retouching. That means that it has a large set of image processing functions, and image enhancing tools. Few of them, however, are of any use for the sciences. On the other hand, not being intended for quantification, it lacks even the most basic measurement tools. Despite what Russ may write in his introduction to his image analysis book [33], we think that software designed for photo retouching is a bad choice for image analysis in the sciences. We cited only what is probably the most popular photo retouching software, but several others exist. Clearly, this discussion holds true for other similar programs, such as GIMP [37].

ImageJ [35] is a popular image analysis package specifically designed for the life sciences, although it may be used for any image analysis application. It can be installed on any computer platform, and used in conjunction with several third-party software. Being in the public domain, it is easy to expand and adapt to new applications, and this has been thoroughly done thanks to the help of the user's and developer's communities. Plugins have been written to make it communicate with virtually any popular image analysis software. It is probably one of the most flexible and complete choices, also thanks to the huge amount of plugins available (the starred columns in Table 5.1 mean that plugins exist to align and deconvolve images). Recently, ImageJ2 [32] was released, a completely new program that shares the same philosophy and is backward compatible with ImageJ. On top of that, Fiji [34] was built as a software distribution whose core is ImageJ2 but coming with a number of well-written and well-documented plugins. For those who are familiar with the GNU/Linux world, we may compare Fiji to a distribution, whereas ImageJ2 can be regarded as the core system itself. Although Fiji supplies several of state-of-the-art processing and analysis tools, and others can be added or created, it preserves the barren look of ImageJ. Image analysis might sometimes prove complex with this austere platform.

Matlab [38] is a prototyping platform with a powerful built-in scripting language, originally designed for matrix manipulation. It comes with several additional packages for various engineering and scientific computing applications, including image processing and analysis. Image analysis is executed by writing appropriate scripts. Matlab thus requires basic programming skills, although scripts for several applications and needs are publically available, thanks to the Matlab developers' community. Unfortunately, execution of functions is usually slow, although methods to speed it up exist. On the other hand, it is a powerful and extremely flexible piece of software. Nonetheless, we do not recommend Matlab for the standard user.

Octave [26] is a free clone of Matlab. As such, it shares the same advantages and drawbacks. Its scripting language corresponds closely to that of Matlab, but the common functions in its library often use a different syntax. That means that some scripts do not work properly, and thus Octave cannot take full advantage of the plethora of scripts written for Matlab.

Imaris [19] is a widely-used commercial suite, specifically designed for microscopy. A core program can be extended with useful add-ons for image alignment, deconvolution, and others. It supplies important processing and measurement functions, as well as an effective 3D visualization and segmentation tools. Imaris supports communication with ImageJ2 and Fiji, as well as with Matlab. ImageJ2 plugins and Matlab scripts can be seamlessly imported and added to Imaris menus, to be used in the same fashion as built-in functions. These capabilities make Imaris flexible and able to take advantage of software developed by the ImageJ2 and the Matlab developers' communities. Although Imaris is less powerful than ImageJ2, its graphical interface appears more pleasant and professional. Imaris is not currently available on Mac platforms.

CellProfiler [9] is a versatile, free, public-funded platform addressed to biologists, specifically designed for 2D cell image analysis. A stand-alone data analysis program, CellProfiler Analyst [20] is also available to analyse large, high-dimensional datasets output from image analysis. Despite its narrow application field, this software can be fruitfully employed for other microscopy applications. The official website [9] carefully lists third-party software that can interact with the input or output produced by CellProfiler.

Olympus Stream [28] is a strongly workflow-oriented application tailored to Olympus microscopes. It guides the user from acquisition, through processing, analysis and measurement, up to the creation of reports and presentations. Although conceived for the material science, it can also be used for food analysis (after all, food is just a special kind of material). Stream is a non-expandable, self-contained software solution that can only send its output to commercial office software (presentation) or database managers (storage and retrieval of data). Its capabilities are comparable to those of ImageJ2 but, again, its graphical interface appears more pleasant and professional, and its workflow philosophy might prove a plus for some users. Stream is only compatible with Windows platforms.

MetaMorph [25] is an image analysis package for the life sciences, with built-in deconvolution functions. Like Imaris and Olympus Stream, it has several high-end, easy-to-use analysis functions. However, its specialization for cell image analysis makes it less suitable for food image analysis. Although problems addressing image analysis of globular structures might benefit from this software, processing algorithms are calibrated on cells, as well as the workflow. MetaMorph is only available for Windows platforms.

Leica LAS Image Analysis software [22] is a proprietary workflow-oriented application for Leica systems. Like Olympus Stream, it is a closed solution that brings you from acquisition to presentation. It also shares the same drawbacks: non-expandable, not communicating with other software, except by sending its output to multimedia presentation software and little more. A specific solution is also available for the life sciences [23] but, again, it is too specialized to cell image analysis. These products are only distributed for Windows platforms.

In the previous list, we did not concentrate on imaging and inspection of large, high-dimensional datasets. Our discussion was limited to generic 3D-viewing capabilities, because the latter topic is somehow more popular. Interested readers may find this web-page useful [21].

Finally, we wish to mention OMERO [5], a software platform for image sharing, annotation, and storing in a central repository. Although no image analysis capabilities are natively supported, results from processing and analysis can be easily shared in an integrated environment. OMERO is thus a good platform for collaborative projects.

Now, if you feel confused, do not worry. This was exactly our purpose. That is, we meant to show the reader that each package has its own pros and cons and that the right choice is up to the microscopist. Our selection was chosen to suggest that you should first clarify your needs, before choosing any software tool for your research. A wrong choice is not necessarily disastrous. However, time might be wasted in learning to use software that does not really help you in every stage of your work. So, our message is, first clarify your requirements, than you choose. We hope that the discussion above can be a good starting point for that.

5.3 Applications to Microscopy for Dairy Science

There are several applications of image analysis, some of which we might not have envisioned so far. Discussing all cases in the scientific literature would require a book on its own. Here, it will suffice to focus on a few of the most popular examples in food science. In order to reach a larger audience, we mostly stick to simple functions supported by the majority of software packages. Nonetheless, it is instructive to take a look to more sophisticated approaches, as well.

Before proceeding, let us introduce a little bit of basic image processing terminology, that we will be using hereafter. We name original images the micrographs acquired with your microscope, without any further processing. Original images may be in full color, or come in shades of grey. The word grey-level is often used in the image processing community to refer to the latter. Please note that grey-level is not the same as black-or-white. The latter is only used for truly black-and-white images, i.e., for images whose pixels may be saturated black or saturated white, with no intermediate shades. These images are referred to as binary or binarized. The process of transforming a grey-level or color image into a binary one is called binarization. It can be obtained by segmentation, a process by which the image is partitioned into homogeneous, non-overlapping regions. Thresholding is a special case of segmentation, where we are interested in just two regions, background and foreground, obtained by choosing a threshold value on image intensities: pixels whose values fall below the threshold are converted to black, whereas those with higher values are shifted to white. Notice that these regions are not required to be connected, i.e., the foreground might contain several distinct connected components (e.g., the pores in Figure 5.4(b)). Binary images may be obtained from full-color images by decoupling color channels, e.g., when using different fluorescent markers.

Whatever your images and objectives might be, the first step in every image analysis session is pre-processing. That is, images should be suitably processed before segmentation and thresholding. Sure enough, microscope images are affected by acquisition noise. This noise may cause severe segmentation error if not properly handled. Modern microscopes are equipped with sophisticated acquisition software that compensate for noise. Most denoising methods, however, introduce some form of blurring or distortion on pixel values. In the search for optimal image quality, engineers are compelled to balance between denoising and fidelity. That is why they often opt for non-invasive denoising, such as fusing multiple acquisitions, leaving some noise in acquired images. Denoising is thus of vital importance for quantitative analysis.

There are several models of noise, the most common being white and speckle noise. Without going into detail, it suffices to say that all image processing packages are equipped with at least basic blurring (e.g., Gaussian) and despeckling (e.g., median) filters, respectively for white and speckle noise. Grey-level morphological filters can be subsequently applied to denoised images to enhance interest features, before binarization. Moreover, binary morphology can be employed to further enhance images after binarization, in order to remove small structures, probably due to residual noise or segmentation errors. More on mathematical morphology can be easily found among the plethora of resources on the web. In the following we assume that this pre- and post-processing procedures are operated on all images.

5.3.1 Porosity

The microstructure of porous materials significantly affects their macroscopic properties. This holds true for food as well [3]. Fruit, baked products, foams, soft candies, all exhibit pronounced porous structures. Studying the extent, shape and distribution of pores is thus of primary importance in food science. The analysis of porous compact structures has a long history and several applications in material sciences, e.g., in the study of porous rocks.

One of the most important measures involving pores is *porosity*, i.e., the volume occupied by pores over the total volume. It is easily computed on binarized images by counting the ratio of foreground pixels over the total image pixels. Easy as may be, this operation often leads to unreliable results due to several reasons, the most important being segmentation errors (more on this in Section 5.4). The procedures described above can reduce the possibility of encountering severe processing errors.

Pore counting is another commonly used measure. Again, it is easily obtained by counting the number of connected regions in the foreground of a binarized image. After images have been binarized and processed, the number of foreground objects should represent our interest features, i.e., pores. All software packages for image analysis are equipped with counting functions. Be careful to not confuse background with foreground, a mistake more common that one may think.

Some applications may require distinction between "large" and "small" pores. In this case, the user can choose a threshold over pore area, related to the application (i.e., what is large and what is small depends on your working hypotheses). The number of pores in each class can then be easily computed.

A few programs support such an operation, but it can be easily implemented as a macro or plugin, if your software supports scripting and you have some basic knowledge of macro programming. One possible way of spotting large pores is to iterate morphological opening on the binarized images, as many times as the required radius. Small pores will be removed from the images, leaving only large ones. Counting what remains gives the number of large pores. Obviously, small pores amount to the total number of pores minus the number of large pores.

Other applications may require the computation of some summary measure on pore area, such as mean or median. Others may take advantage from area distribution, e.g., to test one's hypothesis that pores come in just two representative sizes (say, large or small). In this case, the count of pores may be plotted against their size. If our hypothesis is true, the plot should show two modes corresponding to large and small pores, respectively. All these operations require that your software is capable of analysing distributions and generating reports. Most programs are.

5.3.2 Fat Globules

The microstructure of the fat phase is studied in several fields. Interest attributes of fat obviously depend on the application at hand. A simplification may be attempted by observing that fat usually exhibits a different morphology in different contexts, such as round droplets (e.g., in emulsions, such as milk and mayonnaise), crystals (e.g., in ice cream, chewing gum), or in a sort of amorphous phase (e.g., liquefied in cheese). Feature shape can thus be studied to exclude non-round objects when we are sure that only droplets are to be found. Conversely, in cheese micrographs it could be of interest to count round fat droplets against amorphous ones, e.g., to quantify fat melting in cheese production.

From an image analysis standpoint, studying fat globules is not any different from quantifying pore structure. After binarization, the background of our binary image might well represent a compact matrix as well as the aqueous phase. The only remarkable difference is in image pre-processing and feature detection, mostly due to the differences in input images. Quantification and analysis involve counting, shape measurements, quantification of the distribution of features (fat, in this case). Hence, the very same approaches can be fruitfully employed in both problems, provided that good segmentations are obtained from thresholding or other equivalent techniques.

The distribution of objects in a region has attracted the attention of the scientific community to the point that an autonomous research field devoted to the study of *spatial patterns* [11]. Computing measurements such as homogeneity or other distribution-related measures, might prove harder than one might expect. A simple way to start with is to partition the image using a regular grid. In order to make things easy, we begin with a 2×2 grid, i.e., we partition our image into four equally-sized quadrants. Then, for each quadrant we count the number of pores, their total area, their mean area, or whatever measure we might be interested in. We may say that fat globules are evenly distributed if pore count for each quadrant is more or less close to the average count. Conversely, if a quadrant contains much more pores with respect to other quadrants, there is a clear concentration (or lack) of pores in that quadrant. On a finer scale, we can employ $N \times N$ grids where N is, for example, 8 or more. We could then compute the standard deviation σ of the (somehow normalized) pore count with respect to the mean value. A large value of σ suggests that the imaged sample is not homogeneous.

Such reasoning can be followed for pore area, as well. Actually, there is no limit to the measures we can compute with such a method. A successful example of the application of this method is at the core of our analysis algorithm for milk fat, in confocal images [18]. Unfortunately, image analysis software packages do not support such peculiar methods. Moreover, scripting languages released with most software are not powerful enough to allow easy implementations of refined algorithms, like that above. The support of skilled programmers might be necessary. Another option might be to rely on very specialized software packages, if any. However, if the algorithm used is not known (such as in most commercial software) one cannot be sure of what the software really does. Remote as it may appear, there is always a possibility that some results are more due to the mathematical formulation of image analysis operations, than to real effects of the observed phenomena. We think that public research should only rely on perfectly reproducible results, even for image analysis.

5.3.3 Microbial Cells

The analysis of microbial cells has several applications in dairy science. Bacteria, yeasts, molds and other microorganisms may strongly affect complex evolving ecosystems, such as cheese and other dairy products. A fundamental information that can be derived from color micrographs of stained cells, is the count (and possibly the distribution) of dead and living bacteria. This can be easily done by decoupling color channels, i.e., splitting the image into its color components (e.g., red-green-blue). Each component appears as a grey-level image and can be processed as such. Hence, if living cells are colored with a green dye and red is used to stain dead bacteria, the green and red color channels can be independently binarized in order to detect cells. Then, every interest measure can be computed on each class as above. The image analysis chain is thus the same as in the examples above, except that a color-channel splitting operation must be run beforehand. This operation is provided by every image processing package.

If cells are large enough in our images, morphological analysis can be used to classify bacteria on the basis of their shape. For example, round cells might correspond to isolated *cocci*, whereas elongated ones might suggest the presence of *bacilli*. Shape descriptors used for pores or other image features can be employed for cells, as well. For example, one could measure *roundness*, *aspect ratio*, and *solidity*, as defined in [33]. The meaning of roundness is obvious. Aspect ratio measures the ratio between main diameters of each feature. Solidity measures the ratio between the actual area and the area delimited by a rubber band closely fitting the feature shape. Clearly, round cells should show proper values of roundness, whereas form factor should indicate that there is no substantial difference between the main diameters. Solidity should show that the two areas are close, since a round shape should be almost perfectly circular. On the other hand, elongated cells have significantly different diameters. Moreover, if the shape is somehow curled, such as for *streptococci*, the enclosing rubber band will delimit a larger area with respect to the actual area of the feature. This is just one way to analyse shapes. Many more are possible, and not all of them are equally fit for your application. One should try different combinations of measures in the literature before resting on a particular set of them. As for the last case in Section 5.3.2, knowing the exact mathematical definition of the shape measures used is of great importance, both to clearly understand resulting measurements and to correctly communicate them.

Spatial distribution of cells can be of interest to detect distribution gradients of useful or dangerous populations. The very same algorithm described in Section 5.3.2 can be fruitfully applied to bacterial populations.

Now that we have gone through a few possible applications, we encourage readers to experiment a little bit with their brand-new software. After you have been practising on your images for a while, you will be ready for the last stage of our journey.

5.4 Image Analysis and Quantitative Measurement

In microscopy, image processing has traditionally been used to prepare images for better visual evaluation. Only recently, scientific communities concerned with food and biology research have begun to appraise a proper use of image analysis, implying more than plain qualitative description. Quantitation and measurement have been recognized as a valuable support for experimentation. Far from being a new research field, quantitative analysis has long been studied by the image processing community, being the core and purpose of image analysis itself.

Image analysis, however, is a dangerous scientific tool if not properly handled. Unreliable results may come from its misuse, leading to inaccurate or even wrong conclusions. Great care should be taken when quantitative image measurements are used as a reference to evaluate experimental results. Nonetheless, most published research takes measurement correctness for granted (measurements are output by computers, after all) unaware of whether they are reliable or not. Computers, however, only output manipulated data, regardless of whether the input is correct or not. Potential measurement errors can pop up at every turn. Unconditional faith in computer output may turn out to be a great mistake.

Microscopists should be well aware of that. Being able to decide if quantitative results can be safely used (i.e., if measurement error is acceptable) should be regarded as a basic skill for image analysis users. There are probably a few research tools for which the importance of experimental assessment is more undervalued than image analysis. In the following pages, we will try to give the reader good reasons to agree with us. At the same time, we will overview a few good practices to design sound image analysis experiments.

5.4.1 Image Analysis Basics

Micrographs used for quantitative analysis should be calibrated against known values. That is, the scale of pixels on the specimen must be set. This problem is out of the scope of this chapter, so we assume that the reader knows how to calibrate micrographs. Here, we just point out that no calibration is needed if only dimensionless measurements are to be taken (e.g., porosity).

A further issue with input micrographs is image format. There is virtually a format for each applica-tion. A few popular formats, however, suffice for most applications. The only feature we need to know about them is whether they are "lossy" or "lossless". The latter type includes all those formats, such as PNG and TIFF, that perfectly preserve image pixels as they were imaged. Conversely, lossy formats discard some information to gain better performance, at the expenses of a little quality loss. JPEG is one such format, specifically designed for photography. Since image processing may amplify defects and artifacts introduced by lossy methods, these formats should not be used in scientific analysis.

Successful image analysis requires input micrographs to be conveniently processed before any measurement can be taken. Accurate detection of structural features is an important pre-requisite for reliable quantitative analysis of sample morphology [33]. The measurements computed on the detected features are deeply influenced by the accuracy of feature detection itself, especially when shape descriptors are involved. It is this processing phase, then, that is mostly responsible for measurement errors. That is why microscopists should learn to distinguish between good and bad detection outcomes.

Before proceeding with our discussion, it is worth taking a brief look to relevant image analysis steps.

5.4.1.1 Feature Detection

A digital image is a rectangular array of luminous points called *pixels*. Although signals reaching the sensor (i.e., light or any other form of radiation) spread over a number of adjacent sensor elements due to imperfections of lenses [14], pixels are easily treated as isolated units. Recovering the shape of an object in a digital image amounts to grouping all pixels representing that object. Each group can be composed of one or more connected image regions, e.g., representing a phase of the imaged specimen (see Figure 5.1(b)).

Whatever measurement is being calculated on digital images, one cannot overlook the intrinsic pixel-wise nature of image features. Approximation errors cannot be avoided, because of the sampling process undergoing the digitization of imaged micrographs. We have no means to recover the true shape of a feature once it has been digitized and is represented as a group of pixels in our micrograph. Moreover, approximation errors cannot be exactly computed (otherwise we could undo the sampling process and recover the true shape) and are difficult to estimate, in most cases. Nonetheless, several image descriptors (e.g., area, perimeter, roundness, ...) rely on features' shape.

Image analysis is thus carried out as a two-step process. First, the image is processed in order to mark relevant features (*feature detection*). Then, a series of measurements is taken on the pixels of marked image regions (*quantitative analysis*).

Relevant features are detected by partitioning image pixels into well-defined, non-overlapping, distinct regions (see Figure 5.1(b)). This process, known as image *segmentation*, is one of the open problems in the image processing community [14]. Strictly

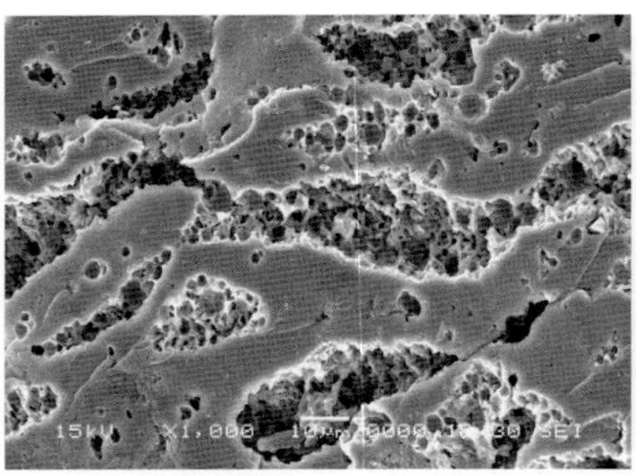

(a) SEM micrograph.

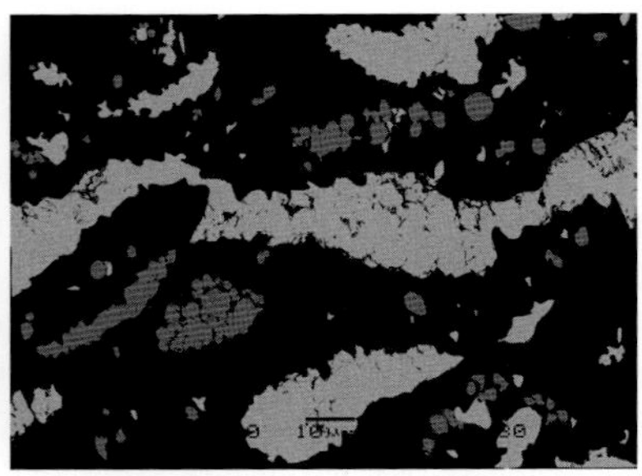

(b) Segmented image with three classes (two plus background).

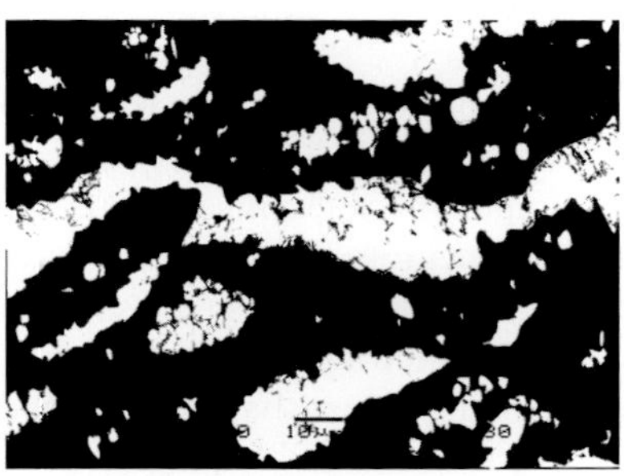

(c) Binarization (two classes: foreground and background).

Figure 5.1 An SEM micrograph of a cheese sample. Examples of segmentation are shown with three and two classes (binarized). (*See color plate section for the color representation of this figure.*)

speaking, a segmentation is partition of the set of image pixels, in the usual definition used in mathematics. That is, a division of image pixels into non-overlapping sets, such that no pixel is left out. The goal of a successful segmentation is to accurately mark one region for each interest feature in the micrograph. Each region should closely cover the image area occupied by the corresponding feature.

Unfortunately, no general-purpose solution exists and no satisfactory solution is available for most applications. Human visual abilities are still unmatched. This is probably the main reason why users are rarely satisfied with segmented images. For the same reason, microscopists should be warned against uncritical acceptance of segmentation output.

The most common segmentation scheme used in the analysis of food micrographs is thresholding: a scalar value is used to partition a greyscale image into two regions containing lighter and darker colors, respectively. A two-level labeling is obtained, partitioning the input image into background and foreground (Figure 5.1(c)). Relevant features are taken as coherent blobs of foreground pixels. This approach, called *binarization*, is the simplest form of segmentation and it is clearly naive. In fact, more sophisticated and effective methods exist. However, it is fast and intuitive for the operator and it often gives acceptable results.

Simple as it might appear, image segmentation involves carefully-designed algorithms of varying complexity, often requiring user intervention (e.g., manual thresholding). Every algorithm has its own peculiarities, making it more or less suited for a specific application. Without going through technicalities, microscopists should be able to understand how segmentation methods work and what output should be expected from them.

5.4.1.2 Quantitative Analysis

Taking measurements is quite easy provided that one knows what to measure. Given the features detected with segmentation, measures like area can be simply computed by counting pixels. Similarly, perimeter can be defined as the count of pixels in the outline of the feature. One should keep in mind that both these measurements are approximations of real measurements due to the discrete nature of digital images. More sophisticated approximations also exist [33], but these are sufficient for most applications.

On the other hand, other kind of measurements are not fully captured by such a simple geometric formulation. This problem often arises with more elusive or ambiguous measures, such as *density*. Take for example a CLSM image of bacteria (Figure 5.2). How can we measure concentration of live (green) cells? There is no simple, well-established mathematical formulation to define what we perceive as spatial density. Spatial patterns have long been studied [11] but no straightforward, general-purpose solution is available for most problems. As a consequence, different mathematical descriptors are often used to capture one natural phenomenon, such as the geometrical distribution in Figure 5.2. Here, a visual representation is shown of two different measures to compute density of live cells, as described in [18]. The two measures produce different representations and different numerical descriptions of the same phenomenon. Which is better depends on the specific applications and, sometimes, on a particular image.

Even worse, some mathematical definitions can fail to capture the common meaning microscopists assign to a measure. This can be the case of *directionality*, i.e., the orientation of a feature or, on a larger scale, the preferred orientation of features in a sample. The cheese sample in Figure 5.1 can help to visualize this measure. Here, features are

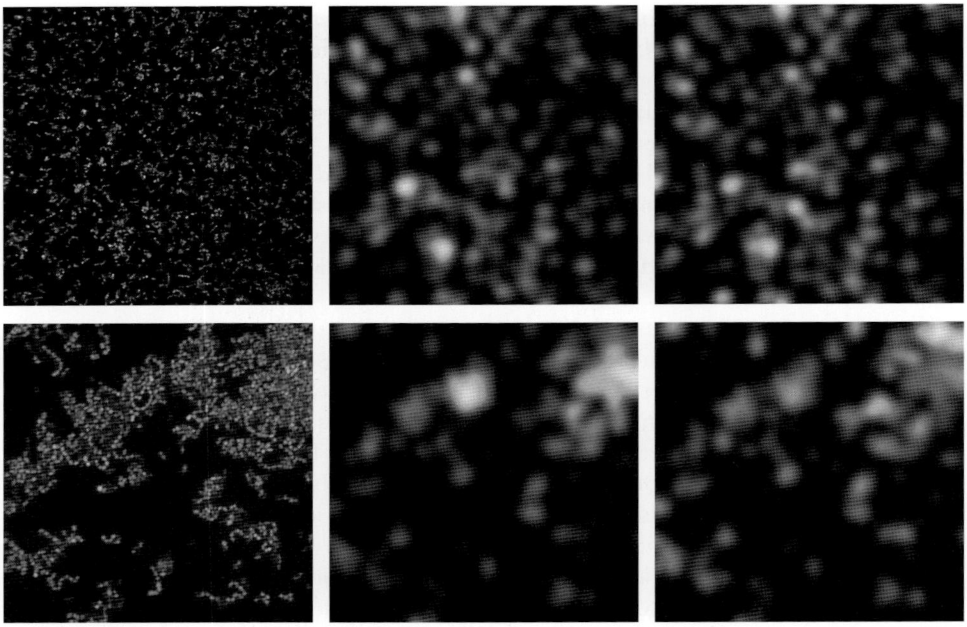

Figure 5.2 Confocal laser scanning micrographs of live (green) and dead (red) cells. Visual representations of two different measures for density are shown in the middle and right columns. (*See color plate section for the color representation of this figure.*)

mainly oriented along two axes, one being parallel to the horizontal axis of the image, the other about 45° counter-clockwise with respect to that. That is probably an informal definition most microscopists can agree upon. In more formal terms, the directionality of a geometric feature is often defined as the direction of the main axis of its shape, i.e., the angle this axis forms with the horizontal axis. Following a naive implementation [17], the main axis is taken as the line between the two farthest points in the shape. To see why this definition fails, observe the shapes presented in Figure 5.3. For each

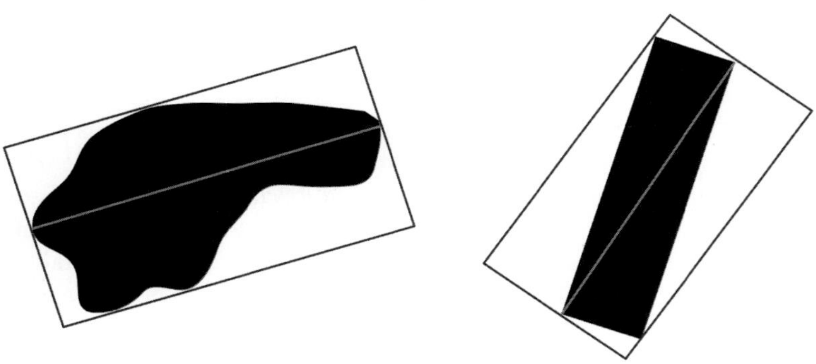

Figure 5.3 Two image features with *directionality* traced as the lines linking the farthest points in their boundaries. The example on the right shows a failure of this simple definition. (*See color plate section for the color representation of this figure.*)

feature (black shapes) directionality is represented as a purple line. The directionality computed for the shape on the left reasonably corresponds with perceived orientation. This does not happen for the feature on the right. Most people (if not all) would say that a rectangular shape is oriented along its longest side. However, the line linking the farthest points in a rectangle is one of its diagonals. Hence, our naive definition takes a diagonal of the rectangle as the main axis. Notice that which one of the two diagonals depends on arbitrary mechanisms of the particular computer program used.

One might think that the case presented is a very peculiar one. This is not so. Most small features are made by squares of 2×2 to 4×4 pixels. Simply ignoring those features is not a solution. In fact, our mathematical definition of directionality imposes a main (again, arbitrary) orientation even on round shapes. If a sample contains several round or small shapes, a distribution of directionality might be unreliable. We do not want to go further into details. Our interest, here, is highlighting possible dangers of simply-stated definitions like ours above.

Finally, we give a last note on possible ambiguities in the meaning we associate with an expression. Take for example *minimum diameter*. If the shape is an ellipse or a rectangle, one would probably take the shortest axis or side, respectively. More complex shapes, however, do not offer such a simple interpretation. One of the first definitions that might come to one's mind is to take the shortest side of a rectangle closely bounding the shape. Another possible definition is again the shortest side of the rectangle oriented with respect to the main direction of the shape. Both definitions are legitimate and describe two possible ways of thinking about our informal notion of minimum diameter. As you probably expect, it turns out that those measures give quite different numerical measurements. In the shape on the right of Figure 5.3, the former definition returns the length of the shortest side of the rectangular shape, whereas the latter one measures the length of the shortest side of the bounding rectangle, which is about twice as long. Notice that this is not the same problem as the one about directionality presented above. In that case, the issue was about the translation from an informal meaning to a mathematical definition. In this last case, the problem is about coherence in the informal notion itself. Image analysts are often forced to cope with ambiguities in the common interchange language used by microscopists, before trying to write down a mathematical definition.

5.4.2 Common Pitfalls

In previous section, we gave a few examples of common issues in image analysis. That was just a small sample of widely-used measures but the described issues are quite common and concern most, if not all possible measurements taken on digital images. A comprehensive survey of image analysis measures is out of the scope of this chapter (see [33] for a good guide). Nonetheless, it is instructive to take a closer look at common problems arising in quantitative analysis of digital images. The reader should become aware of possible limitations of image analysis tools and should learn to be careful and knowing in using them. Before proceeding with our discussion, it is useful to review the issues portrayed above and state them in more precise terms.

1) **Poor image segmentation**, caused by the difficulties image analysis researchers face to replicate vision tasks easily accommodated by the human visual system (e.g.,

object recognition and grouping). This issue often frustrates users who expect that a computer algorithm should at least match their simplest vision capabilities. That is rarely so and probably will not be for several years to come.

2) **Bad approximation of geometric measures** due to the intrinsically discretized nature of digital images. At the finest grain, digital images are composed by square luminous dots, called pixels. Every shape is thus represented with jagged borders. This complicates even the computation of simple measures such as area or perimeter, causing different algorithms to compute largely different approximations.
3) **Difficulty in defining elusive measures**. Some characteristics are difficult to translate into formal mathematical expressions for computer programs. The difficulty arises because image analysis specialists are not able to highlight those geometric peculiarities, if any exists, that fully and exactly describe the informal meaning people assign to a term.
4) **Wrong mathematical definition of apparently simple measures**. This is a problem involving the mathematical representation of informally-defined characteristics of feature shapes. Here, people largely agree on informal terminology, as well as meaning. The informal notion is clearly stated, suggesting a straightforward mathematical encoding. Nonetheless, such simple definitions may turn out to be completely wrong in some cases.
5) **Ambiguity of terminology**, arising from the different informal meaning people assign to measures. In contrast to the previous issue, here people agree on terminology but give different practical meaning to the same term. Whatever choice is used in a computer program will always disappoint some people. Efforts put into finding a common meaning for a term might impair intuition. Clearly, this is a problem in the target scientific domain more than an image analysis issue.

Image segmentation is the first and more critical step in the analysis chain. Segmentation errors can cause severe measurement errors, often leading to useless results. Hence, poor segmentation often prevents successful quantification. To see why this is so, let us delve a little into a few technical issues about binarization, i.e. two-phase segmentation.

A binary (B/W) image is often obtained setting a threshold value on possible pixel values. All image pixels whose value falls below the given threshold are set to black, whereas the remaining are set to white. The threshold can be chosen by hand or with the help of an automatic algorithm. Either case is equivalent for the following discussion.

Consider the SEM micrograph in Figure 5.4(a). A Ragusano cheese sample is shown, where the compact matrix is made of proteins, whereas pores and cracks represent whey or fat [24]. Figure 5.4(b) and Figure 5.4(c) show the result of thresholding with different threshold values. When a high threshold is used (Figure 5.4(b)) most image pixels will be mapped to black (background). Only a few objects are thus visible whereas several features might have been lost (i.e., erroneously marked as background). Conversely, low thresholds (Figure 5.4(c)) cause most pixels to be rendered as white (foreground). Most objects can be correctly detected but a large part of (noisy) background is probably marked as foreground. This trade-off between proper detection of objects and noise suppression in the background is common to all global thresholding techniques, but also to most segmentation algorithms. Manual selection of threshold values is more a trial-and-error art than a rational quantitative evaluation.

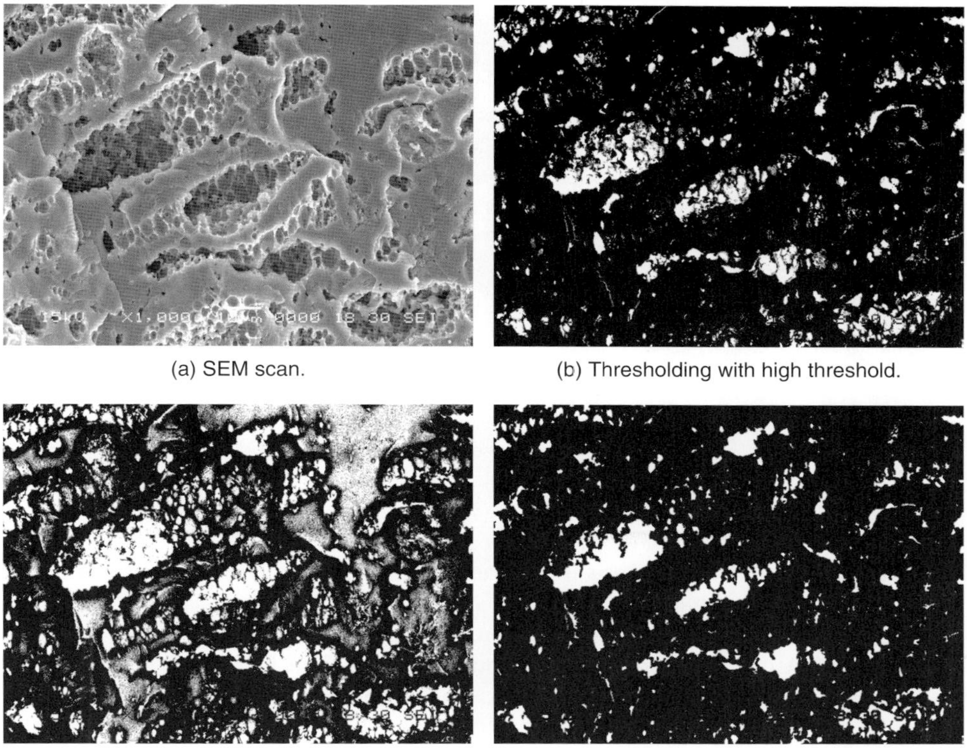

(a) SEM scan. (b) Thresholding with high threshold.

(c) Thresholding with low threshold. (d) A more sophisticated thresholding algorithm tailored to SEM imagery (see [17]).

Figure 5.4 A scanning electron microscope (SEM) image of Ragusano cheese. (*See color plate section for the color representation of this figure.*)

The close-ups in Figures 5.5–5.7 can help to make this point clearer. The pore structure in Figure 5.5(b) is accurately marked only in its deepest part, whereas thresholding fails to detect the half-hollow part on the right. With a different choice of the threshold level (Figure 5.5(c)) that part is correctly marked. However, a large part of the protein matrix is marked as foreground. A closer look to Figure 5.5(a) reveals that this part corresponds to a depression in the compact protein structure, probably due to sample preparation techniques. Hence, uneven surfaces may baffle thresholding algorithms.

Acquisition noise may also cause bad segmentation results. Basically, noise distorts pixel colors and makes smooth surface appear rough. If this roughness is not smoothed out before thresholding, noise artifacts may appear on the binary image. Such artifacts are present in the close-ups in Figures 5.5(c)–5.7(c), resulting in poor segmentations. Luckily, *denoising* has long been studied [14] and several effective solutions are provided by all image analysis software tools.

Noise suppression was used before thresholding in Figure 5.4(d), together with some other clever image processing, before and after thresholding [17]. Figure 5.5(d) shows that most of the noise has been suppressed, while retaining good detection results. This

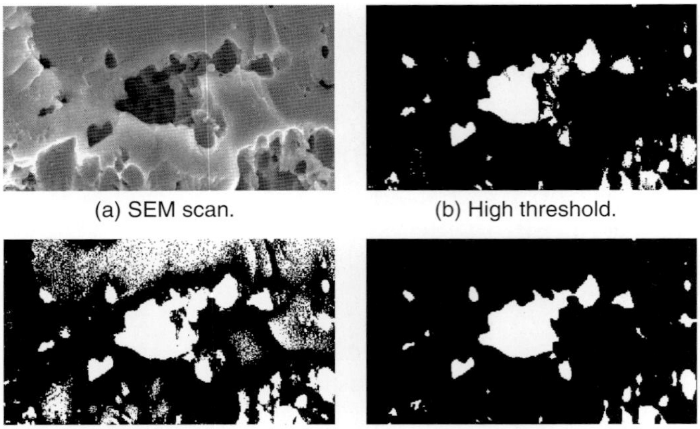

(a) SEM scan. (b) High threshold. (c) Low threshold. (d) Thresholding algorithm erported in [17].

Figure 5.5 Close-up of the SEM scan in Figure 5.4.

does not happen in the close-ups in Figures 5.6(d) and 5.7(d). Here, most part of the features go largely undetected.

Why is that so? What makes the algorithm successful in one case and causes its failure in the others? A first suggestion comes from a closer look at Figure 5.5: a dark hole is deeply cut through a much brighter matrix. The brightness of these two image regions is consistently different, and so are their pixel values. Thus, an intensity level, T, can be easily found so that most matrix pixels are brighter than T (and marked as foreground) and most pore pixels are darker than T (marked as background). A perfect input for

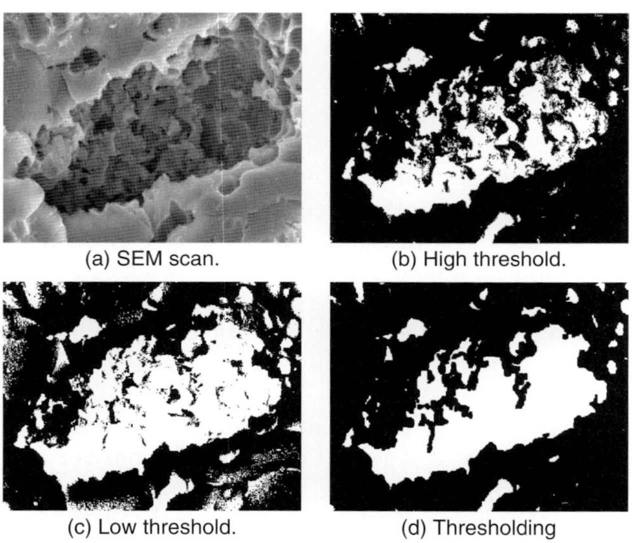

(a) SEM scan. (b) High threshold. (c) Low threshold. (d) Thresholding algorithm erported in [17].

Figure 5.6 Close-up of the SEM scan in Figure 5.4.

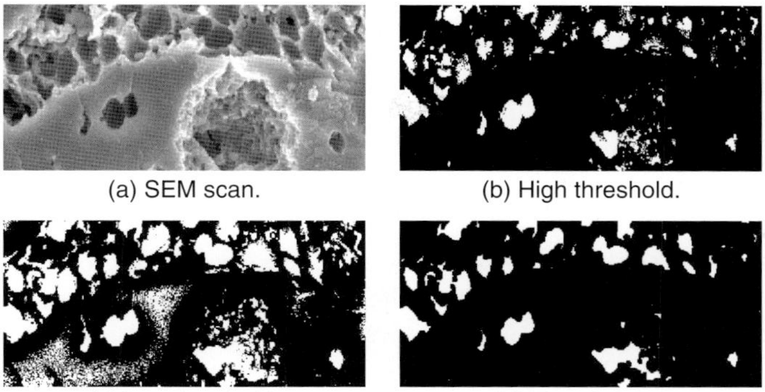

(a) SEM scan. (b) High threshold.

(c) Low threshold. (d) Thresholding algorithm erported in [17].

Figure 5.7 Close-up of the SEM scan in Figure 5.4.

thresholding algorithms. These beautifully-arranged pixel values do not appear in other close-ups. Most features are not significantly lighter or darker than the flat matrix. This violates an important assumption of thresholding algorithms on the distribution of pixel values: foreground and background should have well-separated brightness pixel values. In a word, one region should be dark and the other bright. Since the close-ups in Figures 5.6 and 5.7 do not conform to this assumption, there is no means to detect these pores by using thresholding alone.

The message, here, is that one often has wrong expectations on what an image processing algorithm does. In the example reported, the problem is simply stated: distinguish compact matrix from pores. This task is straightforward for a human operator, but it is very hard for a computer program. In fact, it turns out to be one of the most difficult unsolved problems in computer vision. There is a gap between this plain request and the series of simple mathematical operations onto image pixels that should implement that request. In the case of thresholding, the problem is solved by assuming that the image shows only two interest regions (background and foreground) with considerably different brightness characteristics. This is not quite the same goal the user had in mind. Understanding what kind of processing an algorithm does is thus essential to choose the right tool for one's needs and to foresee results. Basically, there is no general-purpose computer vision algorithm capable of solving all segmentation tasks. Hence, one should learn how to interpret segmentation errors and find more adapt algorithms for the task at hand. That is, try and understand what to expect from a segmentation software tool.

Now, the question is how good my computer vision algorithm is to solve my segmentation problem. Is my segmentation good enough for successful quantitative analysis? Unfortunately, there is no clear-cut answer to the question whether a segmented image is good enough for further analysis. The quality of a segmentation is strongly related to your application, i.e., to the type of analyses you intend to perform on the segmented micrograph. Let us make this point clear with a simple example. Suppose we want to compute porosity of the sample in Figure 5.4(a). If we use the binarized image in Figure 5.4(b), we will underestimate the true value since several pores are incorrectly marked as background, i.e. compact matrix. Conversely, if the binary image in Figure 5.4(c) is used, porosity will be largely overestimated due to the large matrix areas marked as pores.

As a further example, consider the SEM micrograph in Figure 5.8(a). Figure 5.8(b) shows a labeled binary image in which pores have been manually marked by a microscopist. Here, porosity is about 0.163 (i.e. 16.3%) and the number of pores is 206. The segmentation in Figure 5.8(c) shows a relevant number of small spots, mainly due to acquisition noise and matrix surface roughness. In this case, porosity and the number of pores amount, respectively, to 0.114 and 9744. The total area on the micrograph surface occupied by those spots has a small impact on the total area occupied of pores. Porosity has been underestimated by about 30%. In some experimental settings, our segmentation is probably good enough to compute porosity. On the other hand, these spots artificially increase the number of pores and constitute a non-negligible fraction of the total count of pores. In fact, the number of pores is 43 times higher than the correct value. This measurement is thus useless because it is affected by too much high an error. So, the quality of the segmentation in Figure 5.8(c) is too low to reliably estimate the number of pores. Is there any way to detect such an error, without knowing the exact pixel distribution? Figure 5.8(d) shows the same binarization where features smaller than 5 pixels have been dropped. Now, porosity and the number of pores are, respectively, to 0.104 and 1270. About 87% of pores are thus smaller than 5 pixels and

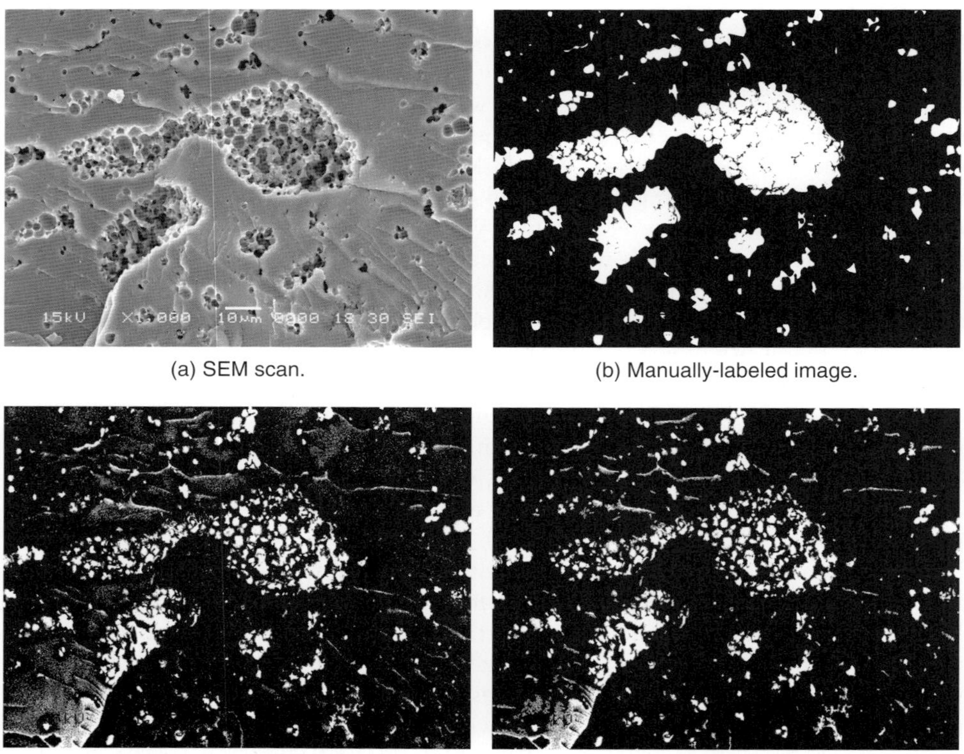

(a) SEM scan.

(b) Manually-labeled image.

(c) Thresholding.

(d) Thresholding cleared from tiny features.

Figure 5.8 A scanning electron microscope (SEM) image of Ragusano cheese.

amount to less that 1% of total area. Hence, they are most probably due to acquisition noise. A good share of alleged pores are thus due to failure in suppressing noise during segmentation. Although the number of pores is still largely overestimated, the simple trick used in Figure 5.8(d) suggests that much better results can be obtained from a sound use of image analysis tools.

These examples should make clear that care must be taken when selecting image analysis procedures for an application. There is a plethora of segmentation algorithms. Unfortunately, most of them are out of reach for the majority of users. Either they are not available in (commercial as well as free) image analysis software packages, or they require fine tuning of parameters whose meaning is obscure for all but image analysis specialists (and sometimes for them, as well). A number of standard thresholding algorithms are released with the ImageJ software [34, 35]. The reader is encouraged to experiment with those methods for a better understanding of this discussion. A detailed discussion on segmentation and image pre-processing is out of the scope of this chapter. Interested readers are referred to the book by Russ [33] and, for more technical details, to the image processing reference by Gonzalez and Woods [14].

When acceptable segmentation results are obtained, measurements can be computed on the features detected by the segmentation algorithm. Troubles, however, have not come to an end. The first problem to deal with is approximation. Digital images are regular grids of pixels. Each pixel can take just one value in a range of possible colors, however large. This implies that the observed object (i.e., the incoming signal, being it light, electrons, or whatever) must be *sampled* into a square grid of finite-sized dots, each corresponding to a pixel. This spatial approximation has a strong influence on shape and size measurements taken on image features. Figure 5.9(a) depicts a circular feature digitized using a regular grid of square pixels, resulting in the pixel arrangement of Figure 5.9(b). Digitization inevitably introduces an approximation whenever shape measurements are taken on a digitized feature. The extent of the approximation error partly depends on the representation chosen to approximate the desired measure. Two such approximations, *city-block* and *pixel-center*, for the computation of perimeter and area are shown in Figures 5.9(c) and 5.9(d), and other exist [14, 33]. The city-block approximation (Figure 5.9(c)) is obtained by walking along the outer edges of pixels, represented as squares in a regular grid. The pixel-center representation (Figure 5.9(d)) approximates the shape boundary by connecting pixels centers. If after calibration pixel width is set, e.g., to 1 μm, the radius of the circular feature is 4 m and its perimeter is about 25.1 μm. The two approximations in Figure 5.9 imply quite different measurements. If perimeter is computed using the city-block approximation, we get a measurement of 32 μm. If the pixel-center approximation is used, we obtain 23.3 μm. Clearly, the former approximation largely overestimates perimeter, whereas the latter shows a better estimate. The approximation error also depends on the feature size. Let us approximate circle area by simply counting the number of pixels involved, as shown in Figure 5.9(b). For a radius of 4 pixels the area is overestimated by about 3.4%. The error drops down to 0.01% for a radius of 175 pixels. This reduction originates from the fact that approximation error affects only the outline of the feature. Since area grows with radius faster than perimeter, error weighs less and less as feature size grows. Clearly, there is no means to determine the exact geometric measurements of a shape once it has been digitized. *Sub-pixel* details are irreparably lost. Approximation error cannot even be computed (otherwise, approximation could be reversed to find the exact value). Bounding

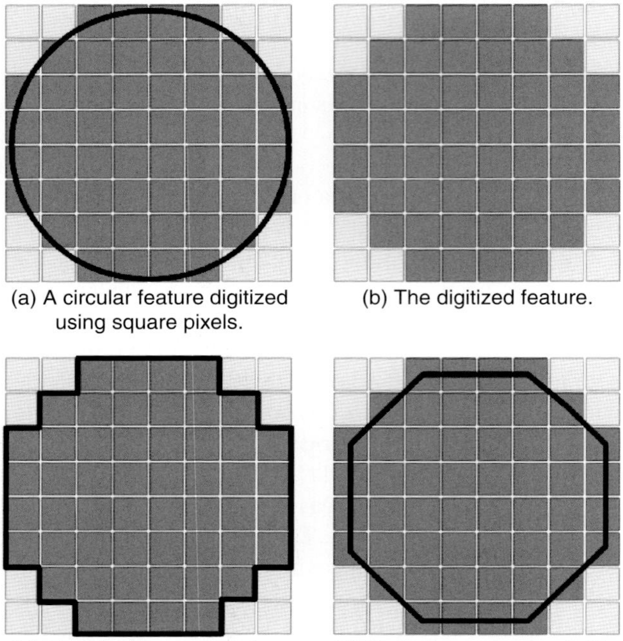

(a) A circular feature digitized using square pixels.
(b) The digitized feature.
(c) *City-block* representation.
(d) Perimeter approximated by connecting pixel centers

Figure 5.9 Approximation of perimeter of a digitized circle.

or estimating approximation error is thus of primary importance to evaluate measurement error. Digital length estimation has attracted a wide interest [10, 39] and is an active area of research [6, 8]. Unfortunately, most software packages implement simple algorithms and do not tell much about approximation. That is why microscopists should avoid computing measurements of small features whenever possible (see the example above), or look for better algorithms.

Approximation is not the only factor affecting measurement quality. Good measurements are of little use if mathematical definitions implemented in computer programs do not represent what users meant. Some measures that are easily defined by microscopists on qualitative grounds are intrinsically difficult to translate into formal definitions. An example of such issue is *density*, interpreted as the concentration of features in certain regions of an image. Two tentative representations are shown in Figure 5.2 for two different images. Although the meaning of "density" and "concentration" are perfectly clear to microscopists, none of those representations seems to fully capture that meaning.

This issue is more common than might be believed. Most image analysis users find frustrating that a simple measure they have in mind cannot be computed as expected. Likewise, image analysis experts are often frustrated because they find no suitable mathematical formulation for user needs. However, this is often a false problem. Slightly different definitions may be no less informative than the desired measure, and much easier to encode in mathematical formulation. Namely, alternative measures can often be devised that are less elusive and enable microscopists to assess the outcome of an experiment equally well. This is the case of Figure 5.10. The same CLSM images as in Figure 5.2 are shown together with one of our two implementations of density. The central column reports a different measure, not trying to reproduce the informal definition of density. For each pixel, we simply computed the percentage of cellular material in a wide circular region around the pixel. The processed images depict percentages as more or less bright tones of orange. Clearly, this measure is not modeled on what most microscopists would define as density. Nonetheless, this mathematical formulation probably produces a more informative measure, for the scope of the experiment, with respect to both attempts to model "density". That is, a better measure has been found when we stopped insisting on perfectly modeling a specific qualitatively-defined measure.

The lesson, here, is that insisting on reproducing informal measures in our computer tools may prove to be deleterious. A slight shift of our requirements is often more fruitful, provided that one looks at the greater picture of the experiment. Involving image analysis experts in early stages of experimental design is a first step in this direction. A first advantage of this practice is to avoid unrealistic expectations from quantitative image analysis. Team working can also be the key to fruitful design of image analysis tools that are more effective and often tailored to the experiment at hand. The interested reader may find it useful to take a closer look to a couple of recent such successful cases of team working [13, 18, 36]. In both cases, the authors did not limit themselves to

Figure 5.10 Confocal laser scanning micrographs of Figure 5.2. In the right and middle columns are shown, respectively, visual representations of a measure for density (right) and a different measure (middle) not trying to reproduce density (see text). (*See color plate section for the color representation of this figure.*)

use available tools with a trial-and-error approach. With the active involvement of image analysis experts, they carefully chose or devised measurement tools specifically designed to capture the image features of interest for the experiment.

Even the more careful image analysis experts, however, may occasionally be baffled by seemingly trivial definitions. In Section 5.4.1.2 we discussed the case of *directionality* showing how disastrous outcomes can be generated by shallow mathematical definitions, such as in Figure 5.3. We will not return on that, mostly because this is more an issue for image analysis experts than for microscopists. Nonetheless, microscopists should carefully analyse the outcome of quantitative analyses produced by computers. Often, following the tracks of dubious or unexpected behaviors can lead to the discovery of errors and bugs, both in mathematical definitions or in the computer program itself. Microscopists should learn to trust less blindly computer results (computers are programmed by human beings, after all) and to rely more on their intuition. Image analysis should not be used to justify one's assumptions but to prove or disprove a model or a reasoning about a phenomenon.

Finally, ambiguities may arise when people assign different meaning to informal terms. Lack of coherence in terminology is a further cause of frustration, since people often expect a different behavior from some image analysis tools. Mathematicians solve these ambiguities by carefully defining each term using unambiguous formalisms. Computer programs often do not put efforts on that, to avoid confusing users. Again, designing experiments with the help of image analysis experts often prevents these ambiguities from messing things up.

5.4.3 Misuse and Wrong Interpretation of Image Analysis Results

Previous discussion should have convinced the reader that image analysis is no silver bullet. You should face the truth that quantitative analysis on bad segmentations is pointless. Analyzing rubbish can only produce rubbish. "Garbage in, garbage out" is a piece of truth every computer science student learns taking basic programming classes. There is no reason to think that this should not apply to image analysis, as well as to other aspects of living. Nonetheless, most people tend to accept unquestioningly computer output even in business and decision-making.

Back to our topic, if your program fails to deliver a good segmentation you should refrain from using quantitative analysis on your data. Justifying one's assumptions using image analysis is tempting but it is just wishful thinking, not rigorous experimentation. Hence, carefully evaluating the results of segmentation is a fundamental step in quantitative analysis, and one that is up to the microscopist. Similarly, all other problems possibly arising in subsequent quantification should be carefully accounted for. That is why we think that a good knowledge of image analysis basics should be a standard tool in the toolbox of microscopists involved in systematic quantitative analysis.

Now, what can happen on subsequent analysis if garbage is produced in one step of the chain? Segmentation errors often have the strongest impact on quantification, segmentation being the first step. The point, here, is to which extent segmentation errors can affect measurements. In Section 5.4.2, we argued that the effect of segmentation errors depends on the kind of analyses to be run on the resulting image. The image in Figure 5.4(c), however, is unlikely to be of any use. Confronted with a binary image manually labeled by an expert microscopist, wrongly-labeled pixels amount to more the 25% of the whole image. The number of pores is overestimated by as much as 3560%, whereas the error on porosity is about 2%. Is this good or bad? Clearly, one cannot seriously take such an overestimate on the number of pores. On the other hand, the error in porosity does not look that bad at all. Here, what happens is that false positives (i.e., background pixels that are marked as foreground) and false negatives (i.e., foreground pixels marked as background) cancel out. We argue that this is not enough to trust such a measurement. A simple figure of merit, the ratio of misclassified pixels, can help to understand why. In our example, false positives and false negatives amount to more that 25% of total pixels, but are well-balanced as to produce a relatively accurate measurement of porosity. Other cases might be encountered where false positives and false negatives are not that balanced, resulting in measurements that largely deviate from the correct value. This is quite a good reason to move to other segmentation techniques. A thorough discussion on evaluation criteria of segmentation algorithms is beyond the scope of this chapter. The interested reader may want to read refs [15, 29].

Wrong mathematical definitions are another source of possibly disastrous results. Figure 5.3 shows a common case that can distort directionality measurements. The rose plots in Figure 5.11 show the distribution of directionality computed from the features in Figure 5.4(d). Plots have been drawn on pruned datasets, where small features are not used for computation. Basically, the directionality of small features is not represented. Features smaller than 2, 5, or 20 pixels (Figures 5.11(a), 5.11(b), and 5.11(c), respectively) were removed from the corresponding plot. The shape of the plot changes accordingly. The first plot shows a dominant direction around 45° that does not appear in the other plots. This suggests that the corresponding spike is the result of directionality computed on tiny features, especially prone to approximation error. In particular, the contribution of features from 2 to 5 pixels to the 45° direction is overwhelming. Clearly, misleading results are obtained if no special attention is put on the analysis of computed data. This can potentially happen with every measure, since definition errors are hard to spot. Are we without hope? Not at all. This very example indicates a possible way out. By simply plotting a number of graphs excluding increasingly larger features, just like we did in Figure 5.11, one can find a point when the succession of plots does not show significant variation, anymore. If such a point exists, one can be reasonably confident

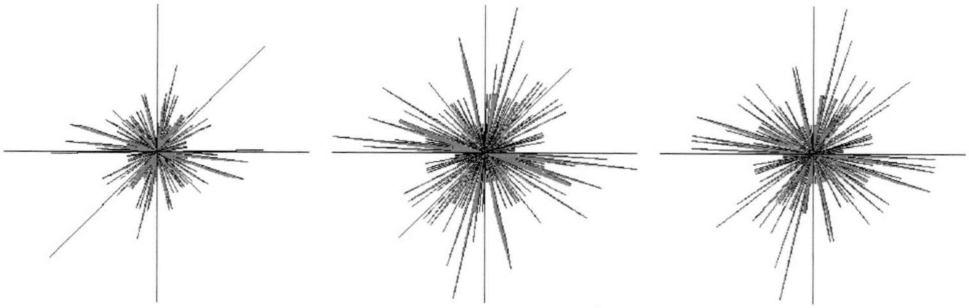

(a) Features larger that 2 pixels. (b) Features larger that 5 pixels. (c) Features larger that 20 pixels.

Figure 5.11 Directionality distribution computed on features in Figure 5.4(d). Small features of varying size were excluded from computation in each plot.

that the plot reproduces a real situation, without confounding factors due to the pixel-wise nature of digital shapes. The question whether pruning small features is acceptable depends on the application at hand. If small features are not a relevant contribution to the microstructural arrangement of the sample, then it should not be a concern. Conversely, if the experiment involves small features, care must be taken before any pruning is made.

This example shows how a little experimentation with image analysis results can foster a more correct interpretation of data. Most definition errors amount to a failure to imagine how approximation can impact on simple geometry. This issue pops up frequently when one translates a definition from continuous domains to discrete ones, just like in the case of our definition of directionality. That is why most flawed definitions ultimately amount to approximation errors, even if they are due to wrong mathematical assumptions, whereas approximation errors are often intrinsic to the problem. Hence, solutions able to cope with approximation errors can be shared to spot both types of error, and the way round. The solution presented above for directionality is such an example. It is also useful for the perimeter approximation problem in Section 5.4.2.

5.4.4 Good Practices

In previous sections potentially disastrous scenarios have been shown as a consequence of misuse of, and blind trust in image analysis. If we somehow scared you, we now try to win your forgiveness by suggesting a few tips for sound experimentation with image analysis. This section is thus devoted to guide the reader through some good practices enabling microscopists to avoid pitfalls without too much effort. A few examples have been already discussed at the end of previous section. Here, we will try to be more systematic.

Before proceeding, it is useful to retrace the path that led us here. Our discussion concentrated on quantitative analysis of digital micrographs, an emerging practice in contemporary food microscopy.

Qualitative inspection is outside the scope of this chapter, since it can be carried out without the help of image analysis. The standard analysis procedure, if such a thing exists, is carried out in two distinct steps: feature detection (i.e., segmentation and grouping) and quantitative measurement of image features. Both steps require careful experimental design and assessment of results. Sure enough, image analysis does not come for free. A number of things can go wrong and the microscopist should learn how to spot possible errors and to cope with them.

Hence, we pointed out a few image analysis issues, often arising on the way: poor image segmentation, bad approximation of geometric measures, difficulty to define elusive measures, wrong mathematical definition of measures, ambiguity of terminology. Image segmentation has probably the strongest in influence on subsequent analysis. On the other hand, detecting poor segmentation results is easy and often amounts to a thorough visual inspection of the segmented image. Leaving aside terminology problems, that are more bound to the applicative domain, all other listed issues occur when taking measurements on the segmented image. They are both due to approximation, arising from sampling and digitization, and to the definition of mathematical descriptors used as quantitative measures. Blind trust in computer output can lead to wrong conclusions, if the output is not carefully checked. Approximation errors and flawed definitions can be insidious and hard to spot, invalidating all subsequent analysis.

Correct interpretation of quantitative results is thus of paramount importance. Microscopists should always ask themselves whether numerical results show peculiarities of the observed phenomenon (i.e., of features of input images) or of the method used to describe it. The directionality example in Figure 5.11 is one of such cases. In this sense, interpretation of image analysis experiments resembles the interpretation of statistical analyses: they both produce huge amount of data, plots, graphs, outliers, that must be correctly explained and used to test experimental hypotheses.

Microscopists should thus learn to design sound image analysis experiments, in order to avoid possible catastrophic errors in the analysis chain, and to spot and circumvent them when they occur. Dubious conclusions should not be considered a good service to science. Hence, let us turn to a few good practices to make our discussion more concrete.

Every microscopist knows that a good experimental design requires careful sampling. The number and distribution of samples should be chosen in order to ensure that a representative selection of material is used for experimentation. There is no reason not to apply the same rationale to take micrographs on the selected specimens. That is, for each specimen you should carefully choose your images and their number. On the other hand, most microscopists wander around the specimen in search for beautiful pictures. Beware of bias. Are you sure that this procedure guarantees a representative sampling of your specimen? Qualitative analysis would probably benefit from carefully-chosen examples, conveying more information and able to better communicate results. Conversely, quantitative analysis may suffer from consistent bias if images are not taken at random positions. Randomly-chosen image sets have more chance to be a representative sampling of the specimen. That is, random is better.

If micrographs are acquired at random positions, their distribution in the sample may be ignored provided that enough images are taken. Our problem is thus simplified to that of finding the minimum number of images that give a representative "picture" of the sample. Several studies on the minimum *Representative Volume Element* have been published in the material science literature. Focusing on the search of a single volume, these methods are only useful for reasonably homogeneous materials. Food materials do not possess such repeatable structure (that is why a number of micrographs are taken for each experiment). Hence, these methods cannot be easily applied to food microscopy [31]. Attempts to employ the same concept of minimum volume were made in [30] and [31] to find the minimum representative image size, using a statistical bootstrapping strategy. However, this is just one aspect of the problem. Once we know the minimum size for our images, we still do not have a clear indication on how many images we need. Moreover, the minimum image size is representative of the image set at hand, but not necessarily of the whole sample. We might be so unfortunate as to capture an image selection that does not fully represent the sample. The information we would get on minimum image size would not tell us much. This case is not improbable (think, for example, to the microstructure of stretched cheeses, such as that shown in Figure 5.4(a)). The problem thus remains open.

Assuming that a good selection of images is available, microscopists should not underestimate the importance of making objectives clear. Wait, wait. Obviously, we do not refer to the experimental hypothesis being tested. Rather, we mean that people approaching image analysis often have vague objectives in mind. Even worse, some people just compute a number of measures without too much pondering on whether they are really useful or not for their experiment. That is a bad habit. You should make clear what you expect from image analysis. What is my program supposed to do with my images? What answers am I looking for? What do I expect when processing my image? Vague objectives translate into vague requests. Like Deep Thought taught us in The Hitchhiker's Guide to the Galaxy [1], it is good questions that make good answers. And *clear* is one aspect of *good*. What is, then, the difference between a clearly-stated image analysis objective and a vague one? Let us make a few examples. You tell which one is which. Here we go:

Problem 1: Compute the number of large and of small pores in a SEM cheese image, such as the one in Figure 5.4(a).

Problem 2: Compute the number of dead and of living cells in a CLSM image, such as those in Figure 5.2.

Problem 3: Compute the directionality of pores in the SEM image of Figure 5.2.

Problem 4: Compute the main directionality of pores in the SEM image of Figure 5.2.

Problem 5: Plot the directionality of the 10% largest pores in the SEM image of Figure 5.2.

Problem 6: Partition pores in Figure 5.4(a) into two sets, containing respectively small and large pores, such that size variation is minimized among the elements of the same set, and maximized among elements of different sets; then compute the number of pores in each set.

Can you figure out which problems propose vague issues? Problem 1 is clearly one of such ill-defined questions. How do you define large and small? That is, how much pixels should a pore contain to be counted as large? Alternatively, how do we automatically compute the size threshold to discriminate small from large pores? On the other hand, Problem 6 clearly states how large and small pores should be set apart. A computer program could partition pores by clustering them into two bins, computing k-means on pore size values. Then, it could return the size of each cluster. A clear-cut definition led us to a simple and unambiguous computer implementation. What about Problem 2? This is quite a lucky case, where informal notions directly translate into operative definitions: dead and living cells have a different color. Problems 3, 4, and 5 show an increasing level of clarity and accuracy. Provided that we are given a good mathematical descriptor for directionality, how do we compute directionality in Problem 3? Should we show a list or output a summary measure, such as mean or median? Problem 4 partially solves this ambiguity. Stating that we are interested in *main* directionality, it suggests that we should use some sort of summary measure. On the other hand, Problem 5 clearly states that we are interested in examining a plot of various directionality values, and that we are interested only in those of larger structures. The choice of 10% is somewhat arbitrary, but this is not a limitation since other values can be tried in further image analysis experiments, if a fine tuning is useful. Clearly, one cannot always hope to be asked such well-defined questions as are posed in Problem 6. We hope, however, that these examples convinced the reader that microscopists should put some effort into making their objectives as clear as they can.

5.4.5 Image Analysis in your Lab

In the previous section we highlighted a few good practices that can help microscopists to design sound image analysis experiments. Now, we try to sketch a sort of image analysis protocol; not a protocol in the usual sense, i.e., a procedure, a sequence of well-defined steps to carry out a lab experiment. Rather, it is a sort of basic outline for image analysis experiments in the context of food science, but also of life sciences and material science. In order to make it more familiar to people used to lab protocols, let us organize our image analysis protocol into ordered steps:

Step 1: Choice of the image dataset.
Step 2: Definition of clear image analysis objectives.
Step 3: Definition of interest features.
Step 4: Feature detection.
Step 5: Visual assessment of segmentation results.
Step 6: Choice of mathematical descriptors.
Step 7: Assessment of results.
Step 8: Statistical analysis of computed data.

Let us discuss in more detail each step. **Step 1** concerns the selection of the set of images that will be used for image analysis. One should carefully select images so as to be a representative sample of the specimen. A sound analysis of a very peculiar image sample is not a good description of the analyzed phenomena and, as such, is of little use. A representative sample is one that conveys enough useful information for the experiment. This involves a correct choice of the number of images per specimen, and varied selection of imaging viewports. In particular, the number of images should be chosen such that it is large enough to show all possible variation of the observed phenomena and, at the same time, small enough to make the experiment feasible and to avoid gathering too much duplicated information. As for the selection of viewports (i.e., positions of the specimen table under the lens) it should be varied enough to avoid overlapping and to avoid any bias in the choice of imaged features. This means that microscopists should not indulge into walking around the specimen in search for beautiful pictures or for pictures that perfectly illustrate the phenomenon one wants to demonstrate. This would probably direct you straight to a biased choice. Roaming randomly around the specimen to take pictures is the best way to avoid any bias. Remember that arbitrary is quite different from random.

In preceding sections, we have discussed in some detail about the importance of clearly defining objectives (**step 2**) and of carefully selecting image features (**step 3**). Here it suffices to highlight once more the difference between the two. Defining image analysis objectives amounts to answering the question of what properties of our specimen we want to investigate. At this stage, all we need to know is what attributes of our specimen we want to quantify, that are visible in our images. After objectives have been clearly stated, we should ask ourselves what specimen features describe those properties and what image features represent those properties. That is, once we define our general objectives, we look for specimen characteristics that expose those properties (step 3). Choosing image features that represent interest specimen characteristics is a consequential operation. Let us make this difference clear with an example. Suppose we want to test our hypothesis that, because fat is hydrophobic, despite of high temperatures in the production of a particular cheese, fat never melts into non-spherical aggregates[1]. We have acquired a large dataset of SEM images like that in Figure 5.1(a). Here, we may distinguish two phases: protein compact matrix and pores. Suppose we measured in an earlier experiment the percentages of whey, fat and matrix. Our image analysis objective (step 2) is to prove that the average percentage of fat in our dataset matches that measured in the other experiment. Our interest features (step 3), then, are pores that are due to fat. Due to our experimental hypothesis that fat never melts, we can assume that fat droplets stay spherical, that is fat pores appear as oval shapes in our images. Hence, our interest image features are (chains of) round pores.

Feature detection (**step 4**) has been extensively discussed in Section 5.4.2. We have shown a number of examples where segmentation failed catastrophically. Learning to select the proper segmentation method from those offered by your program may be painful. A good choice, though, will save you time later on, helping you to avoid coping with noisy data. Basically, segmentation errors may act as a confounding effect added to the phenomenon you are studying. As a first step, you might try to apply all segmentation algorithms in you program to a few images and see what happens. This leads us directly to **step 5**, that is to the assessment of segmentation results. A visual inspection of segmented images usually gives important clues on which algorithms perform better. In fact, several bad-performing algorithms might be immediately excluded at a first glance. Did our segmentation algorithm correctly marked our interest features? Is the segmentation too noisy? These two very basic questions establish a fundamental test-bed for your feature detection algorithms. If you are versed in image analysis machinery, visual inspection might also help you to devise a better pre-processing procedure, e.g., to reduce noise or to better highlight your interest features. In our cheese example above, we could run a sequence of morphological opening-closing operations on grey-level images before feature detection. This operation may highlight elliptic features corresponding to fat droplets. Do not underestimate the importance of visual assessment. Although it does not suffice to tell good from bad results, it often gives good clues on what could possibly have gone wrong. This, in turn, can help you to select a more appropriate pre-processing/segmentation approach. Steps 4 and 5 can be thought as an iterative trial-and-error practice.

Next, we come to the choice of mathematical descriptors (**step 6**) that will compute significant measurements from segmented images. Two aspects should be accounted for when looking for descriptors: *relevance* and *feasibility*. Clearly, descriptors must be relevant both to the problem at hand and to the interest features, defined in step 3. If we want to find the elliptical footprints of fat droplets in our SEM cheese images, geometrical features that refer to roundness and perimeter smoothness are appropriate, whereas descriptors measuring elongation, curl/straightness, or counting the number of inner

1 Notice that this is a purely illustrative example, not demanding any scientific ground.

holes are probably not fit our problem. It is less obvious that descriptors should be feasible, i.e., it should be safe to compute them on our segmented images. Does it still make sense to compute perimeter smoothness if our detected image features show highly jagged borders, due to segmentation noise? Would those measurements be informative or would rather they describe image noise? We have no choice, here, but to rely on roundness alone. Hence, although some descriptors might seem more or less equivalent for our purposes, noise or other segmentation errors could compel us to choose one or another. That is why feasibility should not be overlooked. Notice how visual inspection can be of great value even when we choose mathematical descriptors. Do not limit yourself to choose existing descriptors. If your mathematical skills are strong enough, and you cannot find descriptors that meet your needs in the literature, test yourself and devise your own descriptors. Even to those readers who do not feel like trying anything new, however, we recommend to gain a good understanding of the mathematics they use in their experiments.

Assessment of results in **step 7** is not a repetition of that in step 5. The latter is just a visual assessment of segmentation images, with the aim of avoiding misleading numerical results in subsequent steps. On the other hand, in step 7 we should assess numerical results themselves. That is, are these results reasonable? Do they give a meaningful description of observed phenomena? If the total size of detected fat droplets is not consistent to what we observe, we should investigate what might have gone wrong. Notice that this is not the same as saying that the ratio of fat area is more that 100% or less than 0%. That would just be a software bug. Here, we refer to legal measurements that do not agree with other observations. In our example, we just referred to visual inspection but any observation can be used as a reference. If we had access to precise chemical analyses, for example, we could compare the total ratio of image area occupied by fat footprints to the fat ratio detected in our lab. Unfortunately, these analyses are seldom available. Hence, we should be able to devise more clever tests to assess our image analysis results. The directionality example in Section 5.4.3 is one such example. Figure 5.11(a) shows a substantial prevalence of features oriented at 45° counter-clockwise with respect to the x-axis. We found that this is true for all our images. This made us suspicious. We then arbitrarily picked up an image (Figure 5.4(d)) and, from visual observation, found that this measurement did not match the main orientations, of about 30% and along the x-axis, we observed. Looking at our mathematical definition of directionality, we found that the orientation of rectangular features lies along the diagonal, rather than along their medial axis (that is why it is important to understand the maths being used). We argued that, because of that, small features, such as 2×2 pixels, would have shown a preferred orientation at 45°. Our hypothesis was that an overwhelming number of small features might strongly contribute to that 45° orientation. We tested this hypothesis by removing smaller features from calculations and plotted new diagrams (Figures 11(b) and 11(c)). Now, preferred directionalities pop-up as expected. Hence, we found that small features, probably due to segmentation noise, were responsible for a potentially disastrous outcome. Once we removed them and assessed results again on other images, we can be more confident that measurements are correct for the whole image set. This is just an example of how results can be validated, and not necessarily an easy one. Unfortunately, there is no such thing as a general validation procedure. No silver bullet. Experience, hard thinking, and occasionally the help of expert image analysts are the only weapons microscopists have.

Although it is not essential and, in some respect, it is not even "canonical" to image analysis, we added statistical analysis in this outline (**step 8**) because we think that it is a useful addition to our image analysis protocol. It is useful in several respects. First of all, in order to make our analysis more robust to all possible errors that might have piled up along the way (segmentation error, mathematical definition of descriptors, ...). Basically, errors can be easily treated as outliers in a statistical setting, especially if some degree of data loss can be tolerated. If for example we are interested in computing a few main directions in the example above, we might simply use standard statistical tools to find modes in a histogram reporting the population weight for each of a few quantized directions (the rose plot in Figure 5.11(b) is one such histogram). Principal Component Analysis (PCA) and Factorial Analysis [27] were used in [18] to group CLSM images of milk fat into two classes, of strong and mild milk treatments. Several descriptors were used to describe various aspects of the geometry of milk fat droplets dispersed in milk. The method proved extremely robust to severe segmentation errors and to flawed mathematical definitions (one descriptor was immediately discarded since it gave confounding results) that might not have been otherwise discovered. One more reason to use statistical analysis is to prove, disprove, or at least give more strength to one's assumptions and hypotheses. In the same work, a Bayesian learning [7] framework was used to make statistical inference on new images. Namely, a model was created for each class (strongly-treated and weakly-treated milk) from the statistical distributions of descriptor measurements on a set of training images, where the class was known to the program. These models were used to infer the class of new images, for which the class was unknown to the program. More than 90% of images were correctly classified [18]. These impressive results simultaneously proved that the chosen mathematical descriptors were relevant as well as feasible, and that there is a proven, statistically relevant difference between the two kinds of milk treatment. Again, these are just a few examples. Devising new ways to prove your hypotheses using image analysis and statistics is up to your imagination and creativity.

5.5 Conclusions

Well then, this is the end of our journey through image analysis. Rapid and short as it was, it was hard indeed, I guess, for most readers. Do not worry if you did not grasp every single word of it. If you had, you probably wasted time on something you already knew. If you found our guide helpful for choosing your favorite image analysis software, if you have a clearer idea on how your images and data should be analyzed, well, that is more than we hoped for.

You may have found our last section a bit hard to chew. It may be comforting to you to know that it was even harder for us to write and, most of all, to go through our own experiences in this field before being invited to write this chapter. Do not be afraid to read it through again and again, if you think it useful. You might find new suggestions and bits of useful information you had overlooked before. This is what happens to us when reading a good guide. We hope this is at least a fair one.

Acknowledgments

I am grateful to Nicoletta Fucà, former CLSM microscopist at CoRFiLaC, for the many stimulating discussions in her stay at CoRFiLaC and for carefully revising the manuscript.

References

1 D. Adams. The Hitchhiker's Guide to the Galaxy. Del Rey, 1979.
2 Adobe. Photoshop. Website: http://www.photoshop.com/.
3 J.M. Aguilera. Why food microstructure? Journal of Food Engineering, 67(1-2):3-11, 2005. IV Iberoamerican Congress of Food Engineering (CIBIA IV).
4 J.M. Aguilera and D.W. Stanley. Microstructural principles of food processing and engineering. Aspen, second edition, 1999.
5 C. Allan, J.-M. Burel, J. Moore, C. Blackburn, *et al.* OMERO: flexible, model-driven data management for experimental biology. Nature Methods, 9(3):245–253, March 2012.
6 P. Bhowmick and B.B. Bhattacharya. Digital straightness, circularity, and their applications to image analysis. In V.E. Brimkov and R.P. Barneva, editors, Digital Geometry Algorithms, volume 2 of Lecture Notes in Computational Vision and Biomechanics, pages 247–299. Springer, 2012.
7 C. Bishop. Pattern recognition and machine learning. Springer, New York, USA, 2006.
8 A.M. Bruckstein. Digital geometry in image-based metrology. In V.E. Brimkov and R.P. Barneva, editors, Digital Geometry Algorithms, volume 2 of Lecture Notes in Computational Vision and Biomechanics, pages 3–26. Springer, 2012.
9 A.E. Carpenter, T.R. Jones, M.R. Lamprecht, C. Clarke, *et al.* CellProfiler: image analysis soft-ware for identifying and quantifying cell phenotypes. Genome Biology, 7(10):R100, 2006. Website: http://www.cellprofiler.org/.
10 D. Coeurjolly and R. Klette. A comparative evaluation of length estimators of digital curves. IEEE Transactions on Pattern Analysis and Machine Intelligence, 26(2):252–258, 2004.
11 P.J. Diggle. Statistical Analysis of Spatial Point Patterns. Oxford University Press, 2nd edition, 2003.
12 Free Software Foundation. General Public License. Website: http://www.gnu.org/licenses/license-list.html.
13 N. Fucà, C. Pasta, G. Impoco, M. Caccamo and G. Licitra. Microstructural properties of milk fat globules. International Dairy Journal, 31:44–50, 2013.
14 R.C. Gonzalez and R.E. Woods. Digital Image Processing. Prentice-Hall, Inc., Upper Saddle River, NJ, USA, 3rd edition, 2006.
15 D.S. Hochbaum, C. Lyu and E. Bertelli. Evaluating performance of image segmentation criteria and techniques. EURO Journal on Computational Optimization, 1(1–2):155–180, 2013.
16 ImageMagick Studio LLC. ImageMagick. Website: http://www.imagemagick.org/.

17 G. Impoco, S. Carrato, M. Caccamo and L. Tuminello. Quantitative analysis of cheese microstructure using SEM imagery. In SIMAI 2006, Minisymposium: Image Analysis Methods for Industrial Application, May 2006.

18 G. Impoco, N. Fucà, C. Pasta, M. Caccamo and G. Licitra. Quantitative analysis of nanostructures' shape and distribution in micrographs using image analysis. Computers and Electronics in Agriculture, 84:26–35, June 2012.

19 Bitplane Inc. Imaris. Website: http://www.bitplane.com/imaris/imaris.

20 T.R. Jones, I.H. Kang, D.B. Wheeler, R.A. Lindquist, A. Papallo, D.M. Sabatini, P. Golland and A.E. Carpenter. CellProfiler Analyst: data exploration and analysis software for complex image-based screens. BMC Bioinformatics, 9(1), 2008. Website: http://www.cellprofiler.org/.

21 Laboratory for Optical and Computational Instrumentation (LOCI). Resource page on high-dimensional large dataset visualization software. Website: http://loci.wisc.edu/outreach/3d-viz.

22 Leica Corporation. Leica LAS. Website: http://www.leica-microsystems.com/products/microscope-software/software-for-materials-sciences/details/product/leica-las-image-analysis/.

23 Leica Corporation. Leica LAS X. Website: http://www.leica-microsystems.com/products/microscope-software/software-for-life-science-research/las-x/.

24 W.R. McManus, D.J. McMahon and C.J. Oberg. High-resolution scanning electron microscopy of milk products: A new sample preparation procedure. Food Structure, 12(4):475–482, 1993.

25 Molecular Devices Corporation. MetaMorph. Website: http: //www.moleculardevices.com/systems/metamorph-research-imaging/metamorph-microscopy-automation-and-image-analysis-software.

26 Octave community. Octave. Website: https://www.gnu.org/software/octave/.

27 G.W. Oehlert. A First Course in Design and Analysis of Experiments. W.H. Freeman and Co., 2000. Also available at: http://users.stat.umn.edu/gary/book/fcdae.pdf.

28 Olympus Corporation. Olympus Stream. Website: http://www.olympus-ims.com/en/microscope/stream/.

29 S. Philipp-Foliguet and L. Guigues. Multi-scale criteria for the evaluation of image segmentation algorithms. Journal of Multimedia, 3(5), 2008.

30 C. Ramírez, J.C. Germain, and J.M. Aguilera. Image analysis of representative food structures: application of the bootstrap method. Journal of Food Science, 74(6):R65–72, 2009.

31 C. Ramírez, A. Young, B. James and J.M. Aguilera. Determination of a representative volume element based on the variability of mechanical properties with sample size in bread. Journal of Food Science, 75(8):E516–E521, 2010.

32 C. Rueden, J. Schindelin, M. Hiner, B. DeZonia, L. Kamentsky and K. Eliceiri. Imagej2. Website: http://imagej.net/ImageJ2, 2015.

33 J.C. Russ. Image Analysis of Food Microstructure. CRC Press, 15 November 2004.

34 J. Schindelin, I. Arganda-Carreras, E. Frise, V. Kaynig, *et al.* Fiji: an open-source platform for biological-image analysis. Nature Methods, 9(7):676–682, July 2012.

35 C.A. Schneider, W.S. Rasband, and K.W. Eliceiri. NIH Image to ImageJ: 25 years of image analysis. Nature Methods, 9(7):67–75, July 2012.

36 J.V.C. Silva, D. Legland, C. Cauty, I. Kolotuev and J. Floury. Characterization of the microstruc-ture of dairy systems using automated image analysis. Food Hydrocolloids, 44:360–371, 2015.

37 The GIMP team. GIMP, the GNU Image Manipulation Program. Website: http://www.gimp.org/.

38 The MathWorks Inc. Matlab. Website: http://it.mathworks.com/products/matlab/.

39 V. Toh, C.A. Glasbey and A.J. Gray. A comparison of digital length estimators for image features. In 13th Scandinavian Conference on Image Analysis, SCIA 2003, pages 961–968, June 29–July 2 2003.

40 I.T. Young. Not just pretty pictures: Digital quantitative microscopy. Proceedings of the Royal Microscopical Society, 31(4):311–313, 1996.

6

Microstructure of Milk

Michael H. Tunick

Center for Food & Hospitality Management, Drexel University, 101 N. 33rd Street, Philadelphia, PA 19104, USA

6.1 Components of Milk

Bovine (cow) and caprine (goat) milk are composed of 87–88% water, 3–4% fat, 3–4% protein, 4–5% lactose, and <1% minerals and vitamins (USDA, ARS 2014). The minerals and vitamins will not be discussed here and the lactose is not directly viewable by microscopy since it is dissolved in the water. The fat and protein, however, may be imaged by scanning electron microscopy (SEM), transmission electron microscopy (TEM), and confocal laser scanning microscopy (CLSM). Bacteria and other cells are found in raw milk and may be observed under light microscopes.

This chapter will describe the microstructure of fat, protein, and cells in milk, and will also cover concentrated milk and changes in milk as it undergoes the digestive process. Results of studies of bubaline (buffalo), camelid (camel), equine (horse), ovine (sheep), and human milk are also included.

6.2 Fat

6.2.1 Fat Globules

The fat in raw bovine milk consists of globules ranging from 1 to 20 μm in diameter and enclosed by a milkfat globule membrane (MFGM). The MFGM is formed when the surface membrane of mammary secretory cells surrounds the globule as it is secreted from the cell into the lumen. Homogenization reduces the globule size to <1 μm and strips off the MFGM. The globule contains about 98% triacylglycerols, and its size and shape vary by species. A study by El-Zeini (2006) showed that bubaline milk fat globules average 8.7 μm in diameter, with bovine and ovine milk fat globules averaging 3.8 μm and caprine and camelid around 3.1 μm. The sphericity (perfect sphere – 1) of the bovine and ovine milk globules was 0.9, followed by camelid at 0.8, bubaline at 0.6, and caprine at 0.33. The average diameter of human milk fat globules was found by Zou *et al.* (2012) to be 5.75 ± 0.81 μm in colostrum, 4.74 ± 0.33 μm in mature milk, and

Microstructure of Dairy Products, First Edition.
Edited by Mamdouh Mahmoud Abdel-Rahman El-Bakry, Antoni Sanchez, and Bhavbhuti M. Mehta.

4.01 ± 0.76 μm in transitional milk. The authors noted that variations from diet and genetics affect the results.

6.2.2 Milkfat Globule Membrane

The MFGM is a thin bilayer of surface-active material containing about 90% proteins and lipids, including phospholipids and cholesterol. The MFGM surface appears to consist of liquid-ordered and liquid-disordered phases depending on the orientation of the fatty acid chains: long and symmetric chains are more tightly packed and result in a liquid-ordered phase that is likely rich in sphingomyelin and cholesterol (Gallier *et al.*, 2010). The large dye molecules utilized in CLSM prevent tight packing and partition into the liquid-disordered phase, which consists of the phospholipids, leaving the liquid-ordered regions relatively dark. The distribution of the phases may be seen in Figure 6.1.

The microstructure of fat in milk changes with processing (Ong *et al.* 2010). In Figure 6.2a, CLSM images of raw bovine milk show fat globules stained with Nile Red, some of which are surrounded by a thin layer of protein (stained with FCF Fast Green). Homogenization (Figure 6.2b) and ultrafiltration (UF; Figure 6.2c) strip off much of the MFGM, which allows casein to adsorb to the globule. The MFGM remains when cheesemilk is not homogenized (Figure 6.2d). Lopez (2005) obtained light microscopic images and size distributions of fat globules in pasteurized (Figure 6.3a), ultra-high temperature pasteurized (UHT; Figure 6.3b), and raw full-fat (Figure 6.3c) bovine milk.

Fluorescence microscopy performed by Evers *et al.* (2008) revealed that the microstructure of MFGM varies among globules of the same species, and probably between species as well. They noted heterogeneity within and among globules, with some areas being fluorescent and others not. Figure 6.4 shows variations in these areas in ovine (Figure 6.4a and 4b) and human milk (Figure 6.4c) stained with fluorescent conjugates

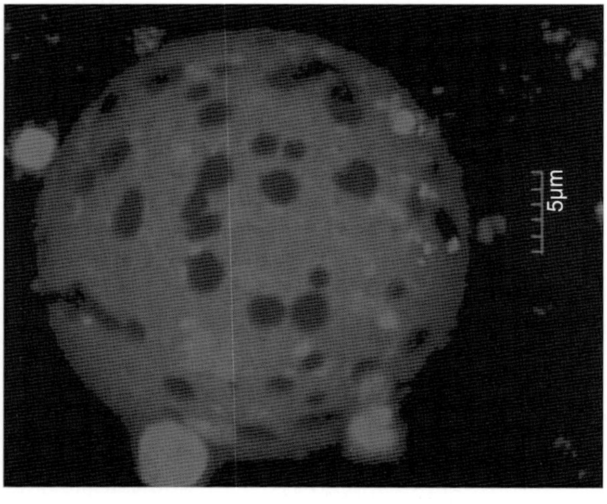

Figure 6.1 Confocal laser scanning micrographs of a milk fat globule in raw cream from bovine milk, showing red liquid-disordered and dark liquid-ordered phases. From Gallier *et al.* (2010), reprinted with permission from the *Journal of Agricultural and Food Chemistry*, copyright 2010 American Chemical Society. (*See color plate section for the color representation of this figure.*)

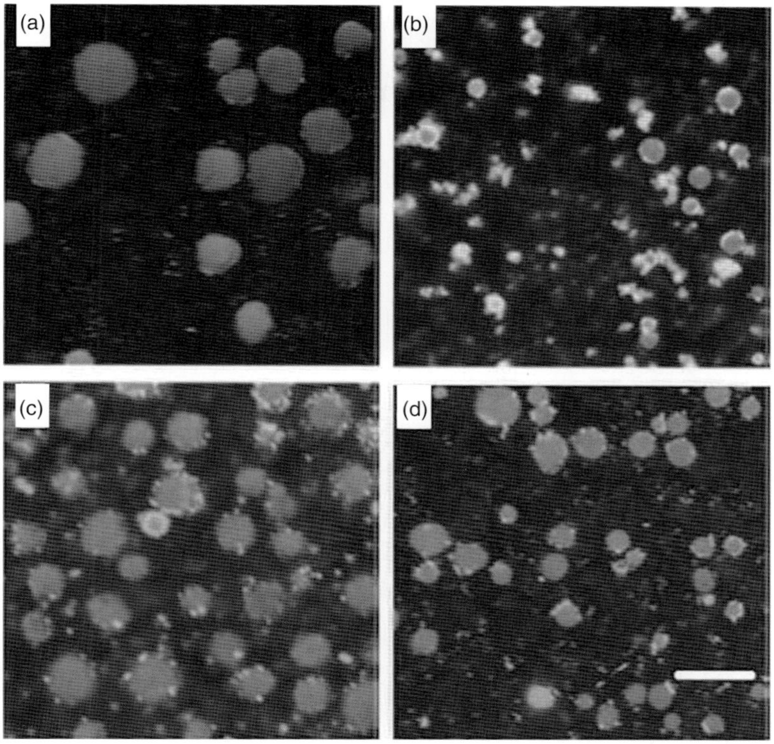

Figure 6.2 Confocal laser scanning micrographs of (a) raw bovine milk; (b) pasteurized and homogenized milk; (c) raw ultrafiltered retentate; and (d) pasteurized and standardized cheesemilk stained so that fat appears as red and protein as green. Samples were diluted 50 times in simulated milk ultrafiltrate solution. Bar at lower right, 10 μm. Adapted from Ong *et al.* (2010), reprinted with permission from the *Journal of Food Science*, copyright 2010 Wiley. (*See color plate section for the color representation of this figure.*)

of lectin wheat germ agglutinin, which binds to fat globules. The areas where it did not bind were apparently without a bilayer membrane.

MFGMs and casein micelles of equine milk were compared to those of bovine and human milk by Welsch *et al.* (1988). The surface of equine (Figure 6.5a) and human (Figure 6.5c) MFGMs was covered by a glycocalyx (glycoprotein-polysaccharide coat) consisting of branched filaments about 0.5 μm long; no glycocalyx was present in bovine milk (Figure 6.5b). The equine fat globules themselves averaged 2.75 μm in diameter and exhibited a smooth and uniform surface. Zou *et al.* (2012) did not observe microstructural differences resulting from stage of lactation in human MFGM.

6.2.3 Cream

The buoyancy of fat globules in raw milk causes them to rise to the surface and form a layer of cream. Figure 6.6 shows SEM images of cream made by Kalab (1985) using two procedures, encapsulating in agar, fixing, and dehydrating; and rapidly freezing, coating with gold, and imaging at low temperature.

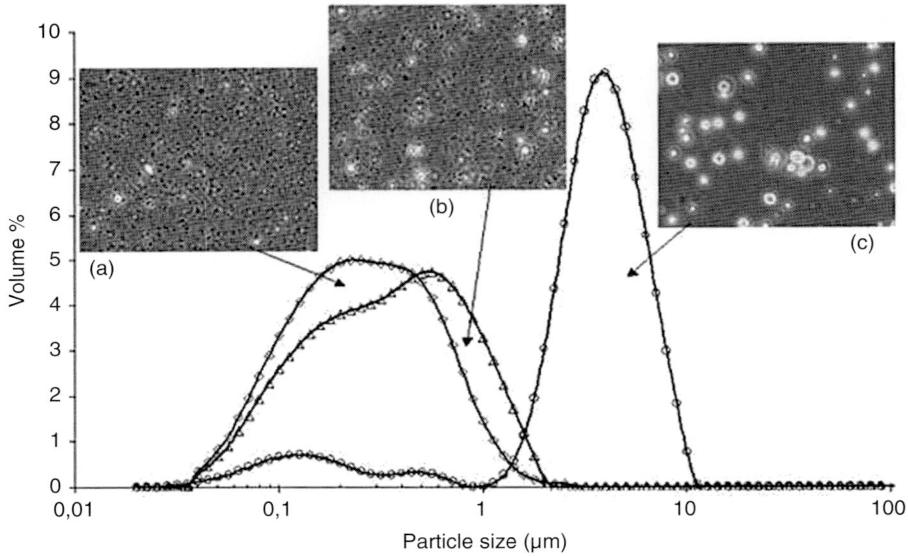

Figure 6.3 Light microscopic images and size distributions (measured by laser light scattering) of milk fat globules in (a) pasteurized; (b) UHT; and (c) raw full-fat bovine milk. From Lopez (2005), reprinted from *Reproduction Nutrition Development*, http://rnd.edpsciences.org, with permission from EDP Sciences, copyright 2005.

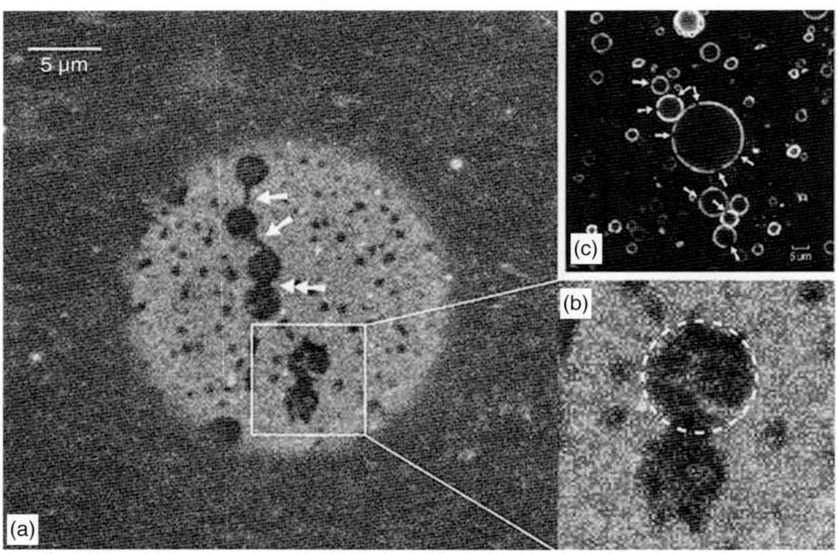

Figure 6.4 (a) Confocal laser scanning micrograph (CLSM) showing part of the membrane of an ovine fat globule. Non-fluorescent areas of different sizes were mostly circular. Some large non-fluorescent areas near the top were connected to each other (arrows). (b) Magnified section showing smaller non-fluorescent areas appearing to combine into larger non-fluorescent areas. The expected final circular non-fluorescent area is marked with a dashed circle. (c) CLSM of human fat globules. Arrows point to non-fluorescent membrane areas. From Evers *et al.* (2008), reprinted from *International Dairy Journal*, copyright 2008, with permission from Elsevier.

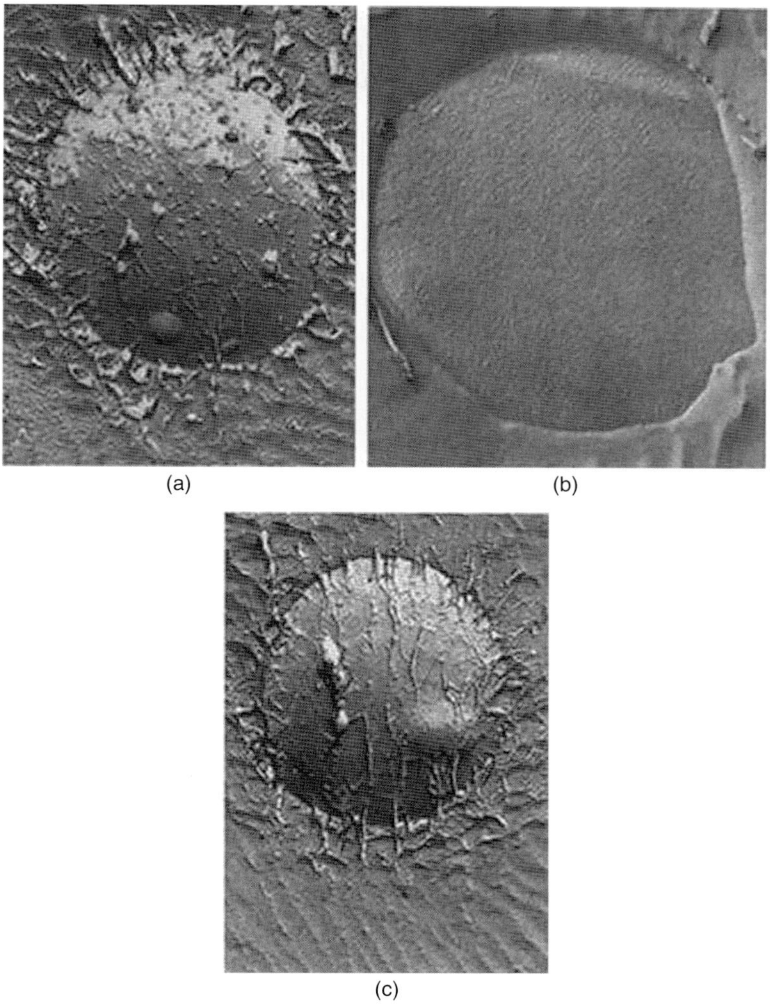

Figure 6.5 Freeze-etching preparations showing the surface of (a) human; (b) bovine; and (c) equine milk fat globules. Surface filaments were present on the human and equine milk globules but not on the bovine globules. Magnifications: (a) 19,000×, (b) 28,300x, (c) 18,600×. From Welsch *et al.* (1988), reprinted from *Histochemistry*, copyright 1988, with permission from Springer.

A layer of cream may form atop UHT milk during storage since heating above 65°C results in the association of β-lactoglobulin with casein micelles, fat globules, and κ-casein, which apparently causes the globules to link together (Hillbrick *et al.*, 1999). Researchers at Utah State University used TEM to compare the microstructure of milk that was not homogenized (Figure 6.7a) with milk that underwent UHT either before or after homogenization, and with direct or indirect heating. They also localized the positions of various proteins within the sample matrix by immunolabeling. Little difference was seen between the four UHT treatments in fat globule size, and casein-fat globule clusters were identified at higher homogenization pressures, such as those in Figure 6.7b (Hillbrick *et al.*, 1999).

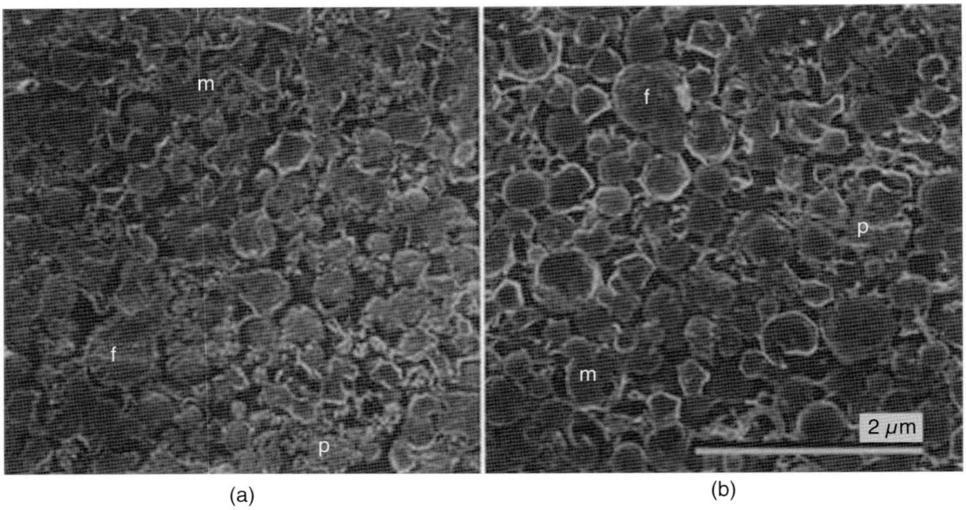

Figure 6.6 Scanning electron micrographs of cultured cream (58% fat) from bovine milk. (a) Cream encapsulated in agar gel, fixed with glutaraldehyde, post-fixed with imidazole-buffered OsO_4, dehydrated in ethanol, freeze-fractured, critical-point dried, and examined at room temperature. (b) Cream rapidly frozen in Freon 12 at −150°C, freeze-fractured in liquid nitrogen, coated with gold while frozen, and examined at 100°C. Key: f, fat globules; m, fat globule membrane fragments; p, protein. From Kalab (1985), reprinted from *Journal of Dairy Science*, copyright 1985, with permission from Elsevier.

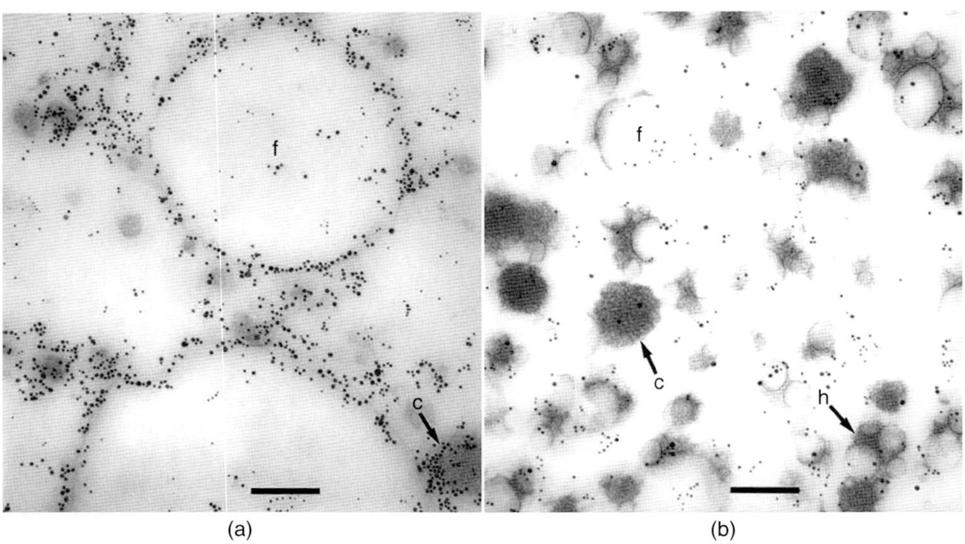

Figure 6.7 Transmission electron micrographs of ultra-high temperature pasteurized (UHT) bovine milk. (a) Not homogenized, immunolabeled for κ-casein (small black dots); (b) homogenized at 34.5 MPa before UHT with indirect heat, immunolabeled for α-lactalbumin (small black dots). Key: c, casein micelle; f, fat globule; h, casein-fat globule homogenization cluster. Bar = 0.2 μm. From Hillbrick *et al.* (1999), reprinted from *LWT-Food Science and Technology*, copyright 1999, with permission from Elsevier.

6.3 Protein

6.3.1 Types of Protein

About 82% of the proteins in bovine milk are classified as caseins, with the remainder being whey proteins. The latter include β-lactoglobulin (3.2 g/L milk), α-lactalbumin (1.2), immunoglobulins (0.8), bovine serum albumin (0.4), and lactoferrins (0.2), and are dissolved as monomers in the aqueous phase of milk, the whey (De Wit, 1998). The casein forms the solid phase, the curd, and in bovine milk is composed of approximately 38% α_{s1}-casein, 10% α_{s2}-casein, 34% β-casein and 15% κ-casein (Fox *et al.*, 2000). Caprine milk contains similar amounts of α_{s2}- and κ-casein but its casein is composed of 42–64% β-casein and 4–26% α_{s1}-casein (Clark and Sherbon, 2000).

6.3.2 Casein Micelles in Bovine Milk

The caseins in milk are in the form of micelles, which are colloidal aggregates dispersed in the whey. About 94% of a micelle consists of casein, with the rest being colloidal calcium phosphate (CCP). Kappa-casein, the only one of the four caseins that is soluble in calcium, resides on the micellar surface and helps to stabilize the other three caseins (Farrell *et al.*, 2006). The micelle is held together by hydrogen bonding, electrostatic and hydrophobic interactions, and CCP linked to the casein molecules (Ahmad *et al.*, 2009).

Casein micelles in bovine milk are generally spheroidal and are 50–600 nm in diameter, with an average hydrodynamic radius of 116 nm (Glantz *et al.*, 2010). Micelles provide nonfat milk with its opalescent appearance, fluidize the casein molecules, and solubilize the calcium and phosphate (Farrell and Thompson, 1988). Caseins comprise about 2.5% of the total weight of milk, but their highly hydrated micelles comprise about 10% of the volume; only 15% of micellar water is bound to the protein, with the rest being trapped within the structure (McMahon and Oommen, 2008).

Electron microscopy of casein micelles in milk has greatly progressed since the first published image by Nitschmann (1949). Researchers have a choice of techniques to obtain the images. In our laboratory, for example, samples are fixed with glutaraldehyde to crosslink proteins and prevent further biological activity. Thin sections are prepared, and uranyl acetate and lead citrate are both added to provide maximum contrast using TEM. Micelles in bovine milk, along with an individual micelle, were imaged in this way by Qi (2009) and are shown in Figure 6.8.

A cold field-emission ultra-high resolution scanning electron microscope has been used at the University of Guelph to examine the surface of casein micelles. The image in Figure 6.9 revealed tubular structures about 20 nm in diameter protruding from the surface (Dalgleish *et al.*, 2004). The authors hypothesized that κ-casein is present at the ends of the tubes since there would not be enough of this protein to cover the entire micelle surface. Later work in that laboratory has led to the model of a sponge-like structure of CCP and casein clusters linked together (Dalgleish and Corredig, 2012).

6.3.3 Casein Micelles in Caprine Milk

Caprine milk contains 2.2–2.8% casein, and the amount of α_{s1}-casein varies from 0.23–0.28% in LaMancha, Nubian, Saanen, and Nubian-Alpine crosses to 0.06–0.10% in

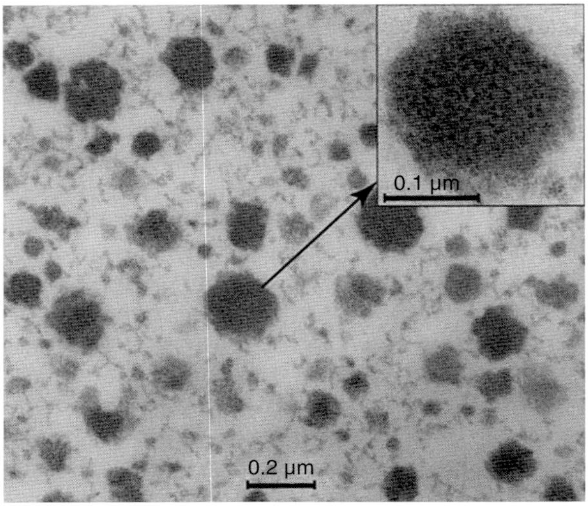

Figure 6.8 Transmission electron micrograph of bovine milk casein micelles. Inset at upper right depicts a micelle at increased magnification. From Qi (2009), reprinted from *Dairy-Derived Ingredients: Food and Nutraceutical Uses*, copyright 2009, CRC Press, Boca Raton, FL.

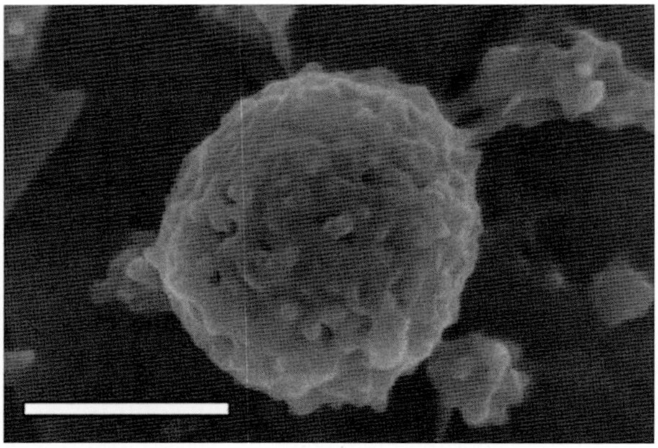

Figure 6.9 Micrograph of a bovine milk casein micelle applied to a polished carbon surface, fixed with glutaraldehyde and OsO_4, dehydrated, and critical point dried. Bar = 200 nm. From Dalgleish *et al.* (2004), reprinted from *International Dairy Journal*, copyright 2004, with permission from Elsevier.

Alpine, Oberhasli, Toggenburg, and Saanen-Alpine crosses (Clark and Sherbon, 2000). In a TEM study of goats bred for a high level (0.47%) of α_{s1}-casein in their milk compared with goats that do not produce α_{s1}-casein at all, Pierre *et al.* (1995) found that the micellar diameters were 199 nm and 280 nm, respectively, though the appearances were otherwise similar (Figure 6.10). They suggested that α_{s1}-casein helps regulate the size of micelles.

High-pressure processing of milk inactivates microorganisms as well as improving rennet and acid coagulation in cheese production. Pressurization of caprine milk was investigated by Law *et al.* (1998), who found that pressure treatment at 20 and 45°C

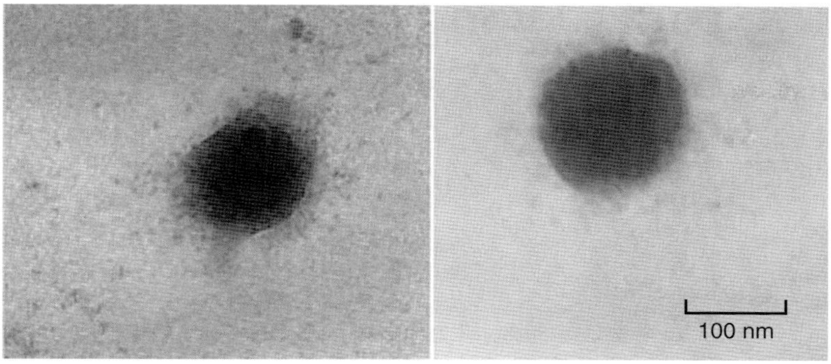

Figure 6.10 Transmission electron micrographs of casein micelles in caprine milk containing (left) 0.47% and (right) 0% α_{s1}-casein. Adapted from Pierre *et al.* (1995), reprinted from *Le Lait*, http://lait.dairy-journal.org, with permission from EDP Sciences, copyright 1995.

caused disruption of micelles into smaller particles, but treatments between 300 and 350 MPa at 45°C led to the formation of very large micelles with almost no κ-casein (Figure 6.11). It was thought that the pressure and temperature conditions affected solubilization of CCP and therefore the structure of the micelles.

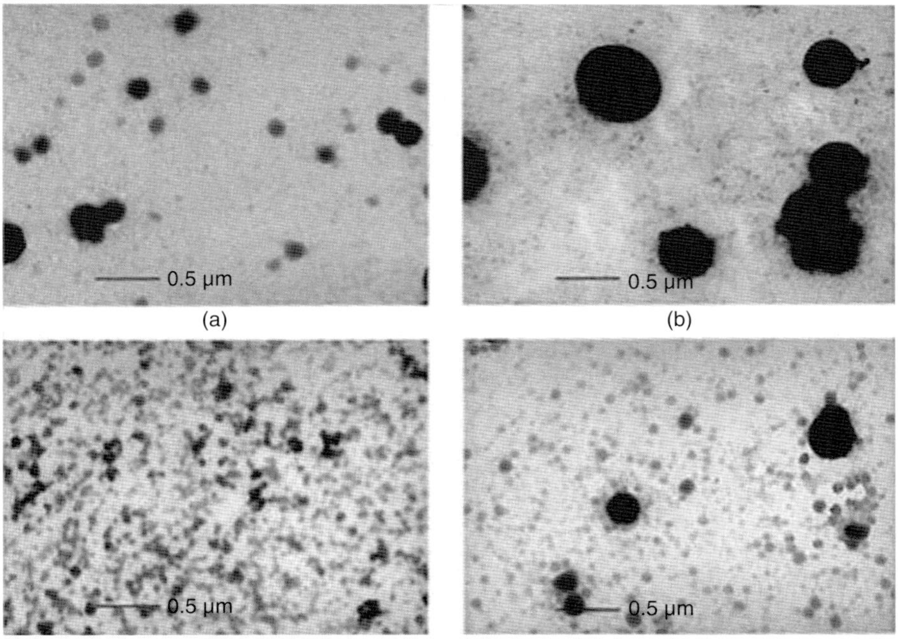

Figure 6.11 Transmission electron micrographs of casein micelles in skim caprine milk pressurized at (a) 300 MPa for 10 min at 20°C; (b) 350 MPa for 10 min at 45°C; (c and e) 500 MPa for 10 min at 20°C; (d) 500 MPa for 10 min at 45°C; and (f) not pressurized. From Law *et al.* (1998), reprinted with permission from the *Journal of Agricultural and Food Chemistry*, copyright 1998 American Chemical Society.

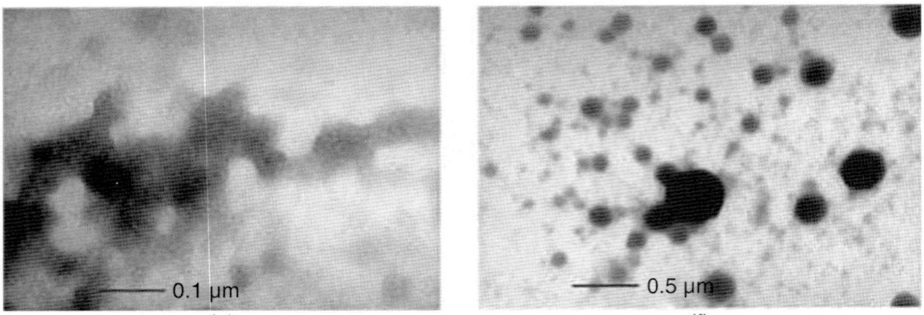

Figure 6.11 (Continued)

6.3.4 Casein Micelles in Milk of Other Species

Researchers in our laboratory have examined the microstructure of casein micelles in human milk (Carroll *et al.*, 1985). The size distribution ranged from 20 to 104 nm, with an average of 42.4 nm for milk obtained at 8 mo of lactation. The appearance of the micelles was similar to that of bovine milk (Figure 6.12).

Casein micelles in ovine milk have not been investigated beyond a determination of the average diameter, 193 nm (Park *et al.*, 2007). Casein micelles have been studied in milk from camels, however. TEM has shown that the majority of casein micelles in camelid milk range from 28 to 240 nm in diameter, with a maximum frequency of 56–70 nm (Ali and Robinson, 1985). Approximately 80% of the micelles have diameters <40 nm, but these values represent <10% of the total mass of casein. The volume distribution curve has a maximum between 260 and 300 nm (Farah, 1993).

6.3.5 Micelle Structure

Scientists do not agree on the exact structure of casein micelles, and much microstructural research on milk is devoted to this question. A model in which submicelle

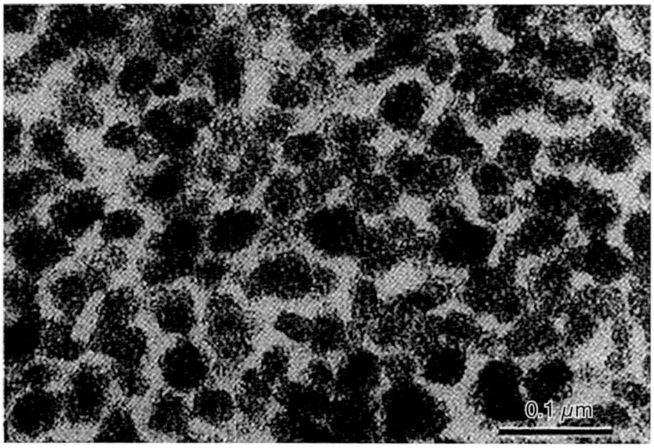

Figure 6.12 Transmission electron micrograph of ultrathin section of casein micelles from full-fat human milk at 6 mo lactation. The sample was concentrated by centrifugation and negatively stained with 1% phosphotungstic acid. From Carroll *et al.* (1985).

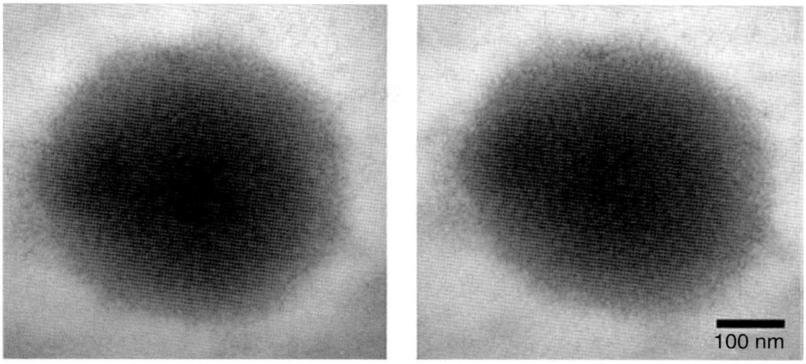

Figure 6.13 Stereo-pair of transmission electron micrograph of a casein micelle from bovine milk, stained with uranyl oxalate. From McMahon and Oommen (2008). Reprinted from *Journal of Dairy Science*, copyright 2008, with permission from Elsevier.

aggregates are linked by CCP to form micelles was proposed by Morr (1967) and seemed to be supported by TEM images such as those by Knoop *et al.* (1979), which revealed a raspberry-like micelle structure. In a second model proposed by Holt (2004), β-casein phosphopeptides bind to CCP to form nanoclusters, which then are cross-linked by the α_S-caseins or aggregated by non-covalent interactions (Dalgleish and Corredig, 2012). Horne (1998) suggested a dual-bonding model, with bonding of hydrophobic regions and with linkage of hydrophilic regions containing phosphoserine clusters to clusters of CCP.

Electron microscopy has not settled the issue. Researchers at Utah State University pointed out that artifacts may be generated by structural changes during preparation of samples, metal coatings, surface tension, and contrast settings on the instrument, which may have led previous researchers to misinterpret their results (McMahon and Oommen, 2008). Their stereo-pair image of a micelle (Figure 6.13) gives a two-dimensional TEM output a more three-dimensional appearance. Misleading aspects of a single image, such as increased scattering at the thicker center of the sample and artifacts resulting from high magnification, are removed by stereo pairing. Their results led them to interpret nanoclusters as interlocking lattices with calcium bridging, hydrogen bonding, and hydrophobic and other interactions (McMahon and Oommen. 2008). Further developments in microscopic instrumentation and techniques may be required to determine how casein micelles are assembled.

6.4 Bacteria and Somatic Cells

Pasteurization removes potentially harmful bacteria from milk. Milk is sterile inside an uninfected udder but is an excellent medium for bacterial invasion outside of it. Various microscopic studies have uncovered the structure of bacteria that tend to contaminate milk. For example, Johne *et al.* (1989) used light microscopy and TEM to examine *Staphylococcus aureus* in cows with mastitis, finding that the bacteria secreted exopolysaccharides that assisted with adherence.

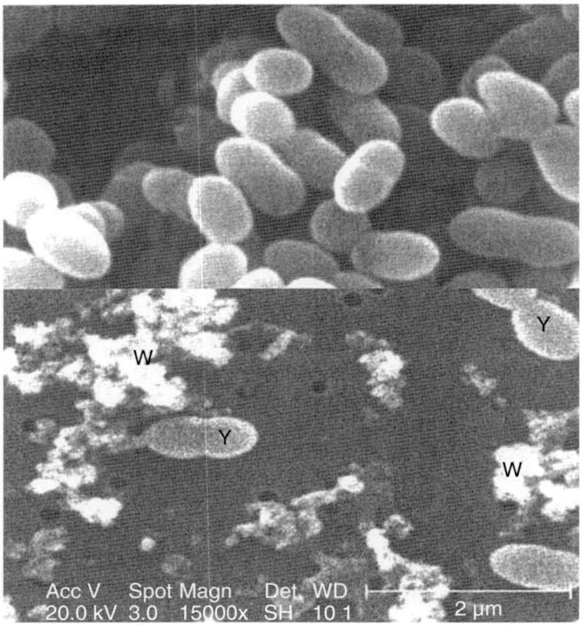

Figure 6.14 Scanning electron micrographs of *Lactococcus lactis* in saline solution (top) before and (bottom) after treatment by pulsed electric field. Key: W, changed cell structure; Y, apparently unchanged cell. From Michalac *et al.* (2003), reprinted with permission from the *Journal of Food Processing and Preservation*, copyright 2003 Wiley.

Raw and UHT bovine milk inoculated with *Bacillus cereus, Lactococcus lactis,* and *Pseudomonas fluorescens* was subjected to pulsed electric field (PEF) treatment by Michalac *et al.* (2003). The protein content, pH, electrical conductivity, viscosity, density, color, and particle size of the milk were unaffected by PEF, but SEM showed that many of the three bacteria tested experienced cell membrane breakdown and loss of structure (Figure 6.14), an indication that PEF will inactivate some spoilage bacteria without altering milk quality.

Somatic cells are white blood cells generated by the animal to ward off bacterial infection and may also include secretory cells, especially in caprine milk. Somatic cell count (SCC) is commonly used to evaluate the quality of milk. SCC, measured as number of cells per milliliter of milk, varies with stage of lactation, breed, season, and other noninfectious factors, but increases if a bacterial infection is present in the mammary gland. A typical SCC in milk is $<100{,}000\ mL^{-1}$. The International Dairy Federation recommends methylene blue staining and direct microscopy for estimation of SCC, and a DNA-specific methyl green-pyronin Y stain is available (Raynal-Ljutovac *et al.*, 2007). Live somatic cells will fluoresce under ultraviolet light if the milk is membrane filtered and the cells are stained with acridine orange. Epifluorescent electron microscopy has been used to visualize and enumerate somatic cells (Pettipher *et al.*, 1980).

6.5 Concentrated Milk

SEM and TEM studies of concentrated milk were conducted by Vélez-Ruiz *et al.* (1998) and Vélez-Ruiz and Barbosa-Cánovas (2000). The microstructure was affected by the solid content and the time-temperature relationship during homogenization

(Figure 6.15). Large complexes of fat globules and protein were formed after evaporation, and the casein micelles dissociated into many small protein particles. The MFGM became thicker as the percentage of solids increased, the casein micelles took on more irregular shapes, and the fat globules exhibited a greater interaction with the micelles. The authors attributed the slower flow of concentrated milk to the increase in solids content and the formation of the fat-protein complexes (Vélez-Ruiz and Barbosa-Cánovas, 2000).

High hydrostatic pressure processing of concentrated milk caused the casein to disintegrate and micelles to fuse, producing a relatively uniform gel. The casein network of concentrated milk containing 39.0% solids was denser than those of 12.9% and 26.1% solids samples, and the size of the fat globules was considerably reduced (Vélez-Ruiz *et al.*, 1998).

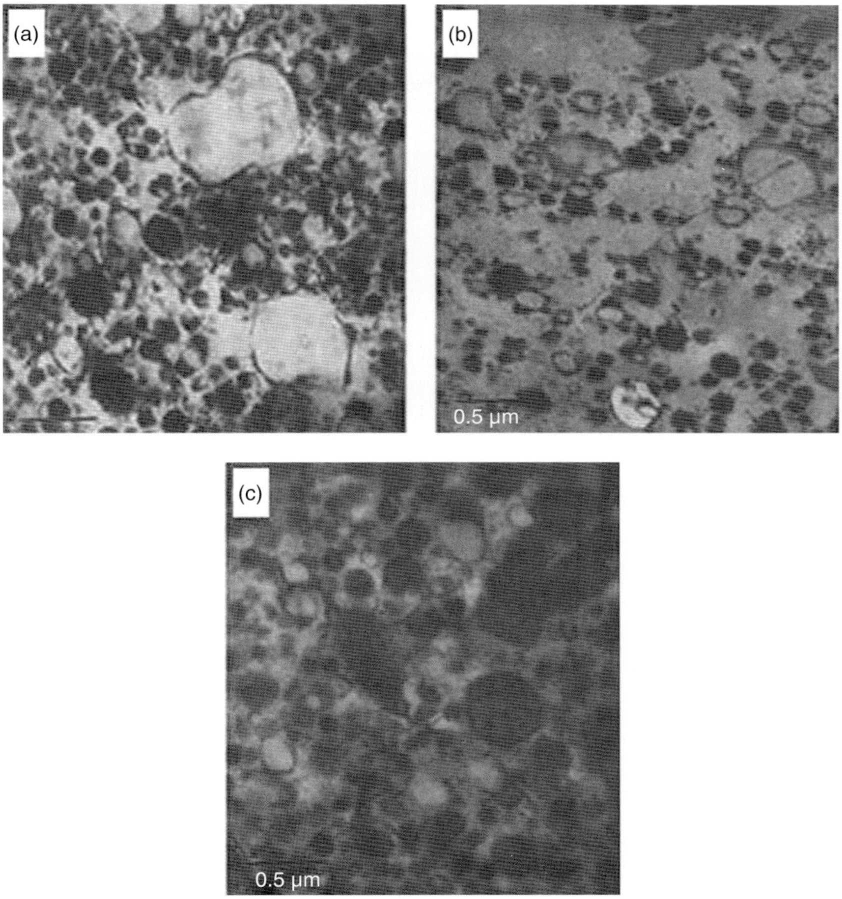

Figure 6.15 Transmission electron micrographs of concentrated milk at 30,000×, showing casein micelles (black) and fat globules (white enclosed structures). (a) 14.6% solids; (b) 26.6% solids; (c) 45.3% solids. Adapted from Vélez-Ruiz and Barbosa-Cánovas (2000), reprinted with permission from *LWT-Food Science and Technology*, copyright 2000 Wiley.

6.6 Digested Milk

Humans digest food with gastric, pancreatic, and intestinal enzymes, releasing the nutrients so they can be absorbed. In laboratory studies on dairy foods, gastric enzymes may be simulated with pepsin diluted in HCl, pancreatic enzymes by pancreatin (a mixture of amylase, pepsin, and trypsin), and intestinal enzymes by bile extract (Hur *et al.*, 2011). When using simulated gastric fluid, Ye *et al.* (2011) observed flocculation of fat globules within 10 min (Figure 6.16). They attributed this effect to aggregation of casein micelles, which either linked the globules or trapped them in casein aggregates.

Our laboratory has researched the effects of processing – pasteurization, UHT, and homogenization – on digestibility of milk. Undigested raw milk (Figure 6.17a) consisted of fat globules (stained red), 1–10 μm in diameter, in an uninterrupted protein matrix (stained green). Digestion of homogenized milk reduced the diameter of fat globules while limiting their size distribution. The protein matrix broke into pieces in simulated gastric fluid (pepsin in HCl; Figure 6.17b) and was mostly disintegrated before the addition of simulated intestinal fluid (bile extract and pancreatin), allowing the fat droplets to aggregate (Figure 6.17c). The fat globules and remaining protein combined into small

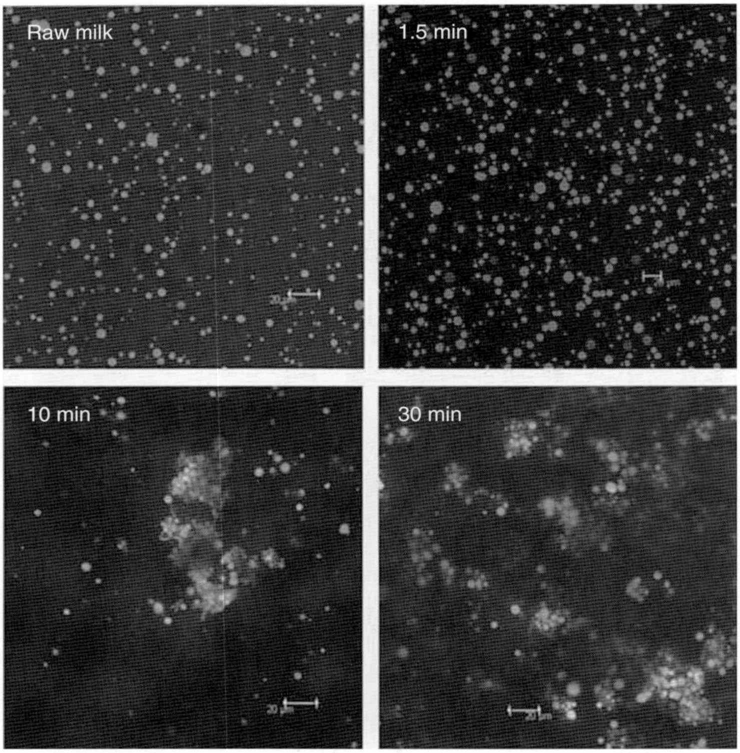

Figure 6.16 Confocal laser scanning micrographs of raw bovine milk 0, 1.5, 10, and 30 min after addition of simulated gastric fluid. Fat appears as red/orange and protein as green. Bars = 20 μm. From Ye *et al.* (2011). Reprinted from *Journal of Dairy Science*, copyright 2011, with permission from Elsevier. (*See color plate section for the color representation of this figure.*)

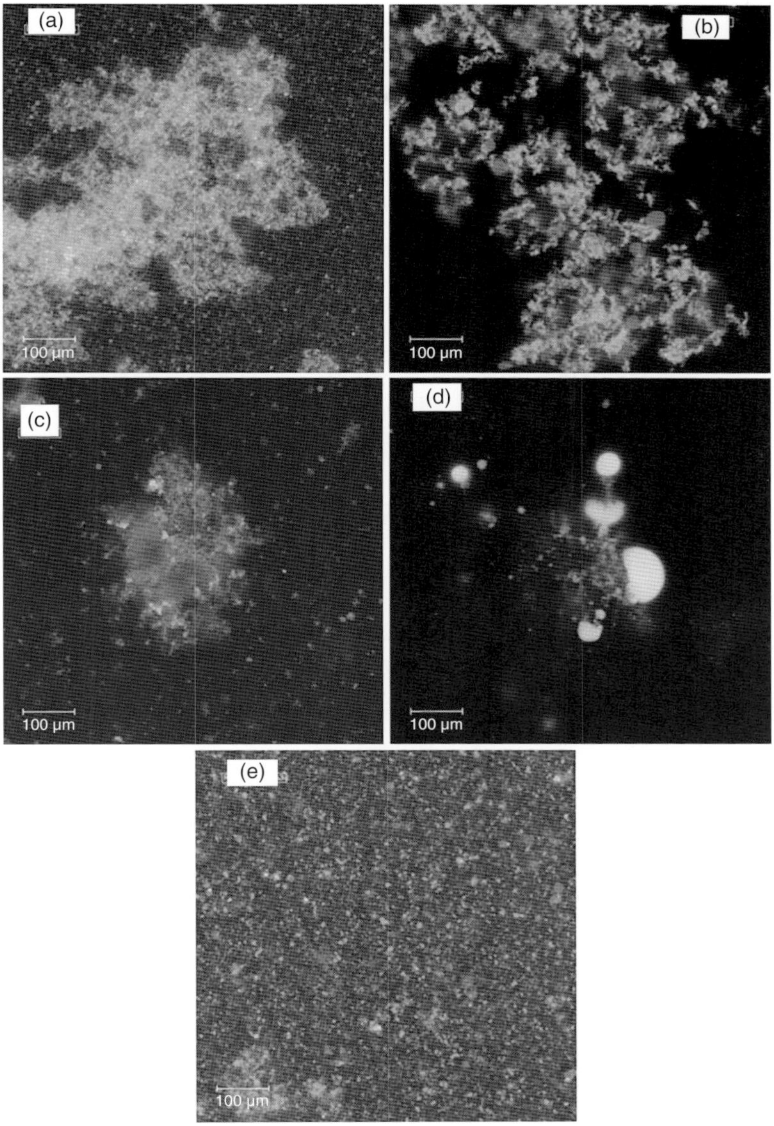

Figure 6.17 Confocal laser scanning micrographs of homogenized bovine milk during simulated digestion. (a) Raw milk before digestion; (b) 60 min after addition of simulated gastric fluid; (c) immediately after addition of simulated intestinal fluid; (d) 120 min later; (e) homogenized and ultra-high temperature pasteurized milk before digestion. Fat appears as red/orange/yellow and protein as green. Bars, 10 μm. (*See color plate section for the color representation of this figure.*)

structures of yellow appearance during 120 min of intestinal digestion (Figure 6.17d). Homogenization and UHT of the milk before digestion caused the protein matrix to quickly disintegrate (Figure 6.17e), leading to somewhat evenly-spaced pieces of protein with fat globules distributed throughout. The microstructures during the subsequent digestion were similar to those of undigested milk, with the formation of fat-protein complexes.

6.7 Conclusion

Light microscopy, SEM, TEM, and CLSM have all been employed to investigate the structural characteristics of milk. Species, treatment, and other factors affect the appearance of fat, protein, and bacteria milk under the microscope. Even today, aspects such as the construction of casein micelles are not completely understood and products such as buffalo milk have not been examined thoroughly. In the coming years, current and perhaps newer techniques will answer the questions still being posed about milk.

Acknowledgments

The author thanks Joseph Uknalis for Figure 6.17, Dina Grimes for processing the illustrations, and Diane L. Van Hekken for helpful comments. Mention of trade names or commercial products in this publication is solely for the purpose of providing specific information and does not imply recommendation or endorsement by the U.S. Department of Agriculture. USDA is an equal opportunity provider and employer.

References

Ahmad, S., Piot, M., Rousseau, F., Grongnet, J.F. and Gaucheron, F. (2009) Physico-chemical changes in casein micelles of buffalo and cow milks as a function of alkalinisation. *Dairy Science and Technology* **89**, 387–403.

Ali, M.K. and Robinson, R.K. (1985) Size distribution of casein micelles in camels' milk. *Journal of Dairy Research* **52**, 303–307.

Carroll, R.J., Basch, J.J., Phillips, J.G. and Farrell, H.M., Jr. (1985) Ultrastructural and biochemical investigations of mature human milk. *Food Microstructure* **4**, 323–331.

Clark, S. and Sherbon, J.W. (2000) Alpha$_{s1}$-casein, milk composition and coagulation properties of goat milk. *Small Ruminant Research* **38**, 123–134.

Dalgleish, D.G. and Corredig, M. (2012) The structure of the casein micelle of milk and its changes during processing. *Annual Review of Food Science and Technology* **3**, 449–467.

Dalgleish D.G., Spagnuolo, P.A. and Goff, H.D. (2004) A possible structure of the casein micelle based on high-resolution field-emission scanning electron microscopy. *International Dairy Journal* **14**, 1025–1031.

De Wit, J.N. (1998) Nutritional and functional characteristics of whey proteins in food products. *Journal of Dairy Science* **81**, 597–608.

El-Zeini, H.M. (2006) Microstructure, rheological and geometrical properties of fat globules of milk from different animal species. *Polish Journal of Food and Nutrition Sciences* **15**, 147.

Evers, J.M., Haverkamp, R.G., Holroyd, S.E., Jameson, G.B., Mackenzie, D.D. and McCarthy, O.J. (2008) Heterogeneity of milk fat globule membrane structure and composition as observed using fluorescence microscopy techniques. *International Dairy Journal* **18**, 1081–1089.

Farah, Z. (1993) Composition and characteristics of camel milk. *Journal of Dairy Research* **60**, 603–626.

Farrell, H.M., Jr., Malin, E.L., Brown, E.M. and Qi, P.X. (2006) Casein micelle structure: What can be learned from milk synthesis and structural biology? *Current Opinion in Colloid and Interface Science* **11**, 135–147.

Farrell, H.M., Jr. and Thompson, M.P. (1988) Caseins as calcium binding proteins. In: *Calcium Binding Proteins.* M.P. Thompson (ed.) CRC Press, Boca Raton, FL, pp. 117–137.

Fox, P.F., Guinee, T.P., Cogan, T.M. and McSweeney, P.L.H. (2000) *Fundamentals of Cheese Science*, pp. 45-53. Aspen Publishers, Gaithersburg, MD.

Gallier, S., Gragson, D., Jiménez-Flores, R. and Everett, D. (2010) Using confocal laser scanning microscopy to probe the milk fat globule membrane and associated proteins. *Journal of Agricultural and Food Chemistry* **58**, 4250–4257.

Glantz, M., Håkansson, A., Lindmark Månsson, H., Paulsson, M. and Nilsson, L. (2010) Revealing the size, conformation, and shape of casein micelles and aggregates with asymmetrical flow field-flow fractionation and multiangle light scattering. *Langmuir* **26**, 12585–12591.

Hillbrick, G.C., McMahon, D.J. and McManus, W.R. (1999) Microstructure of indirectly and directly heated ultra-high-temperature (UHT) processed milk examined using transmission electron microscopy and immunogold labelling. *LWT-Food Science and Technology* **32**, 486–494.

Holt, C. (2004) An equilibrium thermodynamic model of the sequestration of calcium phosphate by casein micelles and its application to the calculation of the partition of salts in milk. *European Biophysics Journal* **33**, 421–434.

Horne, D.S. (1998) Casein interactions: casting light on the black boxes, the structure in dairy products. *International Dairy Journal* **8**, 171–177.

Hur, S.J., Lim, B.O., Decker, E.A., and McClements, D.J. (2011) *In vitro* human digestion models for food applications. *Food Chemistry* **125**, 1–12.

Johne, B., Jarp, J. and Haaheim, L.R. (1989) *Staphylococcus aureus* exopolysaccharide in vivo demonstrated by immunomagnetic separation and electron microscopy. *Journal of Clinical Microbiology* **27**, 1631–1635.

Kalab, M. (1985) Microstructure of dairy foods. 2. Milk products based on fat. *Journal of Dairy Science* **68**, 3234–3248.

Knoop, A.M., Knoop, E. and Wiechen, A. (1979) Sub-structure of synthetic casein micelles. *Journal of Dairy Research* **46**, 347–350.

Law, A.J.R., Leaver, J., Felipe, X., Ferragut, V., Pla, R. and Guamis, B. (1998) Comparison of the effects of high pressure and thermal treatments on the casein micelles in goat's milk. *Journal of Agricultural and Food Chemistry* **46**, 2523–2530.

Lopez, C. (2005) Focus on the supramolecular structure of milk fat. *Reproduction Nutrition Development* **45**, 497–511.

McMahon, D.J. and Oommen, B.S. (2008) Supramolecular structure of casein micelles. *Journal of Dairy Science* **91**, 1709–1721.

Michalac, S., Alvarez, V., Ji, T. and Zhang, Q.H. (2003) Inactivation of selected microorganisms and properties of pulsed electric field processed milk. *Journal of Food Processing and Preservation* **27**, 137–151.

Morr, C.V. (1967) Effect of oxalate and urea upon ultracentrifugation properties of raw and heated skim-milk casein micelles. *Journal of Dairy Science* **50**, 1744–1751.

Nitschmann, H. (1949) Elektronenmikroskopische Grössenbestimmung der Calciumcaseinatteilchen in Kuhmilch. *Helvetica Chimica Acta* **32**, 1258–1264.

Ong, L., Dagastine, R.R., Kentish, S.E. and Gras, S.L. (2010) The effect of milk processing on the microstructure of the milk fat globule and rennet induced gel observed using confocal laser scanning microscopy. *Journal of Food Science* **75**, E135–E145.

Pettipher, G.L., Mansell, R., McKinnon, C.H. and Cousins, C.M. (1980) Rapid membrane filtration-epifluorescent microscopy technique for direct enumeration of bacteria in raw milk. *Applied and Environmental Microbiology* **39**, 423–429.

Park, Y.W., Juárez, M., Ramos, M. and Haenlein, G.F.W. (2007) Physico-chemical characteristics of goat and sheep milk. *Small Ruminant Research* **68**, 88–113.

Pierre, A., Michel, F. and Le Graet, Y. (1995) Variation in size of goat milk casein micelles related to casein genotype. *Lait* **75**, 489–502.

Qi, P.X. (2009) Studies of the biological function and structure of casein micelles, and future implications. In: *Dairy-Derived Ingredients: Food and Nutraceutical Uses.* M. Corredig (ed.) CRC Press, Boca Raton, FL, pp. 147–169.

Raynal-Ljutovac, K., Pirisi, A., De Crémoux, R. and Gonzalo, C. (2007) Somatic cells of goat and sheep milk: Analytical, sanitary, productive and technological aspects. *Small Ruminant Research* **68**, 126–144.

USDA, ARS (2014) *USDA National Nutrient Database for Standard Reference, Release 27, [Online]*, Available: http://www.usda.ars.gov/ba/bhnrc/ndl.

Vélez-Ruiz, J.F. and Barbosa-Cánovas, G.V. (2000) Flow and structural characteristics of concentrated milk. *Journal of Texture Studies* **31**, 315–333.

Velez-Ruiz, J.F., Swanson, B.G. and Barbosa-Canovas, G.V. (1998) Flow and viscoelastic properties of concentrated milk treated by high hydrostatic pressure. *LWT-Food Science and Technology* **31**, 182–195.

Welsch, U., Buchheim, W., Schumacher, U., Schinko, I. and Patton, S. (1988) Structural, histochemical and biochemical observations on horse milk-fat-globule membranes and casein micelles. *Histochemistry* **88**, 357–365.

Ye, A., Cui, J. and Singh, H. (2011) Proteolysis of milk fat globule membrane proteins during in vitro gastric digestion of milk. *Journal of Dairy Science* **94**, 2762–2770.

Zou, X.Q., Guo, Z., Huang, J.H., Jin, Q.Z., Cheong, L.Z., Wang, X.G. and Xu, X.B. (2012) Human milk fat globules from different stages of lactation: a lipid composition analysis and microstructure characterization. *Journal of Agricultural and Food Chemistry* **60**, 7158–7167.

7

Microstructure of Cheese Products

Bhavbhuti M. Mehta

Dairy Chemistry Department, SMC College of Dairy Science, Anand Agricultural University, Anand, Gujarat, India

7.1 Introduction

Milk is one of the important food sources especially for the nourishment of human beings. Due to its high moisture content (~84–87%), milk deteriorates very rapidly. Therefore, since ancient times, people have used various techniques to preserve milk and milk solids. Cheese is one of the techniques used to enhance the shelf-life of milk. A great variety of cheeses has been consumed by human beings. In spite of differences in appearance, texture and flavor, all have in common at least several initial steps in cheese preparation that is the coagulation of milk by a bacterial starter culture and a proteolytic enzyme (e.g. rennet), and the removal of whey from curd (Kalab, 1977). Cheese is a complex soft matter. Cheese is composed of a protein matrix that entraps fat, water, minerals and dispersed bacterial colonies and their metabolites. The formation and changes in the structure of cheese begin from the vat during manufacture, and continue throughout ripening and cold storage (Burdikova *et al.*, 2015).

In Cheddar cheese, the drained curd particles are left on the bottom of a heated jacketed vat and the curd particles start to fuse with each other due to a high fat content and the heat. The scaled curd is cut into slabs which are piled on top of each other. The initially granular curd forms a fibrous texture in the top slabs, as the curd tends to slowly flow down. By reversing the position of the slabs in the pile, all slabs are subjected to this flow and this process in known as cheddaring. The curd is milled/cut into pieces followed by the addition of salts, and then pressed into a block for subsequent ripening (Kalab, 1977). Under electron microscopy, individual aggregated and fibrous forms of casein micelles gradually change to a uniform mass of partially degraded protein. The partly ruptured fat globule membrane with aggregated fat globules, the distribution of lactic acid bacteria, and small crystals of calcium lactate and phosphate are also observed. However, the curd granule surfaces contain less fat and more protein than the interior parts of the granules (Kimber *et al.*, 1974; Brooker *et al.*, 1975; Kalab, 1977).

Cheese is obtained by the precipitation of casein, either by using enzymes, action of lactic acid bacteria or direct acidification to the milk. Then, the curd which is so obtained can be consumed directly or after maturation. Depending upon the type of

Microstructure of Dairy Products, First Edition.
Edited by Mamdouh Mahmoud Abdel-Rahman El-Bakry, Antoni Sanchez, and Bhavbhuti M. Mehta.

coagulants and the extent of maturation, the microstructure of the cheese will differ. The structure of the curd granule is considered as macrostructure, whereas structural components that can only be seen by microscopy are called microstructure (Kalab, 1979). In general, the microstructure of cheese has a complex arrangement of protein matrix in which fat and other constituents are entrapped. Cheese microstructure is the spatial arrangement of the protein (mainly casein micelles) that join together into clusters and chains to form a viscoelastic protein network throughout which moisture, fat globules, minerals and bacteria are dispersed (Kalab *et al.*, 1995; Lucey *et al.*, 2003). To create a desirable texture, understanding the microstructure of cheese is necessary. Microstructure is one of the major controlling factors of functional properties and textures such as firmness, elasticity, crumbliness, cohesiveness, and so on of cheese. The functional properties of cheese depend on how cheese structure is produced and develops, how they interact and change during ripening and requires understanding of where the components of cheese are located in relation to each other (McKenna, 2003). They also affect the physico-chemical, transport and nutritional properties of cheese (Aguilera and Stanley, 1999; Joshi *et al.*, 2004; Impoco *et al.*, 2007).

Various techniques like scanning electron microscopy (SEM), transmission electron microscopy (TEM) and light microscopy are considered as older ones, while confocal laser scanning microscopy (CLSM) is considered as newer a one; they are used to study microstructure of different cheeses. The individual components such as casein micelles and fat globules, casein–fat interactions, state of water (free, bulk, or bound) as well as changes which these components undergo either alone or by interactions with each other or with additional ingredients such as stabilizers, thickeners, starter culture, etc., during the manufacture process are studied using these techniques (Everett and Auty, 2008). SEM provides detailed topographic images of the cheese surface. SEM differentiates fats, proteins, salt crystals and individual bacterial cells/colonies. Standard electron microscopy cannot provide complete 3D information. CLSM with proper fluorescence staining provides three-dimensional visualization of components as well as showing live and dead bacteria (El-Bakry and Sheehan, 2014; Burdikova *et al.*, 2015). However, to obtain detailed information of individual lipid constituents, distribution of trace components and calcium lactate crystals, etc, advanced microscopic techniques are needed. Recently, Burdikova *et al.* (2015) used advanced microscopic techniques like confocal Raman spectroscopy, fluorescence lifetime imaging (FLIM) and non-linear microscopy (two-photon excitation and second harmonic generation imaging microscopy) to study localized physico-chemical properties of cheese matrices.

7.2 Factors Affecting the Development of Microstructures in Cheeses

The structure of cheese is a complex. Moreover, within the same variety of cheese, differences in the structure of cheese are observed. The differences depend on compositional factors and their changes during processing steps and storage conditions (Lawrence *et al.*, 1987; Watkinson *et al.*, 2001). In order to understand what decides the microstructure of cheese, it is necessary to have an understanding of the physico-chemical changes that occur during cheese processing and ripening. The basic steps in cheese manufacturing are: acidification; coagulation; dehydration (cutting, cooking, stirring, pressing, salting); shaping (molding, pressing); and salting. In some varieties of cheese, ripening also has to be carried out. Basically cheese making is a dehydration process in which six to twelve times concentrations of fat and casein are obtained, depending on the variety of cheese to be prepared. The amount of water is regulated and monitored by the various steps during cheese manufacture. The presence of moisture, salt content and microflora regulate the biochemical changes that can be occurred during ripening stage (Holsinge *et al.*, 1995).

The natural cheeses are prepared by the use of rennet enzyme to coagulate the casein micelles and acid is produced by the addition of starter culture into milk. Formation of gel requires a reasonable time. Rennet acts on κ-casein, reduces the steric and charge repulsion, and casein micelles are aggregated into clusters that form a system-spanning network surrounding the fat globules. Serum is also entrapped between and/or within the pores of these aggregates. When the gel has attained enough firmness, it is cut using knives and most of the serum is lost in the form of whey. However, some of the serum still remains in casein aggregates. To increase the rate of syneresis of the curd particles, cooking/heating with continuous stirring of the curd is carried out, and then it is subsequently held in the vat. This facilitates the collisions between the curd particles and increases the syneresis. The curd particles fuse together until it is prevented by stirring in the vat (e.g. in stirred curd-type cheese). In the cheddaring stage, more fusion occurs due to lower pH (<6.0) values. With a further decrease in the pH to <5.8, a fibrous or chicken breast type of structure is developed. This is mainly due to solubilization of colloidal calcium phosphate (CCP) cross-links and the increased mobility of casein. The curd is then cut into small pieces and dry salting is applied; they are then hooped followed by pressing. The pressing helps the formation of bonds between the curd pieces. The fat globules do not remain in spherical form, and partly coalesce due to deformation during molding and pressing (Lucey, 2007).

Recently, Ong *et al.* (2017) have shown the development of typical the microstructure in the initial rennet coagulum in Cheddar cheese, observed using confocal laser scanning microscopy depicted in Figure 7.1.

Traditionally Mozzarella cheese was made using buffalo milk in Italy. Now it is also made from cow's milk in many parts of the world. To obtain desirable functional properties in Mozzarella cheese, curd is heated and stretched. This is known as the pasta filata process. In Mozzarella, a pasta filata type of cheese, the curd undergoes a cooking and stretching process. The stretching process is carried out in hot water (>70°C). These processes give a plastic appearance to the curd and a fibrous structure. The pool of serum and fat are present in the fibrous structure. However, as aging progresses, the cheese becomes softer, less fibrous (but more porous) and more flowable when melted. The amount of serum decreases due to proteolysis and solubilization of calcium, whereas the release of free oil increases (Kiely *et al.*, 1993; Lucey, 2007). Processed cheese is prepared by heating and mixing the blends of natural cheese with emulsifying salts (mostly citrates and sodium phosphates). The emulsifying salts have the ability to chelate calcium and disperse caseins. The dispersed caseins are present in the surrounding fat globules. This helps in the reduction in formation of free oil when processed cheeses are reheated in other food preparations (Lucey, 2007).

The various factors responsible for the formation of microstructure in cheeses are described below.

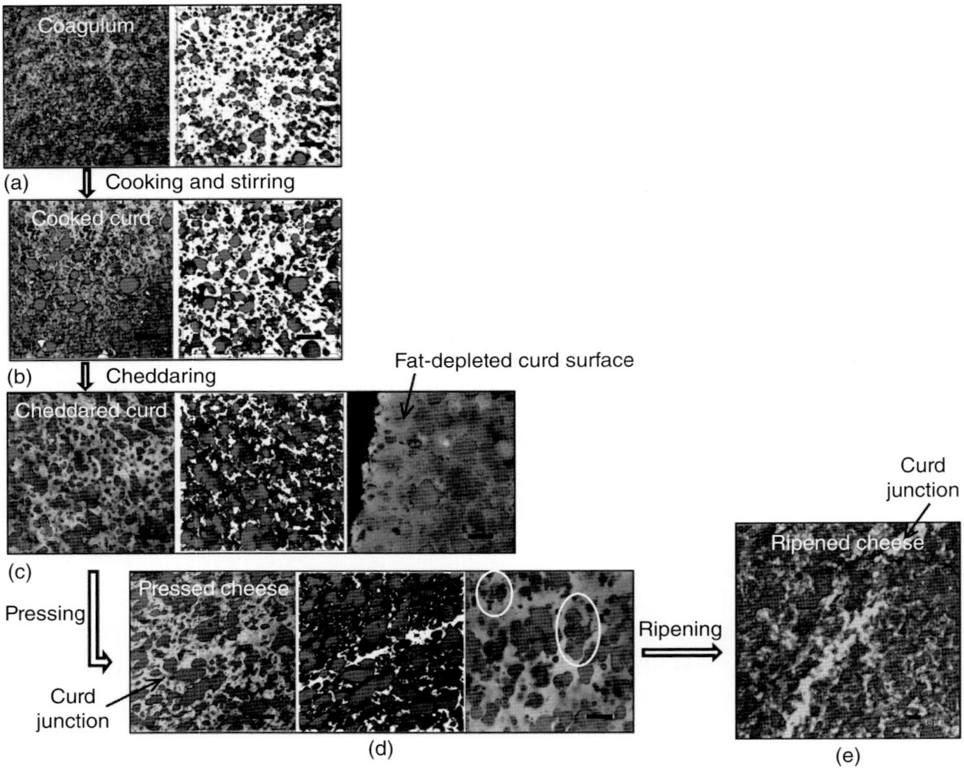

Figure 7.1 Development of microstructure in the initial rennet coagulum through to Cheddar cheese observed using confocal laser scanning microscopy (CSLM). (a) Rennet coagulum; (b) cooked curd; (c) cheddared curd; (d) pressed cheese; and (e) aged Cheddar cheese. Nile red-stained fat appears *red*, fast green FCF-stained protein appears *green*, and the round spots highlighted by the circle in the pressed cheese indicate bacteria. The scale bar is 10 μm in length. Ong *et al.* (2017): Copyright © 2017 by Elsevier Ltd. Reprinted/Reproduced with permission. (*See color plate section for the color representation of this figure.*)

7.2.1 Addition of Calcium Chloride

To influence the coagulation process and increase cheese yield, calcium chloride in the range of 0–0.2 (or as high as 0.5) g/kg of milk is added into the cheese milk. Calcium also affects the rate of proteolysis in cheese. The low calcium and phosphorous content causes higher residual rennet activity. The extent of removal of solubilized calcium and phosphate in whey during cheese making is generally depend on pH of whey draining. The whey draining typically occurs at a pH of 6.2–6.3 for Cheddar cheese. The crumbly texture was obtained in Cheshire cheese with a low draining pH (Lucey and Fox, 1993). The greater proteolysis and reduction in springiness was found in Mozzarella cheese upon lower of draining pH from 6.40 to 6.15 (Yun *et al.*, 1995). Ong *et al.* (2013) studied the effect of calcium chloride addition (200–600 mg/L) on the microstructure and composition of Cheddar cheese. The gel was less porous; cooked curd has denser protein network and retention of more fat observed when 300 or 600 mg/L calcium chloride added. The cheddared curd and cheese contained more micro-pores than cheeses with lower or no calcium addition. These micro-pores were considered possible channels by which fat escaped during pressing. Soodam *et al.* (2015) studied the effect of calcium chloride addition and draining pH on the microstructure pf full fat Cheddar cheese during ripening using both confocal laser scanning microscopy and cryo-scanning electron microscopy. They added varies levels of calcium chloride (0, 100 or 300 mg/kg milk) into the milk. For the cheese prepared without additional calcium at a draining pH of 6.0, the cheese porosity increased with ripening time. The number of protein vertices in the microscopy images declined, indicative of protein solubilization. However, these changes became less significant upon further increase in the amount of $CaCl_2$.

7.2.2 Rennet Coagulation

The first step in the development of protein matrix in the cheese is the rennet coagulation. The calf chymosin specifically hydrolyzes the Phe_{105}–Met_{106} bond in κ-casein. This type of Cheddar cheese is a more compact, sponge-like protein matrix. The proteases from other sources like chickens, pigs, plants, bacteria and fungi have some impact on cheese microstructure. This type of Cheddar cheese has larger cavities and a more fibrous protein matrix and softer curd with less cheese yield (Eino *et al.*, 1976b). Eino *et al.* (1976b) studied the comparison of microstructures of Cheddar cheese curd manufactured with calf rennet, bovine pepsin, and porcine pepsin. They observed that the aggregated protein materials in curd formed with rennet were fused into large compact masses forming a sponge like structural network, whereas the protein materials in curd produced with both bovine pepsin and porcine pepsin formed a fibrous and more open framework. The SEM micrographs of the curd made with rennet display the final stage of gel formation, whereas the fibrous structure of the curds made with both bovine and porcine pepsin may show incomplete gel formation. The formation of the loose and fibrous structure produced with bovine and porcine pepsin might be due to differences in the nature and proteolytic specificities of the pepsins as compared to chymosin (Eino *et al.*, 1976b).

Rennet induced cheese curd comprises of highly concentrated matrix comprising of (a) fused *para*-casein micelles; (b) fat in the form of globules, coalesced globules and/or pools are encased within the casein network; (c) dissolved solutes like soluble salts, lactose, enzymes, peptides, lactic acid entrapped within *para*-casein micelles; and (d) starter and non-starter lactic acid bacteria, located mainly at the fat-protein interface (Laloy *et al.*, 1996; Fox *et al.*, 2017).

The polymer network of fused *para*-casein molecules have hydrophobic cross-links of uncharged amino acid residues as well as calcium bridges formed by divalent calcium ions which bind to glutamate, aspartate and serine phosphate groups. Protein concentrations are responsible for a degree of aggregation as well as volume fraction of *para*-casein particles in the network. The Cheddar, Emmental and Edam contain relatively high protein and hence the volume fraction is higher as compared to Blue and Feta cheese. The residues of chymosin and plasmin hydrolyze the *para*-casein molecules leads to softening of the cheese during ripening (Fox *et al.*, 2017).

7.2.3 Acid-Coagulation

In acid coagulate cheese; fat globules and moisture are entrapped in a continuous protein network, similar to rennet coagulated cheese. However, microstructure of acid cheese differs from rennet induced cheese. In acid cheese (a) the polymer network is comprised of casein (instead of calcium phosphate *para*-casein); (b) the lower calcium bridge is due to solubilization of colloidal calcium phosphate at low pH (4.5 to 4.8) and drains along with the cheese whey; (c) there is a lower degree of fusion and volume fraction of casein particles due to high moisture and less casein concentration; and (d) fat globules are generally less coalesced compared to rennet curd cheese (Fox *et al.*, 2017).

7.2.4 Coagulation Temperature

The coagulation of milk is one of the first steps in the manufacture of cheese. The coagulation temperature, duration of coagulation, type of coagulants and concentration of coagulants affect the final quality of cheese. Depending on the type of cheese being produced, the coagulation temperature varies from 20°C to 40°C. The coagulation temperatures used for Cheddar cheese manufacture range from 31°C to 33°C. The cleavage of κ-casein by rennet is faster due to the increase in the coagulation temperature, but the rate of casein micelle aggregation is significantly affected upon change in the coagulation temperature. This reduces the coagulation time. However, too high coagulation temperature (>36°C) inactivates the rennet and coagulation time increases. The way in which casein micelles are aggregated and rearrange themselves in the gel affect the microstructure of the cheese (Euston *et al.*, 2002; Ong *et al.*, 2011).

Euston *et al.* (2002) used TEM and found that the increase in concentration of rennet and coagulation temperature from 25°C to 35°C produced larger and more densely packed aggregates in the network structure of Feta-, Danbo-, Gouda- and Camembert-type cheeses. A similar study was also conducted by Wium *et al.* (2003) in Feta cheese using TEM. They observed that increasing the chymosin amount and increasing the coagulation temperature from 25°C to 35°C, enhanced the rearrangement of casein networks leading to compact casein micelles aggregates and a coarse protein network structure in Feta cheese.

The plant coagulants and microbial rennet were used to produce skim milk gel over different coagulation temperature (from 25°C to 35°C) and characterized by CLSM. The skim milk gel prepared from plant coagulants were less influenced by the changes in coagulation temperature as compared to microbial rennet suggesting use of plant origin coagulants in the production of cheeses with a wider range of coagulation temperatures (Esteves *et al.*, 2003). The casein matrix was more compact and smaller size of pores was observed using SEM in low-fat Iranian white cheese formed from milk coagulated at 41.5°C in comparison with that of less temperature (34°C and 37°C) by Madadlou *et al.* (2006).

Ong *et al.* (2011) investigated the effect of different coagulation temperature (27°C, 30°C, 33°C and 36°C) on the microstructure of gel, curd and cheese samples during the manufacture of full fat Cheddar cheese using CLSM. The gel obtained at 27°C has fine interconnected regular protein network whereas an irregular, coarse and less continuous protein network was observed at 36°C. The fat content was higher in cheese when the coagulation temperature was 27°C and 30°C as compared to higher temperatures

(33°C and 36°C). At a higher temperature (36°C), aggregation of casein micelles increased with thicker protein strands, reduced the capability of the cooked curd to retain fat during whey draining, resulting in less occupied fat (Ong *et al.*, 2011).

7.2.5 Syneresis

Syneresis is the process of whey expulsion induced by gel contraction. The casein matrix formed by enzymatic coagulation undergoes restructuring or rearrangement during syneresis, which leads to the formation of a more compact gel structure. The expulsion of whey from the porous casein network is an essential step in cheese making. Syneresis is responsible for controlling the moisture content in cheese as well as whey soluble components like lactose, minerals, β-lactoglobulin and α-lactalbumin. Depending on the variety of cheese, syneresis essentially concentrates the fat and casein of milk by a factor of 6–12. The rate and extent of syneresis play an important key role in structural properties of cheese, regulate the growth of bacteria, activity of enzymes, ripening and finished quality of cheese. Syneresis can occur spontaneously. The syneresis is not only a physical process, but it also occurs as a result of a series of chemical interactions that can be stimulated by cutting and agitation of curd. The gel of aggregated para-casein micelles produced by rennet has large pores that contain whey and fat globules. The surfaces of micelle clusters can form junctions with other micelles and provide a driving force for microsyneresis (Walstra *et al.*, 1985, 2006). Walstra *et al.* (1985) found that the rearrangement of the *para*-κ-casein micelle network is the main cause of syneresis.

Syneresis is highly dependent on temperature. However, factors like composition of milk, pH, amount of added $CaCl_2$, firmness of curd and size of curd grains, as well as mechanical treatment, agitation, stirring, and external pressure also influence the removal of whey during cheese making. Geng *et al.* (2011) worked on dynamic visualization and microstructure of syneresis of cheese curd during mechanical treatment. They have developed a set-up for the dynamic visualization of syneresis of cheese curd during mechanical treatment. The group of authors observed that under quiescent conditions, the degree of curd shrinkage significantly increased with increasing the temperature from 32°C to 40°C. Under TEM of curd grain, they established a relationship between frequency of compression, resultant morphology of paracasein clusters and rate of curd shrinkage.

7.2.6 Salting

All cheeses are salted but the concentration varies markedly. Salt diffuses from the surface to the center in Cheddar type cheese. The dry salt is applied on surface of the Blue variety of cheese. The salt dissolves in the surface moisture, causing the outward movement of moisture. Addition of dry salt causes shrinkage of the curd and a rapid rate of release of whey containing calcium and phosphate within a few minutes of pressing (Lawrence *et al.*, 2004). Salted cheeses have more homogeneous and smaller fissures compared with unsalted cheeses (Madadlou *et al.*, 2007). Higher salt concentration enhancing protein–water interaction causes a swelling of the protein phase in cheese (Everett, 2007).

The health risks (like hypertension, kidney stones, osteoporosis, and many more) associated with increased NaCl consumption is a serious issue for the cheese industry.

Hence, there is a high demand for NaCl-reduced cheeses. However, poor texture and other undesirable properties are associated with a simple reduction of NaCl in cheeses. Therefore, partial substitution of NaCl with other salts (e.g. potassium chloride, KCl) is suggested to be an alternative method to reduce salt in cheese without adverse effects on cheese quality. There was a non-significant effect on the texture of Nabulsi and Halloumi cheeses when prepared from partial substitution of NaCl with KCl (Ayyash *et al.* 2011). Ayyash *et al.* (2013) investigated the effect of partial NaCl substitution with KCl on microstructure of low-moisture Mozzarella cheese. Environmental scanning electron microscopy images showed the compact and homogeneous structure of low-moisture Mozzarella cheese upon storage; however, no significant difference was observed among the cheeses prepared from NaCl or NaCl/KCl mixtures indicating the use of KCl as an alternative salt to NaCl for production of low-moisture Mozzarella cheese.

7.2.7 Ripening

Some cheeses, mainly Acid- and Acid-Rennet-Curd Cheese varieties are consumed fresh. However, most cheese varieties undergo a period of ripening, i.e. curing/maturation. The time length of ripening is inversely related to the moisture content of the cheese. The period of ripening varies from 2 weeks to 2 years depending on the variety of cheese. During ripening, various biochemical changes take place that affect the flavor, aroma and texture of the mature cheese. The various factors like coagulant, indigenous milk enzymes (e.g. lipoprotein lipase and plasmin), starter bacteria and secondary microflora as well as their enzymes, moisture content in cheese, NaCl, pH, residual lactose, ripening temperatures, etc. affect the biochemical changes of cheese. Protein and fat are mainly affected during ripening of cheese. Upon maturation of cheese, partial hydrolysis of *para*-casein takes place, and hydration of *para*-casein increases. Fat globules coalesce and form fat pools in the cheese matrix.

Pierre *et al.* (1999) studied the microstructure of soft goat cheeses (Sainte-Maure variety) at different ripening stages. The cleared differences were observed on cheese microstructures at the beginning and end of the ripening period. In the early stage of ripening the structure of cheese was loose as a result of proteolysis that developed in the aqueous phase. A rapid decrease in the water content occurred due to evaporation at the cheese surface and increased the total solids. The proteolytic rate in the cheeses decreased due to lower content in solvent water, and cheese microstructure became more compact with very few cavities remaining between the strands.

The molecular diffusion is a very important characteristic in the manufacture of certain dairy products. In the preparation of cheese, the coagulation of caseins forms a protein gel structure. The gel entraps the aqueous phase (like minerals and whey proteins), fat, and microbes. The protein gel structure, also known as a protein network, works as a sieve. The gel strands obstruct the diffusing molecules. The denser gel structure leads to greater impediment (Floury *et al.*, 2012; Chapeau *et al.*, 2016). The diffusion phenomena are crucial during the ripening of cheese. Recently Chapeau *et al.* (2016) have for the very first time used the Fluorescence Recovery After Photo bleaching (FRAP) technique to find out the relationships between molecular diffusion and the (soft- and hard-) cheese composition and/or its microstructure throughout ripening. The authors used Camembert-type and Swiss-type cheese in the study. The molecular diffusion is characterized by diffusion coefficients. Diffusion coefficients of a dextran were measured in these cheese varieties with different ripening times. Diffusion of the dextrans was mainly dependent on the fraction of a "free" aqueous phase present in the cheese, which is closely linked to cheese-making technology and the ripening stage. The authors concluded that the findings would be useful in the study of the pore size of the protein network, as well as the finding of microstructural changes during the ripening of different types of cheese.

Cheese has long a shelf-life and it can, therefore, be considered to be an excellent delivery vehicle for green tea (*Camellia sinensis*) antioxidants. Green tea has a phenolic content, especially catechins that can serve as antioxidants. It has medical and nutritional properties. There is growing interest in using green tea catechins (e.g. (+)-catechin) in dairy products (Rashidinejad *et al.*, 2013). Recently Rashidinejad *et al.* (2016) reported the effect of (+)-Catechin on the microstructure of full-fat cheese during ripening. The effect of two concentrations of (+)-catechin (250 and 500 ppm) in free form on the structure of mature full-fat cheese (90 day) was studied by transmission electron microscopy (TEM). The mature cheese without (+)-catechin has a homogeneous pattern with regular spacings of milk fat globules. The composite structures of casein micelles in the control cheese are stabilized by electrostatic and hydrophobic interactions, salt bridges, and hydrogen bonding. The cheese containing (+)-catechin has a different electron-dense microstructure. It has a highly disrupted structure. TEM results showed that (+)-catechin can cause destabilization of MFGM, so that fat is released into the protein matrix, and the microstructure of the cheese fortified with these phenolic compounds appears more like butter than cheese. This supports the association between cheese components and green tea catechins during the ripening period, which may affect the feasibility of cheese as a potential delivery vehicle for these antioxidants as observed by Rashidinejad *et al.* (2016).

7.2.8 Homogenization and High Pressure Treatments

Homogenization of cheese milk reduces the size of fat globules, and produces whiter cheese as well as retaining more moisture in the cheese. The larger, more irregularly shaped and small discrete fat globules are observed in Cheddar cheese when prepared from homogenized milk (Guinee *et al.*, 2000). Homogenization increases the surface area of the fat globules and about 25% of the casein micelles are adsorbed on newly created fat globules which cover some 75% of the surface area of the fat globules. The fat globules, now, act almost as large casein micelles in homogenized milk. Thus, fat globules participate in the casein matrix as copolymers. The adsorption of casein on fat leads to reduction in the density of micellar junction zones in the cheese and hence the casein matrix does not form compact structure and retains more moisture. The rate of aggregation of casein micelles during coagulation is slow. The curd fusion is also poor with a weaker curd structure in homogenized milk (Everett, 2007).

High pressure (HP) for milk has been considered as an alternative or additional technology to thermal treatment. The non-covalent bonds within biological matter are perturbed by HP, whereas amino acids, vitamins and flavor compounds remain unaffected by HP. However, HP may induce denaturation, aggregation and gel formation of the macromolecules (O'Reilly *et al.*, 2001). HP treatment not only inactivates undesired microorganisms and/or enzymes in milk, but also affects the milk constituents themselves and final quality of the ripened cheese. HP alters the physico-chemical properties of casein micelles and whey proteins either by reinforcing or weakening intramolecular bonds (Pereira *et al.,* 2009). The pressurization of milk at 300–400 MPa for 30 min caused a decrease of 7.5–15% in protein loss in whey and an increase of 14–20% in curd weight (Lopez-Fandino *et al.*, 1996). HP treatment alters the gelation time and improves coagulation characteristics. It promotes acceleration of cheese ripening by inducing lipolytic and proteolytic activities. It facilitates release of Ca, P and αs_1- and αs_2-caseins and that can lead to destabilization of micelles (Huppertz *et al.*, 2006; Lopez-Fandino, 2006; Pereira *et al.*, 2009).

Optical and electronic microscopy studies of HP treated milk were carried out by Capellas *et al.* (2001) and Kheadr *et al.* (2002). HP treatments of milk reduce the average size of casein micelles that give a closer-packed structure in cheese and produces firmer curds. The formation of casein–casein and/or casein–fat complexes give a finer structure that strengthens the cheese matrix (Lopez-Fandino *et al.*, 1996; Needs *et al.*, 2000; Kheadr *et al.*, 2002). The protein structure can be changed by high pressure (200–400 MPa for 20 min) that can increase the fluidity and flowability as well as reducing melting time upon heating of the cheese (e.g. Cheddar and Mozzarella). However, structural features of the cheese were not affected when applied pressure was 50–200 MPa with longer processing times, i.e. up to 82 h when studied by confocal laser scanning microscopy (Pereira *et al.*, 2009).

Needs *et al.* (2000) reported the differences in the microstructure of curds formed from untreated and HP-treated (600 MPa) milks. Gels made from untreated milk contained coarser strands of whole micelles and the network appeared with larger interstitial spaces whereas gel obtained from HP-treated milk have a dense network of fine strands, which appeared to be continuous over long distances. These differences were due to disruption of casein micelles and denaturation of β-Lg. O'Reilly *et al.* (2000) used a confocal scanning light microscopy technique to observe the microstructural changes in cheese. They did not find any gross structural changes in commercial Cheddar cheese when exposed to 50 MPa for 3 days at 25°C. Capellas *et al.* (1997) pressurized goats' milk cheese 450 MPa at 10°C for 5–30 min. More continuous/homogeneous protein matrix, fat globules lose their circularity, and larger pools of fat were observed than in non-pressurized cheeses. In post-pressurization of Cheddar cheese curd, a more continuous microstructure was observed even after 35 s of pressure treatment (Torres-Mora *et al.*, 1996).

Saldo *et al.* (2002c) did not assess changes in the microstructure of Garrotxa cheese at 400 MPa for 5 min at 14°C on 3 days post-pressurization and 60 days after pressure treatment upon ripening. However, the changes in cheese color after HHP treatments could be related to differences in microstructure between treated and untreated cheese (Martınez-Rodrıguez *et al.*, 2012). O'Reilly *et al.* (2002b) applied 400 MPa pressure on a low-moisture Mozzarella cheese for 20 min. The HHP treatments accelerated age-related changes in microstructure and protein hydration characteristics. Serrano *et al.* (2004, 2005) used moderate pressure treatments (345 and 483 MPa up to 7 min) on unripened milled or stirred-curd Cheddar cheese. These treatments have modified its microstructure and textural properties, which resulted in accelerated shredability.

7.2.9 Evaporation and Ultrafiltration Treatments

Cheese can be prepared from concentrated milk. The yield of the cheese prepared from unconcentrated milk is only 10% and the remaining 90% is drained out in the form of whey. To eliminate the volume of whey and increase the yield of cheese by incorporating more whey proteins into the cheese, concentrated milk has been tried out. The concentration can be obtained from the evaporation of milk under vacuum and from an ultrafiltration technique. The lactose and soluble minerals become concentrated in evaporated milk and microstructure of reduced-fat Cheddar cheese prepared from such milk becomes rough and rippled with a crumbly texture, indicating incomplete curd fusion (Anderson and Mistry, 1994).

Ultrafiltration is another technique for concentrating the milk solids content. The whey proteins act as an inert filler in the protein network structure that binds additional water and provide a smooth texture to the cheese (Everett, 2007). Upon ultrafiltration of milk, lactose is partially concentrated which can lead to too much lactic acid production during cheese manufacture. Addition of extra water in the retentate overcomes this problem. Everett and Jameson (1993) observed that inert whey proteins disrupt the casein matrix responsible for production of softer Cheddar cheese when made from the ultrafiltration of milk. However, Camembert and Feta have been made successfully using UF milk (Green, 1985). The partial homogenizing effect occurs when milk is pumped across a membrane during UF that reduces the size of fat globules. Casein micelles coated the extra surface area of the fat globules and altered the structure of the cheese matrix (Green *et al.*, 1983). Cheddar cheese manufactured from UF milk contains higher protein and moisture content, but less fat compared to conventionally made Cheddar. Cheddar cheese manufactured from UF milk showed thicker protein fibers, coarser protein network and contained more segregated fat globules (Green *et al.*, 1986; Green, 1987). Similarly a coarse protein network was also found in Urfa, a white-brined cheese from Turkey (Ozer *et al.*, 2003) and Feta cheese when UF milk was used (Wium *et al.*, 2003). Kuo and Harper (2003) found larger void spaces in the protein matrix when Feta cheese was prepared from milk protein concentrate.

Ong *et al.* (2013) studied the microstructure of full fat Cheddar cheese made with ultrafiltered milk retentate. In this study, Cheddar cheeses were made from cheese-milk with or without low concentration factor ultrafiltration retentate (LCUFR) addition using a protein concentration of 3.7%–5.8% w/w. The microstructure of the gel was observed using cryo-SEM. After the addition of rennet, gel was formed and contained an aggregation of casein micelles in which fat globules were entrapped. The authors did not find qualitative differences between the microstructure of the gels made using cheese-milk with ~3.7%–4.8% w/w milk protein, but a denser structure with smaller pores were noticed in the gel which was made using 5.8% w/w milk protein. This might be due to the high concentration of protein responsible for decreasing the mean free distance between casein micelles resulting in a densely aggregated network. The casein micelles particles were fused during cooking (at 38°C) and even some bacterial cells were observed within the cooked curd. The bigger pores were observed in curd which was prepared from unconcentrated cheese-milk as compared to those prepared from higher protein concentration. Ong *et al.* (2013) also analyzed the microstructure of the curd after cheddaring and milling. They noticed the micro-pores of about 1 μm in diameter within the milled curd which was made with a lower concentration of milk protein while smaller micro-pores of diameter (<1 μm) were found within the milled curd made from a higher concentration of milk protein. After pressing, the bigger sizes of fat globules within cheeses were found in cheese that was made from a higher concentration of milk protein rather than unconcentrated milk. The microstructure of cheese from cheese-milk with a higher protein concentration was more compact, consistent with the increased hardness but lower cohesiveness (Ong *et al.*, 2013).

7.2.10 Freezing

To extend the shelf life and arresting physico-chemical changes in Mozzarella cheese during ripening, the frozen storage after its manufacture is of commercial interest. Graiver *et al.* (2004) observed that the formation of ice crystals weakens the protein matrix and creates a porous structure upon freezing of Mozzarella cheese. It is known that fast freezing produces smaller ice crystals than the slow freezing. Oberg *et al.* (1992) found that a slow rate of freezing increases the meltability of low-moisture Mozzarella due to the formation of large ice crystals disrupting the protein matrix. The meltability increases upon thawing of the cheese. Water usually bound to the protein fibers of the matrix separate upon freezing and subsequent thawing. Kuo *et al.* (2003) noticed an increase in water self-diffusion coefficient in frozen-stored pasta filata Mozzarella cheese as compared to unfrozen cheese. Due to freezing, separated water was unable to fully recover and be bound by the proteins. Scanning electron microscopy was used to assess the effect of aging on the protein matrix of pasta filata Mozzarella and non-pasta filata pizza cheeses during frozen storage (Kuo and Gunasekaran, 2009). Pores and ruptures of reticular structure were observed in the frozen-stored pasta filata Mozzarella cheese protein matrix, but cracks and clumps of bacteria were shown in frozen-stored non-pasta filata pizza cheese. However, tempering modified the effect of freezing on the protein matrix of Mozzarella cheeses and that can be used to control the quality of the cheese during storage (Kuo and Gunasekaran, 2009).

7.2.11 Fat Replacers

Consumers are becoming more health conscious to avoid fat in their diet. The balance of saturated to unsaturated fatty acids is important in the diet. The production of reduced and low-fat cheese has significantly increased. Upon reduction of fat in foods, the texture and structural properties are altered. Low-fat natural cheeses have a hard and rubbery texture. Fat has the important role in cheese of imparting discontinuity to the protein matrix, so that when fat is removed, the cheese is composed of a homogenous, dense protein network (Rogers *et al.*, 2010). The partial reduction of milk fat modifies the microstructure because the protein matrix becomes more compact and texture more firm, elastic, and chewy in the reduced-fat cheese compared to the full-fat cheese. Emulsified vegetable oils alter the distribution of the fat droplets into the protein network. Whey proteins used as fat replacers in reduced-fat cheese give a porous, less dense and fine protein network to the cheese and close textural characteristics compared to those of full-fat cheese (Lobato-Calleros *et al.*, 1999, 2003, 2007). The SEM showed that the total or partial substitution of the milk fat by emulsified canola oil and/or whey protein concentrate produced fresh cheeses with different structures than that produced from full-milk fat cheese (Lobato-Calleros *et al.*, 2007). When whey protein concentrate predominated, a dense, compact and continuous protein matrix was produced. In contrast, when emulsified canola oil predominated, a loose, more disrupted protein matrix was formed. The microstructures of these different cheeses exhibited differing textural characteristics.

Brown *et al.* (2012) used laser scanning confocal microscopy to determine the location of various starches in the curd and their interaction with the protein matrix. The starches are used in the production of low-fat cheese. Different starches such as modified waxy corn starch, a waxy rice starch, instant tapioca starch, a modified tapioca starch and dextrin were used. The waxy corn, waxy rice, and instant tapioca starches have improved the texture of low fat cheese. These starches were retained more in the curd and interrupt the protein–protein interactions in the protein matrix network. The modified tapioca starch produced a soft non-cohesive gel and interfered with formation of protein structure of the curd, while dextrin did not disrupt the protein network and was not retained in the curd. The microstructure of fresh full fat Tallaga cheese as control and fresh Tallaga cheeses produced by replacing milk fat with sunflower oil/whey protein concentrate (WPC) were studied using SEM (Abd El-Salam, 2015). The total replacement of fat with WPC showed a compact and close network, while total replacement of milk fat with sunflower oil has a compact network with small uniform oil droplets embedded in the protein matrix as compared to full fat cheese. The full fat cheese exhibited an open protein matrix containing milk fat globules of various sizes and forms. The cheese with 50% sunflower oil and 25% whey protein concentrate was more acceptable than that of control cheese.

Reduced fat Cheddar cheese was produced using different concentrations of microbial rennet as well as recombinant camel chymosin. The rennet concentration affected the gel microstructure. The high concentration of rennet produced a more porous gel as most of the rennet was lost to the whey. The residual rennet in the cheese was significantly higher in the cheeses made from camel chymosin compared to Hannilase (a microbial rennet). At the end of the ripening period, the microstructure of the cheeses studied by cryo-SEM showed a thicker protein network in the cheese made with recombinant camel chymosin (Soodam *et al.*, 2015). When the fat content of *pasta filata* cheese is decreased, fiber formation is lost in low-fat Mozzarella cheese. In this cheese, during the stretching and extruding process, protein strands fuse together which leads to increased protein–protein interaction with loss of strand formation/stringiness. The low fat Mozzarella cheese is, therefore, not suitable for the manufacture of string cheese. Oberg *et al.* (2015) studied the use of various polysaccharides (waxy corn starch, waxy rice starch, instant tapioca starch, polydextrose, xanthan gum, and guar gum) as fat mimetics during the stretching and extruding process to increase the stringiness in low-fat Mozzarella cheese. Among the hot (80°C) slurries of various polysaccharides used in this study, xanthan gum functioned best as a fat mimetic and produced low fat string cheese. The xanthan gum-produced cheese showed longer strings with smaller diameters when compared with other low fat string cheeses. Upon examination of low fat string cheese using confocal microscopy, it was found that the polysaccharides were distributed throughout the cheese matrix in a similar way to fat in low-moisture part-skim Mozzarella cheese (Oberg *et al.*, 2015).

7.3 Microstructures of Various Components in Cheese Matrix

There are in general two levels of cheese structure *viz.* macrostructure and microstructure. The macrostructure is visually observed by the naked eye or at <10× magnification using light microscopy. The macrostructure represents eyes, cracks, slits and curd granule into a molded cheese. The microstructure can only be visualized on a micro-scale at >25× magnification using suitable microscopy. It represents the structure within the smallest piece of the cheese in which various casein, fat, minerals, moisture, soluble salts, lactose, lactic acid, and various peptides are distributed through inter-/intra-molecular attractions (Fox *et al.*, 2017). The arrangement of casein particles, fat globules, dispersed water and mineral components on the micron scale is known as the microstructure of cheese (Ong *et al.*, 2011).

7.3.1 Protein in Cheese Matrix

The protein microstructure in cheese mainly develops from casein micelles which are freely present in the milk after coagulation. The temperature and acidity are the factors that can influence formation of the cheese matrix. During processing, casein micelles begin to fuse together and form a dense matrix in which fat globules, free fat, soluble and casein-bound minerals such as calcium, water particles and sodium chloride fractions are all scattered (Everett and Auty, 2008). The fused protein has dimensions of 1–5 μm, with an opening in the matrix of around 1.5 μm in size in Cheddar cheese (Eino *et al.*, 1976a).

The freshly prepared Mozzarella cheese has a large amount of serum/moisture that can be absorbed by the casein matrix upon aging at about ten days. Due to lower temperature during aging, hydrophobic forces of β-casein are lessen that leads to partial disassociation of it from casein matrix and becomes intact casein in the serum portion (Everett, 2007).The serum channels contain minerals, residual serum proteins, fat globules, bacterial cells and water. The serum channels are located in the interstitial casein matrix. Properties such as melting, stretching, cooking and oiling off during baking are derived primarily from the interaction of casein and fat in the cheese matrix. The microstructure of Mozzarella cheese basically exhibited large fat globules uniformly scattered throughout the compact protein matrix with little aggregation.

During the pasta filata stage, the porosity of fused fibrous casein micelles decreases with gradual loss of micellar identity (Oberg *et al.*, 1993). The simultaneous heating and stretching of Mozzarella curd leads to the formation of smooth protein fibers that can align in the direction of stretching. However protein fibers become rough after cooling in brine (McMahon *et al.*, 1993). Kalab (1977) found differences in the shape of the curd granules when microstructure between conventionally stretched and stirred curd Mozzarella were compared. Stretched Mozzarella had a definite curd orientation with no curd junctions and a parallel orientation of protein fibers. These were lacking in stirred curd Mozzarella (Paquet and Kalab, 1988). The fat globules appeared to be evenly distributed in stirred curd Mozzarella, but not in the stretched one. During the maturation of Mozzarella cheese, proteins become more hydrated and swollen (Everett, 2007). The low ionic calcium levels and high salt are responsible for the swelling of proteins (Joshi *et al.*, 2004). Salts leads to solubilization of intact casein. This phenomenon is known as peptization (Guo *et al.*, 1997). The low level of calcium also enhances the peptizing action of sodium chloride. Paulson *et al.* (1998) observed that directly acidified Mozzarella adsorbs more moisture due to less calcium retention than a cultured cheese.

7.3.2 Fat Globule in Cheese Matrix

Fat provides the feel in the mouth, the characteristic flavor and the texture of the cheese. Fat globules soften the texture by disrupting protein fibrous network in the cheese. However, one-third reduction of fat results in too firm a texture of the Cheddar cheese, with poor a flavor (Banks *et al.*, 1989). The fat globules either bind to the cheese matrix or act as inert filler or as a copolymer within the casein matrix. The larger fat globules are bound to the casein matrix and thereby disrupt the structure. The smaller globules are entrapped within the protein void space (Michalski *et al.*, 2004).

Fat can exist in globular form, or it may exist as non-globular free fat. The globular droplets, enveloped by milk fat globular membrane material (MFGM), range from 0.2 to 15 μm with an average of 4 μm. The MFGM is composed of phospholipids, proteins, glycolipids, glycoprotein, triacylglycerols enzymes, cholesterol and other minor components (Lopez *et al.*, 2007; Lopez and Briard-Bion, 2007; Burdikova *et al.*, 2015; Hickey *et al.*, 2015). The high cooking temperature process in a certain variety of cheeses and subsequent pressing of the curds result in coalescence of the fat globules and fracturing of MFGM. This leads to release of free fat and MFGM materials (Lopez and Briard-Bion, 2007). The phospholipids in the MFGM facilitate fat globules to act as emulsifiers and stabilize the water-in-oil and oil-in-water emulsions within the matrix structure of cheese. Moreover, phospholipids also help fat–protein interactions which are essential for the cheese matrix (Pereira *et al.*, 2009; Gallier *et al.*, 2010).

Recently, Lopez *et al.* (2015) reported the role of technological processes on organization of lipids in milk, infant milk formulas and various dairy products (e.g. cheese). Figure 7.2 shows the microstructure of cheeses showing the effect of dairy technological processes on the organization of lipids.

The fat globules become altered during the manufacture of Emmental cheese. Upon coagulation of milk, fat globules are entrapped in a continuous network of casein and subsequently cooking and pressing produces non-globular free fat. On the other hand, fat is located in the aqueous phase entrapped between the elongated protein fibers formed during stretching of the curd in Mozzarella cheese. The globules are individualized in the casein matrix or aggregated in the mold-ripened soft cheeses. In homogenized cream cheeses, the average size of the lipid droplets is around 1 μm. In these cream cheeses, there are interactions between the proteins adsorbed at the lipid/water interface and the proteins in the surrounding phase (Lopez *et al.*, 2015). Moreover, homogenization and/or draining operations given during the manufacture of cheeses, Lopez *et al.* (2015) have also shown various organizations of lipids (Figure 7.3). In their review paper and they concluded the following:

- In non-homogenized, moderately drained, high-moisture Mozzarella cheese and mold-ripened soft cheeses, fat globules of 4 μm in diameter are covered by a biological membrane.

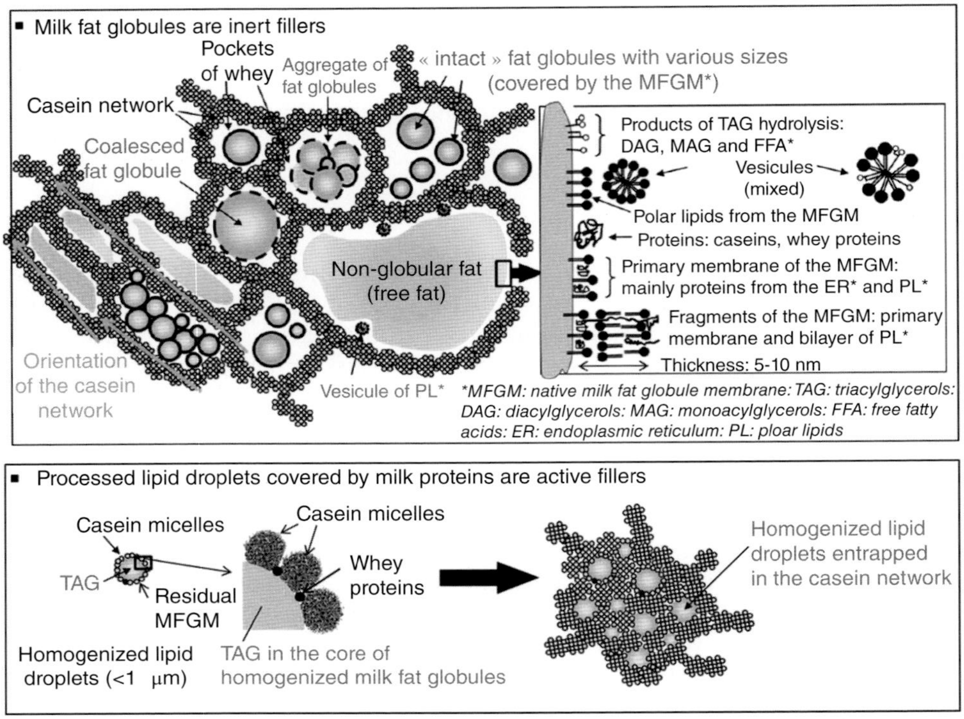

Figure 7.2 Schematic representation of the organization of milk fat in a casein matrix (e.g. cheeses). Lopez *et al.* (2015): Copyright © 2015 by Springer. Reprinted/Reproduced with permission.(*See color plate section for the color representation of this figure.*)

- In non-homogenized, intensely drained, low-moisture Mozzarella, Emmental and Cheddar cheeses, non-globular free fat, several μm in size, is present.
- In fresh, blue and cream cheeses, very small homogenized lipid droplets are covered by milk proteins.

Mozzarella curd undergoes a stretching process resulting in the formation of fat into pools. The fats are present in between void spaces of elongated protein fibers. The fat particles are more spherical in Cheddar, Processed, Cream and other related cheeses.

Figure 7.3 Microstructure of cheeses showing the effect of dairy technological processes on the organization of lipids. Confocal laser scanning microscopy (CLSM) images of (a–d) Emmental cheeses: (a) after coagulation, (b) after pressing of the curd grains; (c), (d) in the final product; (e), (f) high-moisture Mozzarella cheese; (g) low-moisture Mozzarella cheese; (h) mold-ripened soft-type cheese; (i) homogenized cream cheese; (j) homogenized fresh cheese. In all the images, proteins are labeled by using fast green FCF fluorescent dye (green color). Images a–c, e, g, h–j: Total fat is labeled using Nile red (red color). Images d and f: Phospholipids from the milk fat globule membrane are labeled using Rhodamine-DOPE fluorescent dye (red color). Scales bars are in the images. Lopez *et al.* (2015): Copyright © 2015 by Springer. Reprinted/Reproduced with permission. (*See color plate section for the color representation of this figure.*)

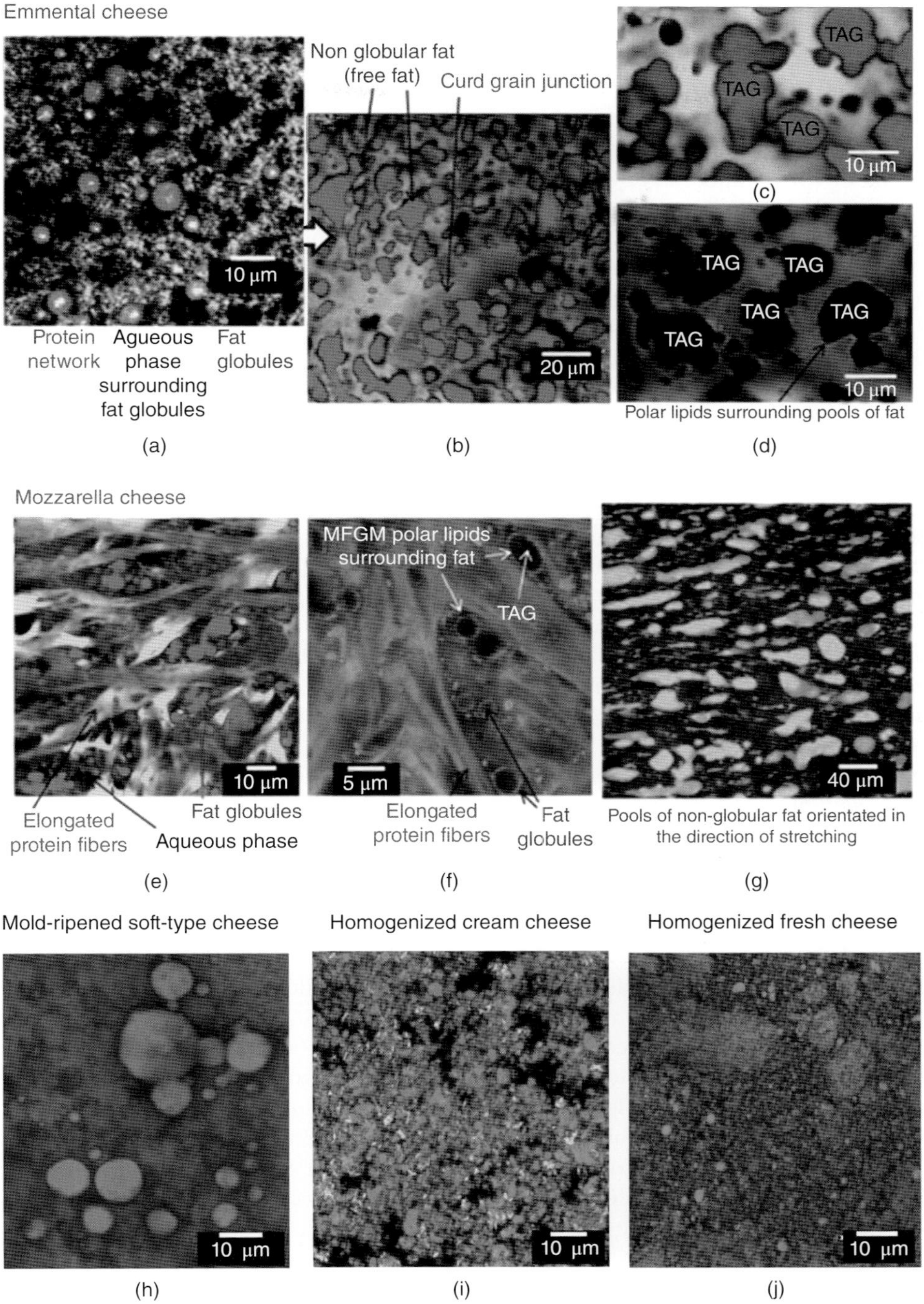
Emmental cheese
10 μm
Protein network
Agueous phase surrounding fat globules
Fat globules
(a)
Non globular fat (free fat)
Curd grain junction
20 μm
(b)
TAG
TAG
TAG
10 μm
(c)
TAG
TAG
TAG
TAG
TAG
10 μm
Polar lipids surrounding pools of fat
(d)
Mozzarella cheese
10 μm
Elongated protein fibers
Fat globules
Aqueous phase
(e)
MFGM polar lipids surrounding fat
TAG
5 μm
Elongated protein fibers
Fat globules
(f)
40 μm
Pools of non-globular fat orientated in the direction of stretching
(g)
Mold-ripened soft-type cheese
10 μm
(h)
Homogenized cream cheese
10 μm
(i)
Homogenized fresh cheese
10 μm
(j)

7.3.3 Calcium in Cheese Matrix

One of the major compositional variables that affects the microstructure of cheese is calcium. Calcium participates in the protein network that forms through milk coagulation (Lucey and Fox, 2015). The ratio of soluble to bound calcium increases upon reduction in pH (<6 to 5.4) of cheese curd. The solubility of water-insoluble calcium phosphate increases when either pH or temperature of the milk decreases. Protein–protein interactions increase in the presence of soluble calcium. However, the presence of citrate will create a complex with the soluble calcium ions, and prevent that interaction with expansion of the protein phase in cheese (Everett, 2007). Pastorino *et al.* (2003) observed more protein phase expansion in Cheddar cheese with added citrate.

During ripening, alteration in physico-chemical changes leads to an increase in hydration of the cheese matrix. The proteolysis, changes in pH and solubilization of CCP increase the hydration of cheese. The physical expansion or swelling of the matrix due to hydration pushes the fat globules closer together causing coalescence of the fat globules forming fat pools, which results in an increase in free fat upon melting (Fox *et al.*, 2000). The mineralization of casein has a major influence on microstructure. The formation of CCP increases matrix density. Ong *et al.* (2013) supplemented cheese milk with Ca at 300 and 600 mg/L and found cheeses with a greater number of micron-sized pores in the structure. Moreover, hardness also increased with >100 mg/L addition levels. This is due to reduced micellar fusion, an increased number of CCP crosslinks in the matrix and increased density of the para-casein matrix. Pastorino *et al.* (2003) injected 40% $CaCl_2$ solution into cheese post-manufacture. They found contraction of the matrix and release of water into pores of the matrix. This was due to increased casein–casein bonding leading to decreased matrix hydration. On the other hand, Joshi *et al.* (2003) reduced the Ca content of cheese milk. Matrix density was thus decreased by allowing more hydration and swelling of the matrix, creating more voids in the microstructure. The decrease in CCP content of cheese during early ripening is responsible for the initial softening of cheese and an increase in its meltability (Lucey *et al.*, 2003; O'Mahony *et al.*, 2006).

Calcium has a significant effect on Mozzarella cheese microstructure. The pH is responsible for partitioning of calcium into the curd and whey in the manufacture of Mozzarella cheese. The average pH for this cheese is above 5.4 and the melted structure of the cheese becomes more fibrous. Incomplete fused particles with poor meltability were found below pH 5 (Kindstedt *et al.*, 2001). Calcium bound to protein in the cheese ultimately affects the microstructure in Mozzarella cheese. The protein fibers are more closely associated through calcium phosphate bridging (Kindstedt and Guo, 1998). There is enhancement of protein–protein interactions due to the increase in the amount of soluble calcium and that leads to a decrease in the association of protein with the water phase. This is responsible for poor meltability of the cheese (Joshi *et al.*, 2003). On the other hand, lower levels of calcium are responsible for decreases in the number of serum pockets. This causes decreased firmness, but increased the meltability.

7.4 Crystals in Cheese Matrix

Crystals of calcium phosphate 20–30 μm in size are found in Cheddar cheese (Brooker *et al.*, 1975). Brooker *et al.* (1975) found white calcium D(−)-lactate crystals around 80 μm size on the surface of cultured ripened cheese. Washam *et al.* (1982) worked on the microstructure of various chemical compounds crystallized on Cheddar cheese using SEM. The tyrosine, sorbic acid, calcium lactate, calcium phosphate and sodium chloride were induced to crystallize on the surface of mild Cheddar cheese and these crystals were observed using SEM. Calcium lactate crystals appeared to form flat and ribbon-like fibers (as bundles or sheets of parallel fibers). The sorbic acid crystals occurred as more thread-like, whereas sodium chloride crystals appeared as individual or fused aggregates of the characteristic cubic crystals. Tyrosine crystals, like calcium lactate and sorbic acid crystals were also fibrous. However, the fiber of tyrosine crystals was much larger and appeared to exist in a very compact, almost solid aggregate. The calcium phosphate crystals have a coral-like appearance (Washam *et al.*, 1982).

When the solubility of Ca^{2+} and lactate ions exceed certain levels after the post-manufacture of cheese, calcium lactate crystals are formed within the cheese matrices. The D-lactate is formed by non-starter lactic acid bacteria. The lower solubility of Ca-D-lactate forms the crystal (Lawrence *et al.*, 2004; Upreti and Metzger, 2006). Crystallization of calcium lactate on the surface of cheese is considered a defect. The crystals can be visualized using X-ray crystallography, but it does not show details about the placement of fat and protein around the crystals. Burdikova *et al.* (2015) used second harmonic generation (SHG) microscopy to visualize calcium lactate crystals placed inside the cheese matrix (Figure 7.4). These authors have also mentioned that a combination of SHG microscopy and CLSM helped in obtaining information about the relationship between crystals and the surrounding environment, such as location of fat, protein, phospholipids, and live/dead bacteria in three dimensions.

7.5 Starter Bacteria in Cheese Matrix

Starter bacteria are important to cheese manufacture and ripening. They acidify milk during the initial fermentation and during subsequent ripening release of enzymes into the cheese matrix resulting in the breakdown of proteins, amino acids and fatty acids

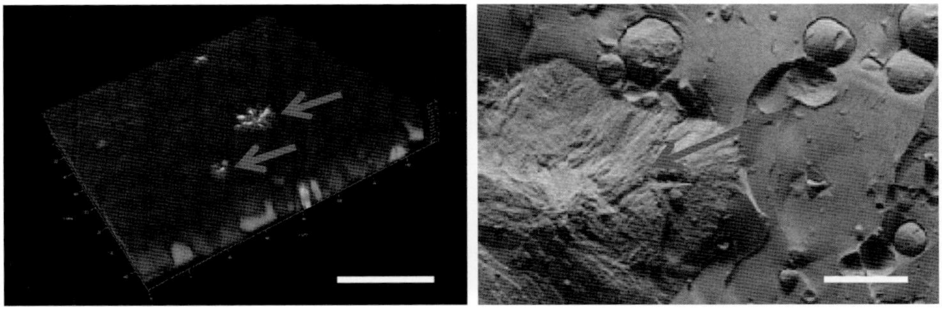

(a) (b)

Figure 7.4 Calcium lactate crystals inside the Cheddar cheese matrix. (a) (red) two-photon fluorescence of protein matrix stained with Nile blue; (green) two-photon fluorescence of fat stained with Nile blue; (grey) SHG signal from the calcium lactate crystals (purple arrows); scale bar 50 μm. (b) Freeze-fracture cryo-SEM image showing the details of fat droplets in protein matrix and a single calcium lactate crystal (purple arrow); scale bar 5 μm. Burdikova *et al.* (2015): Copyright © 2015 by Springer. Reprinted/Reproduced with permission. (*See color plate section for the color representation of this figure.*)

(Hickey *et al.*, 2015). The amphipolar property of the phospholipids in MFGM makes fat–casein interaction possible in the cheese matrix. The bacterial colonies are generally located at the fat–casein interface in contact with MFGM due to the hydrophobic nature of lactic acid bacteria (Lopez *et al.*, 2006; Pereira *et al.*, 2009; Gallier *et al.*, 2010). However, Jeanson *et al.* (2011) showed bacterial colonies to be randomly distributed and static within the cheese matrix. Romeih *et al.* (2012) and Martinovic *et al.* (2013) used CLSM combination with SEM to localize colonies of *L. lactis* and *Lb. casei* within the protein matrices of the Cheddar cheese. Exopolysaccharide-producing starter cultures can be added to milk. The exopolysaccharides have the ability to bind water and help to increase the ratio of water to protein in low-fat Mozzarella cheese which improves meltability (Everett, 2007).

For complete information about the location of microorganisms in cheese, readers are advised to refer Chapter 13 of this book.

7.6 Microstructure of Selected Varieties of Cheeses

Every cheese variety has different characteristic structural features. The microstructures of selected cheeses are discussed at below.

7.6.1 Processed Cheese

Processed cheese is prepared by heating and mixing the blends of different stages of matured hard and semi-hard natural cheeses with melting salts/emulsifying salts (mostly citrates and sodium phosphates) and eventually, other dairy or non-dairy ingredients. Processed cheese varieties include creamy processed cheese spreads as well as firm and sliceable products usually sold in slices, blocks, or in grated form.

The proper selection of aged cheeses as raw material is very important as it significantly affects the microstructure, texture and rheology properties of the final product. The long-term matured cheese varieties like Cheddar or Emmental produces processed cheese having a firm, long body and good sliceability (Mulsow *et al.*, 2007). The emulsifying salts (1–3%) help in enhancing protein hydration, emulsification of free fat, binding/masking the calcium, increasing pH, the buffering capacity and structure formation. The emulsifying salts generally used are sodium salts of phosphoric acid and citric acid, including sodium citrates, sodium ortho-, pyro- and polyphosphates as well as various phosphate blends. These salts convert insoluble para-casein into hydrated sodium para-caseinate in the presence of heat and action of shearing during processing. The para-caseinate has a higher water-binding capacity and stabilizing oil-in-water emulsion. It has the ability to chelate calcium which is attached to the casein and replace the monovalent sodium of the emulsifying salt that helps in the dispersion of caseins. The dispersed caseins present in surrounding fat globules. This helps in the reduction of free oil when Processed cheeses are reheated in other food preparations (Caric *et al.*, 1985; Lucey, 2007; Mulsow *et al.*, 2007). The dispersed calcium–paracaseinate complex interacts with fat via hydrophobic interactions. During the cooling stage of the Processed cheese, the dispersed caseinate matrix forms “flocs”. The flocs then interact to form a uniform, closely knit gel network (Figure 7.5).

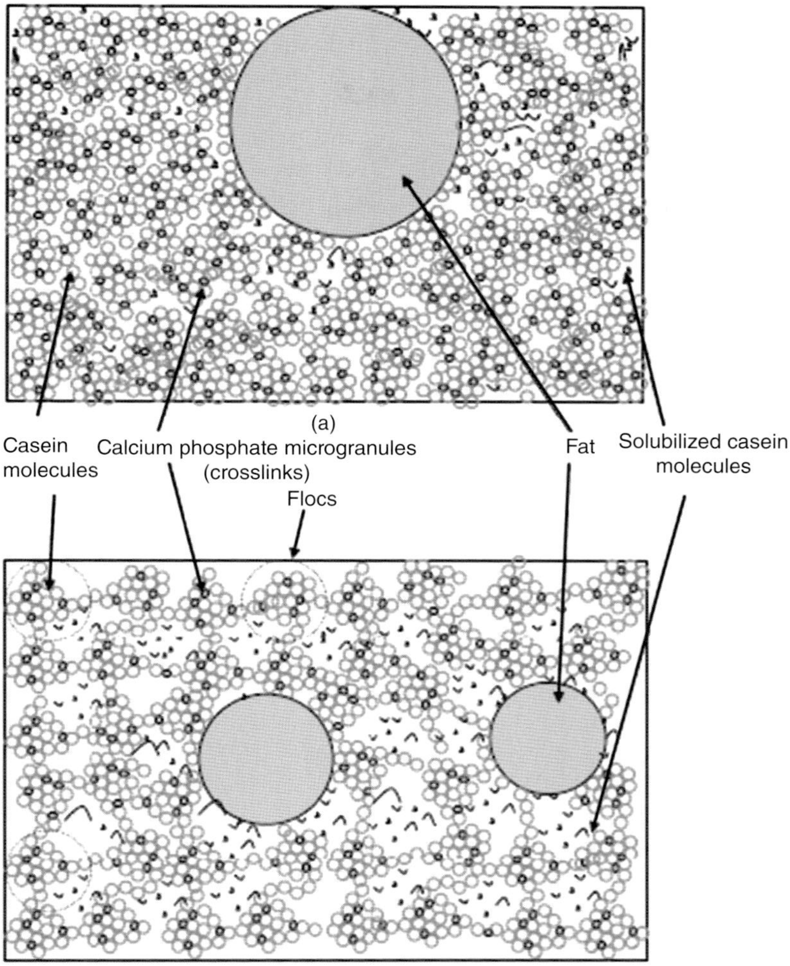

Figure 7.5 Schematic microstructure of (a) natural cheese and (b) processed cheese. Kapoor and Metzger (2008): Copyright © 2008 by Wiley. Reprinted/Reproduced with permission. (*See color plate section for the color representation of this figure.*)

The microstructure of Processed cheese essentially consists of a fat phase (1 to 5 μm size of fat globules) evenly dispersed in a partially dispersed casein gel network (Heertje *et al.*, 1981). Therefore, Processed cheese can be described as a stable oil-in-water emulsion. On the other hand, fat in natural cheese is under-emulsified and the fat phase as well as the water phase is supported by a network of water-insoluble calcium–paracaseinate phosphate complexes (Shimp, 1985; Kapoor and Metzger, 2008). The microstructure of Processed cheese resembles, to some extent, the microstructure of natural cheese (Caric *et al.*, 1985). However, there are several differences that can be studied using light microscopy and by electron microscopy. Caric *et al.* (1985) have concluded following differences:

7.6.1.1 Curd Granules and Fat

Natural cheese is made by pressing curd granules. The curd granules have insoluble calcium caseinate and fat droplets, and become a homogeneous mass. The name of the sites at where granules are fused with each other are known as curd granule junctions. The junctions have typical patterns. Due to milling and pressing of curd, Cheddar cheese has complex patterns compared with stirred-curd cheeses. The fat globules are depleted in the curd granule junctions. Most of the fat globules have their fat globule membranes preserved in natural cheese.

In the case of processed cheese, the fat globule membranes and curd granule junctions become extinct due to various processes such as heating, melting and stirring of the cheese. Several micrometers in diameter of fat particles are formed during melting out the fat. The insoluble protein present in natural cheese becomes partially solubilized by melting/emulsifying salts, which leads to the formation of a smooth and homogeneous mass. The salts increase the natural emulsifying properties of the cheese proteins, and the fat disperses in the form of minute globules.

7.6.1.2 Occurrence of Crystals

Natural cheese contains calcium phosphate and lactate crystals. Calcium phosphate crystals (up to 30 μm in diameter) are abundant. The calcium lactate crystals (up to 30 μm in diameter) are irregular in shape having curved, needle-like crystals. Swiss cheese also contains calcium tyrosinate crystals.

The calcium reacts with sodium phosphates in cheese and form insoluble calcium phosphate crystals which can withstand processing of cheese and are found in the finished product. Anhydrous phosphates absorb water and form large aggregates. Sodium citrate produces a smaller number and smaller size of calcium aggregates in Processed cheese than the sodium phosphate used as an emulsifying salt in this cheese preparation. Small white crystals of calcium citrate occasionally develop on the surface of Processed cheese that can be prevented by eliminating citrate from the emulsifying agent.

7.6.2 Cheese Analogs

Cheese analogs are substitutes for Processed cheese. Cheese analogs are prepared using blend of dairy ingredients (milk fat, milk proteins, protein fractions like casein and caseinates), or products in which dairy components are partly or entirely replaced by non-dairy ingredients (mainly vegetable oils like soy bean oil, peanut oil, palm kernel oil, corn oil, or coconut oil) (Shaw, 1984).

Lobato-Calleros *et al.* (1998) studied the microstructure of cheese analogs using SEM which were made from different fats. They indicated that:

> The size, morphology, and distribution of the fat droplets varied with fat type. The soybean oil analog showed a more spherical and smaller fat droplet sizes. It has the greatest number of individually dispersed small fat droplets surrounded by a finely interwoven protein network. The soybean fat cheese analog formed aggregated amorphous structures. It is characterized by a non-defined shaped fat mass with sporadically projecting protein filaments, giving the impression of an agglomeration of unmelted fat covered by a blanket of protein. The butterfat

analog had a bigger fat droplet size than the soybean oil analog but a more homogenous fat droplet size distribution, forming three dimensional clusters of fat droplets held together and covered by protein molecules. The cheese analog made with fat blends showed great variability in their fat droplets size and distribution. The protein matrix and the fat droplet distribution affected the cheese analog structure and morphology. The cheese analog containing the soybean fat-soybean oil (50:50) blend had an open, porous structure with spaces occupied by fat droplets with an unhomogeneous size distribution. The cheese analog made with butterfat and soybean oil (50:50) showed a densely packed agglomeration of fat droplets surrounded by protein filaments, whereas the blend in which butterfat predominated over soybean fat and soybean oil showed an open, porous structure reminiscent of the soybean fat-soybean oil (50:50) blend, but with a greater homogeneity in the particle size distribution.

Cunha *et al.* (2010) studied the microstructure of Requeijãocremoso (spreadable processed cheese analog), a traditional product of Brazil. They substituted part of the dairy cream (25% and 50%) with hydrogenated vegetable fat. The size of the fat globules increased, and the uniformity of their distribution in the protein matrix decreased, with an increase in the percentage of vegetable fat used as observed under SEM. The microstructure of Mozzarella cheese analog with the addition of modified casein and inulin at several ratios were studied using SEM, and it was found that the increasing proportion of inulin in the modified casein: inulin ratio produced a more uniform cheese matrix with cavities in Mozzarella cheese analog. A higher inulin ratio gave a smaller cavity, indicating that inulin produced a more effective fat emulsification and binds more free water (Hakim *et al.*, 2016).

7.6.3 Feta Cheese

Hassan *et al.* (2002) observed even distribution of fat with a larger number of smaller globules in Feta cheese made with a non-capsule-forming culture compared to that made with the capsule-forming culture. Non-fat cheese had a compact structure when made with a non-capsule-forming culture which contrasted with the open structure observed in cheese made with the capsule-forming culture.

7.6.4 Domiati Cheese

Fresh Domiati cheese has a soft texture, but becomes firmer when stored in brine. The internal structure of the fresh Domiati cheese has a spherical framework of casein aggregates held together by bridges, and that the fat globules are embedded within these aggregates. During storage of the cheese in brine, the casein aggregates dissociate into smaller spherical particles forming a loose structure (Hofi *et al.*, 2001). In UF Domiati cheese, the protein matrix is denser and made of larger protein aggregates, and the whey proteins are also included in the matrix (Hofi *et al.*, 2001).

7.6.5 Fresh Cheese

Fresh cheese has a diverse group of varieties. In the processing of traditional fresh cheese, destabilization of pasteurized, homogenized milk is carried out through acidification via fermentation using lactic acid bacteria, which leads to aggregation of the milk protein. The concentration of the milk protein can be carried out by centrifugation. The curd is pasteurized to inactivate the lactic acid bacteria and is usually homogenized before the product is packed. The microstructure of a fresh cheese can be described as a protein matrix filled with fat droplets. That means an acid protein-stabilized oil-in-water emulsion. The protein aggregates into clusters of droplets, but surplus protein tends to aggregate onto these clusters as well (Guinee *et al.*, 1993).

Fresh cheese can also be prepared by coagulation of milk, cream or whey via acidification, a combination of heat and acid, or acidification with a small quality of rennet. Fresh cheese, generally, is consumed once manufacturing is complete. The acid coagulated milk/cream based fresh cheeses are Quark, Fromagefrais, Neufchâtel, Labneh, Chakka, etc. Cottage cheese is an example of a rennet/acid-coagulated milk/cream based cheese, whereas Ricotta, Paneer, Mascarpone and Quescoblanco come under the acid/heat coagulated milk/cream based fresh cheeses. The whey-based fresh cheeses are Ricottone, Mysost and Gudbrandsdalsost (Fox *et al.*, 2017).

Heertje *et al.* (1985) monitored the changes in casein micelles during acidification using freeze-fracture electron microscopy. The casein micelles retained their shape and integrity without major changes at pH values ≥5.9. However, due to solvation of β- and κ-caseins, casein micelles become more porous at pH ~5.5. Upon a further fall in pH towards 5.2, all the CCP is solubilized and a greater number of caseins dissociate from the micelles. On a further reduction in pH towards the isoelectric point of the casein (~4.6), the casein aggregates fuse into a larger size and separate by large void spaces, forming a casein gel/network. Upon scanning electron microscopy, individual gel strands are found to be composed of casein aggregates that look like beads in a necklace and with a lower calcium bridge (Fox *et al.*, 2017).

7.6.6 Cream Cheese

Cream cheese is a soft fresh acid-coagulated cheese product, which is acidified by mesophilic lactic acid starter culture. Cream cheese is obtained by fermenting high fat homogenized milk and then some of the whey is separated out. The homogenized fat globules covered by caseins and whey proteins are the active contributors to the cheese matrix (Lucey, 2007). Cream cheese contains protein aggregates (20 to 1000 μm in size) containing casein, heat denatured whey proteins and small fat globules (Kalab, 1985; Sainani *et al.*, 2004). However, more clusters and larger sized fat globules are also found in Cream cheese prepared from high fat or high salt. In some Cream cheese, a gritty mouthfeel is detected due to the compact aggregate of proteins. Cream cheese prepared from homogenized milk has more dispersed, smaller fat globules, less free oil with more elastic properties and fat which interacts more strongly with the casein matrix. The processing conditions influenced the structure development in cream cheese. High fat content (12% versus 0%), higher inoculum level (2% versus 1%), higher incubation temperature (26°C versus 20°C), and higher homogenization pressure (250 bar versus 100 bar) gave acid gels with a firmer texture, while lower pH at breaking the gels (pH 4.7 versus pH 5.1) gave acid gels with a firmer texture (Phadungath, 2005).

7.6.7 Mold-Ripened Cheeses

Camembert, Roquefort, Stilton, Brie, Blue and Gorgonzola cheeses are mold-ripened cheeses in which mold can be added and is present either on the surface or interior of the cheese to obtain a desired texture and flavor of the final product. Stilton, Roquefort, Gorgonzola, and blue cheeses contain a mold in the interior of the cheese. Blue cheese contains less than 1 μm in the size of channels that exist within the protein matrix. More mold growth can be observed in the channels. Camembert and Brie are characterized by a surface flora layer (Everett, 2007). The proteins are broken down by surface flora present on Camembert cheese, and by the release of ammonia and carbon dioxide. This can increase the pH of the exterior surface and consequently leads to precipitation of calcium phosphate from the aqueous phase. It promotes a sandy mouthfeel. The calcium ion gradient occurs between the surface and bulk of the cheese (Brooker, 1987). This leads to migration of calcium ions from the interior outwards and causes reduction in the casein crosslinking resulting in a softening of the cheese. The rate of plasmin-induced proteolysis increases during increase in pH, and is also responsible for softening the cheese (Pereira *et al.*, 2009). Washam *et al.* (1979) investigated the growth of mold (*Penicillium roqueforti*) during maturation using scanning electron microscopy of commercial Blue cheese. A porous structure, many crevices and some ungerminated conidia appeared in the internal structure 5 days after the manufacture of the cheese. After 13 days, most of the conidia are germinated. Dense masses of mycelia with enumerable spores characterized a 21-day-old cheese. Upon further maturation for 34 and 44 days, these mycelial masses have undergone extensive degeneration (Washam *et al.*, 1979).

7.6.8 Cheese Powder

Cheese powder is usually used as a food ingredient. It improves the flavor and mouthfeel of food products. Cheese powder is used in crackers, chips, crisps and relish for ready-to-eat products, salads and fresh vegetables, as a flavor–odor additive. A certain amount of aroma is inevitably lost during drying in the production of cheese powder. Hence, aged cheese as a starting material is generally used. In cheese powders, either a single cheese variety or blends of various cheeses, or a blend of cheese with other dairy ingredients (i.e. skimmed milk solids, whey, lactose, etc.), are normally used. The cheese slurry is concentrated to 35–45% total solids and homogenized followed by spray-drying.

The effect of trisodium citrate and sodium polyphosphate added to the cheese suspension prior to its spray drying on the microstructure of the final powdered cheese was studied by Chojnowski *et al.* (2006). Cheese powder particulate microstructure was analyzed using a scanning electron microscope. The cheese powder prepared using 3% trisodium citrate had a particle diameter ranging from 2.9 to 50 μm. Powder particles were irregular in shape and had numerous creases. The diameters of the particles of cheese powder with addition of 2% trisodium citrate ranged from 5.8 to 42.3 μm. They were more spherical in shape and had pores all over their surface. The particulate diameters ranging from 3.8 to 43.3 μm were found in cheese powder with a 2% addition of sodium polyphosphate. A certain number of creases and numerous cracks were observed on the particulate surface, as well as a slight aggregation capacity of cheese powder particles. The addition of sodium polyphosphate at the rate of 2.5% resulted in particulate diameters varying from 4.8 to 46.2 μm. The powder particles aggregated into larger clusters. Thus, addition of trisodium citrate caused more visible deformation of cheese powder particulate microstructure, while the use of sodium polyphosphate caused the formation of greater clusters of powder particles (Chojnowski *et al.*, 2006). Further Chojnowski *et al.* (2006) estimated the presence of milk fat using fluorescence microscopy in cheese powder. They found that milk fat was present in both the powder particulate surface structure and the capillary pores and fissures. Fat can be enclosed in non-damaged powder particles, which constituted the majority in the microstructure of the examined powders.

Mounir *et al.* (2011) used an instant controlled pressure process to produce cheese powder. The cheese powder has porous microstructure particles that improve reconstitution, solubility, and functional characteristics. A porous structure revealed the presence of alveoli in the cheese powder. The SEM of white cheese powders produced at different atomization pressures (294 kPa, 441 kPa and 588 kPa) were studied by Koca *et al.* (2015). The SEM showed that high atomization pressure (588 kPa) resulted in larger numbers of small particles in the powder. Wrinkled particles of different sizes were also noticed.

7.7 Cheese Matrix and Digestion

The food matrix acts as a nutrient-release regulator as the matrix disintegrates during the digestion process in the human body (Turgeon and Rioux, 2011). Thus, the food matrix influences the entire digestion process. The physico-chemical state of the cheese matrix affects the kinetics of digestion. The complex matrix of cheese consists of a casein network with open spaces occupied by both fat and serum. The protein contributes to hardness, and fat and water contribute to smoothness. Each cheese variety, from soft to hard, has typical structural characteristics. Processing of cheese affects the texture and macronutrient organization and that modulate the kinetics of cheese disintegration and nutrient release in the gastrointestinal (GI) tract (Gunasekaran and Ak, 2003; Fang *et al.*, 2016). The *in vitro* studies on kinetics of cheese matrix degradation and fat release for different textures of mild Cheddar, aged Cheddar, light Cheddar, and Mozzarella were investigated by Lamothe *et al.* (2012). The aged Cheddar cheese matrix was easily degraded as compared to light Cheddar cheese. The light Cheddar cheese has a firm and cohesive cheese texture which limits cheese degradation (Lamothe *et al.*, 2012). Fang *et al.* (2016) studied the *in vitro* digestion of commercial cheeses with different textural properties. They found that at the end of gastric digestion, elastic cheeses (Mozzarella) were less disintegrated when compared with ripened and soft cheeses with a high fat content (Camembert, aged Cheddar). Fang *et al.* (2016) suggested that the adjustment of cheese processing parameters, resulting in structural and textural changes to cheese, can be used to modulate the digestion of cheese proteins. Cheeses with a higher protein content disintegrate less and release fewer nutrients during digestion.

The impact of different levels of calcium on *in vitro* digestion of Cheddar cheese was studied by Ayala-Bribiesca *et al.* (2016). They found that Cheese hardness increased with higher calcium content resulting in a slower disintegration during *in vitro* digestion. They also found that higher calcium levels led to faster lipolysis during the first half of intestinal digestion, possibly by enhancing lipase activity. The protective effect of the matrix on lipolysis was observed for very high calcium cheese towards the end of digestion. They concluded that the food matrix and microstructure could help to control the release of lipid nutrients from cheese.

7.8 Conclusions

Cheese structure is heterogeneous. The microstructure of cheese can change depending on the cheese processing technology. The constituents present in the cheese matrix are in the form of solid (paracasein), liquid (obtained from residual whey) and either solid or liquid (fat). The structural organization of lipids in the cheese matrix can be easily observed using advanced microscopy techniques. The visualization of calcium lactate crystals placed inside the cheese matrix is easy with the help of second harmonic generation (SHG) microscopy. There are different electron microscopy techniques used to study the microstructure of cheese products. Standard electron microscopy cannot provide complete 3D information. CLSM with fluorescence staining helps in three-dimensional visualization of various components as well as bacterial cells present in the cheese matrix. Moreover, advanced microscopic techniques such as confocal Raman spectroscopy, FLIM and non-linear microscopy are used to study various properties of cheese matrices. The proper selection of aged cheese as raw material is very important, as it significantly affects the microstructure of processed cheese. There are thousands of varieties of cheese that can be produced, and all cheeses have their own microstructures. The understanding of the cheese matrix and microstructure can help to control the release of nutrients from cheese during the digestion process.

References

Abd El-Salam, B.A. (2015). Effect of milk fat replacement with vegetable oil and/or whey protein concentrate on microstructure, texture and sensory characteristics of fresh soft cheese. *International Journal of Dairy Science,* **10**(3): 117–125.

Aguilera, J.M. and Stanley, D.W. (1999). *Microstructural Principles of Food Processing and Engineering.* Springer, US.

Anderson, D.L. and Mistry, V.V. (1994). Reduced fat Cheddar cheese from condensed milk. 2. Microstructure. *Journal of Dairy Science,* **77**: 7–15.

Ayala-Bribiesca, E., Lussier, M., Chabot, D., Turgeon, S.L. and Britten, M. (2016). Effect of calcium enrichment of Cheddar cheese on its structure, in vitro digestion and lipid bioaccessibility. *International Dairy Journal,* **53**: 1–9.

Ayyash, M.M., Sherkat, F., Francis, P., Williams, R.P. and Shah, N.P. (2011). The effect of sodium chloride substitution with potassium chloride on texture profile and microstructure of Halloumi cheese. *Journal of Dairy Science,* **94**: 37–42.

Ayyash, M.M., Sherkat, F. and Shah, N.P. (2013). Effect of partial NaCl substitution with KCl on the texture profile, microstructure, and sensory properties of low-moisture Mozzarella cheese. *Journal of Dairy Research,* **80**: 7–13.

Banks, J.M., Brechany, E.Y. and Christie, W.W. (1989). The production of low fat Cheddar-type cheese. *Journal of the Society of Dairy Technology*, **42**: 6–9.

Brooker, B.E. (1987). The crystallization of calcium phosphate at the surface of mold-ripened cheeses. *Food Microstructure*, **6**: 25–33.

Brooker, B.E., Hobbs, D.G. and Turvey, A. (1975). Observations on microscopic crystalline inclusions in Cheddar cheese. *Journal of Dairy Research*, **42**: 341–348.

Brown, K.M., McManus, W.R. and McMahon, D.J. (2012). Starch addition in renneted milk gels: Partitioning between curd and whey and effect on curd syneresis and gel microstructure. *Journal of Dairy Science*, **95**: 6871–6881.

Burdikova, Z., Svindrych, Z., Hickey, C., Wilkinson, M.G., Auty M. A.E., Samek, O., Bernatova, S., Krzyzanek, V., Periasamy, A. and Sheehan, J.J. (2015). Application of advanced light microscopic techniques to gain deeper insights into cheese matrix physico-chemistry. *Dairy Science and Technology*, **95**: 687–700.

Capellas, M., Mor-Mur, M., Sendra, E. and Guamis, B. (2001). Effect of high-pressure processing on physico-chemical characteristics of fresh goats' milk cheese (Mato). *International Dairy Journal*, **11**: 165–173.

Capellas, M., Mor-Mur, M., Trujillo, A.J., Sendra, E. and Guamis, B. (1997). Microstructure of High Pressure Treated Cheese Studied by Confocal Scanning Light Microscopy. In: *High Pressure Research in the Biosciences and Biotechnology*. Heremans, K. (ed.) Leuven University Press, Leuven, pp. 391–394,

Caric, M., Gantar, M. and Kalab, M. (1985). Effects of emulsifying agents on the microstructure and other characteristics of process cheese: a review. *Food Microstructure*, **4**: 297–312.

Chapeau, A.L., Silva, J.V.C, Schuck, P., Thierry, A. and Floury, J. (2016). The influence of cheese composition and microstructure on the diffusion of macromolecules: A study using Fluorescence Recovery After Photo bleaching (FRAP). *Food Chemistry*, **192**: 660–667.

Chojnowski, W., Fornal, J., Dec, B. and Blaszczak, W. (2006). Effect of some technological parameters during drying of ripening cheese on its physico-chemical properties and microstructure. *Polish Journal of Food and Nutrition Sciences*, **15**: 47–50.

Cunha, C.R., Dias, A.I. and Viotto, W.H. (2010). Microstructure, texture, colour and sensory evaluation of a spreadable processed cheese analog made with vegetable fat. *Food Research International*, **43**: 723–729.

Dunshea, F.R., Ostrowska, E., Ferrari, J.M. and Gill, H.S. (2007). Dairy proteins and the regulation of satiety and obesity. *Australian Journal of Experimental Agriculture*, **47**(9): 1051–1058.

Eino, M.F., Biggs, D.A., Irvine, D.M. and Stanley, D.W. (1976a). Microstructure of Cheddar cheese: sample preparation and scanning electron microscopy. *Journal of Dairy Research*, **43**: 109–111.

Eino, M.F., Biggs, D.A., Irvine, D.M. and Stanley, D.W. (1976b). Comparison of microstructures of Cheddar cheese curd manufactured with calf rennet, bovine pepsin, and porcine pepsin. *Journal of Dairy Research*, **43**: 113–115.

El-Bakry, M. and Sheehan, J. (2014). Analyzing cheese microstructure: a review of recent developments. *Journal of Food Engineering*, **125**: 84–96.

Esteves, C.L.C., Lucey, J.A., Hyslop, D.B. and Pires, E.M.V. (2003). Effect of gelation temperature on the properties of skim milk gels made from plant coagulants and chymosin. *International Dairy Journal*, **13**: 877.

Euston, S.R., Piska, I., Wium, H. and Qvist, K.B. (2002). Controlling the structure and rheological properties of model cheese systems. *Australian Journal of Dairy Technology*, **57**: 145–152.

Everett, D.W. and Auty, M.A.E. (2008). Cheese structure and current methods of analysis. *International Dairy Journal*, **18**: 759–773.

Everett, D.W. (2007). Microstructure of natural cheeses. In: *Structure of Dairy Products*. Tamime, A. (ed.) Blackwell Publishing, Oxford, pp. 170–209.

Everett, D.W. and Jameson, G.W. (1993). Physico-chemical aspects of Cheddar cheese made from ultrafiltered milk. *Australian Journal of Dairy Technology*, **48**: 20–29.

Fang, X., Rioux, L-E., Labrie, S. and Turgeon, S.L. (2016). Commercial cheeses with different texture have different disintegration and protein/peptide release rates during simulated *in vitro* digestion. *International Dairy Journal* **56**: 169–178.

Floury, J., Madec, M.-N., Waharte, F., Jeanson, S. and Lortal, S. (2012). First assessment of diffusion coefficients in model cheese by fluorescence recovery after photo bleaching (FRAP). *Food Chemistry*, **133**: 551–556.

Fox, P.F., Guinee, T.P., Cogan, T.M. and McSweeney, P.L.H. (2017). *Fundamentals of Cheese Science*, 2nd edn. Springer, New York, pp. 475–532.

Fox, P.F., Guinee, T.P., Cogan, T.M. and McSweeney, P.L.H. (2000). *Fundamentals of Cheese Science*. Aspen Publishers, Gaithersburg, MD.

Fox, P.F., Guinee, T.P., Cogan T.M. and McSweeney P.L.H. (2017). Fresh cheese products: principals of manufacture and overview of different varieties. In: *Fundamentals of Cheese Science*, 2nd edn. Fox, P.F., Guinee, T.P., Cogan T.M. and McSweeney, P.L.H (Eds). Springer New York, pp. 543–588.

Gallier, S., Gragson, D., Jiménez-Flores, R. and Everett, D. (2010). Using confocal laser scanning microscopy to probe the milk fat globule membrane and associated proteins. *Journal of Agricultural Food Chemistry*, **58**: 4250–4257.

Geng, X.L., van den Berg, F.W.J., Bager, A.N. and Ipsen, R. (2011). Dynamic visualization and microstructure of syneresis of cheese curd during mechanical treatment. *International Dairy Journal*, **21**: 711–717.

Graiver, N.G., Zaritzky, N.E. and Califano, A.N. (2004). Viscoelastic behavior of refrigerated and frozen low-moisture Mozzarella cheese. *Journal of Food Science*, **69**: 123–128.

Green, M.L. (1985). Effect of milk pretreatment and making conditions on the properties of Cheddar cheese from milk concentrated by ultrafiltration. *Journal of Dairy Research*, **52**: 555–564.

Green, M.L. (1987). Effect of manipulation of milk composition and curd-forming conditions on the formation, structure and properties of milk curd. *Journal of Dairy Research*, **54**: 303–313.

Green, M.L., Langley, K.R., Marshall, R.J., Brooker, B.E., Willis, A. and Vincent, J.F.V. (1986). Mechanical properties of cheese, cheese analogs and protein gels in relation to composition and microstructure. *Food Microstructure*, **5**: 169–180.

Green, M.L., Marshall, R.J. and Glover, F.A. (1983). Influence of homogenization of concentrated milks on the structure and properties of rennet curds. *Journal of Dairy Research*, **50**: 341–348.

Guinee, T.P., Auty, M.A.E., Mullins, C., Corcoran, M.O. and Mulholland, E.O. (2000). Preliminary observations on effects of fat content and degree of fat emulsification on the structure-functional relationship of Cheddar-type cheese. *Journal of Texture Studies*, **31**: 645–663.

Guinee, T.P., Pudja, P.D. and Farkye, N.Y. (1993). Fresh acid-curd cheese varieties. In: *Cheese: Chemistry, Physics and Microbiology*. Fox, P.F. (Ed.) Chapman and Hall, *London, Chapter* **13**, pp. 363–419.

Gunasekaran, S. and Ak, M.M. (2003). Cheese texture. In: *Cheese Rheology and Texture.* Gunasekaran, S. and Ak, M.M. (Eds.) CRC Press Boca Raton, FL, pp. 299–329.

Guo, M.R., Gilmore, J.A. and Kindstedt, P.S. (1997). Effect of sodium chloride on the serum phase of Mozzarella cheese. *Journal of Dairy Science*, **80**: 3092–3098.

Hakim, L., Thohari, P.I., Evanuarini, H. and Manab, A. (2016). Physical and chemical properties of Mozzarella cheese analog microwavable. *International Journal of ChemTech Research*, **9**(7): 171–181.

Hassan, A.N., Frank, J.F. and Corredig, M. (2002). Microstructure of Feta Cheese made using different cultures as determined by confocal scanning laser microscopy. *Journal of Food Science*, **67**(7): 2750–2753.

Heertje, I., Boskamp, M.J., van Kleef, F. and Gortemaker, F.H. (1981). The microstructure of processed cheese. *Netherlands Milk and Dairy Journal*, **35**: 177–179.

Heertje, I., Visser, J. and Smits, P. (1985). Structure formation in acid milk gels. *Food Microstructure*, **4**: 267–278.

Hickey, C.D., Sheehan, J.J., Wilkinson, M.G. and Auty, M.A.E. (2015). Growth and location of bacterial colonies within dairy foods using microscopy techniques: a review. *Frontiers in Microbiology*, **6**: 99.

Hofi, M.A., Hofi, A.A., El-Kanany, Y.M. and El-Beialy, A.R. (2001). Microstructure of Domiati cheese manufactured by ultrafiltration. In: *Proceedings 8th Egyptian Conference of Dairy Science and Technology*, pp. 269–282.

Holsinger, Y.H., Smith, P.W. and Tunick, M.H. (1995). Overview: cheese chemistry and rheology. In: *Chemistry of Structure-Function Relationships in Cheese.* Malin, E.L. and Tunick, M.H. (Eds). Springer Science+Business Media, LLC, New York, pp. 1–6.

Huppertz, T., Fox, P.F., de Kruif, K.G. and Kelly, A.L. (2006). Highpressure-induced changes in bovine milk proteins: a review. *Biochimica et Biophysica Acta*, **1764**: 593–598.

Impoco, G., Carrato, S., Caccamo, M., Tuminello, L. and Licitra, G. (2007). Quantitative analysis of cheese microstructure using SEM imagery. *Communications to SIMAI Congress*, ISSN 1827-9015, *Vol.* **2**, pp 1–10. DOI: 10.1685/CSC06096

Jeanson, S., Chadoeuf, J., Madec, M.N.S., Floury, A.J., Brocklehurst, T.F. and Lortal, S. (2011). Spatial distribution of bacterial colonies in a model cheese. *Applied Environmental Microbiology*, **77**: 1493–1500.

Joshi, N.S., Muthukumarappan, K. and Dave, R.I. (2004). Effect of calcium on microstructure and meltability of part skim Mozzarella cheese. *Journal of Dairy Science*, **87**: 1975–1985.

Joshi, N.S., Muthukumarappan, K. and Dave, R.I. (2003). Understanding the role of calcium in functionality of part skim Mozzarella cheese. *Journal of Dairy Science*, **86**: 1918–1926.

Joshi, N.S., Muthukumarappan, K. and Dave, R.I. (2003). Effect of calcium on physico-chemical properties of fat-free Mozzarella cheese. *Journal of Food Science*, **68**: 2289–2294.

Kalab, M. (1977). Electron microscopy in dairy research. *MSC-SMC Bulletin, pp.* 4–10. (www.researchgate.net/publication/276205007).

Kalab, M. (1977). Milk gel structure. *VI. Cheese texture and microstructure. Milchwissenschaft* **32**: 449–458.

Kalab, M., Allan-Wojtas, P. and Miller, S.S. (1995). Microscopy and other imaging technique in food structure analysis. *Trends in Food Science and Technology*, **6**: 177–186.

Kalab, M. (1979). Scanning electron microscopy of dairy products: an overview. *Scanning Electron Microscopy*, III: 261–272.

Kalab, M. (1985). Microstructure of dairy foods. 2. Milk products based on fat. *Journal of Dairy Science*, **68**: 3234–3248.

Kapoor, R. and Metzger, L.E. (2008). Process Cheese: Scientific and Technological Aspects – A Review. *Comprehensive Reviews in Food Science and Food Safety*, **7**: 194–214.

Kheadr, E.E., Vachon, J.F., Paquin, P. and Fliss, I. (2002). Effect of dynamic high pressure on microbiological, rheological and microstructural quality of Cheddar cheese. *International Dairy Journal*, **12**: 435–446.

Kiely, L.J., Kindstedt, P.S., Hendricks, G.M., Levis, J.E., Yun, J.J. and Barbano, D.M. (1993). Age related changes in the microstructure of Mozzarella cheese. *Food Structure*, **12**: 13–20.

Kimber, A.M., Brooker B.E., Hobbs, D.G. and Prentice, J.H. (1974). *Journal of Dairy Research*, **41**: 389–396.

Kindstedt, P.S. and Guo, M.R. (1998). A physico-chemical approach to the structure and function of Mozzarella cheese. *Australian Journal of Dairy Technology*, **53**: 70–73.

Kindstedt, P.S., Zielinski, A., Almena-Aliste, M. and Ge, C. (2001). A post-manufacture method to evaluate the effect of pH on Mozzarella cheese characteristics. *Australian Journal of Dairy Technology*, **56**: 202–207.

Koca, N., Erbay, Z. and Kaymak-Ertekin, F. (2015). Effects of spray-drying conditions on the chemical, physical, and sensory properties of cheese powder. *Journal of Dairy Science*, **98**: 2934–2943.

Kuo, C.J. and Harper, W.J. (2003). Effect of hydration time of milk protein concentrate on cast Feta cheese texture. *Milchwissenschaft*, **58**: 283–286.

Kuo, M.I. and Gunasekaran, S. (2009). Effect of freezing and frozen storage on microstructure of Mozzarella and pizza cheeses. *LWT – Food Science and Technology* **42**: 9–16.

Kuo, M.I., Anderson, M.E. and Gunasekaran, S. (2003). Determining effects of freezing on pasta filata and non-pasta filata Mozzarella cheeses by nuclear magnetic resonance imaging. *Journal of Dairy Science*, **86**: 2525–2536.

Laloy, E., Vuillemard, J.C., El Soda, M. and Simard, R.E. (1996). Influence of the fat content of Cheddar cheese on retention and localization of starters. *International Dairy Journal*, **6**: 729–740.

Lamothe, S., Corbeil, M.-M., Turgeon, S. L. and Britten, M. (2012). Influence of cheese matrix on lipid digestion in a simulated gastro-intestinal environment. *Food and Function*, **3**: 724–731.

Lawrence, R.C., Creamer, L.K. and Gilles, J. (1987). Texture development during cheese ripening. *Journal of Dairy Science*, **70**: 1748–1760.

Lawrence, R.C., Gilles, J., Creamer, L.K., *et al.* (2004). *Cheddar cheese and related dry-salted cheese varieties*, 3rd edn. Elsevier Academic Press, London.

Lobato-Calleros, C., Reyes-Hernández, J., Beristain, C.I., Hornelas-Uribe, Y., Sánchez-García, J.E. and Vernon-Carter, E.J. (2007). Microstructure and texture of white fresh cheese made with canola oil and whey protein concentrate in partial or total replacement of milk fat. *Food Research International*, **40**: 529–537.

Lobato-Calleros, C., Velázquez-Varela, J., Sánchez-García, J. and Vernon-Carter, E.J. (2003). Dynamic rheology of Mexican manchego cheeselike products containing canola oil and emulsifier blends. *Food Research International,* **36**: 81–90.

Lobato-Calleros, C., Vernon-Carter, E.J., Sánchez-García, J. and García-Galindo, H.S. (1999). Textural characteristics of cheese analogs incorporating fat replacers. *Journal of Texture Studies,* **30**: 533–548.

Lobato-Calleros, C., Vernon-Carter, E.J. and Hornelas-Uribe, Y. (1998). Microstructure and texture of cheese analogs containing different types of fat. *Journal of Texture Studies,* **29**: 569–586.

Lopez, C. and Briard-Bion, V. (2007). The composition, supramolecular organization and thermal properties of milk fat: a new challenge for the quality of food products. *Lait* **87**: 317–336.

Lopez, C., Camier, B. and Gassi, J.-Y. (2007). Development of the milk fat microstructure during the manufacture and ripening of Emmental cheese observed by confocal laser scanning microscopy. *International Dairy Journal,* **17**: 235–247.

Lopez C, Maillard MB, Briard-Bion V, Camier B, and Hannon J.A. (2006). Lipolysis during ripening of Emmental cheese considering organization of fat and preferential localization of bacteria. *Journal of Agricultural Food Chemistry,* **54**: 5855–5867.

Lopez, C., Cauty, C. and Guyomarc'h, F. (2015). Organization of lipids in milks, infant milk formulas and various dairy products: role of technological processes and potential impacts-Review. *Dairy Science and Technology,* **95**: 863–893.

Lopez-Fandino, R. (2006). High pressure-induced changes in milk proteins and possible applications in dairy technology. *International Dairy Journal,* **16**: 1119–1131.

Lopez-Fandino, R., Carrascosa, A.V. and Olano, A. (1996). The effect of high pressure on whey protein denaturation and cheesemaking properties of raw milk. *Journal of Dairy Science,* **79**: 929–936.

Lucey, J.A. and Fox, P.F. (2015). Importance of calcium and phosphate in cheese manufacture: a review. *Journal of Dairy Science,* **76**: 1714–1724.

Lucey, J.A. and Fox, P.F. (1993). Importance of calcium and phosphate in cheese manufacture: A review. *Journal of Dairy Science,* **76**(6): 1714–1724.

Lucey, J.A. (2007). Microstructural approaches to the study and improvement of cheese and yogurt products. In: *Understanding and Controlling the Microstructure of Complex Foods* McClements, D.J. (Eds). Woodhead Publishing Limited and CRC Press LLC, USA, pp. 600–621.

Lucey, J.A., Johnson, M.E. and Horne, D.S. (2003). Invited review: Perspectives on the basis of the rheology and texture properties of cheese. *Journal of Dairy Science,* **86**: 2725–2743.

Madadlou, A., Khosroshahi, A., Mousavi, S.M. and Djome, Z.E. (2006). Microstructure and rheological properties of Iranian white cheese coagulated at various temperatures. *Journal of Dairy Science,* **89**: 2359–2364.

Madadlou, A., Khosrowshahi, A., Mousavi, M.E. and Farmani, J. (2007). The influence of brine concentration on chemical composition and texture of Iranian white cheese. *Journal of Food Engineering,* **81**: 330–335.

Madadlou, A., Khosrwoshahi, A. and Mousavi, M.E. (2005). Rheology, microstructure and functionality of low-fat Iranian White cheese made with different concentrations of rennet. *Journal of Dairy Science,* **88**: 3052–3062.

Martınez-Rodrıguez, Y., Acosta-Muniz, C., Olivas, G.I., Guerrero-Beltran, J., Rodrigo-Aliaga, D. and Sepulveda, D.R. (2012). High hydrostatic pressure processing of cheese. *Comprehensive Reviews in Food Science and Food Safety*, **11**: 399–416.

Martinovic, A., Moe, K., Romeih, E., Aideh, B., Vogensen, F.K., Ostlie, H. and Skeie, S. (2013). Growth of adjunct *Lactobacillus casei* in Cheddar cheese differing in milk fat globule membrane components. *International Dairy Journal*, **31**: 70–82.

McMahon, D.J., Oberg, C.J. and McManus, W. (1993). Functionality of Mozzarella cheese. *Australian Journal of Dairy Technology*, **48**: 99–104.

Michalski, M.C., Camier, B., Briard, V., Leconte, N., Gassi, J.Y., Goudedranche, H., Michel, F. and Fauquant, J. (2004). The size of native milk fat globules affects physico-chemical and functional properties of Emmental cheese. *Lait*, **84**: 343–358.

Mounir, S., Halle, D. and Allaf, K. (2011). Characterization of pure cheese snacks and expanded granule powders textured by the instant controlled pressure drop (DIC) process. *Dairy Science and Technology*, **91**: 441–455.

Mulsow, B.B, Jaros, D. and Rohm, H. (2007). Processed cheese and cheese analogs. In: *Structure of Dairy Products.* Tamime, A. (Ed.) Blackwell Publishing, Oxford, pp. 210–235.

Needs, E.C., Stenning, R.A., Gill, A.L., Ferragut, V. and Rich, G.T. (2000). High-pressure treatment of milk: effects on casein micelle structure and on enzymic coagulation. *Journal of Dairy Science*, **67**: 31–42.

O'Mahony, J. A., McSweeney, P.L.H. and Lucey, J.A. (2006). A model system for studying the effects of colloidal calcium phosphate concentration on the rheological properties of Cheddar cheese. *Journal of Dairy Science*, **89**: 892–904.

O'Reilly, C.E., Murphy, P.M., Kelly, A.L., Guinee, T.P., Auty, M.A.E. and Beresford, T.P. (2002). The effect of high pressure treatment on the functional and rheological properties of Mozzarella cheese. *Innovative Food Science and Emerging Technology*, **13**: 3–9.

O'Reilly, C.E. Kelly, A.L. Murphy, P.M. and Beresford, T.P. (2001). High pressure treatment: applications in cheese manufacture and ripening-Review. *Trends in Food Science and Technology*, **12**: 51–59.

O'Reilly, C.E., O'Connor, P.M., Murphy, P.M., Kelly, A.L. and Beresford, T.P. (2000). The effect of exposure to pressure of 50 MPa on Cheddar cheese ripening. *Innovative Food Science and Emerging Technology*, **1**: 109–117.

Oberg, C.J., McManus, W.R. and McMahon, D.J. (1993). Microstructure of Mozzarella cheese during manufacture. *Food Structure*, **12**: 251–258.

Oberg, C.J., Merrill, R.K., Brown, R.J. and Richardson, G.H. (1992). Effects of freezing, thawing, and shredding on low moisture, part-skim Mozzarella cheese. *Journal of Dairy Science*, **75**: 1161–1166.

Oberg, E.N., Oberg, C.J., Motawee, M.M., Martini, S. and McMahon, D.J. (2015). Increasing stringiness of low-fat Mozzarella string cheese using polysaccharides. *Journal of Dairy Science*, **98**: 4243–4254.

Ong, L., Dagastine, R.R., Kentish, S.E. and Gras, S.L. (2011). Microstructure of milk gel and cheese curd observed using cryo scanning electron microscopy and confocal microscopy. *LWT – Food Science Technology*, **44**: 1291–1302.

Ong L., Dagastine, R.R., Auty, M.A.E., Kentish, S.E. and Gras, S.L. (2011). Coagulation temperature affects the microstructure and composition of full fat Cheddar cheese. *Dairy Science and Technology*, **91** (6): 739–758.

Ong, L. Lawrence, R.C. Gilles J., *et al.* (2017). Cheddar cheese and related dry-salted cheese varieties. In: McSweeney, P.L.H, Fox, P.F, Cotter, P.D. and Everett, D.W. (Eds.) *Cheese: Chemistry, Physics and Microbiology*, 4th edn. Academic Press, Oxford, pp. 829–863.

Ong, L., Dagastine, R.R., Kentish, S.E. and Gras, S.L. (2013). The effect of calcium chloride addition on the microstructure and composition of Cheddar cheese. *International Dairy Journal*, **33**: 135–141.

Ong, L., Dagastine, R.R., Kentish, S.E. and Gras, S.L. (2013). Microstructure and composition of full fat Cheddar cheese made with ultrafiltered milk retentate. *Foods*, **2**: 310–331.

Ozer, B.H., Robinson, R.K. and Grandison, A.S. (2003). Textural and microstructural properties of urfa cheese (a white-brined Turkish cheese). *International Journal of Dairy Technology*, **56**: 171–176.

Paquet, A. and Kalab, M. (1988). Amino acid composition and structure of cheese baked as a pizza ingredient in conventional and microwave ovens. *Food Microstructure*, **7**: 93–103.

Pastorino, J., Hansen, C.L. and McMahon, D.J. (2003). Effect of sodium citrate on structure-function relationships of Cheddar cheese. *Journal of Dairy Science*, **86**: 3113–3121.

Pastorino, J., Ricks, N.P., Hansen, C.L. and McMahon, D.J. (2003). Effect of calcium and water injection on structure-function relationships of cheese. *Journal of Dairy Science*, **86**, 105–113.

Paulson, B.M., McMahon, D.J. and Oberg, C.J. (1998). Influence of sodium chloride on appearance, functionality, and protein arrangements in nonfat Mozzarella cheese. *Journal of Dairy Science*, **81**, 2053–2064.

Pereira C.I., Gomes, A.M.P. and Xavier Malcata, F. (2009). Microstructure of cheese: Processing, technological and microbiological considerations. *Trends in Food Science and Technology*, **20**(5): 213–219.

Phadungath, C. (2005). Cream cheese products: A review. *Songklanakarin Journal of Science and Technology*, **27**(1): 191–199.

Pierre, A., Michel, F., Graët, Y.L. and Berrier, J. (1999). Soft goat cheeses at different ripening stages: cheese structure, composition and non solvent water. *Lait*, **79**: 489–501.

Rashidinejad, A., Birch, E.J. and Everett, D.W. (2016). Effects of (+)-catechin on the composition, phenolic content and antioxidant activity of full-fat cheese during ripening and recovery of (+)-catechin after simulated in vitro digestion. *Antioxidants*, **5**(3): 29.

Rashidinejad, A., Birch, E.J., Sun-Waterhouse, D. and Everett, D.W. (2013). Effects of catechin on the phenolic content and antioxidant properties of low-fat cheese. *International Journal of Food Science and Technology*, **48**: 2448–2455.

Rogers, N.R., McMahon, D.J., Daubert, C.R., Berry, T.K. and Foegeding, E.A. (2010). Rheological properties and microstructure of Cheddar cheese made with different fat contents. *Journal of Dairy Science*, **93**: 4565–4576.

Romeih, E., Moe, K. and Skeie, S. (2012). The influence of fat globule membrane material on the microstructure of low-fat Cheddar cheese. *International Dairy Journal* **26**: 66–72.

Sainani, M.R., Vyas, H.K. and Tong, P.S. (2004). Characterization of particles in cream cheese. *Journal of Dairy Science*, **87**, 2854–2863.

Saldo, J., Sendra, E. and Guamis, B. (2002). Colour changes during ripening high-pressure-treated hard caprine cheese. *High Pressure Research*, **22**: 659–663.

Serrano, J., Velazquez, G., Lopetcharat, K., Ramırez, J.A. and Torres, J.A. (2004). Effect of moderate pressure treatments on microstructure, texture, and sensory properties of stirred-curd Cheddar shreds. *Journal of Dairy Science*, **87**: 3172–3182.

Serrano, J., Velazquez, G., Lopetcharat, K., Ramırez, J.A. and Torres, J.A. (2005). Moderately high hydrostatic pressure processing to reduce production costs of shredded cheese: microstructure, texture and sensory properties of shredded milled curd Cheddar. *Journal of Food Science*, **70**: 286–293.

Shaw, M. (1984). Cheese substitutes: threat or opportunity? *Journal of the Society of Dairy Technology*, **37**: 27–31.

Shimp, L.A. (1985). Process cheese principles. *Food Technology*, **39**: 63–64, 66, 68, 70.

Soodam, K., Ong, L., Powell, I B, Kentish, S.E. and Gras, S.L. (2015). Effect of calcium chloride addition and draining pH on the microstructure and texture of full fat Cheddar cheese during ripening. *Food Chemistry*, **181**: 111–118.

Soodam, K., Ong, L., Powell, I.B., Kentish, S.E. and Gras, S.L. (2015). Effect of rennet on the composition, proteolysis and microstructure of reduced-fat Cheddar cheese during ripening. *Dairy Science and Technology*, **95**(5): 665–686.

Torres-Mora, M.A., Soeldner, A., Ting, E.Y., *et al.* (1996). Early microstructure changes in Cheddar cheese and the effects of high pressure curd processing. In: *Book of Abstracts of I.F.T. Annual Meeting*, pp. 9, New Orleans, LA, USA.

Turgeon, S.L. and Rioux, L.-E. (2011). Food matrix impact on macronutrients nutritional properties. *Food Hydrocolloids*, **25**: 1915–1924.

Upreti, P. and Metzger, L.E. (2006). Influence of calcium and phosphorus, lactose, and salt-to-moisture ratio on Cheddar cheese quality: manufacture and composition. *Journal of Dairy Science*, **89**: 420–428.

Walstra, P., van Dijk, H.J.M. and Geurts, T.J. (1985). The syneresis of curd. 1. General considerations and literature review. *Netherlands Milk and Dairy Journal*, **39**: 209–246.

Walstra, P., Wouters, J.T.M. and Geurts, T.J. (2006). *Dairy Science and Technology*, 2nd edn. CRC Press, Boca Raton, FL, pp. 603–608.

Washam, C.J., Kerr, T.J. and Hurst, V.J. (1982). Microstructure of various chemical compounds crystallized on Cheddar cheese. *Journal of Food Protection*, **45**(7): 594–596.

Washam, C.J., Kerr, T.J. and Todd, R.L. (1979). Scanning electron microscopy of blue cheese: mold growth during maturation. *Journal of Dairy Science*, **62**: 1384-1389.

Watkinson, P., Coker, C., Crawford, R., Dodds, C., Johnston, K., McKenna, A. and White, N. (2001). Effect of cheese pH and ripening time on model cheese textural properties and proteolysis. *International Dairy Journal*, **11**: 455–464.

Wium, H., Pedersen, P.S. and Qvist, K.B. (2003) Effect of coagulation conditions on the microstructure and the large deformation properties of fat-free Feta cheese made from ultrafiltered milk. *Food Hydrocolloids*, **17**: 287–296.

Yun, J.J., Barbano, D.M., Kindstedt, P.S. and Larose, K.L. (1995). Mozzarella cheese: impact of whey pH at draining on chemical composition, proteolysis, and functional properties. *Journal of Dairy Science*, **78**(1): 1–7.

8

Microstructural Aspects of Yogurt and Fermented Milk

P.H.P. Prasanna[1], *C.S. Ranadheera*[2], *and J.K. Vidanarachchi*[3]

[1] *Department of Animal and Food Sciences, Faculty of Agriculture, Rajarata University of Sri Lanka, Anuradhapura, 50000, Sri Lanka*
[2] *Advanced Food Systems Research Unit, College of Health and Biomedicine, Victoria University, Werribee Campus, Werribee, VIC 3030, Australia*
[3] *Department of Animal Science, Faculty of Agriculture, University of Peradeniya, Peradeniya, 20400, Sri Lanka*

8.1 Yogurt and Fermented Milk: An Overview

Yogurt is fermented milk where the fermentation has been carried out with lactic acid producing microorganisms. Fermented milk is a milk product obtained by fermentation of milk and/or products derived from milk, where the fermentation involves the action of microorganisms and results in coagulation and a reduction in pH (Food Standards Code of Food Standards Australia New Zealand, 2014). Fermentation is one of the oldest methods practiced by the humans to preserve milk with an extended shelf life (Weerathilake *et al.*, 2014; Balthazar *et al.*, 2017). The origin of fermented milk products dates back to 6000 BC, when the Neolithic people in Central Asia transformed from a status of a food hunting people to a food producer/gathering society, where they began milking their domesticated farm animals with the establishment of a fixed dwelling place. The fermented milk products were discovered accidentally when they stored milk in sheepskin bags, and this evolved over centuries into today's commercial yogurt and fermented drinking milk manufacture. These have different commercially available varieties with a range of flavors, forms and textures designed to meet consumers' varying desires (Weerathilake *et al.*, 2014). Fermented milk including yogurt is mainly produced by using cow's milk. However, milk from other diary animals such as buffalo, goats and sheep is also used and milk from mares and camels is used in certain regions of the world (Hashim *et al.*, 2009; Ranadheera *et al.*, 2012; Balthazar *et al.*, 2017). Although the exact origination of milk fermentation is not clear (Weerathilake *et al.*, 2014), it is accepted that the evolution process of various fermented milk products has taken place in different regions of the world (Table 8.1).

Yogurt is arguably the most popular fermented dairy product in the world and has great consumer acceptability due to its nutritional and therapeutic properties. It is the food produced by culturing one or more of the optional dairy ingredients such as cream, milk, partially skimmed milk, and skimmed milk, used alone or in combination with a

Microstructure of Dairy Products, First Edition.
Edited by Mamdouh Mahmoud Abdel-Rahman El-Bakry, Antoni Sanchez, and Bhavbhuti M. Mehta.

Table 8.1 Fermented products originating in different countries/regions of the world.

Region	Country or region of origin	Traditional name of the fermented milk product
Europe	Turkey	Jugurt/Eyra/Ayran
	Balkans	Kisselmleka/Naja/Yaourt
	Balkan mountains	Urgotnic
	Greece	Yiaourti
	Italy	Cieddu
	Sicily	Mezzoradu
	Sardinia	Gioddu
	Hungary	Tarho/taho
	Finland	Viili
	Scandinavia	Filmjolk/Fillbunke/Filbunk/Surmelk/Taettem
	Iceland	Skyr
	Yugoslavia	Gruzoviz
	Portugal	Iogurte
Eurasia	Russia	Donskaya/Varenetes/Kurugna/Ryzhenka/Guslyanka
	Turkestan	Busa
	Transcaucasia (South Caucasian state was once extended across the modern-day countries of Armenia, Azerbaijan, and Georgia)	Katyk
	Armenia	Mazun/Matzoon, Matsun, Matsoni, Madzoon
Middle East and Asia	Lebanon and some Arab countries	Leban/Laban
	Egypt and Sudan	Zabady/Zabade
	Iran and Afghanistan	Mast/Dough/Doogh
	Iraq	Roba/Rob
	India	Dahi/Dadhi/Dahee
	Mongolia	Tarag
	Nepal	Shosim/Sho/Thara
	Sri Lanka	Meekiri/Deekiri (Curd from buffalo milk)

Adapted and modified from Tamime and Robinson (1999); da Cruz *et al.* (2007); Weerathilake *et al.* (2014); Ekanayake (2016).

characterizing bacterial culture that contains the lactic acid-producing bacteria, *Streptococcus thermophiles* and *Lactobacillus delbrueckii* subsp. *bulgaricus*. One or more of the other optional ingredients such as vitamins and concentrated skimmed milk, non-fat dry milk, buttermilk, whey, lactose, lactalbumins, lactoglobulins, or whey modified by partial or complete removal of lactose and/or minerals can also be included. Most of these ingredients are mainly added to increase the non-fat solids content of the final product and should be included in the culturing process (FDA, 2013). Yogurt, before the addition of bulky flavors should contain not less than 3.25% milk fat and not less than 8.25% milk solids not-fat with a titratable acidity of not less than 0.9%, expressed as lactic acid. The mixture may be homogenized and will be pasteurized or ultra-pasteurized prior to the addition of the bacterial culture. Flavoring ingredients may be added after pasteurization or ultra-pasteurization.

The nutritional composition of yogurts and fermented milk can be varied depending on various factors such as the strains of starter culture and adjunct cultures used in the fermentation, type of milk used (whole , semi or skimmed milk), species from which milk is obtained (bovine, goat, sheep, camel, mare), type of milk solids, solid non-fat, sweeteners and fruits added before fermentation, manufacturing process as well as the length of the fermentation process (Ranadheera *et al.*, 2012; Weerathilake *et al.*, 2014; Ranadheera *et al.*, 2016) (Table 8.2). These factors may influence the microbiological properties, physico-chemical characteristics and sensory attributes of these products (Ranadheera *et al.*, 2012; Ranadheera *et al.*, 2016).

Yogurt and fermented milk can be categorized into various varieties and types. One of the major categories is based on the type of starter culture and adjunct cultures used in the fermentation process: standard culture yogurts those made with *S. thermophiles* and *L. delbrueckii* subsp. *Bulgaricus* and bio- or probiotic yogurts. Probiotic are the live

Table 8.2 Nutritional composition of different types of yogurts and fermented milk (serving size 100 g).

Component	Whole milk yogurt	Low fat yogurt	Non-fat yogurt	Greek style yogurt	Drinking yogurt	Frozen yogurt (non-fat with low-calorie sweeteners
Energy (kcal)	79	56	54	113	62	104
Protein (g)	5.7	4.8	5.4	5.7	3.1	4.4
Carbohydrate (g)	7.8	7.4	8.2	4.8	13.1	19.70
Fat (g)	3.0	1.0	0.2	10.2	Trace	0.80
Thiamin (mg)	0.06	0.12	0.04	0.12	0.03	0.04
Riboflavin (mg)	0.27	0.22	0.29	0.13	0.16	0.18
Niacin (mg)	0.2	0.1	0.1	0.1	0.1	0.2
Vitamin B6 (mg)	0.10	0.01	0.07	0.01	0.05	–
Vitamin B12 (mg)	0.2	0.3	0.2	0.2	0.2	0.09
Folate (μg)	18	18	8	6	12	12
Carotene (μg)	21	Trace	Trace	Trace	Trace	–
Vitamin D (μg)	0	0.1	Trace	0.1	Trace	0
Potassium (mg)	280	228	247	184	130	339
Calcium (mg)	200	162	160	126	100	159
Phosphorus (mg)	170	143	151	138	81	129

Adapted and modified from The Dairy Council (2013); USDA, 2013; Weerathilake *et al.*, (2014).

microorganisms which, when administered in adequate amounts, confer a health benefit on the host (FAO/WHO, 2002). There is an increasing trend in using various probiotic cultures in producing yogurt and fermented drinking milk products due to their potential health benefits (Ranadheera *et al.*, 2014). However, such final product should contain at least 10^6–10^8 cfu/mL or g of live probiotic microorganisms at the time of consumption (Ranadheera *et al.*, 2010). Recently a term called "Live and Active Cultures" has been introduced which refers to the living microorganisms including both standard yogurt cultures and probiotic cultures present in the final product. Yogurt and fermented milk products can be categorized according to various other factors such as the physical and chemical nature, added flavors and post-incubational processes. In addition to these categories and classifications, the yogurt and fermented drinking milk products available in the market are in a wide variety of flavors, textures and forms that suit a vast array of palates and meal occasions. These can be consumed either as a snack, dessert or as part of a meal (Weerathilake *et al.*, 2014).

8.2 Yogurt and Fermented Milk: Production Technologies

Figure 8.1 illustrates the main processing steps involved in manufacturing of fermented milk and popular yogurt types and these steps include the standardization of milk (fat and protein content), homogenization, milk heat treatment, incubation/fermentation at thermophilic temperatures (40–45°C), cooling, and storage (Lee and Lucey, 2010. These steps and ingredients used in manufacturing contribute to the yogurt and fermented milk microstructure.

Milk is the main ingredient used in yogurt and fermented milk manufacturing and other main ingredients include sweeteners, stabilizers, flavors, and bacterial cultures. Based on the type of milk used the final product will be varied in their composition. For example, whole milk is used for full fat/regular yogurt, partially skimmed milk is used for low fat yogurt and skimmed milk is used for non-fat yogurt. Cream/butter fat is used to adjust the fat content whereas ingredients such as skimmed milk powder, whey protein concentrate or milk protein concentrates are used to elevate the total solid content of the yogurt mix. Stabilizers are usually added to the mix in order to increase the body and texture giving an increase in firmness, and preventing unacceptable whey separation/syneresis which leads to low consumer acceptability; this also aids in uniform distribution of ingredients. In addition, sweeteners are added to improve the flavor and sensory characteristics (Weerathilake *et al.*, 2014).

Milk is often mixed with skimmed milk and cream to standardize or adjust the fat content to the desired level. Various types of milk powders, including non-fat dry milk, whey protein concentrates, or milk protein concentrate, can be blended with the milk using a powder dispersion unit and can also adjust the total solid content (Lee and Lucey, 2010). Many commercial yogurt products have milk solids contents of 14–15% and the minimum milk solids not-fat content required in standards or regulations in many countries ranges from 8.2 to 8.6% (Tamime and Robinson, 1999) while fermented drinking milk is often have less total solid contents (Ranadheera *et al.*, 2016). However, Codex Alimentarius Commission recommends yogurt should have a minimum protein content of 2.7% and a maximum fat content of 15% (Codex Alimentarius Commission, 2010).

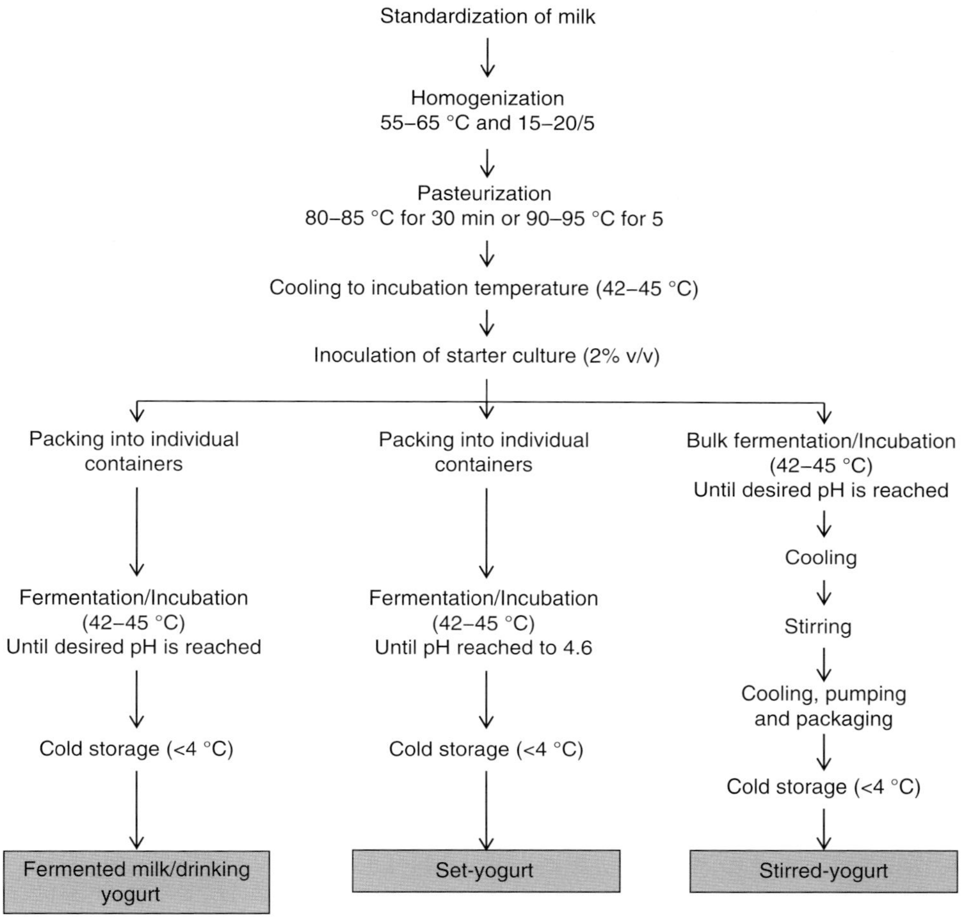

Figure 8.1 General manufacturing procedure for set-, stirred and drinking yogurt Adapted and modified from Lee and Lucey (2010); Weerathilake *et al.* (2014); Ranadheera *et al.* (2016).

In yogurt manufacturing, stabilizers such as pectin and gelatine are added to the yogurt mix in order to attain the characteristic properties of the final product such as texture, mouth feel, appearance, viscosity and also to prevent the whey separation or syneresis (Tamime and Robinson, 1999; Lee and Lucey, 2010). However, both over-stabilization and under-stabilization may cause quality defects as the over-stabilization results in a "jelly-like" springy body of yogurt, whereas the under-stabilization causes a weak "runny body" or unacceptable whey separation (Lee and Lucey, 2010; Weerathilake *et al.*, 2014). However, in the manufacturing of fermented milk, addition of stabilizers may be not necessary as it is consumed as a dairy drink.

After standardization the mixture is subjected to homogenization. Homogenization is considered as an important processing step that reduces the diameter of fat globules to less than 1–2 μm, and ensures uniform distribution throughout the food matrix. Thus, it results in no distinct creamy layer on surface of the yogurt, which helps to increase the consumer acceptability and improves consistency of the yogurt (Lee and

Lucey, 2010; Weerathilake *et al.*, 2014). Homogenization is accomplished by using a homogenizer or viscolizer where the milk is forced through small openings at a high pressure in which the fat globules are broken up due to the shearing forces. Generally milk is homogenized using pressures of 10–20 and 5 MPa in first and second stages, respectively for over 10–17 min. Recently, ultra-high pressure homogenization has been introduced, mainly for commercial yogurt manufacture leading to an increase in yogurt firmness and water holding capacity compared to that of the conventional homogenization process (Serra *et al.*, 2008 and 2009). Ultra-high pressure causes whey protein denaturation as well as partial disruption of the casein micelles (Lee and Lucey, 2010); thus it may affect the microstructure and the nutritional properties of the final product.

The homogenized and standardized yogurt mixture is then subjected to heat treatment. Generally heat treatment of the yogurt mixture is considered as an essential step in the yogurt manufacturing process, because it greatly influences the microstructure and physical properties of yogurt (Lee and Lucey, 2010). There are many advantages of heat treatment as it will destroy the unwanted microorganisms present in the milk or yogurt mixture which can potentially interfere with the controlled fermentation process; provide less competition for starter culture microorganisms and also ensure the food safety aspects of the final product; it will denature the whey proteins that will give the final product a better body and texture; and will release the compounds in milk that stimulate growth of the starter culture microorganisms (Weerathilake *et al.*, 2014). Additionally, it will help some ingredients to achieve the required state to form gels and a protein lattice that affects the final texture and viscosity of the product while aiding in removing dissolved oxygen in the yogurt mixture, assisting the growth and survival of oxygen sensitive starter cultures (Lee and Lucey, 2010; Weerathilake *et al.*, 2014). Heat treatment is a continuous- or batch-process involving heating of milk to a relatively high temperature for a predetermined time period. The time–temperature combinations for the batch heat treatments that are usually employed in the commercial yogurt manufacture include 85° for 30 min and 90–95°C for 5 min (Tamime and Robinson, 1999; Weerathilake *et al.*, 2014). However, very high temperature–short time (100°C to 130°C for 4 to 16 seconds) or ultra-heat temperature (UHT) (140°C for 4 to 16 seconds) are also sometimes used as alternative heat treatment procedures (Lee and Lucey, 2010). These time and temperature combinations are used as a mandatory requirement to destroy the unnecessary pathogenic microorganisms and enzymes present in yogurt mixture.

The heat treated yogurt mixture is then cooled to 42–45°C prior to the addition of yogurt starter culture bacteria (Ranadheera *et al.*, 2012). Usually the recommended starter culture concentration is about 2% (v/v). This temperature range is optimal for the thermophilic microorganisms used in the yogurt starter culture. The typical standard yogurt culture consists of *S. thermophilus* and *L. delbrueckii* subsp. *bulgaricus* in 1:1 ratio (Lee and Lucey, 2010; Weerathilake *et al.*, 2014). Generally, the inoculation of starter cultures takes place in a sealed hygienic stainless steel vessel. However, the place of the fermentation process can be varied depending on the types of yogurt: set and stirred yogurts. It usually occurs in individual containers and in large hygienic stainless steel vats known as bulk fermentation in set- and stirred yogurt manufacturing processes, respectively. The incubation temperature is maintained and monitored at optimal level throughout the fermentation process for a few hours (2.5–3 h) until the pH and acidity reached their desired levels prior to discontinuing the fermentation process by rapid cooling (usually pH of 4.5–4.6 and titratable acidity of 0.3% expressed as percentage of lactic acid for set-yogurts). During the fermentation process, due to the metabolic activity of the lactic acid bacteria used, lactose converts into lactic acid which coagulates milk proteins along with the production of certain volatile compounds that give its characteristic texture development, flavor and aroma (Weerathilake *et al.*, 2014). In addition to starter cultures, probiotics can also be included into the yogurt mixture (Ranadheera *et al.*, 2012). Fermented dairy drinks can also be produced through fermentation using only probiotic microorganisms (Ranadheera *et al.*, 2016).

In order to stop the fermentation process and thereby stop further undesirable acid development, the product is usually blast chilled to refrigerated temperatures (<10°C) when the yogurt or fermented milk has reached the desired pH (usually around 4.5). In the manufacture of set-yogurt, yogurts are directly transferred to a cold store or blast chilled in cooling tunnels while in the manufacture of stirred-yogurt, cooling is first performed by agitating the coagulum in the jacketed fermentation vat in order to produce smoothened product before filling the containers (Lee and Lucey, 2010; Weerathilake *et al.*, 2014). Agitation is not required in the case of fermented drinking milk (Ranadheera *et al.*, 2016).

8.3 Microstructure of Yogurt and Fermented Milk

In general, the structure of foods greatly affects their various properties including texture, functionality, appearance and organoleptic properties. The casein micelles exist as individual entities in milk. Lactic acid produced by the starter culture microorganisms reduces the pH of the milk and destabilizes the casein micelles, which become subsequently linked to each other in the form of clusters and chains. The clusters and chains are part of a protein network matrix in which the liquid phase of milk is immobilized and a three-dimensional gel matrix is formed (Kaláb *et al.*, 1983; Aguilera and Stanley, 1999). Microstructure has a major impact on the texture and other physical properties of acid milk gels (Lee and Lucey, 2010).

The microstructure of acid milk gels has been observed using various techniques such as electron microscopy and confocal scanning laser microscopy. Accordingly, the acid gels consist of a coarse particulate network of casein particles linked together in clusters, chains and strands. The network has pores or void spaces in which the aqueous phase is confined (Lee and Lucey, 2010). In fat-containing products, the presence of fat globules obscures the finer details of pores and strands. The diameter of these pores varies considerably, with larger pores in gels made at a high gelation temperature (usually 80°C) usually consisting of a more cross-linked and branched protein structure, with small pores compared with milk heated at low temperature. Stirred yogurt has very large clusters of caseins presumably created by the collisions and shearing during the mixing process (Lee and Lucey, 2006; Lee and Lucey, 2010). The characteristic three-dimensional gel matrix of set-yogurt is therefore no longer visible in stirred products. Although particle size is occasionally reported for stirred yogurt, it should be recognized that there are no individual particles; rather there are weakly associated clusters of proteins that make up the weak network. Stirring and dilution during the particle size measurement process disrupt the weak network and create the particle (Lee and Lucey, 2010). The way in which casein micelles may interact during gel formation due to pH drop depends on many factors. These factors include, but are not limited to: the heat treatment and various processing steps involved in yogurt and fermented milk production; bacterial strains and the ratio of starter culture microorganisms, *S. thermophiles* and *L. delbrueckii* subsp. *Bulgaricus*; rate of acid development and amount of lactic acid developed; incubation temperature; total solids; presence of additives (thickening agents), etc. (Kaláb *et al.*, 1983). Hence, all these factors could greatly influence the microstructure of yogurt and fermented milk.

8.4 Factors Influencing Microstructure of Yogurt and Fermented Milk

Different factors influencing microstructure of yogurt and fermented milks are shown in Figure 8.2.

8.4.1 Effects of Type of Milk on Structure

Yogurt and fermented milks are basically produced by using cow, goat and sheep milk around the world. There are differences in composition of sheep's milk from that of cow and goat milk. Sheep milk contains higher amount of total solids, fat, proteins, lactose and ash (Hadjipanayiotou, 1995). In addition, different types of milk have different technological and functional properties such as heat stability, rennet clotting and gelation properties (Pesic *et al.*, 2012). The main differences between cow's milk and milk from

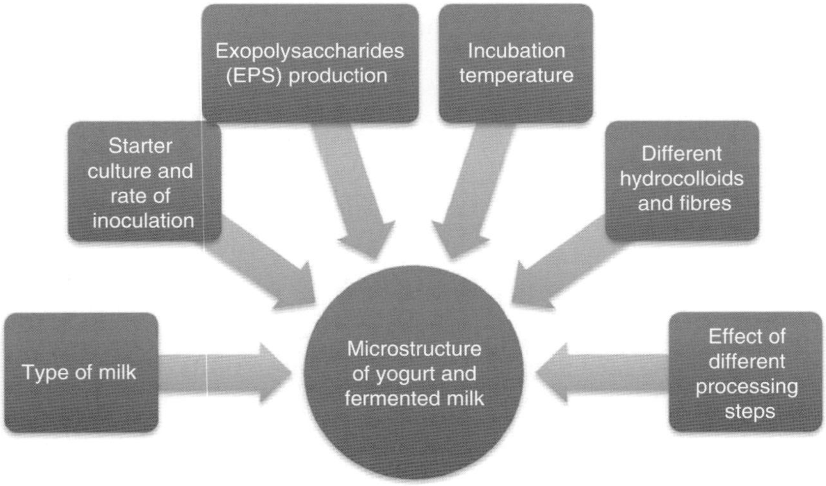

Figure 8.2 Factors that could influence on microstructure of yogurt and fermented milk.

small ruminants are due to the different proportion of the different kind of caseins such as αs_1-casein, αs_2-casein and κ-casein (Raynal-Ljutovac *et al.*, 2008). In the case of goat milk, it is considered to be a poorer source of αs_1-casein which plays an important role in milk coagulation. In addition, cow's milk is different from small ruminant milk, in terms of size of casein micelles and fat globules. Caprine casein micelles are reported to have higher mineralization, lower hydration level and higher dispersion level than cow milk micelles. These differences lead to changes in the coagulation process of cow's milk from those of coagulation of milk from small ruminants, leading to microstructural changes of coagulum (Vargas *et al.*, 2008).

Many studies have reported differences in yogurt made from the milk of different species of animals. Domagała (2009) studied yogurt made from cow, sheep and goat milk. The author reported that goat milk yogurt had a protein matrix of casein micelles connected with thick chains and larger void spaces and large clusters of protein. Sheep milk yogurt was observed to have rarely clustered casein micelles with smaller and more regular void spaces compared to those of goat and cow milk yogurt. Moreover, the sheep milk yogurt was ranked to produce a stronger gel-resisting deformation, compared to that of cow and goat milk yogurt. In another study, cow milk yogurt gel was compared with buffalo milk yogurt, and a more irregular protein network with a greater number of serum pores were observed in buffalo yogurt compared to cow milk yogurt (Nguyen *et al.*, 2014). Similarly, formation of a less compact structure formation in goat milk yogurt than cow milk yogurt was reported by Vlahopoulou *et al.* (1994). Smaller casein micelles and fat globules in goat milk yogurt than cow milk yogurt were reported by Karademir *et al.* (2002) and they further reported that the gel of goat milk yogurt was less stable than that of cow milk yogurt. In a separate study, Nguyen *et al.* (2014) studied buffalo and bovine yogurt and observed a weaker protein network which was porous and irregular.

8.4.2 Rate of Inoculation Level and Starter Culture Composition on Microstructure of Yogurt and Fermented Milk

8.4.2.1 Rate of Inoculation

The rheological properties and structural characteristics of acid gels are affected by different starter concentrations (Lee and Lucey, 2004). Kristo *et al.* (2003) reported that rheological properties of fermented milk products were affected by different concentrations of inoculum. In another study, yogurts prepared with 0.5, 1, 2, 3, or 4% (w/w) inoculation rates of yogurt bacteria (*Streptococcus thermophilus* and *Lactobacillus delbrueckii* subsp. *bulgaricus*) were shown to affect microstructure of yogurt. At a lower inoculation rate, yogurt gel showed larger pores in micrographs, while a uniformly distributed protein network with more cross links was observed at higher inoculation rates. This is due to formation of a weaker gel at a lower inoculation rate, which is vulnerable to whey separation and rearrangement of the gel (Lee and Lucey, 2004).

8.4.2.2 Culture Composition

The type of culture has a major impact on physical properties of fermented milk products. (Hassan *et al.*, 1996). Probiotic bacteria are commonly used in fermented dairy products. Fermented products containing probiotic bacteria must maintain the characteristic textural and sensory attributes (Boylston *et al.*, 2004). Therefore, it is quite common to use probiotic bacteria with other types of bacteria suited for the fermentation of the specific product. In the case of yogurt, probiotic *Bifidobacterium* and *Lactobacillus* strains are generally mixed with *Streptococcus thermophilus* and *Lactobacillus delbrueckii* subsp. *bulgaricus* to achieve the desired flavor and texture (Saarela *et al.*, 2000). Therefore, selection of appropriate culture is very important to have the desired textural and rheological properties of fermented milk products (Penna and Barbosa-Cánovas, 2007). Set-yogurts made using 11 commercially available cultures were shown to be different in microstructure and viscoelastic properties (Rohm and Kovac, 1994). Hassan *et al.* (1995) reported that yogurt made from capsulated bacterial cells had a microstructure with larger pores compared to yogurt with fee cells. Formation of larger pores is due to the action of capsuled cells as nuclei for formation of pores. In another study, Priyashantha *et al.* (2015) studied and reported positive relations with the culture composition and structure of fermented milk gel. However, Oliveira *et al.* (2001) used *Lactobacillus acidophilus* (LA5) and *Lactobacillus rhamnosus* (LC35) in pure culture and in mixed culture with *Streptococcus thermophiles* to produce fermented milks and did not observe any effect of culture composition on the textural properties of fermented products.

8.4.3 Effect of Exopolysaccharide Producing Starter Culture on Microstructure

Exopolysaccharides (EPS) are long chain and high molecular mass biopolymers and they can be covalently linked to bacterial surfaces forming a capsule, non-covalently associated with the surface or totally secreted to the environment (Leivers *et al.*, 2011). These extracellular biopolymers are secreted by many LAB and other bacteria present in fermented foods and human environments such as propionibacteria and bifidobacteria. In-depth understanding of the microstructure of yogurt is considered as a key tool in developing products with desired textural, rheological and organoleptic properties (Laverse *et al.*, 2011). EPS producing LAB produces a yogurt gel which is a well-defined porous web like structure, while EPS negative cultures produce yogurt with a compact homogenous structure. The effect of EPS of LAB on the microstructure of yogurt is mainly due to the incompatibility with the protein aggregates and EPS in the product. This could affect the aggregation prior to gelation and rearrangement of the protein aggregates after the gelation point, hence could result different microstructural properties (Hassan *et al.*, 2003a and 2003b). The final textural properties of EPS containing yogurt depends on structural properties of EPS, including type (ropy or capsular), the degree of ropiness, sugar composition and degree of branching (Purwandari *et al.*, 2007).

A porous structure of yogurt made with EPS producing LAB species was reported in many studies (Amatayakul *et al.*, 2006; Hassan *et al.*, 2003a and 2003b; Kelvin *et al.*, 2005). In addition, the porous structure of yogurt can increase the water holding capacity of the gel structure leading to the lower syneresis of the product. Similarly, the EPS of different LAB species, such as *Lactobacillus delbrueckii* subsp. *bulgaricus* and *Streptococcus thermophilus* (Abbasi *et al.*, 2009; Amatayakul *et al.*, 2006) were reported to produce yogurt with porous structures leading to lower syneresis. Furthermore, the porous structure of yoghurt can withstand mechanical treatments without disturbing its structure, which is considered very important during mechanical handling such as transporting and other related handling practices (Folkenberg *et al.*, 2005).

8.4.4 Incubation Temperature on Structure

Incubation or fermentation temperature has a significant effect on the growth rate of cultures, aggregation rate, molecular interactions, thermal motion, and the type of aggregates formed during the manufacture of fermented milk including yogurt (Wu *et al.*, 2009). In addition, incubation temperature can affect network development, bond mobility, strengthening of bond and rearrangements of bonds after the gelation point (Peng *et al.*, 2010). In general, a temperature of 37–45°C is used for the fermentation of milk during its manufacture, as this is considered the optimum temperature for growth of the commonly used bacterial species such as *Streptococcus thermophilus* and *Lactobacillus delbrueckii* subsp. *bulgaricus* (Purwandari *et al.*, 2007). Higher temperatures during the fermentation process could lead to a lower gel firmness and a weaker protein network with a coarser microstructure (Lee and Lucey, 2003; Wu *et al.*, 2009). This is due to the faster acidification rate at high temperatures, which limits the available time for casein molecules to remain around the gelation point, which helps to develop a fine-stranded network (Peng *et al.*, 2010).

In addition, extensive particle rearrangements occur at higher incubation temperatures, leading to formation of dense clusters of aggregated particles, resulting in a more compact conformation and a contraction of casein micelles (Mellema *et al.*, 2002). Casein particles have higher voluminosity at low temperature, and under the slow acidification process where the pH level changes slowly, the initially weak gel spends more time around the gelation point, leading to a swelling of casein micelles and enhancing inter-cluster strand formation instead of formation of intra-particle bonding (Lucey *et al.*, 1997; Peng *et al.*, 2010).

Lee and Lucey (2004) studied two incubation temperatures at 40°C and 45.7°C to produce yogurt. Yogurt incubated at 40°C was shown to have a more branched and homogenous interconnected protein network compared with yogurt made at 45.7°C, whereas at a higher incubation temperature, yogurt gel had finely dispersed protein clusters. In another study, three incubation temperatures (37°C, 42°C and 45°C) were used to produce yogurt samples (Abbasi *et al.*, 2009). Yogurt made at 37°C was shown to have a regular and compact network with fewer small pores, while yogurt prepared at 42°C and 45°C had larger pores with the irregular network. Similarly, larger pores with a less dense protein network in yogurt produced at 40°C and 46°C were reported by Lee and Lucey (2003). In addition, Nguyen *et al.* (2014) reported that buffalo yogurt prepared at the incubation temperature 43°C had a less dense microstructure with larger pores compared to yogurt made at the incubation temperatures of 37°C and 40°C.

8.4.5 Effect of Different Processing Steps

8.4.5.1 Homogenization of Milk

In yogurt production, milk is generally subjected to homogenization at pressures of 100–200 bar at 55–65°C (Lucey and Singh, 1997). This results in smaller fat globules which are partly or entirely coated with absorbed casein and denatured whey proteins, and these adsorbed proteins have been shown to be incorporated in the gel network during acid induced coagulation of milk leading for better texture and gel firmness (Tamime and Robinson, 2007). Microstructure of buffalo yogurt made from homogenized milk was shown to have a more interconnected protein network with smaller embedded fat globules compared to un-homogenized yogurt samples (Nguyen *et al.*, 2015). In another study by Serra *et al.* (2009), use of ultra-high pressure homogenization in yogurt manufacturing was shown to result in a more homogenous and compact microstructure compared to the control yogurt. Penna and Barbosa-Cánovas (2007) reported that high hydrostatic pressure treated milk produced yogurt having interconnected clusters of densely aggregated protein with uniform size, smooth surfaces and more spherical in shape.

8.4.5.2 Heat Treatment of Milk

Heat treatment of milk is considered as a very important step in yogurt and the fermented milk manufacturing process, and this has been reported to affect the microstructure of yogurt and fermented milk products. Heating of milk induces the denaturation of whey proteins which encourage association of whey proteins with casein micelles through disulfide linkages and hydrophobic interactions. In general, severe heat treatment of milk results in a more branched structured gel compared to a mild heat treatment (Remeuf *et al.*, 2003). Time temperature combination ranging from 80 to 85°C for 30 min to 90 to 95°C for 5 min is considered to be adequate to produce high quality yogurt (Lucey *et al.*, 1998a and 1998b; Soukoulis *et al.*, 2007). Many studies have reported the effect of heat treatment of milk based on the microstructural properties of fermented milk products. A heavily crossed linked and branched protein structure having small pores was observed by Lee and Lucey (2003) in yogurt prepared from milk heated at high temperature (>80°C) compared to that of yogurt prepared with milk heated at low temperature. In another study, yogurt prepared using heat treated milk was shown to have fewer interconnected chains of casein micelles with irregular shapes, which is due the presence of superficial filamentous appendages on the casein micelles that inhibit the fusion of casein micelles (Penna and Barbosa-Cánovas, 2007). Similarly in another study, Remeuf *et al.* (2003) reported a denser and more branched yogurt gel with a severely heated milk base compared with that of a mildly heated milk base. In addition, Harwalkar and Kaláb (1980) reported that acid gel made from unheated milk had larger clusters of proteins than that of gels prepared using heated milk. Heat treatment of milk was shown to produce milk gels with visible cracks and a rough surface, instead of a smooth and unblemished appearance. The large cracks and rough appearance of gels prepared using heated milk is due to structural rearrangements of the network (Lucey *et al.*, 1998a and 1998b).

8.4.5.3 Effect of Stirring

Yogurt and fermented milks are considered to have a higher water content and this water is retained in the structure by the network of the gel. External forces can break the structure of the homogenous gel and result in rearrangements of the structure, leading to expulsion of whey from the coagulum called syneresis (Lee and Lucey, 2010). Stirring during manufacturing of the stirred fermented milk products destroys the network of the coagulum and leads to the formation of microgel particles which are suspended in the milk serum (Nöbel *et al.*, 2014). Hassan *et al.* (2003b) reported that stirring of yogurt resulted in a denser protein network having smaller pores than the unstirred yogurt. Similarly, very large clusters of caseins were observed in stirred yogurt, which are thought to be produced by the collisions and shearing during the stirring of the yogurt mixture (Lee and Lucey, 2006, 2010).

8.4.6 Effect of Addition of Different Hydrocolloids and Fibers on Microstructure

The incorporation of different hydrocolloids has been used to improve viscoelastic and microstructural properties of fermented milk. These hydrocolloids can bind water, react with milk proteins and stabilize the protein network. In addition, they can stop free movement of water (Soukoulis *et al.*, 2007).

Some studies reported that incorporation of different materials can modify the microstructure of yogurt. In one study, Sandoval-Castilla *et al.* (2004) used whey protein concentrate, microparticulated whey protein, and modified tapioca starch as fat replacers in yogurt, and observed that yogurt containing fat replacers had open and less dense microstructure than that of full fat yogurt. They further observed that casein micelles of yogurt containing whey protein concentrate were linked by particle-to-particle attachment in long chains. However, casein micelles of yogurt containing microparticulated whey protein showed a spatial distribution while yogurt with modified tapioca starch had a loose structure with solubilized starch molecules integrated into the casein micelle network. Similarly, low fat yogurt containing whey protein concentrate was shown to yield a compact structure with robust casein particles and larger aggregates than the control non-fat yogurt (Aziznia *et al.*, 2008). In another study, casein and whey protein hydrolysates were shown to modify the microstructure of yogurt which resulted in a highly branched and cross linked gel. In addition, the microstructure of hydrolysate-containing yogurt was observed to be coarser and fluffier. This coarse structure of hydrolysate-containing yogurt may be due to the reduction of thermal denaturation of β-lactoglobulin in solution in the presence of peptide fractions. Therefore, these hydrolysates may prevent the interaction between the surface of casein micelles and denatured β-lactoglobulin, resulting in a lower degree of interaction between β-lactoglobulin and casein micelles (Sodini *et al.*, 2005).

Fibers from plant cell wall materials derived from fruits and vegetables have been shown to improve microstructural properties of yogurt. These materials are generally from by-products of fruit and vegetable processing plants (McCann *et al.*, 2011). Espírito-Santo *et al.* (2013) studied the microstructure of passion fruit fiber-enriched yogurt and observed a higher density of the casein network than that of the control yogurt. Casein network of passion fruit fiber-enriched yogurt had a lower number of pores and a higher number of cross links between strands. However, pineapple peel powder-supplemented yogurt was reported to have a more open network with larger pores, which may be due to thermodynamic incompatibility between polysaccharide of pineapple peel powder and milk proteins (Sah *et al.*, 2016). In another study, the microstructure of dehydrated carrot cell wall powder-supplemented yogurt was shown to have a higher density in protein network compared to that of control yogurt, which may be due to the negative charge of carrot particles at the pH of the yogurt, leading to electrostatic interactions with the protein matrix (McCann *et al.*, 2011). Agave fructans was shown to act as cementing material in the yogurt gel network, resulting in less dense, open and less fused casein micelle aggregates than that of full fat yogurt (Crispín-Isidro *et al.*, 2015). In addition, the same authors used inulin which is a very common ingredient in stirred yogurt. The microstructure of inulin-supplemented yogurt was shown to have an elongated yogurt gel which intermingled into the protein network. They further observed numerous secondary gelled structures at higher concentrations of inulin. This kind of secondary structure formation was observed in yogurt supplanted with inulin by Guggisberg *et al.* (2009). Perhaps inulin may form a gel network of insoluble sub-micron crystalline inulin particles in the water phase.

8.5 Microscopy Methods Used for Analyzing Microstructure of Fermented Milk

Light and electron microscopy are commonly used to examine the microstructure of fermented milk. In general, light microscopy results in images with a limited resolution for thick samples. Electron microscopy provides images with a higher resolution compared to that of light microscopy techniques. However, sample preparation for some electron microscopy techniques may be complex and expensive (Dufour, 2011). Figure 8.3 shows different microscopy techniques which have been used to examine microstructure of fermented dairy products.

8.5.1 Light Microscopy

Bright field light microscopy, polarized light microscopy, fluorescence microscopy and confocal scanning laser microscopy techniques have been used to examine the microstructure of dairy products.

8.5.1.1 Bright Field Light Microscopy

In bright field light microscopy, transmission of illumination is transmitted sequentially through a condenser, the specimen and the objective resulting in the real image which is further adjusted based on the requirement using different techniques. In addition, for clear identification, different structural components are colored using different dyes or stains that will improve the contrast of the samples (Kaláb *et al.*, 1995). Morell *et al.* (2015) observed yogurt using bright field light microscopy and reported to have a grey colored protein network stained with toluidine blue. However, in yogurt containing whey protein, the protein network was observed as explained above and the whey protein network was observed to be a faint blue color. In another study, bright field microscopy was used to examine the gel network of fat-free yogurt, whole yogurt, fat-free with milk solid non-fat, and whole fat with milk solid non-fat yogurt. The size of the protein network of the two yogurt containing milk solid non-fat was shown to be greater than that of yogurt produced without adding milk solid non-fat (Yu *et al.*, 2016) (Figure 8.4).

8.5.1.2 Polarized Light Microscopy

The basic microscope can be converted to polarized light microscopy by attaching polarizing accessories (El-Bakry and Sheehan, 2014). This microscopy technique has been used in microstructural studies related to yogurt and fermented milk. The microstructure of amylomaltase-treated starch supplemented set-yogurt was examined using polarized light microscopy to check the presence or absence of crystalline regions in amylomaltase-treated starch yogurt gel (Kanning *et al.*, 2012) (Figure 8.5). In another study, Aryana *et al.* (2015) used a polarized light microscope to examine the microstructure of resistant starch enriched yogurt. In a separate study, yogurt supplemented

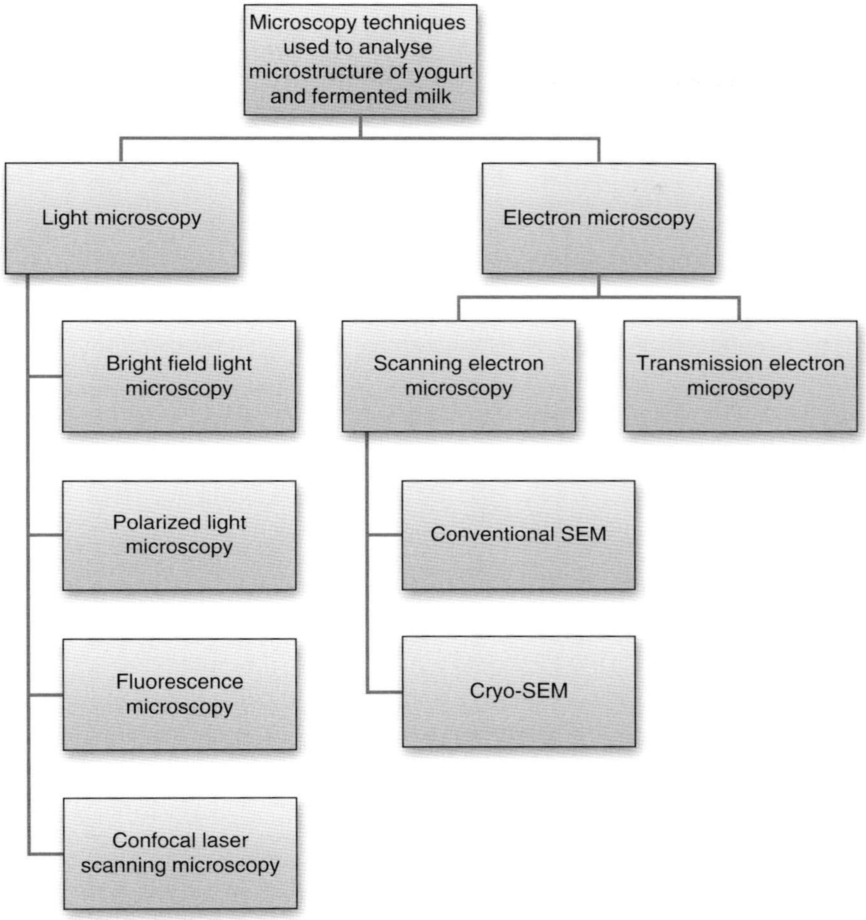

Figure 8.3 Different microscopy techniques used to examine microstructure of yogurt and fermented milk.

with physically modified starch was examined using a polarized light microscope and the samples were stained using toluidine blue (Morell *et al.*, 2015).

8.5.1.3 Fluorescence Microscopy

In fluorescence microscopy, molecules in the specimen absorb light of a specific wavelength and the energy is re-emitted as light of a longer wavelength and lower intensity (fluorescence) (Kaláb *et al.*, 1995). Furthermore, in fluorescence microscopy specific fluorescent dyes are used to stain the components in the specimen. These dyes are basically used to induce fluorescence only in the specific environment (El-Bakry and Sheehan, 2014). Therefore, a sample preparation should be carried out before microstructural analysis of yogurt and fermented milk which is time consuming and it is a main disadvantage associated with this technique. Singh and Kim (2009)

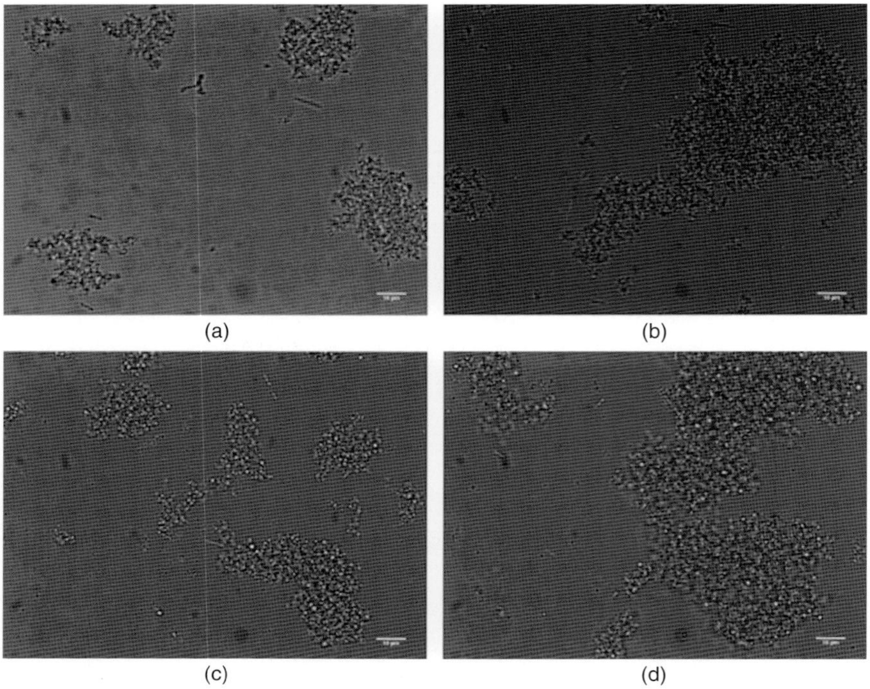

Figure 8.4 Bright field microscopy images of yogurt. (a) Fat free yogurt. (b) Fat free with milk solids non-fat yogurt. (c) Whole fat yogurt. (d) Whole fat with milk solids non-fat yogurt. Adapted from Yu, Wang, and McCarthy (2016).

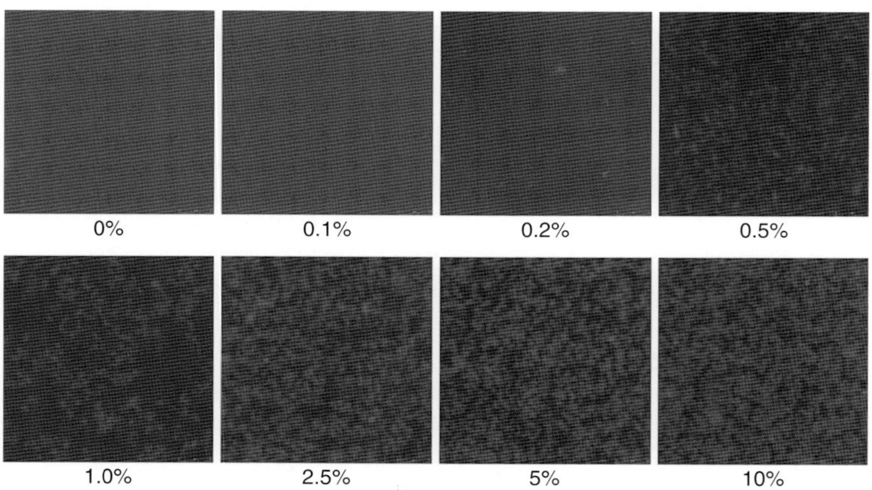

Figure 8.5 Confocal laser scanning microscopy images of 0–10% (w/w) amylomaltase-treated starch in yoghurt serum after 1 d of storage at 5°C. Image size is 62 mm 62 mm. Structures in bright red are amylomaltase-treated starch domains. Adapted from Kanning *et al.* (2012). (*See color plate section for the color representation of this figure.*)

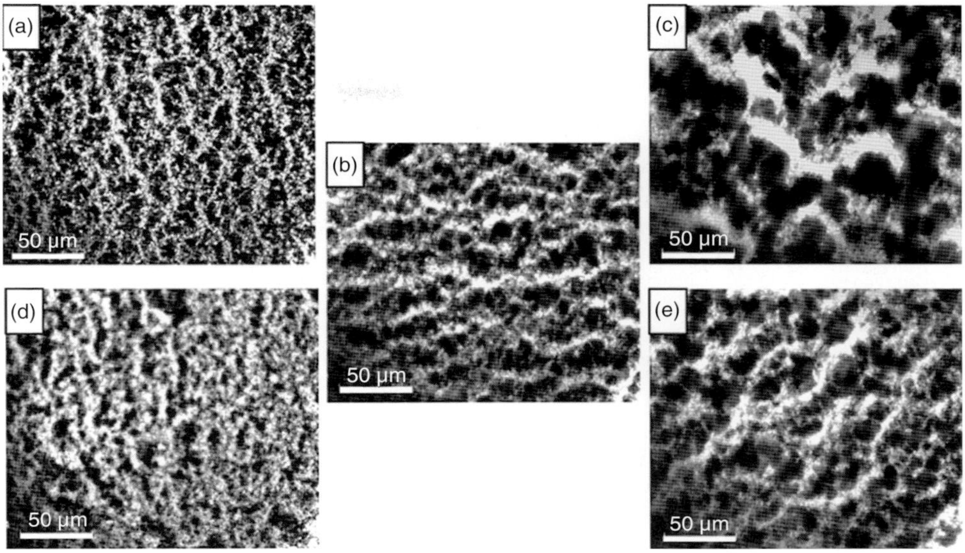

Figure 8.6 Fluorescence images of yogurt gels prepared with the pre-acidification pH of (a) 6.55; (b) 6.10; and (c) 5.65 and incubated for 375 min. Microstructure of yogurt gels prepared with the same pre-acidification pH of 6.10 and incubated for (d) 250 min; (b) 375 min; and (e) 500 min. Scale bar = 50 µm. The protein matrix is white, whereas pores are dark. Adapted from Peng, Horne, and Lucey (2009).

used fluorescence microscopy to examine yogurt gel containing starch–lipid composite. They used 7-Chloro-4-Nitrobenz-2-Oxa-1,3-Diazole (NBD-Cl) to label yogurt. In another study, the microstructure of yogurt gel was examined using fluorescence microscopy by using acridine orange as the fluorescent protein dye (Peng *et al.*, 2009) (Figure 8.6). The microstructure of yogurt gel containing tetrasodium pyrophosphate was examined by using fluorescence microscopy and acridine orange was used as the fluorescent protein dye (Ozcan *et al.*, 2008).

8.5.1.4 Confocal Laser Scanning Microscopy

Confocal laser scanning microscopy (CLSM) has been widely used in microstructural analysis of yogurt and fermented milk. The main advantage of the CLSM is that this technique can be used to produce optical sections through a three-dimensional (3-D) specimen (Dürrenberger *et al.*, 2001). In addition, CLSM can be used to identify different components of a sample by using specific fluorescent labels. CLSM has been used in many structural studies of fermented dairy products. CLSM was used to examine the microstructure of yogurt using the samples stained with Congo red (0.01% in water) and fluorescein isothiocyanate (FITC, 0.025% in dimethyl sulfoxide) (McCann *et al.*, 2011) (Figure 8.7). In another study, acid milk gel was observed using CLSM, and proteins of samples were labeled by using Fast Green CFC dye (Oh *et al.*, 2011). Hassan *et al.* (2003b) examined the microstructure of the protein network in the fermented milk prepared using EPS producing bacteria, and wheat germ agglutinin conjugated with Alexa flour 488 was used to label EPS of samples. In another study, low fat yogurt was observed using CLSM (London *et al.*, 2015). In this study, protein samples were labeled

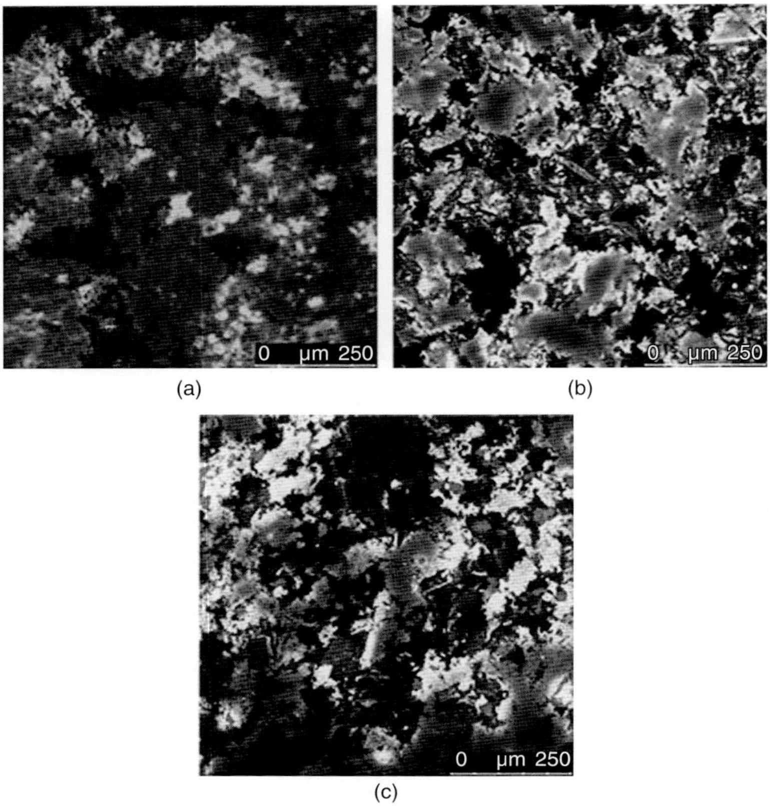

Figure 8.7 CLSM images of the structure of milk–CWP gels containing: (a) 0%; (b) 0.5% and (c) 1.5% rehydrated carrot cell wall fiber particles ($d_{0.5}$ = 71 µm, dry). Adapted from McCann *et al.* (2011). (*See color plate section for the color representation of this figure.*)

using Fast Green FCF fluorescent stain while EPS was labeled using lectin wheat germ agglutinin Alexafluor 555 conjugate. In addition, SYTO9 fluorescent dye was used to label both viable and non-viable bacteria.

8.5.2 Electron Microscopy

Electron microscopy techniques are used to obtain high resolution images, and these result in higher resolution images compared to light microscopy techniques. Electron microscopy can be broadly categorized into two groups, namely scanning electron microscopy (SEM) and transmission electron microscopy (TEM); both techniques have been used to examine microstructure of yogurt and fermented dairy products.

8.5.2.1 Scanning Electron Microscopy

SEM is used to examine the three-dimensional surface of the samples (Kaminskyj and Dahms, 2008). SEM is considered as a very important technique in dairy research which can be used to gain information on microstructure of dairy products which may be related to physical properties (Kaláb, 1993). In SEM, there are two sub-divisions, namely, conventional SEM (used with dried samples) and cryo-SEM (used with frozen samples).

Conventional SEM In conventional SEM, the sample should be dried and then it should be coated with a specific metal which is necessary to prevent a charging that effect on the quality of the micrographs (El-Bakry and Sheehan, 2014). A focused electron beam is used to scan the sample under the vacuum conditions and its enlarged image is formed by processing secondary or beck scattered electrons (Kaláb *et al.*, 1995). The high vacuum conditions used in SEM lead to evaporation of free water in the sample, resulting in distortion of the sample. Furthermore, fat and other volatile compounds in the sample should be fixed or extracted before SEM observation to avoid melting under the electron beam. In general, dairy products contain a high moisture level and fat content leading to extensive sample preparation prior to microscopic observation (Hassan *et al.*, 2003a). Conventional SEM has been used in studying microstructure of yogurt and fermented dairy products. Sandoval-Castilla *et al.* (2004) used SEM to examine microstructure of reduced-fat yogurts containing three different fat replacers. In this study, sample preparation included fixation, and stabilizing unsaturated lipids, followed by dehydration in a series of graded ethanol. In another study, SEM was used to evaluate the structure of goat milk yogurt where sample preparation included fixation, staining and dehydration using an acetone dehydration series (Farnsworth *et al.*, 2006). SEM was successfully used to examine the microstructure of reduced-fat stirred yogurt (Crispín-Isidro *et al.*, 2015). Stirred milk gel was examined using SEM micrographs by Pang *et al.* (2016) (Figure 8.8). In this study, samples were fixed with glutaraldehyde, dehydrated using ethanol and dried with a CO_2 critical point dryer.

Cryo-SEM Cryo-SEM can be easily used to overcome many problems associated with conventional SEM (Kaláb *et al.*, 1995). Cryo-SEM is considered as an ideal technique to

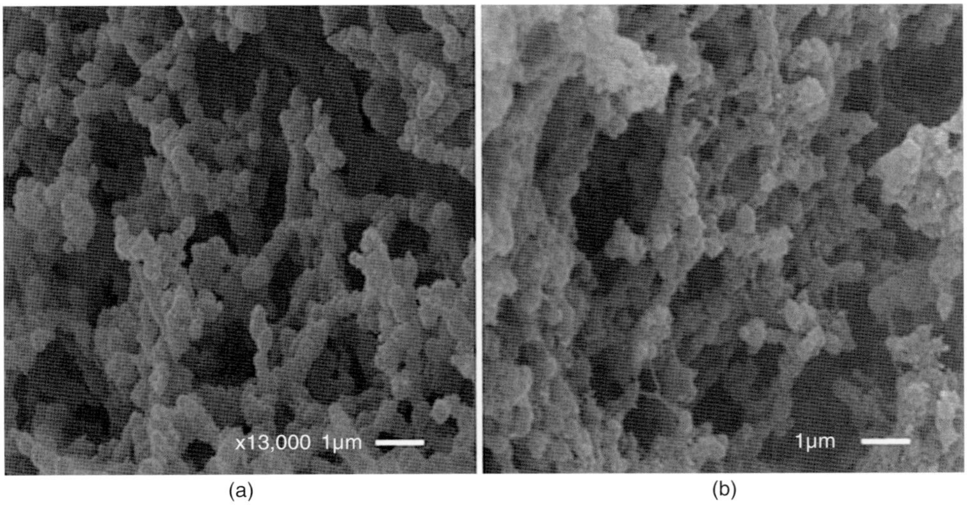

Figure 8.8 SEM micrographs of control milk gel (a) and xanthan and locust bean gum (b). Scale bars in the images are 1 µm. Adapted from Pang *et al.* (2016).

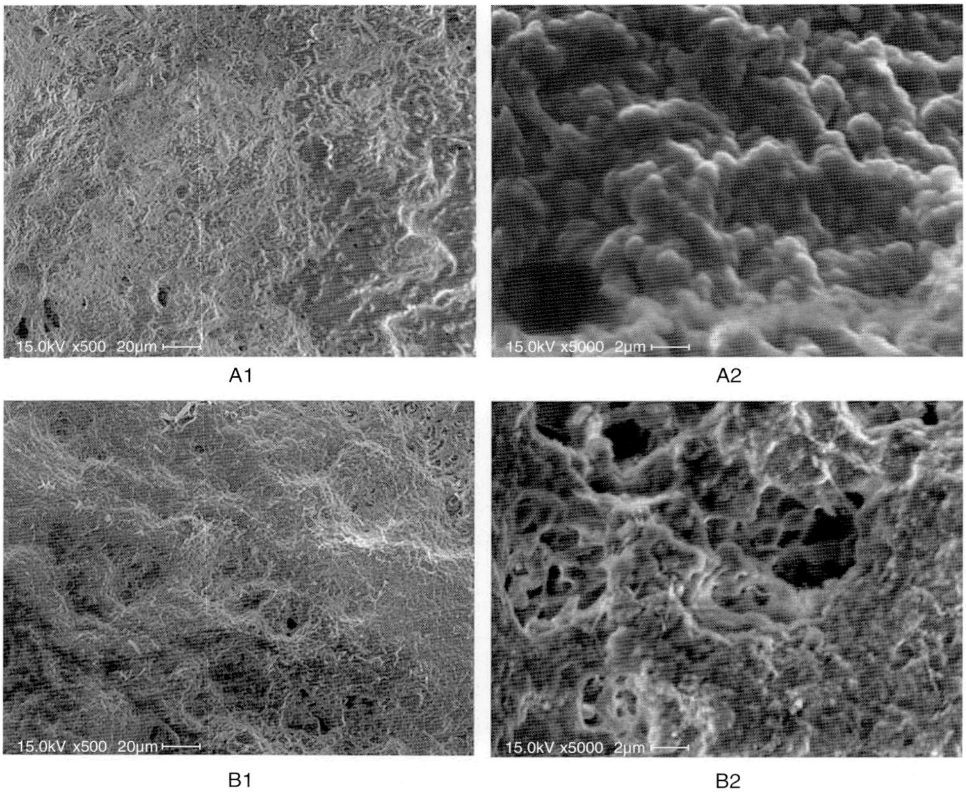

Figure 8.9 Cryo-SEM micrographs of yoghurts from conventionally (A) heated and (B) thermosonicated milk (1.5% fat). (Magnification: A1, B1 ×500; A2, B2 ×5000). Adapted from Riener *et al.* (2009).

examine dairy products such as yogurt that are high in water and fat content, with minimum distortion to samples (El-Bakry and Sheehan, 2014). In this technique, rapid freezing is used to fix samples and they are examined in their frozen hydrated state. The frozen sample is placed in a chamber where it can be cryo-fractured allowing observation of a fresh surface (Hassan *et al.*, 2003a). In one study, the microstructure of yogurt gel was examined using cryo-SEM to evaluate the effect of thermosonication on structural properties of yogurt gel (Riener *et al.*, 2009) (Figure 8.9). In another study, Prasanna *et al.* (2013) examined yogurt produced with EPS producing bifidobacteria using micrographs of cryo-SEM to evaluate the effect of *in situ* EPS production on the microstructure of yogurt. The effect of polyunsaturated fatty acids on microstructural properties of fermented milk was evaluated by using micrographs of cryo-SEM (Martín-Diana *et al.*, 2004). Vargas *et al.* (2008) successfully used cryo-SEM to examine microstructural properties of yogurt produced using cows' and goats' milk.

8.5.2.2 Transmission Electron Microscopy

Transmission electron microscopy (TEM) is used to visualize the internal structure of samples (Figure 8.10). This technique uses electrons which pass through the specimen

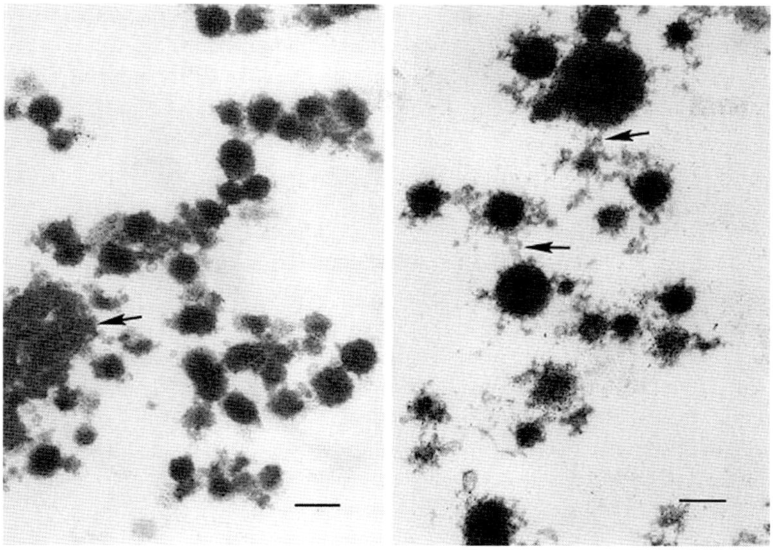

(a) (b)

Figure 8.10 TEM micrographs showing set yogurt gels, cooled and held at 4 mC for 24 h following gelation. Gels were prepared from (a) milk pressure-treated at 600 MPa for 15 min and warmed from 4 to 43°C (amorphous material entrapped within the gel network) and (b) milk heated at 85 mC for 20 min and cooled to 43°C (filamentous projections form long-range bridges between micelles). Scale bar, 200 nm. Adapted from Needs *et al.* (2000).

and the image is formed based on the scattered electrons (El-Bakry and Sheehan, 2014; Kaláb *et al.*, 1995). Tamime *et al.* (1995) used TEM to study the microstructure of yogurt. In this study, yogurt samples were cut into 0.5 mm cubes, post-fixed in veronal acetate buffer, and dehydrated in a series of ethanol. Samples were stained with uranyl acetate and lead citrate solutions and examined by using a transmission electron microscope. In another study, TEM was successfully used to examine the microstructure of acid skim milk gels (Sanchez *et al.*, 2000). In the process of sample preparation, the sample was fixed with glutaraldehyde solution, post-fixed and dehydrated in a series of ethanol followed by staining with razur-II. In addition, TEM was used to examine microstructure of fat free yogurt by Trejo *et al.* (2014).

8.6 Conclusions

Yogurt and fermented milk microstructure influences the texture and other physical properties, and also the organoleptic characteristics of these products. Various factors including the type of milk; rate of inoculation level and starter culture composition; type of culture; the various metabolites including exopolysaccharides produced by the cultures; manufacturing procedures and processing conditions such as incubation temperature; and the inclusion of various food hydrocolloids and fibers, influence the microstructure of yogurts and fermented milk products. The microstructure of yogurts and fermented milk are already well-studied, mainly using various microscopic techniques. However, research and techniques involved in improving microstructure of these dairy products with the ultimate aim of enhancing their consumer acceptability are constantly being improved, and evaluation of microstructure of fermented dairy foods is currently an active area of research.

References

Abbasi, H., Mousavi, M. E., and Ehsani, M. R. (2009). Influence of starter culture type and incubation temperatures on rheology and microstructure of low fat set yoghurt. *International Journal of Dairy Technology, 62*(4), 549–555.

Aguilera, J. M., and Stanley, D. W. (1999). Microstructural principles of food processing and engineering. Springer Science and Business Media.

Amatayakul, T., Sherkat, F., and Shah, N. P. (2006). Physical characteristics of set yoghurt made with altered casein to whey protein ratios and EPS-producing starter cultures at 9 and 14% total solids. *Food Hydrocolloids, 20*(2–3), 314–324.

Aryana, K., Greenway, F., Dhurandhar, N., Tulley, R., Finley, J., Keenan, M., Martin, R., Pelkman, C., Olson, D., and Zheng, J. (2015). A resistant-starch enriched yogurt: fermentability, sensory characteristics, and a pilot study in children. *F1000Research, 4*(139).

Aziznia, S., Khosrowshahi, A., Madadlou, A., and Rahimi, J. (2008). Whey protein concentrate and gum tragacanth as fat replacers in nonfat yogurt: chemical, physical, and microstructural properties. *Journal of Dairy Science, 91*(7), 2545–2552.

Balthazar, C. F., Pimentel, T. C., Ferrão, L. L., *et al.* (2017). Sheep milk: physicochemical characteristics and relevance for functional food development. *Comprehensive Reviews in Food Science and Food Safety, 16*(2), 247–262.

Boylston, T. D., Vinderola, C. G., Ghoddusi, H. B., and Reinheimer, J. A. (2004). Incorporation of bifidobacteria into cheeses: challenges and rewards. *International Dairy Journal, 14*(5), 375–387.

Codex Alimentarius Commission. 2010. Codex standard for fermented milks. Codex Stan 243-2003. [online]. Available at: < www.codexalimentarius.org/input/download/.../400/CXS_ 243e.pdf> Accessed 23 March 2018.

Crispín-Isidro, G., Lobato-Calleros, C., Espinosa-Andrews, H., Alvarez-Ramirez, J., and Vernon-Carter, E. (2015). Effect of inulin and agave fructans addition on the rheological, microstructural and sensory properties of reduced-fat stirred yogurt. *LWT-Food Science and Technology, 62*(1), 438–444.

da Cruz, A. G., de AF Faria, J., and Van Dender, A. G. (2007). Packaging system and probiotic dairy foods. *Food Research International, 40*(8), 951–956.

Domagała, J. (2009). Instrumental texture, syneresis and microstructure of yoghurts prepared from goat, cow and sheep Milk. *International Journal of Food Properties, 12*(3), 605–615.

Dufour, E. (2011). Recent advances in the analysis of dairy product quality using methods based on the interactions of light with matter. *International Journal of Dairy Technology, 64*(2), 153–165.

Dürrenberger, M. B., Handschin, S., Conde-Petit, B., and Escher, F. (2001). Visualization of food structure by confocal laser scanning microscopy (CLSM). *LWT – Food Science and Technology, 34*(1), 11–17.

Ekanayake, S. (2016). Ethnic fermented foods and beverages of Sri Lanka. In: Ethnic Fermented Foods and Alcoholic Beverages of Asia (pp. 139–150). Springer, India.

El-Bakry, M., and Sheehan, J. (2014). Analysing cheese microstructure: A review of recent developments. *Journal of Food Engineering, 125*, 84–96.

Espírito-Santo, A. P., Lagazzo, A., Sousa, A. L. O. P., Perego, P., Converti, A., and Oliveira, M. N. (2013). Rheology, spontaneous whey separation, microstructure and sensorial characteristics of probiotic yoghurts enriched with passion fruit fiber. *Food Research International, 50*(1), 224–231.

FAO/WHO (2002). FAO/ WHO working group report on drafting guidelines for the evaluation of probiotics in food. London, Ontario, Canada, *30*.

Farnsworth, J. P., Li, J., Hendricks, G. M., and Guo, M. R. (2006). Effects of transglutaminase treatment on functional properties and probiotic culture survivability of goat milk yogurt. *Small Ruminant Research, 65*(1–2), 113–121.

FDA (2013). US Food and Drug Administration, Yogurt. 21 CFR 131.200, Code of Federal Regulations. U. S. Dept. of Health and Human Services, Washington, DC.

Folkenberg, D., Dejmek, P., Skriver, A., and Ipsen, R. (2005). Relation between sensory texture properties and exopolysaccharide distribution in set and in stirred yoghurts produced with different starter cultures. *Journal of Texture Studies, 36*(2), 174–189.

Food Standards Code of Food Standards Australia New Zealand (2014). Food Standards Code – Standard 2.5.3.

Guggisberg, D., Cuthbert-Steven, J., Piccinali, P., Bütikofer, U., and Eberhard, P. (2009). Rheological, microstructural and sensory characterization of low-fat and whole milk set yoghurt as influenced by inulin addition. *International Dairy Journal, 19*(2), 107–115.

Hadjipanayiotou, M. (1995). Composition of ewe, goat and cow milk and of colostrum of ewes and goats. *Small Ruminant Research, 18*(3), 255–262.

Harwalkar, V. R., and Kaláb, M. (1980). Milk gel structure. XI. Electron microscopy of glucono-δ-lactone-induced skim milk gels1. *Journal of Texture Studies, 11*(1), 35–49.

Hashim, I. B., Khalil, A. H., and Habib, H. (2009). Quality and acceptability of a set-type yogurt made from camel milk. *Journal of Dairy Science, 92*(3), 857–862.

Hassan, A. N., Frank, J. F., and Elsoda, M. (2003a). Observation of bacterial exopolysaccharide in dairy products using cryo-scanning electron microscopy. *International Dairy Journal, 13*(9), 755–762.

Hassan, A. N., Frank, J. F., Farmer, M. A., Schmidt, K. A., and Shalabi, S. I. (1995). Formation of yogurt microstructure and three-dimensional visualization as determined by confocal scanning laser microscopy. *Journal of Dairy Science, 78*(12), 2629–2636.

Hassan, A., Frank, J., Schmidt, K., and Shalabi, S. (1996). Rheological properties of yogurt made with encapsulated nonropy lactic cultures. *Journal of Dairy Science, 79*(12), 2091–2097.

Hassan, A., Ipsen, R., Janzen, T., and Qvist, K. (2003b). Microstructure and rheology of yogurt made with cultures differing only in their ability to produce exopolysaccharides. *Journal of Dairy Science, 86*(5), 1632–1638.

Kaláb, M. (1993). Practical aspects of electron microscopy in dairy research. *Food Structure, 12*(1), 11.

Kaláb, M., Allan-Wojtas, P., and Miller, S. S. (1995). Microscopy and other imaging techniques in food structure analysis. *Trends in Food Science and Technology, 6*(6), 177–186.

Kaláb, M., Allan-Wojtas, P., and Phipps-Todd, B. E. (1983). Development of microstructure in set-style nonfat yogurt-a review. *Food Structure, 2*(1), 7.

Kaminskyj, S. G., and Dahms, T. E. (2008). High spatial resolution surface imaging and analysis of fungal cells using SEM and AFM. *Micron, 39*(4), 349–361.

Kanning, M. W., van de Velde, F., Alting, A. C., Mulleners, L., Sein, A., van den Akker-Bleumink, B., and Buwalda, P. (2012). Improved creaminess in stirred yoghurt through amylomaltase-treated starch domains. *International Dairy Journal, 27*(1–2), 86–91.

Karademir, E., Atamer, M., Tamucay, B., and Yaman, S. (2002). Some properties of goat milk yoghurts produced by different fortification methods. *Milchwissenschaft, 57*(5), 261–263.

Kelvin, K. T. G., Haisman, R. D., and Singh, H. (2005). Examination of exopolysaccharide produced by *Lactobacillus delbrueckii* subsp. *bulgaricus* using confocal laser scanning and scanning electron microscopy techniques. *Journal of Food Science, 70*(4), M224–M229.

Kristo, E., Biliaderis, C., and Tzanetakis, N. (2003). Modelling of rheological, microbiological and acidification properties of a fermented milk product containing a probiotic strain of Lactobacillus paracasei. *International Dairy Journal, 13*(7), 517–528.

Laverse, J., Mastromatteo, M., Frisullo, P., Albenzio, M., Gammariello, D., and Del Nobile, M. A. (2011). Fat microstructure of yogurt as assessed by x-ray microtomography. *Journal of Dairy Science, 94*(2), 668–675.

Lee, W. J., and Lucey, J. A. (2004). Structure and physical properties of yogurt gels: Effect of inoculation rate and incubation temperature. *Journal of Dairy Science, 87*(10), 3153–3164.

Lee, W. J., and Lucey, J. A. (2010). Formation and physical properties of yogurt. *Asian–Australasian Journal of Animal Sciences, 23*(9), 1127–1136.

Lee, W. J., and Lucey, J. A. (2006). Impact of gelation conditions and structural breakdown on the physical and sensory properties of stirred yogurts. *Journal of Dairy Science, 89*(7), 2374–2385.

Lee, W. J., and Lucey, J. A. (2003). Rheological properties, whey separation, and microstructure in set-style yogurt: Effects of heating temperature and incubation temperature. *Journal of Texture Studies, 34*(5–6), 515–536.

Leivers, S., Hidalgo-Cantabrana, C., Robinson, G., Margolles, A., Ruas-Madiedo, P., and Laws, A. P. (2011). Structure of the high molecular weight exopolysaccharide produced by *Bifidobacterium animalis* subsp. *lactis* IPLA-R1 and sequence analysis of its putative eps cluster. *Carbohydrate Research, 346*(17), 2710–2717.

London, L. E. E., Chaurin, V., Auty, M. A. E., Fenelon, M. A., Fitzgerald, G. F., Ross, R. P., and Stanton, C. (2015). Use of Lactobacillus mucosae DPC 6426, an exopolysaccharide-producing strain, positively influences the techno-functional properties of yoghurt. *International Dairy Journal, 40*, 33–38.

Lucey, J. A., Munro, P. A., and Singh, H. (1998a). Rheological properties and microstructure of acid milk gels as affected by fat content and heat treatment. *Journal of Food Science, 63*(4), 660–664.

Lucey, J. A., Teo, C. T., Munro, P. A., and Singh, H. (1998b). Microstructure, permeability and appearance of acid gels made from heated skim milk. *Food Hydrocolloids, 12*(2), 159–165.

Lucey, J. A., and Singh, H. (1997). Formation and physical properties of acid milk gels: a review. *Food Research International, 30*(7), 529–542.

Lucey, J. A., Van Vliet, T., Grolle, K., Geurts, T., and Walstra, P. (1997). Properties of acid casein gels made by acidification with glucono-δ-lactone. 1. Rheological properties. *International Dairy Journal, 7*(6–7), 381–388.

Martín-Diana, A. B., Janer, C., Peláez, C., and Requena, T. (2004). Effect of milk fat replacement by polyunsaturated fatty acids on the microbiological, rheological and sensorial properties of fermented milks. *Journal of the Science of Food and Agriculture, 84*(12), 1599–1605.

McCann, T. H., Fabre, F., and Day, L. (2011). Microstructure, rheology and storage stability of low-fat yoghurt structured by carrot cell wall particles. *Food Research International, 44*(4), 884–892.

Mellema, M., Walstra, P., Van Opheusden, J., and Van Vliet, T. (2002). Effects of structural rearrangements on the rheology of rennet-induced casein particle gels. *Advances in Colloid and Interface Science, 98*(1), 25–50.

Morell, P., Hernando, I., Llorca, E., and Fiszman, S. (2015). Yogurts with an increased protein content and physically modified starch: rheological, structural, oral digestion and sensory properties related to enhanced satiating capacity. *Food Research International, 70*, 64–73.

Needs, E. C., Capellas, M., Bland, A. P., Manoj, P., Macdougal, D., and Paul, G. (2000). Comparison of heat and pressure treatments of skim milk, fortified with whey protein concentrate, for set yogurt preparation: effects on milk proteins and gel structure. *Journal of Dairy Research, 67*(3), 329–348.

Nguyen, H. T. H., Ong, L., Kentish, S. E., and Gras, S. L. (2015). Homogenization improves the microstructure, syneresis and rheological properties of buffalo yoghurt. *International Dairy Journal, 46*, 78–87.

Nguyen, H. T. H., Ong, L., Lefèvre, C., Kentish, S. E., and Gras, S. L. (2014). The microstructure and physicochemical properties of probiotic buffalo yoghurt during fermentation and storage: a comparison with bovine yoghurt. *Food and Bioprocess Technology, 7*(4), 937–953.

Nöbel, S., Hahn, C., Hitzmann, B., and Hinrichs, J. (2014). Rheological properties of microgel suspensions: Viscoelastic modelling of microstructural elements from casein micelles to fermented dairy products. *International Dairy Journal, 39*(1), 157–166.

Oh, H., Wong, M., Pinder, D., and Anema, S. (2011). Comparison of pressure treatment and heat treatment of skim milk with added starch on subsequent acid gelation of milk. *LWT-Food Science and Technology, 44*(4), 1189–1198.

Oliveira, M., Sodini, I., Remeuf, F., and Corrieu, G. (2001). Effect of milk supplementation and culture composition on acidification, textural properties and microbiological stability of fermented milks containing probiotic bacteria. *International Dairy Journal, 11*(11), 935–942.

Ozcan, T., Lucey, J. A., and Horne, D. S. (2008). Effect of tetrasodium pyrophosphate on the physicochemical properties of yogurt gels. *Journal of Dairy Science, 91*(12), 4492–4500.

Pang, Z., Deeth, H., Prakash, S., and Bansal, N. (2016). Development of rheological and sensory properties of combinations of milk proteins and gelling polysaccharides as potential gelatin replacements in the manufacture of stirred acid milk gels and yogurt. *Journal of Food Engineering, 169*, 27–37.

Peng, Y., Horne, D. S., and Lucey, J. A. (2009). Impact of preacidification of milk and fermentation time on the properties of yogurt. *Journal of Dairy Science, 92*(7), 2977–2990.

Peng, Y., Horne, D. S., and Lucey, J. A. (2010). Physical properties of acid milk gels prepared at 37°C up to gelation but at different incubation temperatures for the remainder of fermentation. *Journal of Dairy Science, 93*(5), 1910–1917.

Penna, A., and Barbosa-Cánovas, G. (2007). High hydrostatic pressure processing on microstructure of probiotic low-fat yogurt. *Food Research International, 40*(4), 510–519.

Pesic, M. B., Barac, M. B., Stanojevic, S. P., Ristic, N. M., Macej, O. D., and Vrvic, M. M. (2012). Heat induced casein–whey protein interactions at natural pH of milk: A comparison between caprine and bovine milk. *Small Ruminant Research, 108*(1), 77–86.

Priyashantha, T. V. H., Pérez, A. Q., Baixauli, R. M., Albalat, M. P. F. and Vidanarachchi, J. K. 2015. Effect of starter culture on structural and sensorial properties of fermented gel. 9th NIZO Dairy Conference: Milk Protein Functionality, 30th September–2nd October 2015, Papendal, The Netherlands.

Prasanna, P. H. P., Grandison, A. S., and Charalampopoulos, D. (2013). Microbiological, chemical and rheological properties of low fat set yoghurt produced with exopolysaccharide (EPS) producing *Bifidobacterium* strains. *Food Research International, 51*(1), 15–22.

Purwandari, U., Shah, N., and Vasiljevic, T. (2007). Effects of exopolysaccharide-producing strains of *Streptococcus thermophilus* on technological and rheological properties of set-type yoghurt. *International Dairy Journal, 17*(11), 1344–1352.

Ranadheera, C. S., Evans, C. A., Adams, M. C., and Baines, S. K. (2012). Probiotic viability and physico-chemical and sensory properties of plain and stirred fruit yogurts made from goat's milk. *Food Chemistry, 135*(3), 1411–1418.

Ranadheera, C. S., Evans, C. A., Adams, M. C., and Baines, S. K. (2014). Effect of dairy probiotic combinations on in vitro gastrointestinal tolerance, intestinal epithelial cell adhesion and cytokine secretion. *Journal of Functional Foods, 8*, 18–25.

Ranadheera, C. S., Evans, C. A., Adams, M., and Baines, S. K. (2016). Co-culturing of probiotics influences the microbial and physico-chemical properties but not sensory quality of fermented dairy drink made from goats' milk. *Small Ruminant Research, 136*, 104–108.

Ranadheera, R. D. C. S., Baines, S. K., and Adams, M. C. (2010). Importance of food in probiotic efficacy. *Food Research International, 43*(1), 1–7.

Raynal-Ljutovac, K., Lagriffoul, G., Paccard, P., Guillet, I., and Chilliard, Y. (2008). Composition of goat and sheep milk products: An update. *Small Ruminant Research, 79*(1), 57–72.

Remeuf, F., Mohammed, S., Sodini, I., and Tissier, J. (2003). Preliminary observations on the effects of milk fortification and heating on microstructure and physical properties of stirred yogurt. *International Dairy Journal, 13*(9), 773–782.

Riener, J., Noci, F., Cronin, D. A., Morgan, D. J., and Lyng, J. G. (2009). The effect of thermosonication of milk on selected physicochemical and microstructural properties of yoghurt gels during fermentation. *Food Chemistry, 114*(3), 905–911.

Rohm, H., and Kovac, A. (1994). Effects of starter cultures on linear viscoelastic and physical properties of yogurt gels. *Journal of Texture Studies, 25*(3), 311–329.

Saarela, M., Mogensen, G., Fonden, R., Mättö, J., and Mattila-Sandholm, T. (2000). Probiotic bacteria: safety, functional and technological properties. *Journal of Biotechnology, 84*(3), 197–215.

Sah, B. N. P., Vasiljevic, T., McKechnie, S., and Donkor, O. N. (2016). Physicochemical, textural and rheological properties of probiotic yogurt fortified with fiber-rich pineapple peel powder during refrigerated storage. *LWT-Food Science and Technology, 65*, 978–986.

Sanchez, C., Zuniga-Lopez, R., Schmitt, C., Despond, S., and Hardy, J. (2000). Microstructure of acid–induced skim milk–locust bean gum–xanthan gels. *International Dairy Journal, 10*(3), 199–212.

Sandoval-Castilla, O., Lobato-Calleros, C., Aguirre-Mandujano, E., and Vernon-Carter, E. (2004). Microstructure and texture of yogurt as influenced by fat replacers. *International Dairy Journal, 14*(2), 151–159.

Serra, M., Trujillo, A. J., Jaramillo, P. D., Guamis, B., and Ferragut, V. (2008). Ultra-high pressure homogenization-induced changes in skim milk: impact on acid coagulation properties. *Journal of Dairy Research, 75*(1), 69–75.

Serra, M., Trujillo, A., Guamis, B., and Ferragut, V. (2009). Evaluation of physical properties during storage of set and stirred yogurts made from ultra-high pressure homogenization-treated milk. *Food Hydrocolloids, 23*(1), 82–91.

Singh, M., and Kim, S. (2009). Yogurt fermentation in the presence of starch–lipid composite. *Journal of Food Science, 74*(2), C85–C89.

Sodini, I., Lucas, A., Tissier, J. P., and Corrieu, G. (2005). Physical properties and microstructure of yoghurts supplemented with milk protein hydrolysates. *International Dairy Journal, 15*(1), 29–35.

Soukoulis, C., Panagiotidis, P., Koureli, R., and Tzia, C. (2007). Industrial yogurt manufacture: monitoring of fermentation process and improvement of fnal product quality. *Journal of Dairy Science, 90*(6), 2641–2654.

Tamime, A. Y., and Robinson, R. K. (1999). *Yoghurt: Science and Technology*. Woodhead Publishing.

Tamime, A. Y., and Robinson, R. K. (2007). *Tamime and Robinson's Yoghurt: Science and Technology (Vol.* 3). Elsevier, Cambridge.

Tamime, A. Y., Kaláb, M., Muir, D. D., and Barrantes, E. (1995). The microstructure of set-style, natural yogurt made by substituting microparticulate whey protein for milk fat. *International Journal of Dairy Technology, 48*(4), 107–111.

The Dairy Council (2013). *The Nutritional Composition of Dairy Products*. London.

Trejo, R., Corzo-Martínez, M., Wilkinson, S., Higginbotham, K., and Harte, F. M. (2014). Effect of a low temperature step during fermentation on the physico-chemical properties of fat-free yogurt. *International Dairy Journal, 36*(1), 14–20.

USDA (2013), US Department of Agriculture, Agricultural Research Service, USDA National Nutrient Database for Standard Reference, *Release 26*.

Vargas, M., Cháfer, M., Albors, A., Chiralt, A., and González-Martínez, C. (2008). Physicochemical and sensory characteristics of yoghurt produced from mixtures of cows' and goats' milk. *International Dairy Journal, 18*(12), 1146–1152.

Vlahopoulou, I., Bell, A. E., and Wilbey, A. (1994). Starter culture effects on caprine yogurt fermentation. *International Journal of Dairy Technology, 47*(4), 121–123.

Weerathilake, W. A. D. V., Rasika, D. M. D., Ruwanmali, J. K. U., and Munasinghe, M. A. D. D. (2014). The evolution, processing, varieties and health benefits of yogurt. *International Journal of Scientific and Research Publications, 4*(4), 1–10.

Wu, S., Li, D., Li, S.-J., Bhandari, B., Yang, B.-l., Chen, X. D., and Mao, Z.-H. (2009). Effects of incubation temperature, starter culture level and total solids content on the rheological properties of yogurt. *International Journal of Food Engineering, 5*(2), 1–19.

Yu, H.-Y., Wang, L., and McCarthy, K. L. (2016). Characterization of yogurts made with milk solids nonfat by rheological behavior and nuclear magnetic resonance spectroscopy. *Journal of Food and Drug Analysis, 24*(4), 804–812.

9

Microstructure of Milk Fat and its Products

Pere Randy R. Ramel and Alejandro G. Marangoni

University of Guelph, Food Science Building, 50 Stone Road East, Guelph, Ontario, N1G 2W1, Canada

9.1 Introduction

Milk fat, together with proteins, is one of the major components of milk at a concentration of around 3-6% (Walstra *et al.*, 1999). Because of its ease of use and unique taste, it has been widely used as a key ingredient in many food products as it provides a characteristic structure and flavor. Its functionality, however, is greatly dependent on its thermal properties (i.e. crystallization and melting behavior). Thus, characterization of the structure of the crystal network formed by milk fat as affected by thermal processes, and different compositional and environmental factors has been studied extensively (Herrera and Hartel, 2000; Shi *et al.*, 2001; Lopez *et al.*, 2001c; Campos *et al.*, 2002; Marangoni *et al.*, 2012).

The thermal behavior of milk fat is very complex due to the extreme heterogeneity of its triacylglycerol composition (Timms 1984; Lopez *et al.*, 2001a,b; Lopez *et al.*, 2005). Triacylglycerols (TAGs), which constitute 98% of milk fat, are composed of a glycerol backbone with fatty acids esterified to its three stereospecific positions (*sn1*, sn2, *sn3*) (Figure 9.1) (Jensen *et al.*, 1991; McGibbon and Taylor, 2006). The type of fatty acids present as well as their distribution on the glycerol backbone determine the properties of the TAG molecule (Jensen *et al.*, 1991; Palmquist, 2009; Tzompa-Sosa *et al.*, 2014). The composition of fatty acids in milk changes greatly with the cow's feed, thus, depending on the season (i.e. summer grass vs winter diet), fatty acids could range from unsaturated to saturated, and short chain to long chain and therefore causing the large variation in TAG molecular species present in milk fat (Shi *et al.*, 2001; Jensen 2002; Bugeat *et al.*, 2011). This diversity in TAG composition in combination with the presence of minor components (2%) in milk fat lead to formation of compound crystals upon crystallization which brings about the broad melting range of milk fat at –40°C to 40°C (Timms, 1980; Shi *et al.*, 2001; Wright and Marangoni, 2002; Wright and Marangoni, 2003; Mazzanti *et al.*, 2004). Besides heterogeneity in composition, TAG crystals also exhibit polymorphism, which is a general property of TAG molecules. This increases the level of complexity even further (Grotenhuis *et al.*, 1999; Sato, 2001).

Microstructure of Dairy Products, First Edition.
Edited by Mamdouh Mahmoud Abdel-Rahman El-Bakry, Antoni Sanchez, and Bhavbhuti M. Mehta.

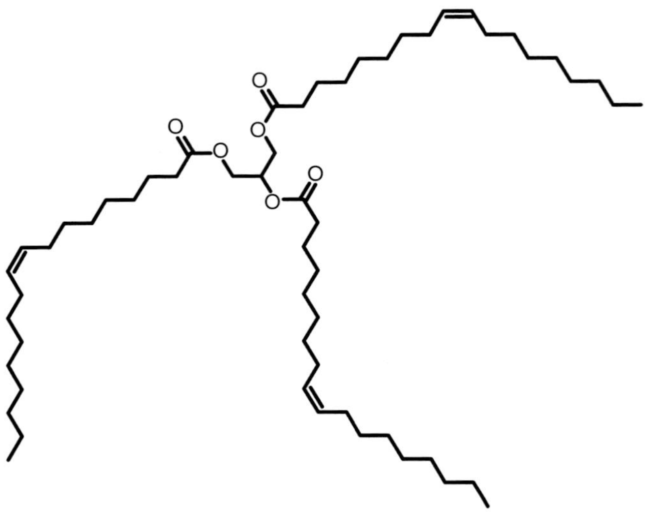

Figure 9.1 Structure of a triacylglycerol molecule.

The two steps involved in the crystallization of TAGs include nucleation and then crystal growth (Marangoni and Wesdorp, 2013b; Sangwal and Sato, 2012). However, before nucleation can take place, supersaturation of TAGs must first be achieved (Sato, 2001; Marangoni *et al.*, 2012). Supersaturation happens when the supercooling or undercooling becomes sufficiently high (i.e. the number of degrees below the melting point of the highest melting TAG) so that the melt becomes supersaturated with a particular TAG with respect to the liquid phase in which they are dissolved (Marangoni *et al.*, 2012). The supersaturation of TAGs is very important as it determines the polymorphism of milk fat (Kloek *et al.*, 2000; Sato 2001; Dibildox-Alvarado *et al.*, 2010; Marangoni *et al.*, 2012). Nucleation then occurs wherein TAG molecules organize to form ordered domains up to a certain critical size (i.e. large enough to not be easily dissolved) (Sangwal and Sato 2012; Marangoni *et al.*, 2012; Marangoni and Wesdorp, 2013b). The nucleation process is essential in the formation of a fat crystal network (Sato, 2001; Martini *et al.*, 2001; Marangoni *et al.*, 2012). When stable nuclei are formed, crystal growth rapidly ensues and the number of nuclei present determines the amount and size of the crystals formed (Kloek *et al.*, 2000; Martini *et al.*, 2001; Marangoni *et al.*, 2006; Marangoni *et al.*, 2012; Marangoni and Wesdorp, 2013b). Depending on the cooling rate, crystallization temperature, as well as other external factors, which will be described in the following sections, milk fat crystallizes in a certain state (i.e. polymorphism) which greatly affects the microstructure and overall quality of dairy products (Grotenhuis *et al.*, 1999; van Aken and Visser, 2000; Lopez *et al.*, 2001a,b; Lopez *et al.*, 2005).

Advances in the methods used for analyzing and characterizing the microstructure formed by milk fat have been made throughout the years and this has led to numerous opportunities for improving the properties of several food products (Peyronel and Campos, 2012). Besides its importance in food manufacturing, increased knowledge of the microstructure of milk fat could also significantly impact consumer health. Nutritionally, milk fat is considered to be unhealthy as it contains high levels of saturated and *trans* fats (Parodi, 2004). Thus, a popular solution nowadays is to blend milk fat or butter with vegetable oils to increase unsaturation which in turn, greatly affects microstructure, thermal and mechanical properties, and sensory perceptions (Smet *et al.*, 2010; Kerr *et al.*, 2011; Bugeat *et al.*, 2011; Kaufmann *et al.*, 2012). By knowing specific effects of these changes on the crystallization behavior of milk fat, effective solutions can be provided.

In this chapter, the different structures formed by milk fat crystal networks are described. In addition, current methods used to characterize these structures, and the effect of compositional and environmental factors on the microstructure are discussed. Lastly, and more importantly, we also elaborate on the impact of the resulting microstructure on the quality (e.g. rheology and texture) of different milk fat products.

9.2 Milk Fat Crystal Structure

The microstructure of milk fat determines, to a large extent, its mechanical, textural and functional properties. Due to the advantages that milk fat provides in numerous food applications, many studies have been conducted on its crystallization behavior, crystal structure formation and how these are affected by compositional changes and different environmental factors.

A range of methods such as various microscopy techniques (e.g. polarized light microscopy (PLM) and scanning electron microscopy (SEM)), differential scanning calorimetry (DSC), pulsed nuclear magnetic resonance (pNMR), and wide-angle (WA-) and small-angle (SA-) powder X-ray diffraction (XRD) are currently being used to analyze the microstructure and thermal behavior of milk fat (Peyronel and Campos, 2012). However, most recently, ultra-small angle X-ray scattering (USAXS) and cryogenic transmission electron microscopy (cryo-TEM) were found to be useful in describing the structure formed by triacylglycerol networks on the nanoscale (Acevedo *et al.*, 2011; Marangoni *et al.*, 2012; Acevedo, 2012).

9.2.1 Mesoscale Structure of Milk Fat

Milk fat crystal structure at the mesoscale (approximately 1-200μm) is usually studied using PLM and SEM (Marangoni *et al.*, 2012). At this scale, fat crystals are described according to their crystal habits (size and shape) which include spherulites, needle-like crystals, platelets, disordered aggregates and fractal-like aggregates (Figure 9.2) (Marangoni and Mcgauley, 2003; Marangoni *et al.*, 2012). Often, along with this microstructural characterization, powder XRD and DSC measurements are also performed to be able to describe the molecular arrangement of TAGs (polymorphism) and the melting behavior of the crystals associated with the particular microstructure. pNMR is also a useful tool to complement these studies and is used to quantify the amount of crystals formed, or solid fat content (SFC).

9.2.1.1 Polymorphism

Polymorphism of milk fat is a result of the various conformations in which TAGs can pack and stack in a lamellar crystal (Sato, 2001; Marangoni *et al.*, 2012). The more compact the TAG packing, the more bonds or interactions they make and the more dense

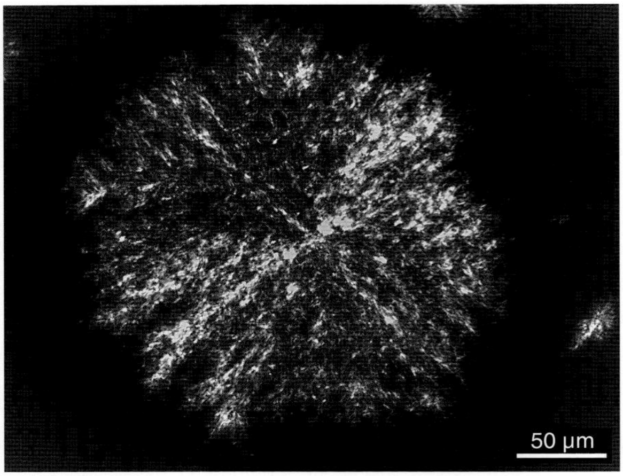

Figure 9.2 Spherulitic crystal growth of milk fat observed under polarized light microscope (PLM).

and stable they are (D'Souza *et al.*, 1990). Generally, three polymorphs are observed in milk fat namely alpha (α), beta prime (β′) and traces of beta (β). Polymorphism refers to the geometrical packing arrangement of the ethylene groups within the long fatty acid hydrocarbon chains within the TAG unit cell. Surprisingly, only three major subcell structures are observed in fats - hexagonal, orthorhombic perpendicular and triclinic (Figure 9.3) (Woodrow and deMan, 1968; Sato, 2001; Lopez *et al.*, 2001a,b; Lopez *et al.*, 2005; Marangoni and Wesdorp, 2013a). According to D'Souza *et al.* (1990) and Larsson (1966), α, β′ and β polymorphs are characterized by d-spacings 4.15Å, 3.8 and 4.2Å, and 4.6Å, respectively. The β polymorph is the most stable and therefore has the highest melting point, while α is the least stable and has the lowest melting point (D'Souza *et al.*, 1990; deMan, 1992). The driving force for TAGs to nucleate into a certain polymorphic form can be described using the concept of activation energies for nucleation (Marangoni *et al.*, 2012). Formation of α crystals requires a lower activation energy of nucleation

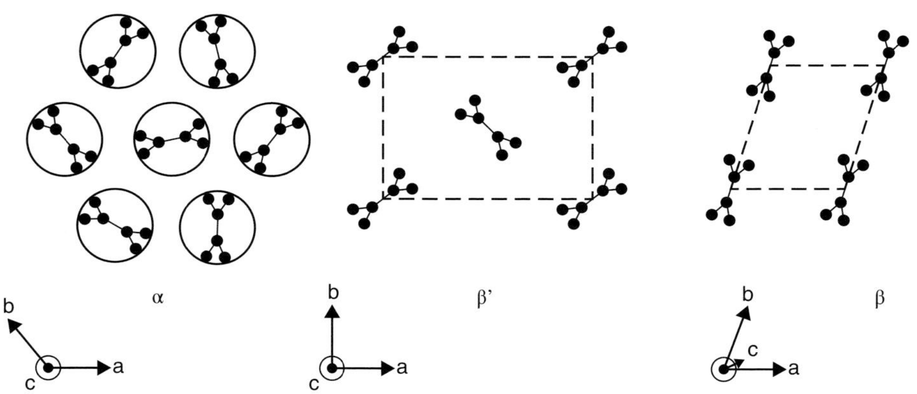

Figure 9.3 Three major subcell structures observed in fats: (a) hexagonal; (b) orthorhombic; perpendicular; and (c) triclinic.

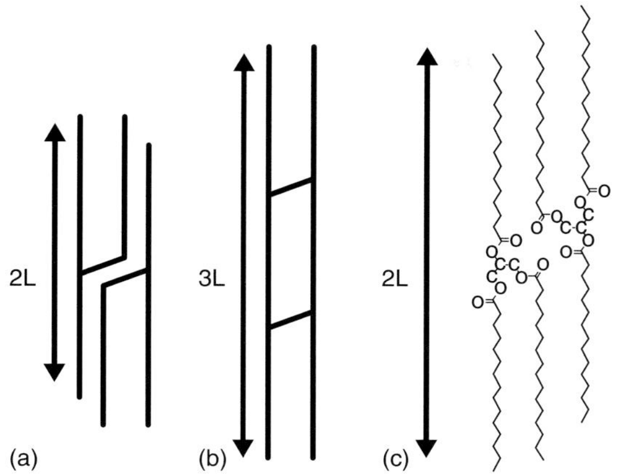

Figure 9.4 Epitaxial packing of triacylglycerols in different chain length configurations: (a) double 2L; and (b) triple, 3L chain lengths.

while β′ and β polymorphs have higher activation energies of nucleation (Marangoni *et al.*, 2012; Sangwal and Sato, 2012; Marangoni and Wesdorp, 2013b). As a consequence, the kinetics of nucleation for the α form are much faster than for the other polymorphs (Grotenhuis *et al.*, 1999; Martini *et al.*, 2001). Thus, more frequently, upon crystallization, the α polymorph appears first which eventually transforms into the β′ or β form, since the α polymorph is thermodynamically unstable (Sato, 2001; Sato and Ueno, 2011). According to Sato (2001), polymorphic nucleation is governed by the Ostwald rule where the metastable form nucleates first under high supersaturation or undercooling conditions (i.e., high cooling rate or lower temperatures), followed by a step-by-step phase change that leads to the formation of more stable crystal polymorphs. It is only when the supersaturation is decreased (i.e. very slow cooling or higher temperatures) that this rule is not followed and the more stable forms nucleate first (Grotenhuis *et al.*, 1999; Sato, 2001; Lopez *et al.*, 2001b; Lopez *et al.*, 2005).

Another aspect of polymorphism is what used to be called at one time, polytypism. This refers to the ability of TAGs to stack epitaxially in different chain length configurations such as double (2L) and triple (3L) chain lengths (Marangoni *et al.*, 2012). Due to the flexibility of the TAG hydrocarbon chains, TAGs can pack into either a tuning fork or a chair conformation (Marangoni *et al.*, 2012). These conformations are characterized by having two fatty acids (*sn*-1 and *sn*-3, and *sn*-1 and *sn*-2, respectively) in the same alignment as that of the glycerol plane. TAGs therefore could stack-up in the lamellae either "back to back" or "seat to seat", depending on the type of TAGs present, which then result to the 2L and 3L chain lengths (Figure 9.4) (Marangoni *et al.*, 2012). These two different structures exist because of chain sorting, that is, when three fatty acid moieties in the TAG structure are very similar, a 2L structure is formed while when there is one or two fatty acids that are different from the other, a 3L structure is formed (van Aken and Visser, 2000; Sato, 2001).

9.2.1.2 Phase Behavior and Fractionation

TAGs in milk fat can be categorized according to their melting points (Timms, 1980; Dimick *et al.*, 1996; van Aken and Visser, 2000). TAGs that have final melting points above

50°C are classified as high melting fraction (HMF), while those melting between 40°C to 30°C as middle melting fraction (MMF), and TAGs melting above 15°C as low melting fraction (LMF) (Timms, 1980). Due to the presence of these fractions, three physical phases may be present during crystallization of milk fat, namely, solid, liquid and semi-solid at certain temperatures (Timms, 1984; Marangoni and Lencki, 1998; Campos *et al.*, 2003). This forms the basis for separating the different fractions since it is known that the solid phase at higher temperatures is mainly composed of HMF and some MMF, while LMF is expected to crystallize at very low temperatures (Timms, 1984; Dimick *et al.*, 1996). Separation of the fractions is performed either by dry fractionation, which utilizes only the difference in the melting points as basis for separating the fractions, or solvent fractionation, which also uses the melting points as basis but the fractions are obtained with the use of a solvent (Dimick *et al.*, 1996; Marangoni and Lencki, 1998; Vanhoutte *et al.*, 2002).

These fractions and their mixtures have been extensively studied for their properties and applications. For example, Cisneros *et al.* (2006) found that the isothermal crystallization of a mixture of HMF and LMF (40:60) at 5°C, follows the Ostwald rule where the α crystals are first formed which transform into the β' polymorph and some traces of β crystals in time. In another study, Marangoni and Lencki (1998) examined the ternary phase behavior of milk fat, in which they found that mixing LMF with either HMF or MMF forms a partial solid solution, while a mixture of HMF and MMF forms an ideally soluble mixed crystal. Furthermore, they found that interaction between LMF and MMF is stronger than that of LMF and HMF, and that between MMF and HMF.

The study of phase behavior is important as it can be used to manufacture the desired structures or polymorph for optimum product structure and quality.

9.2.1.3 Solid Fat Content and Crystallization/Melting Behavior

The amount of crystals formed, or solid fat content (SFC), is frequently measured using pNMR but it could also be determined by the amount of energy released during melting using DSC (Lopez *et al.*, 2001a). Authors often describe the crystallization of milk fat as a two-step process where the first step is characterized by a rapid increase in SFC at a fast rate which indicates the growth of α crystals, followed by a plateau, which is related to the transformation of α crystals into the β'. This process does not generate an increase in SFC as some α crystals are melted to form the more stable β' (Foubert *et al.*, 2006). The second step is then indicated by an increase in SFC due to the growth of β' crystals up until a maximum is reached (Figure 9.5) (Grotenhuis *et al.*, 1999; Foubert *et al.*, 2006; Cisneros *et al.*, 2006; Wiking *et al.*, 2009). By crystallizing a mixture HMF and LMF at 5°C, Cisneros *et al.* (2006) showed that the sudden increase in the SFC during the second step of crystallization is due to the delayed crystallization of the melt into the β' polymorph.

The two-step crystallization process is often characterized and quantified using the Avrami equation given by:

$$\frac{\text{SFC}\,(t)}{\text{SFC}\,(\infty)} = 1 - e^{-kt^n}$$

where SFC (t) is solid fat content as a function of time, SFC (∞) is the limiting SFC, k is the crystallization rate constant; t is time of crystallization; and n is the Avrami

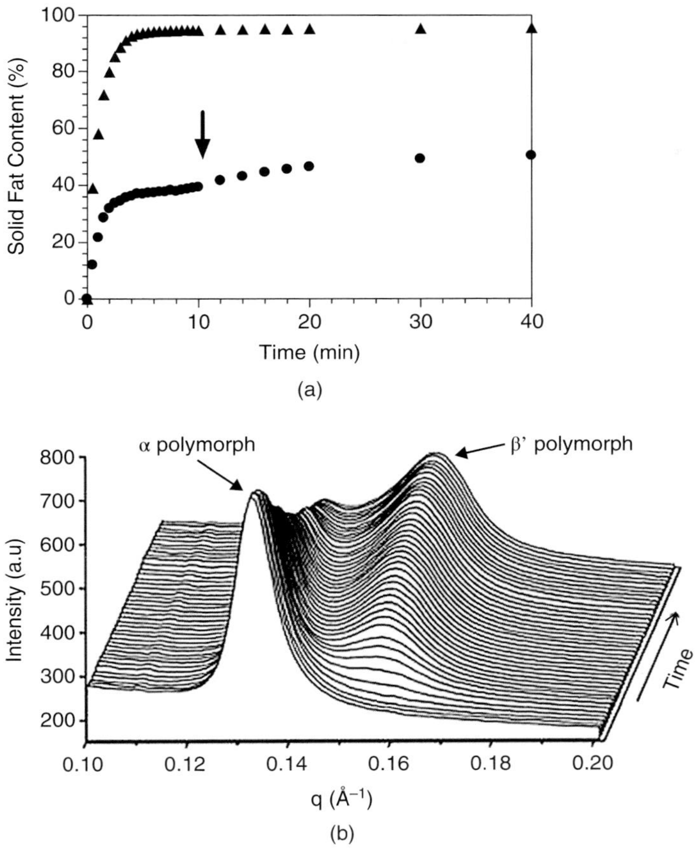

Figure 9.5 Two-step isothermal crystallization of milk fat as observed by (a) a plot showing the increase in solid fat content over time as measured by pulsed nuclear magnetic resonance (pNMR); and (b) evolution of polymorphism (from α to β') in time measured by X-ray diffraction (XRD).

exponent which indicates the crystal growth mechanism and has a value ranging from 0 to 4 (Wright and Marangoni, 2003; Campos *et al.*, 2003).

Due to polymorphism and the presence of the three TAG melting fractions (i.e., HMF, MMF and LMF) in milk fat, a broad melting range is often recorded using DSC upon heating. According to Timms (1980), upon melting of milk fat, three overlapping melting peaks are observed whose enthalpies and peak temperatures vary depending on the thermal treatments that were applied (i.e. crystallization and heating rates; and thermal history). In terms of polymorphism, according to (Lopez *et al.*, 2005), the general pattern for melting a previously slowly cooled milk fat involves the melting of the α 3L crystals first with a final melting point of around 13°C, and then the melting of the β′ crystals with final melting points at 19.5°C and 24°C, and higher melting β′ crystals (i.e. composed mainly of higher melting TAGs) with the final melting point at around 40°C. When fat is subjected to fast cooling, however, a lower final melting point of 35°C for the β′ polymorph is recorded upon melting (Grotenhuis *et al.*, 1999).

9.2.2 Nanoscale Structure of Fat Crystal Networks

An important breakthrough in the characterization of the structures created by triacylglycerol crystal networks has been made recently, and that is, the extraction, isolation and characterization of nanoplatelets formed through the crystallization of blends of tristearin and triolein (Acevedo and Marangoni, 2010b; Acevedo, 2012).

During fat crystallization, growth of the crystals results in the formation of different polycrystals and crystal aggregates. Throughout the years, these polycrystals were believed to be the result of the aggregation of primary crystals that are observable under PLM (Marangoni *et al.*, 2012). However, with the application of shear to reduce the size of the crystals up to the nanometer range, Acevedo and Marangoni (2010a,b) showed that the primary crystals are far smaller in size and involve crystal nanoplatelets (CNPs). CNPs are a result of the stacking of several TAG lamellae (Figure 9.6) that are held together by London-van der Waals forces to form crystalline domains (Acevedo *et al.*, 2011; Acevedo and Marangoni, 2015). The thicknesses of these domains agree with the thickness of the nanoplatelets (Acevedo *et al.*, 2011). Aggregation of these CNPs then therefore result to the formation of different polycrystals which are visualized under PLM and affect the properties of the milk fat macrostructure.

With the use of USAXS, the size of scatterers and their fractal structures could be characterized (Peyronel, Quinn *et al.*, 2014). In contrast to wide angle X-ray Scattering (WAXS) and small angle X-ray scattering (SAXS), which define reflections at angles between $12° < 2\theta < 35°$ and between $1° < 2\theta < 12°$, respectively, USAXS measures at an angle close to $2\theta = 0$ (Peyronel and Campos, 2012). WAXS determines distances between atomic planes (i.e. polymorphism), and SAXS determines distances between TAG lamellae, while USAXS allows for the determination of colloidal sizes (CNP aggregates), surface characteristics and distribution of mass in space (Figure 9.7) (Peyronel and Campos, 2012; Peyronel *et al.*, 2014; Peyronel, Quinn, *et al.*, 2014).

Observation of single nanocrystals under cryo-TEM is challenging because of the presence of liquid oil (Marangoni *et al.*, 2012). Therefore, an extraction method was developed using cold isobutanol to wash-off oil present in the system, either trapped within the crystalline structures or is free. After that, disruption of the crystals down to nanoplatelets is achieved using a rotostator with speeds up to 30,000 rpm. The nanoplatelets are then kept from aggregating using a sonicator. This method allows for imaging under cryo-TEM and in combination with USAXS, certain measurements (e.g. length and thickness) could be confirmed (Figure 9.8) (Acevedo, 2012).

Figure 9.6 Stacking of several triacylglycerol lamellae to form crystal nanoplatelets (CNPs).

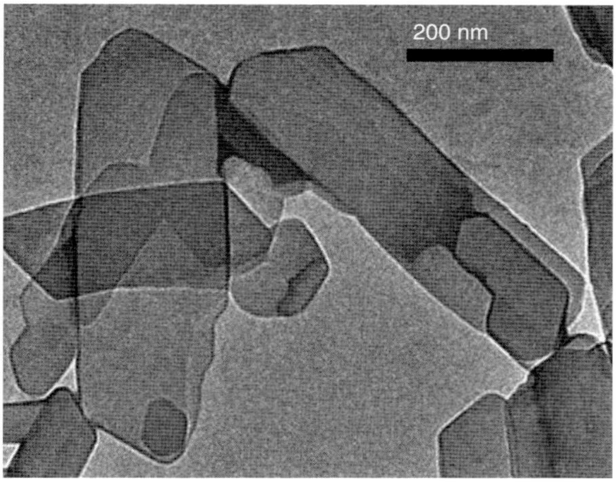

Figure 9.7 Crystal nanoplatelets (CNPs) of tristearin observed under cryogenic transmission electron microscope (Cryo-TEM).

Figure 9.8 Polarized light microscope (PLM) micrographs of anhydrous milk fat crystallized (a) slowly (0.1°C/min) and (b) rapidly (1.0°C/min).

Findings by Acevedo *et al.* (2012), suggest that processing conditions do not greatly affect the nanoscale structure of fat crystal networks as much as they affect the microstructure. As CNPs are the building blocks of larger crystal aggregates, nano-engineering could be key to improving the crystalline structure of products and developing new technologies for various applications (Acevedo and Marangoni, 2010b).

9.3 Effect of Different Factors on the Crystallization Behavior and Microstructure of Milk Fat

Crystallization of milk fat, like other fats and oils, is strongly affected by its composition and various environmental factors. Significant variation in its saturation index, for example, (i.e. saturated fatty acid to unsaturated fatty acid ratio) greatly affects its crystallization behavior (i.e. induction time and polymorphism). Furthermore, as milk fat is added as an ingredient, processing conditions (e.g. mixing, shear emulsification, and etc.) for the manufacture of the products affect its resulting microstructure, and therefore its contributory effect to the quality of the product. Discussed below are some factors that were found to significantly affect milk fat crystal structure.

9.3.1 Processing Conditions

9.3.1.1 Different Crystallization Mechanisms

The study of the kinetics of milk fat crystallization is carried out under isothermal and non-isothermal conditions (Marangoni *et al.*, 2006; Fredrick *et al.*, 2011). Isothermal crystallization is performed by rapidly cooling milk fat down to a target temperature, keeping it at that crystallization temperature and monitoring the changes in the crystalline mass. Here, crystallization temperature is the most important variable. On the other hand, non-isothermal crystallization is carried-out by gradually and continuously decreasing the temperature down to the desired crystallization point. Observed here is the differences in the microstructure at different cooling rates and transitions that occur as the temperature is decreased (Lopez *et al.*, 2001a,b).

9.3.1.2 Crystallization Temperature and Cooling Rate

Crystallization Temperature Stability of the crystals formed during the crystallization process and during storage is dependent on the crystallization temperature. At lower temperatures (i.e. higher supercooling), the transition of less stable to more stable crystal forms (e.g. α to β') is delayed or completely hindered mainly due to the higher ratio of solid to liquid fat. Under these conditions, there is less mobility of molecules, preventing them from rotating or migrating into more stable packing (Grotenhuis *et al.*, 1999; Cisneros *et al.*, 2006). On the other hand, crystallization at higher temperatures results in a faster transition of less stable crystals to more stable crystal forms because more liquid fat is available for the dissolution of TAGs (Grotenhuis *et al.*, 1999; Cisneros *et al.*, 2006; Bugeat *et al.*, 2011).

Subjecting milk fat to fast cooling down to different target crystallization temperatures allowed Lopez et al. (2001a) and Mazzanti *et al.* (2004a) to demonstrate that at crystallization temperatures greater than or equal to 20°C, all α crystals are transformed into β′ with some traces of β, and no α crystals remain. On the other hand, at temperatures between −8 and 20°C, co-existence of α and β′ crystals occurs because transformation of α to β′ is slowed down and at temperatures below or equal to −8C°, only the α form exists.

Fluctuations in the temperature during crystallization was also found to affect the resulting microstructure of milk fat (Buldo *et al.*, 2013). It was found that rise in temperature from 5 to 25°C for short periods of time (temperature cycling) during 8 days of isothermal crystallization resulted to harder, more stable and denser crystals due to sintering (Buldo *et al.*, 2013). Sintering allows TAGs to melt and migrate to fill spaces in the network, thus resulting in a more compact structure.

Cooling Rate Another important factor in determining the resulting microstructure of milk fat during crystallization is cooling rate. Microscopically, at fast cooling rates (>0.5C°/min), smaller crystals are formed while at slow cooling rates (≰0.5C°/min), larger crystals develop (Campos *et al.*, 2002; Wiking *et al.*, 2009). In terms of polymorphism, it was found that when milk fat was subjected to fast cooling during crystallization, the formation of the least stable α crystals was favored (Lopez *et al.*, 2001a; Lopez *et al.*, 2005). This is because during fast cooling, TAGs do not have the time to arrange themselves into more stable forms and therefore, it is more energetically favorable, from a kinetic perspective, for them to self-assemble into a less stable, but more accessible form (Marangoni *et al.*, 2012). On the other hand, at slow cooling rates, milk fat crystallizes into the β′ form because TAGs have time to arrange themselves and exclude TAGs that do not fit into the more compact crystalline lattice (Martini *et al.*, 2001). It is therefore expected that at very slow cooling rates, β crystals would be formed. This was observed by Lopez *et al.* (2001b), however at slow cooling rates ending at high crystallization temperatures, Mazzanti *et al.* (2004a) found that only the β′ form exists because formation of β crystals is dependent on the formation and transition of the α form.

9.3.1.3 Agitation, Shear and Ultrasound

Agitation Agitation of milk fat during crystallization usually occurs during the manufacture of many milkfat-containing food products. Herrera and Hartel (2000) found that when a mixture of HMF and LMF was subjected to high agitation rates during crystallization, many small crystals were formed. On the other hand, when milk fat was crystallized statically, fewer and larger crystals are formed. One of the reasons provided was that during agitation, air bubbles are incorporated in the system which could act as nucleating agents thus, increasing nucleation rate, therefore leading to the formation of numerous small crystals. In another experiment, Vanhoutte *et al.* (2002) suggested that agitation during crystallization of milk fat results in better fractionation due to uniform heat transfer.

Shear-Induced Crystallization Shear is often applied in the processing of products such as chocolate, margarine, spreads and butter to improve their sensorial attributes because mixing improves heat transfer and induces uniformity in the final product. Many authors have described the effects of shear on fat crystallization, especially cocoa butter (Mazzanti *et al.*, 2004b; Campos and Marangoni, 2014). Most recently, Mazzanti *et al.* (2009) studied the crystallization kinetics of milk fat under shear using a detailed synchrotron X-ray and the results showed that high shear rates induces the transition of the α to the β′ polymorph which agrees with a previous study showing phase transitions being accelerated in different edible fats due to application of shear (Mazzanti *et al.*, 2003).

Ultrasound or Ultrasonication Ultrasound application has been widely used in food processing due to its advantages such as improving uniform heat transfer during cooking, drying and freezing; enhancing degassing and filtration of different liquid products; and speeding-up different processes like emulsification and oxidation (Chemat *et al.*, 2011).

With the aim of improving its mechanical properties, high-intensity ultrasound (HIU) at frequencies between 100 kHz and 10 MHz was applied to milk fat during crystallization (Martini *et al.*, 2008; Suzuki *et al.*, 2010). It was found that when HIU is applied, faster crystallization occurs and smaller crystals are formed (Martini *et al.*, 2008; Suzuki *et al.*, 2010). In another study, Frydenberg *et al.* (2013) found that with the application of HIU, transition of α crystals to a denser β′ was faster and that more uniform crystals were formed. Furthermore, HIU treatment for short periods of time were found to increase solid fat content but have detrimental effects when applied for longer periods of time, mainly due to excessive heating.

9.3.2 Composition

9.3.2.1 Minor Components

Milk fat is composed of approximately 96–98% TAGs, with minor components such as monoacylglycerols (MAG), diacylglycerols (DAG), free fatty acids (FFA), cholesterols, glycolipids and other polar lipids amounting to approximately 2–4% (McGibbon and Taylor, 2006). Although very low in concentration, these components were found to have a significant effect on the crystallization kinetics of milk fat (Wright *et al.*, 2000; Wright and Marangoni, 2002 and 2003; Mazzanti *et al.*, 2004a).

By comparing the crystallization behavior of native AMF and milk fat containing only TAGs (MF-TAG), Wright *et al.* (2000) and Mazzanti *et al.* (2004a) showed that the presence of minor components at their natural levels delays the onset of crystallization at high crystallization temperatures and favors formation of α and β crystals at high cooling rates and low crystallization temperatures. Equilibrium SFC and overall mechanical and structural properties of milkfat, on the other hand, remained unaffected (Wright *et al.*, 2000; Wright and Marangoni, 2002 and 2003).

9.3.2.2 Blending with Different Fats and Oils, and Waxes

Blending of different fats and oils with milk fat is usually carried out to improve its functionality and nutritional value. In most cases, they are added to increase the unsaturation level which in turn alters the resulting crystal structure (Smet *et al.*, 2010; Bugeat *et al.*, 2011). For example, Wright *et al.* (2005), showed that adding canola oil to milk fat leads to dissolution of milk fat solids and entrainment of oil in the crystalline structures which results in very soft and oily products. Furthermore, at increasing concentrations, crystallization is delayed and at a 1:8 w/w) ratio (milk fat: canola oil), formation of β crystals is favored. Kaufmann *et al.* (2012), on the other hand, showed that addition of rapeseed oil results in a faster transition of α to β′ form due to increased mobility through an increased ratio of liquid fat.

Addition of wax, in contrast, is usually done to increase the melting point and speed-up hardening of milk fat. According to Kerr *et al.* (2011), adding sunflower oil wax to milk fat results in a faster crystallization and formation of smaller sized crystals. This is because waxes have high melting points which allows them to crystallize at a lower supercooling and therefore serve as seeding nuclei for milk fat crystallization.

9.3.3 In a Dispersed State (Emulsion)

Naturally in milk, fat is present as emulsion droplets (i.e. globules) surrounded by a membrane (phospholipid bilayer), composed of proteins, phospholipids, glycolipids and other materials (McGibbon and Taylor 2006). In products such as whipped cream, butter, and ice cream, this globular property of milk fat is essential because of the structures that it can form through controlled emulsion destabilization.

Destabilization of milk fat occurs because of its lower density than water and aggregation with other globules. Creaming is an example of the former wherein milk fat globules rise-up and concentrate on top of the water phase (skim milk) (Figure 9.9a). On the other hand, coalescence and flocculation are examples of the latter, wherein globules could collide and coalesce or flocculate (Figure 9.9b and c).

The most useful destabilization phenomenon for the products stated above, however, is partial coalescence. When TAGs inside the globules crystallize, needle-like crystal structures are formed. Upon application of shear, these crystals can pierce through and rupture the membrane where they are contained and through that of another globule. Thus the globules become interconnected through these crystals. At favorable temperatures, oil is released which becomes the dispersing phase for the crystals and prevents complete coalescence of the globules. Their shape is therefore partially retained and a network of interconnecting partial globules is formed (Figure 9.10). This is different from flocculation since partial coalescence results in an irreversible aggregation while flocs can easily be broken-up.

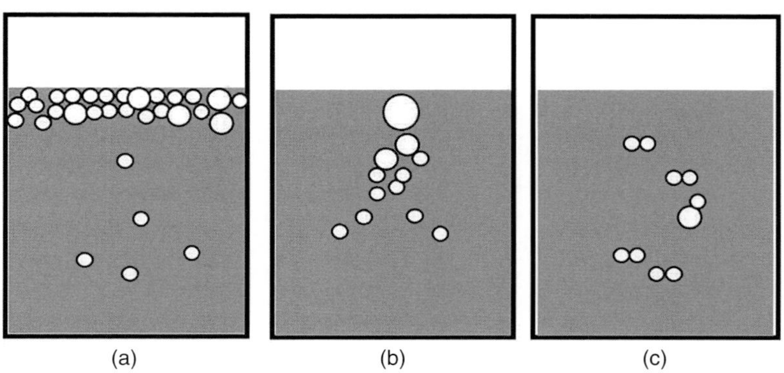

Figure 9.9 Schematic representation of different types of emulsion destabilization in milk: (a) creaming; (b) coalescence; and (c) flocculation.

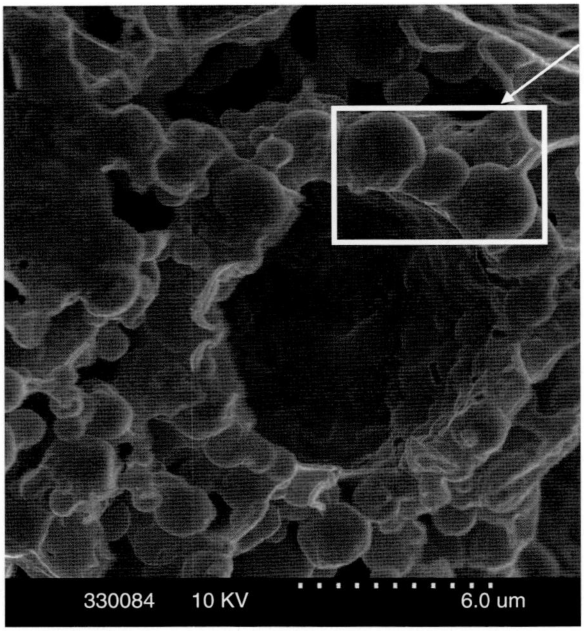

Figure 9.10 Partial coalescence of milk fat globules observed under cryogenic scanning electron microscope (cryo-SEM). Photo courtesy of Dr. Alexandra Smith, University of Guelph.

9.3.3.1 Emulsified State (Cream) vs Bulk State or Anhydrous Milk Fat (AMF)

Crystallization of natural milk fat (cream) is more complicated than that of bulk (without the membrane) due to the variation of TAGs within each globule. Furthermore, in terms of rate of crystallization, that of the emulsified state is slower (Fredrick *et al.*, 2011). According to (Walstra *et al.*, 1999), crystallization of fats is sped up by presence of catalytic impurities which induce heterogeneous nucleation of milk fat. In the bulk state, these catalytic impurities are readily available for nucleation and growth. On the other hand, in the emulsified state, depending on the amount, some of these catalytic impurities are partitioned into globules while some globules will not contain any (Figure 9.11). Thus, nucleation of TAGs in some of these globules will take place heterogeneously, while in globules devoid of these impurities, TAGs would have to nucleate homogeneously which is energetically more difficult, displays a longer induction time and requires a higher degree of supercooling.

In terms of polymorphism, at fast cooling rates, cream and AMF crystallize via the same mechanism where the α polymorph is formed and then transforms into the β' form. Transition of one form to another in cream, however, is slower than that in AMF. Thus, coexistence of the different crystal forms occurs in cream (Lopez *et al.*, 2002). On the other hand, at slow cooling, AMF crystallizes directly into the β' form while cream crystallizes in the α polymorph.

9.3.3.2 Emulsion Droplet Size

Milk fat globules naturally have a size of 1.8 to 15 μm which makes them large enough to favor destabilization as they can easily rise-up or collide with each other (McGibbon

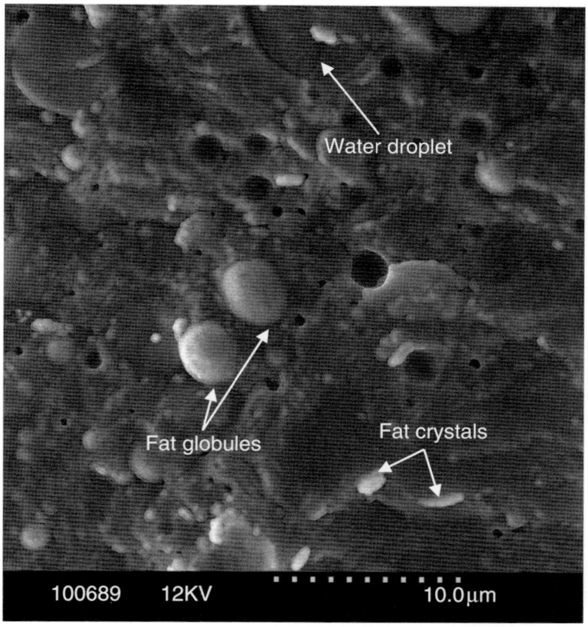

Figure 9.11 Schematic representation of the distribution of catalytic impurities in (a) bulk and (b) globules.

and Taylor, 2006). In many industries where destabilization is undesirable, reduction of emulsion droplets up to 10× is performed.

This is usually carried out by homogenization at high pressures.

The effect of this process on crystallization was then investigated and it was found that the smaller the average emulsion droplet size, the higher supercooling was needed and the slower the crystallization rate. Furthermore, Bugeat *et al.* (2011) found that confinement of long chain TAGs in smaller emulsion droplets enhances the formation of β crystals while decreases that of β′. In homogenous products, prevention of fat clumping is desirable, however, in products that depend on partial coalescence, reduction of droplet size could be a disadvantage.

9.3.3.3 Addition of Emulsifiers

Emulsifiers are usually added to food to improve the stability of oil in water emulsions. By studying the effect of the addition of lactose monolaurate (LML) on the crystallization behavior of milk fat, Wagh *et al.* (2013), concluded that the addition of emulsifiers delays the onset of crystallization, decreases the amount of solid crystals formed and therefore, decreases the viscosity of the system.

9.3.4 In Food Matrices

9.3.4.1 Water-in-Oil Emulsion

Butter is a product made from cream and is a water-in-oil emulsion. As discussed in the previous section, the destabilization process governing crystal network formation in this product is partial coalescence. Two major processes influencing the final microstructure are churning and working. During churning, continuous collision of

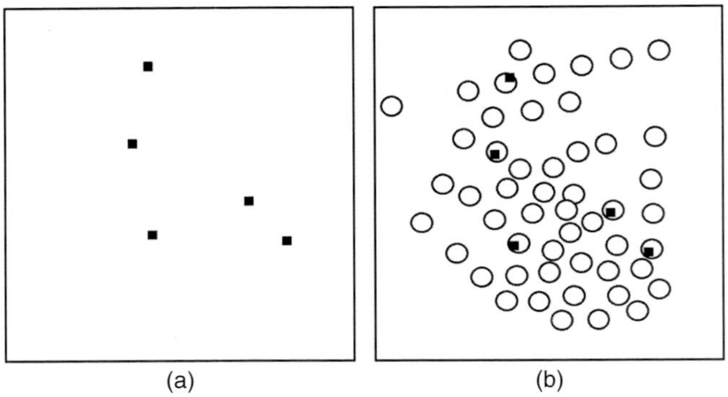

Figure 9.12 Microstructure of butter under cryogenic scanning electron microscope (cryo-SEM). Photo courtesy of Dr. Alexandra Smith, University of Guelph.

fat globules occurs which leads to damage of the fat globule membrane and formation of large butter grains. After this process, large moisture droplets become entrapped which prevents the grains from forming a continuous network. Therefore, working is done to reduce moisture and moisture droplet size. This then allows the formation of a water-in-oil emulsion. Figure 9.12 shows the resulting microstructure of butter.

A substantial amount of liquid fat must be present to act as a sticking agent to keep the network together and to prevent complete coalescence of the globules. This could be achieved by churning at 15–20C° (Walstra *et al.*, 2006).

Studies by Buldo *et al.* (2013), show that before churning, the crystals observed in cream are in α form which transforms readily to β′ upon churning. This result agrees well with those discussed in section 9.3.1.3, where the application of agitation, shear or ultrasound induces faster phase transitions. On the other hand, Ronholt *et al.* (2012), showed that maturing cream prior to churning could result in the formation of β crystals. Furthermore, slow cooling of cream results in the formation of fewer crystals with a wider size distribution while fast cooling yields more uniform crystal size distribution.

9.3.4.2 Foamed Emulsions

Milk fat products that depend on partial coalescence to stabilize air bubbles include whipped cream and ice cream. Whipped cream is created by beating air into the cream, therefore resulting into the entrapment of air bubbles within the milk fat crystal network (Figure 9.13). In this system however, too much liquid fat could cause the foam to collapse, thus, very little is required. Furthermore, large crystals can pierce through the air bubbles, thus, the globules are homogenized to disrupt the membrane and to keep fat crystals small. Reducing the size of the globules, in turn, could result in excessive stabilization of globules and thus a surfactant or emulsifier is usually added. According to Truong *et al.* (2014), reducing fat particle size up to the nano range results in a lower percentage of overrun (i.e., lower amount of air incorporation) and destabilization or complete collapse of foams.

Milk fat greatly contributes to the structure formation in ice cream (Goff and Hartel, 2013). Like whipped cream, homogenization is carried out so that small fat droplets (average size of 0.8 μm) become dispersed throughout the ice cream matrix. Addition

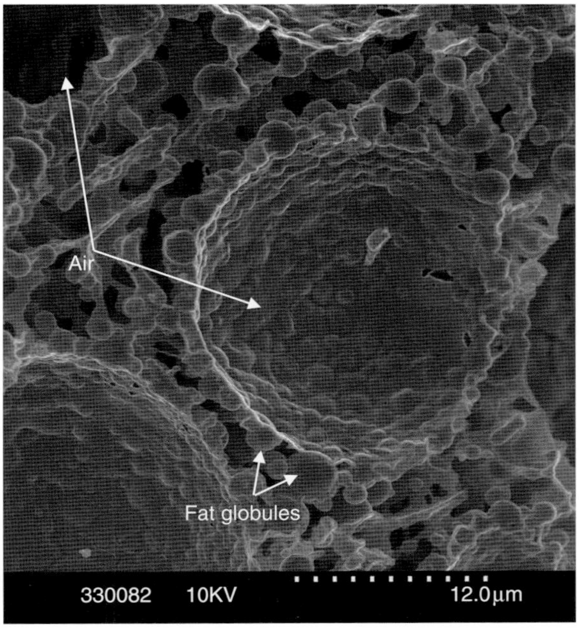

Figure 9.13 Microstructure of whipped cream under cryogenic scanning electron microscope (cryo-SEM). Photo courtesy of Dr. Alexandra Smith, University of Guelph.

of surfactant, or emulsifiers then follows to enhance globule partial coalescence (Goff, 2008). Competitive displacement of the large protein molecules on the surface of fat droplets by smaller molecule surfactants lowers surface tension and therefore favors partial coalescence (Goff and Hartel, 2013). Together with proteins, the network formed by clusters of fat globules stabilize the air bubbles that are incorporated in the ice cream mix during whipping or air injection (Goff, 1997). As the dispersed droplets can form a semi-continuous network throughout the matrix, they are able to cut through the lamellae between air cells, resulting in smaller air bubbles which is important for maintaining the shape of ice cream (Goff and Hartel, 2013).

9.3.4.3 Chocolate

Application of shear during conching of chocolate greatly determines the final quality of the product especially viscosity and flavor development (Afoakwa, 2010). This process, in turn affects its microstructure and polymorphism (Brunello *et al.*, 2003; Campos *et al.*, 2010; Campos and Marangoni, 2014). Milk fat is usually blended with cocoa butter (CB) because of its desirable sensory qualities and lower price. Furthermore, it prevents the extensive hardening of chocolate as found by Martini and Marangoni (2007) that when HMF is added to CB, larger crystals are formed which results in a reduced contact area between crystals and a less compact network. In addition, blending of milk fat with CB was found to prevent bloom formation, a type of defect in chocolate. This is because the presence of milk fat crystals in the matrix delays the transition of the less stable V polymorph to the most VI polymorph which is responsible for formation of white film on the surface of chocolate (Sonwai and Rousseau, 2010).

9.3.4.4 Cheese

Fat in natural cheese strongly influences sensory qualities. During processing, milk fat affects the amount of moisture retained in cheese and therefore determines greatly the texture of the final product. Then, during ripening, it acts as a precursor for several volatiles that provide distinct flavors to different cheese varieties (Guinee and McSweeney, 2006).

Looking into the microstructure of cheese, fat globules can be seen in the spaces or pockets throughout the protein gel matrix formed by caseins. Processing temperatures, however, liquefy most of the milk fat and therefore the tendency for them to coalesce and form large globules at certain areas in the cheese is high. According Lopez *et al.* (2006), aggregation and coalescence of milk fat in Emmental cheese during processing significantly affects its crystallization and melting properties.

Such destabilization also plays a role in the manufacture of processed cheese. As the texture of this product is greatly dependent on milk fat properties, stability of fat within the protein matrix is one of the most important considerations during processing. By homogenizing milk fat, globule size distribution is made more uniform which is followed by emulsifier addition to keep a stable dispersion of fat throughout the product. Processed cheese is stored at refrigeration temperature which causes fat to crystallize and according to Gliguem *et al.* (2011), the thermal history and polymorphism of milk fat affects the viscoelastic properties of processed cheese. Using XRD, it was found that TAGs in processed cheese crystallize in the β' form upon equilibration at 25°C from 4°C. Furthermore, after melting to 60°C and crystallizing down to −10°C, only a single α polymorph is formed (Gliguem *et al.*, 2009). Ultimately, it was found that changes in the viscoelastic and rheological properties of processed cheese as a function of temperature is affected mainly by the polymorphism and thermal properties of milk fat (Gliguem *et al.*, 2009 and 2011).

9.4 Impact of Resulting Microstructure on the Properties of Different Milk Fat Products

As elaborated in the previous sections, different processing conditions result in changes to milk fat microstructure which translate into significant effects on the macrostructure and therefore, the overall quality of milk fat products. In this section, the impact of these changes on certain milk fat-structured products will be described.

9.4.1 Rheology

One of the most important qualities of products made-of or containing milk fat (i.e., milk fat products) that determine their acceptability is texture as it affects mouthfeel, appearance and other sensory attributes. The rheological properties of milk fat products such as hardness, firmness, spreadability and viscoelasticity are of economic importance for the industry especially in butter. Analytical techniques such as penetrometry, compression studies, sectility or the ability of the product to be cut into pieces, extrusion methods, and small deformation studies have been used extensively to characterize the texture of butter after manufacturing (Wright *et al.*, 2001).

The hardness or firmness of milk fat products has been found to be greatly affected by the size of crystals present. When large crystals are present, the product is softer and easier deformed which could be explained by the amount of surface area in which crystals are in contact with other crystals. Due to their large size, only few areas can interact with other crystals thus resulting to a more loose structure. In contrast, when smaller fat crystals are present, the product becomes firmer because the surface area for interaction becomes larger. The force needed to break inter-crystalline interactions and to deform the product therefore increases (Campos *et al.*, 2002; Kerr *et al.*, 2011; Danthine, 2012).

The ratio between solid and liquid fat, i.e., the SFC at specific temperatures also affects the viscoelastic properties of milk fat-containing products. When more liquid fat is present in the system, the product becomes soft and runny while when more solid fat is present, it becomes firm and viscous. As described earlier, due to the complexity of the TAG composition of milk fat, solid and liquid phases could exist at the same time at certain temperatures, thus, a specific balance between these phases could result to a desirable viscoelastic property (Wright *et al.*, 2001; Marangoni *et al.*, 2012).

An important topic in the discussion of the rheological properties provided by fat crystal networks is their fractal nature. Fractals are defined as self-similar across different length scales and this applies to fat crystal networks as it is a product of the aggregation of primary crystals (Marangoni, 2002). The relationship of the shear elastic modulus (G') to the volume fraction of solid fat (Φ) has been used to quantify the microstructure of fat crystal networks, and Narine and Marangoni (1999a) demonstrated that this could be done via the mass fractal dimension (D) of the network,

$$G' \sim \Phi^{m}$$

where G' is the shear elastic modulus, Φ is the particle volume fraction of solid fat and m depends on the fractal dimension.

The fractal dimension is a measure of the order in which solid mass is spatially distributed in the network. The higher the fractal dimensions, the more ordered the distribution (Figure 9.14) (Marangoni, 2002). Furthermore, by examining different fat mixtures (e.g. milk fat, cocoa butter, lard), Narine and Marangoni (1999b) showed that increasing fractal dimensions (D) results to lower shear modulus (G') normalized by a constant γ. A lower G' indicates a harder material (Wright *et al.*, 2001) (Figure 9.15).

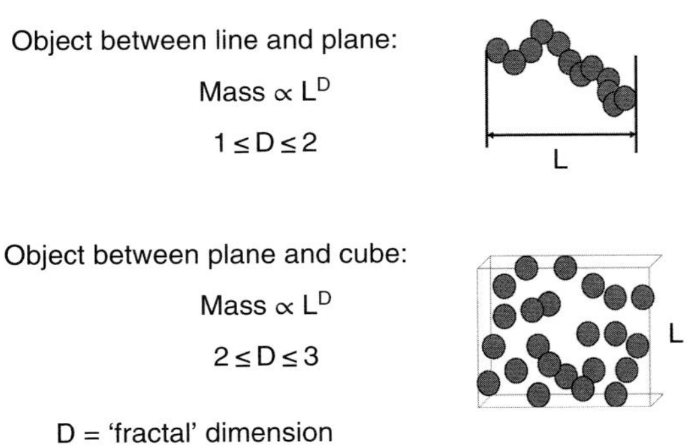

Figure 9.14 Spatial distribution of mass and its corresponding fractality, where D is fractal dimension.

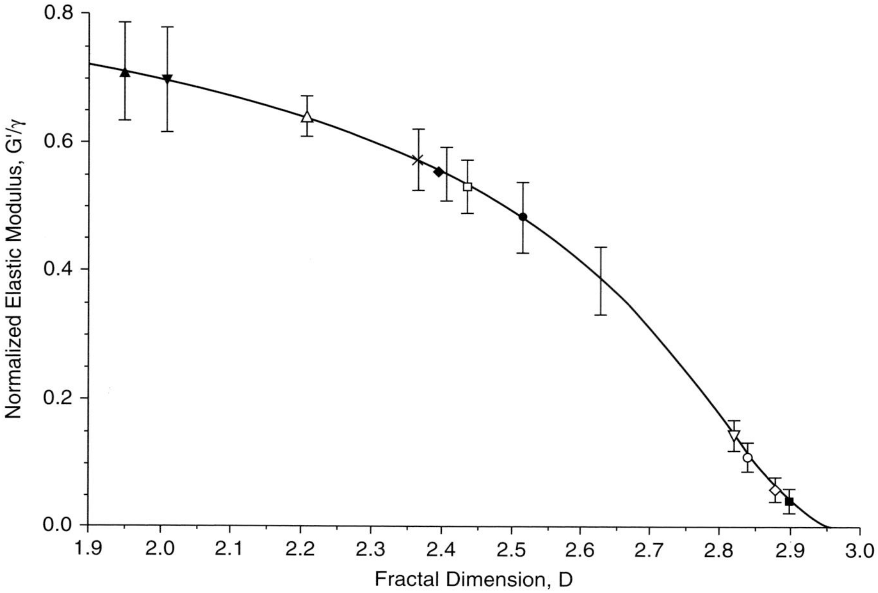

Figure 9.15 The relationship between fractal dimension (D) and shear modulus (G') normalized by a constant γ.

Measurement of fractal dimensions is carried out by image analysis of the micrographs obtained using PLM (Wright *et al.*, 2001; Marangoni and Ollivon 2007). The fractal dimension (D) is defined as,

$$N = \left(\frac{R}{\sigma}\right)^D \quad \mathrm{N} >> 1$$

where N is the number of particles in an object or region of interest, R is linear size of the fractal and σ is linear size of one particle making up the fractal object (Narine and Marangoni, 1999a).

It is important to note that micrographs are a 2-dimensional (2D) representation of the crystal network while it is known that crystal networks extend into a 3-dimensional (3D) plane. According to Tang and Marangoni (2006), to be able to count the 3D fractal dimension of fat crystal networks, counting must be done through image analysis of 3D stacks of the sample.

The box counting fractal dimension of a wax crystal network is strongly related to its oil binding capacity (Blake *et al.*, 2014). It was found that a higher fractal dimension, and therefore a lower porosity of wax-oil gels results in an increased oil binding capacity. Marangoni and Mcgauley (2003), on the other hand, showed that box counting fractal dimension could also be used to quantify changes in microstructure as it indicates nucleation behavior. Fractal dimension was found to be inversely proportional with the Avrami exponent. The use of box counting fractal dimension was also extensively used in making models to study the crystallization behavior of triacylglycerols and properties of colloidal networks (Tang and Marangoni, 2006; Marangoni *et al.*, 2006; Tang and Marangoni, 2008).

9.4.2 Thermal Stability

Most products containing milk fat rely on their thermal stability for their shelf stability and marketability. This is because, when fat melts, the integrity of the product is lost. Thus, much work has been geared towards increasing the melting point of milk fat-containing products. Processing conditions are usually manipulated to induce formation of stable crystals (i.e. to increase melting point), however, a prerequisite for these crystal forms is the presence of symmetric TAGs (van Aken and Visser, 2000; Cisneros *et al.*, 2006). Symmetric TAGs are usually composed of saturated or *trans* fatty acids so that no kinks or bends are found in their structure, thus, allowing close packing with other TAGs. With the growing trend of decreasing and eliminating saturated and *trans* fatty acids in food products, addition of unsaturated fatty acids are believed to be the best solution. However, this could result in a decrease in the melting point of milk fat products as well as decrease in SFC, causing the product to melt easily and to become very oily. Thus, most companies nowadays look for alternatives to increase thermal stability so as to maintain structure such as the use of edible organogels or oleogels. For example, Zulim Botega *et al.* (2013) used wax oleogels as fat replacement in ice cream, while Zetzl *et al.* (2012) used ethycellulose oleogels in frankfurters.

9.4.3 Sensory Qualities

Besides providing richness of flavor to food products, milk fat also adds to their texture and mouthfeel. Especially in ice cream, the network formed by milk fat provides smoothness and creaminess. When too much fat crystal agglomeration is present, the product becomes dry and grainy, while when there is too less, the product becomes too light, smooth and somewhat slimy (Walstra *et al.*, 1999; Goff, 2006).

Polymorphism also affects the texture and mouthfeel of milk fat products. As described in section 2, β crystals are the most stable, followed by β′ and then α. When α crystals dominate the product, it becomes soft. On the other hand, when β′ crystals are present at high concentration, the product becomes more stable, spreadable and elastic, and not easily melted (deMan 1992). Lastly, when β crystals are present, the product is most stable and high-melting, hard and coarse, grainy, and non-spreadable (deMan, 1992; Danthine, 2012).

9.5 Conclusions

The understanding of the structural contribution of milk fat to different milk fat products has been shown to provide numerous opportunities for improving product quality, optimizing manufacturing processes, and providing cost-effective solutions. Milk fat is a very complex system due to the heterogeneity of its composition, however, with the advances in methodologies for characterizing the structure of milk fat at different scales, changes that occur in its crystallization behavior and crystal network formation (e.g. crystal habit, polymorphism) as affected by composition and different processing conditions can be described and explained with precision. The characterization of the nanoscale structure of triacylglycerol networks using USAXS and cryo-TEM brings about new dimensions for analysis and improving the microstructure of milk fat by nano-engineering of primary crystals (i.e. CNPs). The effect of the resulting microstructure of milk fat on product properties such as rheology, thermal stability, and sensory characteristics is as well important in providing a comprehensive description of the relationship between processing conditions and microstructure, and product quality.

References

Acevedo, N., 2012. Characterization of the nanostructure of triacylglycerol networks. In: A.G. Marangoni, ed. *Structure-Function Analysis of Edible Fats*. AOCS Press, Urbana, IL, pp. 5–24.

Acevedo, N.C., Block, J.M. and Marangoni, A.G., 2012. Critical laminar shear-temperature effects on the nano- and mesoscale structure of a model fat and its relationship to oil binding and rheological properties. *Faraday Discussions*, 158, p.171.

Acevedo, N.C. and Marangoni, A.G., 2010a. Characterization of the nanoscale in triacylglycerol crystal networks. *Crystal Growth and Design*, 10, 3327–3333.

Acevedo, N.C. and Marangoni, A.G., 2010b. Toward nanoscale engineering of triacylglycerol crystal networks. *Crystal Growth and Design*, 10(8), 3334–3339.

Acevedo, N.C. and Marangoni, A.G., 2015. Nanostructured Fat Crystal Systems. *Annual Review of Fluid Mechanics*, (November 2014), p.42.

Acevedo, N.C., Peyronel, F. and Marangoni, A.G., 2011. Nanoscale structure intercrystalline interactions in fat crystal networks. *Current Opinion in Colloid and Interface Science*, 16(5), 374–383. Available at: http://dx.doi.org/10.1016/j.cocis.2011.05.004.

Afoakwa, E.O., 2010. *Chocolate Science and Technology*. John Wiley, Oxford.

van Aken, G.A. and Visser, K.A., 2000. Firmness and crystallization of milk fat in relation to processing conditions. *Journal of dairy science*, 83(9), 1919–1932.

Blake, A.I., Co, E.D. and Marangoni, A.G., 2014. Structure and physical properties of plant wax crystal networks and their relationship to oil binding capacity. *JAOCS, Journal of the American Oil Chemists' Society*, 91(6), 885–903.

Brunello, N., McGauley, S.E. and Marangoni, A., 2003. Mechanical properties of cocoa butter in relation to its crystallization behavior and microstructure. *LWT – Food Science and Technology*, 36(5), 525–532.

Bugeat, S. *et al.*, 2011. Enrichment in unsaturated fatty acids and emulsion droplet size affect the crystallization behaviour of milk triacylglycerols upon storage at 4°C. *Food Research International*, 44(5), 1314–1330. Available at: http://dx.doi.org/10.1016/j.foodres.2011.01.003.

Buldo, P., Kirkensgaard, J.J.K. and Wiking, L., 2013. Crystallization mechanisms in cream during ripening and initial butter churning. *Journal of Dairy Science*, 96(11), 6782–91. Available at: http://www.ncbi.nlm.nih.gov/pubmed/24035028.

Campos, R. and Marangoni, A.G., 2014. Crystallization Dynamics of Shear Worked Cocoa Butter. *Crystal Growth and Design*, 14(3), 1199–1210. Available at: http://pubs.acs.org/doi/abs/10.1021/cg4017273.

Campos, R., Narine, S. and Marangoni, A. 2002. Effect of cooling rate on the structure and mechanical properties of milk fat and lard. *Food Research International*, 35(10), 971–981.

Campos, R., Ollivon, M. and Marangoni, A.G., 2010. Molecular composition dynamics and structure of cocoa butter. *Crystal Growth and Design*, 10(1), 205–217. Available at: http://pubs.acs.org/doi/abs/10.1021/cg900853e.

Campos, R.J., Litwinenko, J.W. and Marangoni, A.G., 2003. Fractionation of milk fat by short-path distillation. *Journal of Dairy Science*, 86(3), 735–745. Available at: 10.3168/jds.S0022-0302(03)73654-6.

Chemat, F., Zill-E-Huma and Khan, M.K., 2011. Applications of ultrasound in food technology: Processing, preservation and extraction. *Ultrasonics Sonochemistry*, 18(4), 813–835. Available at: http://dx.doi.org/10.1016/j.ultsonch.2010.11.023.

Cisneros, A. *et al.*, 2006. Polymorphic transformation in mixtures of high- and low-melting fractions of milk fat. *Journal of Agricultural and Food Chemistry*, 54(16), 6030–6033.

D'Souza, V., DeMan, J.M. and DeMan, L., 1990. Short spacings and polymorphic forms of natural and commercial solid fats: A review. *Journal of the American Oil Chemists' Society*, 67(11), 835–843.

Danthine, S., 2012. Physicochemical and structural properties of compound dairy fat blends. *Food Research International*, 48(1), 187–195.

deMan, J.M., 1992. X-ray diffraction spectroscopy in the study of fat polymorphism. *Food Research International*, 25(6), 471–476.

Dibildox-Alvarado, E., Marangoni, A.G. and Toro-Vazquez, J.F., 2010. Pre-nucleation structuring of triacylglycerols and its effect on the activation energy of nucleation. *Food Biophysics*, 5(3), 218–226.

Dimick, P.S., Reddy, S.Y. and Ziegler, G.R., 1996. Chemical and thermal characteristics of milk-fat fractions isolated by a melt crystallization. *Journal of the American Oil Chemists Society*, 73(12), 1647–1652. Available at: <Go to ISI>://A1996VY89600005.

Foubert, I. *et al.*, 2006. Modelling two-step isothermal fat crystallization. *Journal of Food Engineering*, 75(4), 551–559.

Fredrick, E. *et al.*, 2011. Isothermal crystallization behaviour of milk fat in bulk and emulsified state. *International Dairy Journal*, 21(9), 685–695.

Frydenberg, R.P. *et al.*, 2013. Ultrasonication affects crystallization mechanisms and kinetics of anhydrous milk fat. *Crystal Growth and Design*, 13(12), 5375–5382.

Gliguem, H. *et al.*, 2009. Crystallization and polymorphism of triacylglycerols contribute to the rheological properties of processed cheese. *Journal of Agricultural and Food Chemistry*, 57(8), 3195–3203.

Gliguem, H. *et al.*, 2011. The viscoelastic properties of processed cheeses depend on their thermal history and fat polymorphism. *Journal of Agricultural and Food Chemistry*, 59(7), 3125–3134.

Goff, D.H. and Hartel, R.W., 2013. Ice cream structure. In: D.H. Goff and R.W. Hartel, eds. *Ice Cream*. pp. 313–352.

Goff, H.D., 2008. 65 Years of ice cream science. *International Dairy Journal*, 18(7), 754–758.

Goff, H.D., 1997. Colloidal aspects of ice cream – A review. *International Dairy Journal*, 7(6–7), 363–373.

Goff, H.D., 2006. Ice cream. In: P.F. Fox and P.L.H. McSweeney, eds. *Advanced Dairy Chemistry Volume 2: Lipids*. Springer Science, Berlin, pp. 441–450.

Grotenhuis, E. *et al.*, 1999. Polymorphism of milk fat studied by differential scanning calorimetry and real-time x-ray powder diffraction. *Journal of the American Oil Chemists' Society*, 76(9), 1031–1039.

Guinee, T.P. and McSweeney, P.L.H., 2006. Significance of Milk Fat in Cheese. In: P.F. Fox and P.L.H. McSweeney, eds. *Advanced Dairy Chemistry Volume 2: Lipids*. Springer, Berlin, pp. 377–440.

Herrera, M.L. and Hartel, R.W., 2000. Effect of processing conditions on crystallization kinetics of a milk fat model system. *Journal of the American Oil Chemists' Society*, 77(11), 1177–1188.

Jensen, R.G., 2002. The composition of bovine milk lipids: January 1995 to December 2000. *Journal of Dairy Science*, 85(2), 295–350. Available at: http://dx.doi.org/10.3168/jds.S0022-0302(02)74079-4.

Jensen, R.G., Ferris, a M. and Lammi-Keefe, C.J., 1991. The composition of milk fat. *Journal of Dairy Science*, 74(9), 3228–3243.

Kaufmann, N. *et al.*, 2012. Shear-induced crystal structure formation in milk fat and blends with rapeseed oil. *Food Biophysics*, 7(4), 308–316.

Kerr, R.M. *et al.*, 2011. Crystallization behavior of anhydrous milk fat-sunflower oil wax blends. *Journal of Agricultural and Food Chemistry*, 59(6), 2689–2695.

Kloek, W., Walstra, P. and Vliet, T., 2000. Nucleation kinetics of emulsified triglyceride mixtures. *Journal of the American Oil Chemists' Society*, 77(6), 643–652.

Larsson, K., 1966. Classification of glyceride crystal forms. *Acta Chemica Scandinavica*, 20(8), 2255–2260.

Lopez, C. *et al.*, 2006. Milk fat thermal properties and solid fat content in emmental cheese: a differential scanning calorimetry study. *Journal of Dairy Science*, 89(8), 2894–2910. Available at: http://dx.doi.org/10.3168/jds.S0022-0302(06)72562-0.

Lopez, C., Lavigne, F., Lesieur, P., Keller, G., *et al.*, 2001a. Thermal and structural behavior of anhydrous milk fat. 2. Crystalline forms obtained by slow cooling. *Journal of Dairy Science*, 84(11), 2402–2412. Available at: http://dx.doi.org/10.3168/jds.S0022-0302(01)74689-9.

Lopez, C. *et al.*, 2005. Thermal and structural behavior of anhydrous milk fat. 3. Influence of cooling rate. *Journal of Dairy Science*, 88(2), 511–526.

Lopez, C. *et al.*, 2001b. Thermal and Structural Behavior of Milk Fat. *Journal of colloid and interface science*, 240(1), 150–161. Available at: http://www.ncbi.nlm.nih.gov/pubmed/11446797.

Lopez, C., Lavigne, F., Lesieur, P., Bourgaux, C., *et al.*, 2001c. Thermal and structural behavior of milk fat. 1. Unstable species of anhydrous milk fat. *Journal of Dairy Science*, 84(4), 756–766. Available at: http://dx.doi.org/10.3168/jds.S0022-0302(01)74531-6.

Lopez, C. *et al.*, 2002. Thermal and structural behavior of milk fat. 3. Influence of cooling rate and droplet size on cream crystallization. *Journal of Colloid and Interface Science*, 254(1), 64–78.

Marangoni, A.G. *et al.*, 2012. Structure and functionality of edible fats. *Soft Matter*, 8(5), 1275.

Marangoni, A.G., 2002. The nature of fractality in fat crystal networks. *Trends in Food Science and Technology*, 13(2), 37–47.

Marangoni, A.G. and Lencki, R.W., 1998. Ternary phase behavior of milk fat fractions. *Journal of Agricultural and Food Chemistry*, 46(10), 3879–3884. Available at: http://pubs.acs.org/doi/abs/10.1021/jf9801668.

Marangoni, A.G. and Mcgauley, S.E., 2003. Relationship between crystallization behavior and structure in cocoa butter. *Crystal Growth and Design*, 3, 95–108.

Marangoni, A.G. and Ollivon, M., 2007. Fractal character of triglyceride spherulites is a consequence of nucleation kinetics. *Chemical Physics Letters*, 442(4–6), 360–364.

Marangoni, A.G., Tang, D. and Singh, A.P., 2006. Non-isothermal nucleation of triacylglycerol melts. *Chemical Physics Letters*, 419(1–3), 259–264.

Marangoni, A.G. and Wesdorp, L.H., 2013a. Crystallography and polymorphism. In: A.G. Marangoni and L.H. Wesdorp, eds. *Structure and Properties of Fat Crystal Networks*. CRC Press, Boca Raton, FL, pp. 1–25.

Marangoni, A.G. and Wesdorp, L.H., 2013b. Nucleation and crystalline growth kinetics. In: A.G. Marangoni and L. Wesdorp, eds. *Structure and Properties of Fat Crystal Networks*. CRC Press, Boca Raton, FL, pp. 27–100.

Martini, S., Herrera, M.L. and Hartel, R.W., 2001. Effect of cooling rate on nucleation behavior of milk fat–Sunflower oil blends. *Journal of Agricultural and Food Chemistry*, 49(7), 3223–3229.

Martini, S. and Marangoni, A.G., 2007. Microstructure of dairy fat products. In: A. Tamime, ed. *Structure of Dairy Products*. Blackwell Publishing, Oxford, pp. 72–103.

Martini, S., Suzuki, A.H. and Hartel, R.W., 2008. Effect of high intensity ultrasound on crystallization behavior of anhydrous milk fat. *JAOCS, Journal of the American Oil Chemists' Society*, 85(7), 621–628.

Mazzanti, G. *et al.*, 2004a. Effect of minor components and temperature profiles on polymorphism in milk fat. *Crystal Growth and Design*, 4(6), 1303–1309.

Mazzanti, G. *et al.*, 2004b. Novel shear-induced phases in cocoa butter. *Crystal Growth and Design*, 4(3), 409–411.

Mazzanti, G. *et al.*, 2003. Orientation and phase transitions of fat crystals under shear. *Crystal Growth and Design*, 3(5), 721–725.

Mazzanti, G., Marangoni, A.G. and Idziak, S.H.J., 2009. Synchrotron study on crystallization kinetics of milk fat under shear flow. *Food Research International*, 42(5–6), 682–694. Available at: http://dx.doi.org/10.1016/j.foodres.2009.02.009.

McGibbon, A.K.H. and Taylor, M.W., 2006. Composition and structure of bovine milk lipids. In: P. F. Fox and P.L.H. McSweeney, eds. *Advanced Dairy Chemistry Volume 2: Lipids*. Springer Science, Berlin, pp. 1–42.

Narine, S.S. and Marangoni, A.G., 1999a. Fractal nature of fat crystal networks. *Physical Review E*, 59(2), 1908–1920.

Narine, S.S. and Marangoni, A.G., 1999b. Relating structure of fat crystal networks to mechanical properties: A review. *Food Research International*, 32(4), 227–248.

Palmquist, D.L., 2009. Milk fat: Origin of fatty acids and influence of nutritional factors thereon. *Advanced Dairy Chemistry*, 2, 43–92.

Parodi, P.W., 2004. Milk fat in human nutrition. *Australian Journal of Dairy Technology*, 59(1), 3–59. Available at: <Go to ISI>://FSTA:2004-Pa1304.

Peyronel, F., Quinn, B., *et al.*, 2014. Ultra small angle X-ray scattering in complex mixtures of triacylglycerols. *Journal of Physics: Condensed Matter*, 26(46), p.464110. Available at: http://stacks.iop.org/0953-8984/26/i=46/a=464110?key=crossref.ef31620cf45eb7eae5af6ba01058f318.

Peyronel, F. and Campos, R., 2012. Methods used in the study of the physical properties of fats. In: A.G. Marangoni, ed. *Structure-Function Analysis of Edible Fats*. AOCS Press, Urbana, IL, pp. 231–294.

Peyronel, F., Pink, D.A. and Marangoni, A.G., 2014. Triglyceride nanocrystal aggregation into polycrystalline colloidal networks : Ultra-small angle X-ray scattering , models and computer simulation. *Current Opinion in Colloid and Interface Science*, 19, 459–470.

Ronholt, S. *et al.*, 2012. Polymorphism, microstructure and rheology of butter. Effects of cream heat treatment. *Food Chemistry*, 135(3), 1730–1739.

Sangwal, K. and Sato, K., 2012. Nucleation and crystallization kinetics of milk fat. In: A.G. Marangoni, ed. *Structure-Function Analysis of Edible Fats*. AOCS Press, Urbana, IL, pp. 25–78.

Sato, K., 2001. Crystallization behaviour of fats and lipids – a review. *Chemical Engineering Science*, 56(7), 2255–2265.

Sato, K. and Ueno, S., 2011. Crystallization, transformation and microstructures of polymorphic fats in colloidal dispersion states. *Current Opinion in Colloid and Interface Science*, 16(5), 384–390. Available at: http://dx.doi.org/10.1016/j.cocis.2011.06.004.

Shi, Y., Smith, C.M. and Hartel, R.W., 2001. Compositional effects on milk fat crystallization. *Journal of Dairy Science*, 84(11), 2392–2401. Available at: http://dx.doi.org/10.3168/jds.S0022-0302(01)74688-7.

Smet, K. *et al.*, 2010. Crystallization behavior of milk fat obtained from linseed-fed cows. *Journal of Dairy Science*, 93(2), 495–505.

Sonwai, S. and Rousseau, D., 2010. Controlling fat bloom formation in chocolate – Impact of milk fat on microstructure and fat phase crystallisation. *Food Chemistry*, 119(1), 286–297.

Suzuki, a. H. *et al.*, 2010. Altering functional properties of fats using power ultrasound. *Journal of Food Science*, 75(4).

Tang, D. and Marangoni, A.G., 2006. 3D fractal dimension of fat crystal networks. *Chemical Physics Letters*, 433(1–3), 248–252.

Tang, D. and Marangoni, A.G., 2008. Modified fractal model and rheological properties of colloidal networks. *Journal of Colloid and Interface Science*, 318(2), 202–209.

Timms, R.E., 1984. Phase behaviour of fats and their mixtures. *Progress in Lipid Research*, 23(1), 1–38.

Timms, R.E., 1980. The phase behaviour and polymorphism of milk fat, milk fat fractions and fully hardened milk fat. *American Journal of Dairy Technology*, 47–53.

Truong, T. *et al.*, 2014. Effects of emulsion droplet sizes on the crystallisation of milk fat. *Food Chemistry*, 145, 725–735.

Tzompa-Sosa, D.A. *et al.*, 2014. Influence of C16:0 and long-chain saturated fatty acids on normal variation of bovine milk fat triacylglycerol structure. *Journal of Dairy Science*, 97(7), 4542–4551. Available at: http://www.ncbi.nlm.nih.gov/pubmed/24835976.

Vanhoutte, B. *et al.*, 2002. Monitoring milk fat fractionation: Effect of agitation, temperature, and residence time on physical properties. *Journal of the American Oil Chemists' Society*, 79(12), 1169–1176.

Wagh, A., Walsh, M.K. and Martini, S., 2013. Effect of lactose monolaurate and high intensity ultrasound on crystallization behavior of anhydrous milk fat. *JAOCS, Journal of the American Oil Chemists' Society*, 90(7), 977–987.

Walstra, P. *et al.*, 1999. *Dairy Technology*, Marcel Dekker, New York.

Walstra, P., Wouters, J. and Geurts, T.J., 2006. *Butter*. In: *Dairy Science and Technology*. CRC Press, Boca Raton, FL, pp. 467–495.

Wiking, L. *et al.*, 2009. Relations between crystallisation mechanisms and microstructure of milk fat. *International Dairy Journal*, 19(8), 424–430.

Woodrow, I.L. and deMan, J.M., 1968. Polymorphism in Milk Fat Shown by X-ray Diffraction and Infrared Spectroscopy. *Journal of Dairy Science*, 51(7), 996–1000. Available at: http://linkinghub.elsevier.com/retrieve/pii/S0022030268871127.

Wright, A.J., Batte, H.D. and Marangoni, A.G., 2005. Effects of canola oil dilution on anhydrous milk fat crystallization and fractionation behavior. *Journal of dairy science*, 88(6), 1955–1965. Available at: http://dx.doi.org/10.3168/jds.S0022-0302(05)72871-X.

Wright, A.J. *et al.,* 2001. Rheological Properties of Milkfat and Butter Concise Reviews in Food Science. *Journal of Food Science*, 66(8), 1056–1071.

Wright, A.J. and Marangoni, A.G., 2003. The Effect of Minor Components on Milk Fat Microstructure and Mechanical Properties. *Journal of Food Science*, 68(1), 182–186. Available at: http://doi.wiley.com/10.1111/j.1365-2621.2003.tb14137.x.

Wright, A.J. *et al.,* 2000. The effect of minor components on milk fat crystallization. *Journal of the American Oil Chemists' Society*, 77(5), 463–475.

Wright, A.J. and Marangoni, A.G., 2002. Effect of DAG on milk fat TAG crystallization. *Journal of the American Oil Chemists' Society*, 79(4), 395–402.

Zetzl, A.K., Marangoni, A.G. and Barbut, S., 2012. Mechanical properties of ethylcellulose oleogels and their potential for saturated fat reduction in frankfurters. *Food and Function*, 3(3), 327.

Zulim Botega, D.C. *et al.,* 2013. Development of formulations and processes to incorporate wax oleogels in ice cream. *Journal of Food Science*, 78(12), 1845–1851.

10

Microstructure of Ice Cream and Frozen Dairy Desserts

Samantha R. VanWees and Richard W. Hartel

University of Wisconsin-Madison, 1605 Linden Drive, Madison, WI 53706, USA

10.1 Overview of Frozen Desserts

Frozen dairy desserts are a diverse class of products that are popular around the world. Ice cream is perhaps the most well-known product in this class; however, these products vary in composition and processing, resulting in a wide variety of products with unique characteristics.

In the United States, the Code of Federal Regulations (CFR) defines the standards of identity (SOI) of many frozen desserts. Ice cream is defined as having at least 10% milkfat, 20% milk solids, weighing at least 539.2 kg/m^3 (4.5 lb/gal) and having at least 35.6% total solids (Food and Drug Administration, 2016). Products that do not meet all these specifications cannot be labeled and sold as ice cream in the United States, though labeling and composition requirements vary by region and country.

The CFR also defines standards of identity for sherbet, mellorine, and frozen custard, but other frozen dairy desserts such as gelato, frozen yogurt, and milk ices have no regulatory standard of identity. Sherbet is similar to ice cream but contains fewer milk ingredients and has a predominant fruit flavor, with high sugar content (25–30%) and a slightly lower pH. Mellorine is very similar to ice cream, except that some or all of the milkfat has been replaced by vegetable or other animal fats. Frozen custard is also very similar to ice cream, but must contain 1.4% egg yolk solids.

Gelato does not have a standard of identity in the United States, but generally contains less milkfat (3–4%), and more sugar (20–25%) than ice creams and is served at a warmer temperature. It also has, on average, a lower overrun, defined as the percent increase in volume caused by air incorporation during dynamic freezing. Frozen yogurts are made from a mix inoculated with yogurt starter bacteria *Streptocccus thermophilus* and *Lactobacillus delbruckii* subsp. *bulgaricus*, resulting in a slightly acidic flavor and some microstructural differences due to the partial gelation of caseins. Milk ices contain more than 2.5% milkfat and more than 6% milk solids non-fat (MSNF), with no non-dairy protein or fat ingredients.

Microstructure of Dairy Products, First Edition.
Edited by Mamdouh Mahmoud Abdel-Rahman El-Bakry, Antoni Sanchez, and Bhavbhuti M. Mehta.

10.1.1 Ingredients

The formulation of a frozen dessert must abide by all regulatory definitions and deliver the desired physical and sensorial properties of the product. Ingredients in the mix supply fat, MSNF, water, sugars, stabilizers, and emulsifiers.

Milk and cream are used as the base of many ice cream mixes because they contain both milkfat and MSNF in desirable proportions, and are readily available. These ingredients also contain water, some of which will be converted to ice crystals during freezing. Butter, anhydrous milkfat, and twice-separated plastic creams may also be used as a source of milkfat. In some frozen desserts, vegetable or nondairy animal fats and oils can be added to the formulation, provided the product meets all regulatory standards. In this case, the properties of the fats must be considered for adequate fat destabilization, texture, and flavor of the product.

Concentrated or powdered products, such as evaporated milk, sweetened condensed skim milk, nonfat dry milk powder, and whey powder, are added to increase the amount of MSNF and total solids in the mix. The CFR limits the use of whey ingredients in ice cream to contributing less than 25% of the final MSNF of the mix. These products contain predominantly lactose, milk minerals, and proteins, which are important to product texture, structure, and flavor.

Sweeteners are added for taste, and also depress the freezing point of the mix and promote freeze concentration of the serum phase. Sucrose is most commonly used in ice cream because it is readily available, depresses the freezing point, and is soluble at low temperatures. Lactose is naturally present in dairy products and also depresses the freezing point of the mix, but may crystallize at low temperatures. Lactose monohydrate crystals dissolve slowly when consumed, resulting in a sandiness defect. Corn syrup solids may be used to reduce costs, but the body and texture of the frozen dessert is affected. For low-sugar and sugar-free frozen dessert formulations, sugar alcohols, high-intensity sweeteners, and combinations of these may be used to create a product with desirable physical and sensorial characteristics.

Additionally, many frozen desserts contain small amounts of stabilizers (0.2–0.5%) and emulsifiers (0.1–0.2%). These ingredients promote the formation of the initial microstructure, and provide stability during storage. Hydrocolloid stabilizers used may be produced by seaweeds (carrageenan, alginate), vegetables and seeds (guar gum, locust bean gum), bacteria (xanthan gum, gellan gum), animals (gelatin), or other sources (pectin). Modified starches and other macromolecules having similar functionality are also used, but are less common. Historically, gelatin was used as a stabilizer, but today the most common stabilizers are carrageenan, guar gum, and locust bean gum. These gums increase the viscosity of the serum phase and provide stability during storage by inhibiting ice recrystallization. Stabilizers provide a desirable, creamy texture upon consumption, though excessive stabilizer addition results in a heavy, gummy body.

Emulsifiers promote the formation of partially-coalesced fat networks and stabilize air cells. Common emulsifiers used are mono- and diglycerides (0.1–0.2%) and sorbitan esters (0.02–0.04%), such as polysorbate 80. Mono- and diglycerides have greatest function at the air interface and increase the whipping capacity of the mix, while polysorbate 80 promotes the partial coalescence of fat (Goff *et al.*, 1987, 1999; Goff and Jordan, 1989; Warren, 2015). Natural emulsifying ingredients, such as lecithin from egg yolk or soybeans, can also be used for improving frozen dessert quality. Phosphatides in egg yolk lecithin contribute to the unique smoothness and body of frozen custards and homemade ice cream products.

Flavors, colors, variegates, and inclusions are not critical to the microstructure of frozen desserts, but contribute unique flavors and textures. Chocolate mixes are often produced with cocoa powders or syrups, which increase the solids content and marginally increase the fat content, but the overall microstructure of the final product is comparable to non-chocolate products. Variegates and inclusions are often added after the initial freezing step so as not to interfere with the freezing process and structural development, but the addition of these components may affect the storage stability of the product. Products that contain a peanut butter swirl variegate, for example, introduce additional fat to the product, while fruit-based variegates or inclusions must include enough sugar and stabilizer to prevent ice crystals from forming at their surfaces.

10.1.2 Processing

In addition to formulation, the structure of the frozen product is determined by the processing parameters during the major sequential steps: mixing, pasteurization, homogenization, aging, freezing, and hardening (Figure 10.1).

The processing of frozen dessert begins with the mixing step. This gently heats and combines all liquid and powder ingredients, after which the mixture is pasteurized using high-temperature short-time (HTST) or vat pasteurization for safety. Due to the high fat content and viscosity, the frozen dessert mixes are pasteurized at a higher temperature than fluid milk (82 °C for 15 s).

Following pasteurization, the mixture is homogenized at elevated temperatures in a two-stage homogenizer. The homogenization pressure of the first stage is 13.7–17 MPa (2000–2500 psi) to decrease the size of the milkfat globules, while the second stage (3.4 MPa, 500 psi) reduces clustering and results in a stable oil-in-water emulsion with discrete globules approximately 1 μm in diameter. After the mix is pasteurized and homogenized, it is cooled and aged without agitation at 4°C for 10–24 h. During aging, emulsifiers displace proteins from the milkfat globule membrane surface, proteins and stabilizers are hydrated, and milkfat partially crystallizes.

The mix is frozen using a scraped-surface heat exchanger, and can be done in batch or continuous processes. Some ice crystallizes and air is incorporated into the product during this stage, resulting in a frozen foam that contains air, water, fat, and ice. Soft-serve ice creams and gelato are consumed immediately after freezing. Most frozen desserts are frozen to the desired endpoint, packaged into containers, and immediately hardened at −30°C for 1–3 h, depending on the size of the container. Some frozen novelties, such as ice cream bars or sandwiches, and slow-churned style products are extruded

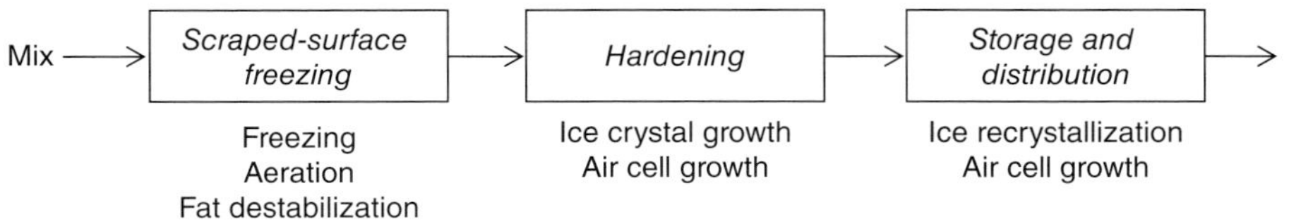

Figure 10.1 Overview of frozen desserts processing with reference to microstructure development.

at very low temperatures after freezing, thus eliminating the need for a hardening step and resulting in the formation of more small ice crystals and air cells compared to conventional hardened products.

The storage of frozen desserts is critical for maintaining a high-quality product. Low temperatures that do not fluctuate are ideal for maintaining the microstructure created during the freezing and hardening process. Temperature cycling results in changes to the ice phase, which also influences the surrounding phases and lowers product quality. Additional storage parameters including light and oxygen exposure, packaging selection, and formulation also influence the shelf stability of the product. Since frozen desserts are pasteurized products that are stored at low temperatures before consumption, it is this quality degradation and not microbiological concerns that determine the shelf life of the product.

10.2 Frozen Dessert Structure

Frozen desserts are multi-phase systems that contain air, water, fat, and ice (Figure 10.2). During the dynamic freezing process, some milkfat globules partially coalesce to form networks, while some remain as discrete globules dispersed in the unfrozen serum phase. Some water in the mix is converted to ice crystals, resulting in a freeze-concentrated serum phase that contains sweeteners, dairy salts, water, and some MSNF. The primary components of frozen dessert structure include the serum phase, ice crystals, fat phase, air cells, and proteins and hydrocolloids.

10.2.1 Serum Phase

Frozen dairy dessert mixes are oil-in-water emulsions stabilized by hydrated proteins and stabilizers following the pasteurization, homogenization, and aging processes. The continuous phase of the emulsion, often called the serum, contains dispersed fat globules and casein micelles. During the dynamic freezing process, the mix is freeze concentrated as some of the water becomes ice. Remaining in the unfrozen serum phase are sweeteners, dairy salts and minerals, some proteins and stabilizers, and water. Many of

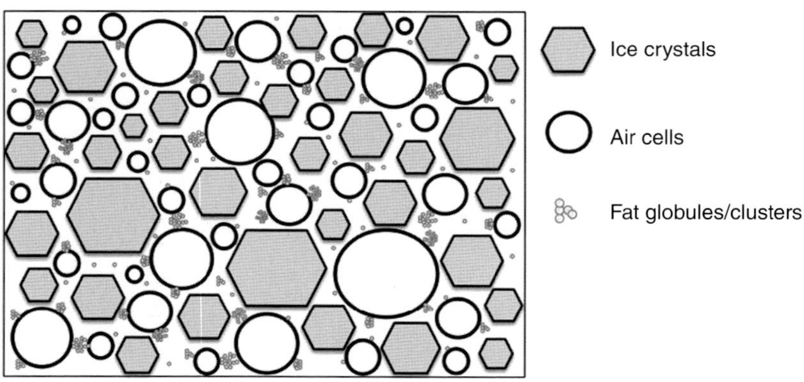

Figure 10.2 Ice cream microstructure (highly schematic).

the properties of this phase depend on mix formulation, freezing conditions, and storage temperatures.

Sweeteners such as sucrose, lactose, dextrose, sugar alcohols, and corn syrups depress the freezing point of frozen dessert mixes, with monosaccharides having the largest effect compared to other saccharides (Smith and Bradley, 1983). The dairy salts present in MSNF also have a large effect on freezing point. Large molecules, such as hydrocolloid stabilizers or starches, generally do not significantly influence the freezing point of the mix because of their low molality. A standard ice cream mix, with 12% fat, 15% sucrose, 11% MSNF, and 0.3% stabilizers and emulsifiers, has a freezing point of approximately –2.5°C (Goff and Hartel, 2013). Increasing the fat content and lowering the sugar or MSNF content results in a higher freezing point, causing more water to be frozen at draw, while increasing the MSNF or sugar content further depresses the freezing point below –2.5°C. The initial freezing point of the mix can be calculated empirically (see Goff and Hartel (2013) for calculations) or determined using a cryoscope or osmometer.

The initial freezing point and the freezing point depression curve determine the amount of water frozen at any temperature, which is important for the physical and sensorial properties of the product. Once freezing begins, the solute concentration of the serum increases as some water becomes ice. This causes a decrease in freezing point of the remaining serum. As freezing continues, the amount of ice increases, as does the concentration of the serum phase (Figure 10.3a). By plotting the freezing point against the concentration of solute, one can calculate the freezing point depression and thus the amount of ice present at a given temperature, as a function of solute concentration, assuming phase equilibrium (Figure 10.3b). Precise knowledge of the freezing point depression curve provides control over the ice and serum phases during processing and storage, and thus product quality.

At typical draw temperatures of –6 to –8°C, approximately 50% of the water exists as ice crystals and 50% remains unfrozen in the serum (Cook and Hartel, 2010). The product temperature decreases to near −18°C during hardening, further concentrating the serum. At this temperature, approximately 75–80% of water is frozen (Hartel, 1996). At –30°C, 90–95% of the water is frozen (Berger *et al.*, 1972).

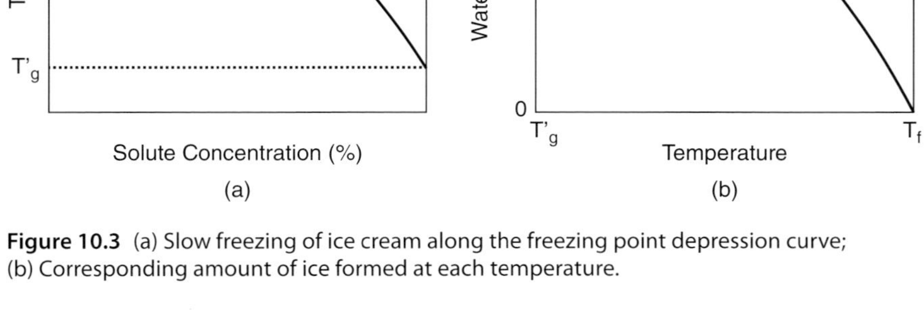

Figure 10.3 (a) Slow freezing of ice cream along the freezing point depression curve; (b) Corresponding amount of ice formed at each temperature.

The ingredients added to a mix also influence the glass transition temperature of the product (Goff, 1992). At this temperature, polymeric molecules change from being a viscoelastic rubber to an amorphous glass, resulting in an increased viscosity as temperatures decrease and ice phase volume increases. For ice cream, the glass transition temperature, T_g', is approximately –40°C, though the exact values of a specific product vary with formulation (Sahagian and Goff, 1995; Goff and Sahagian, 1996; Roos, 2010). The melting onset temperature, T_m', which is a few degrees higher than Tg′, is of more interest to manufacturers, because storage above this temperature allows molecular mobility, and shelf life is reduced because molecular diffusion and growth of ice crystals is enhanced (Roos, 2010). Storage below the glass transition temperature restricts the mobility of water such that there is always some unfrozen water in the product no matter how low the storage temperature. Products stored below –40°C have excellent stability and long shelf life because the molecular mobility of these molecules is severely restricted by the glassy matrix (Park *et al.*, 2015). However, commercial products are generally stored at temperatures well above T_m' and have a higher amount of unfrozen water in the serum. The increased molecular mobility negatively influences the shelf life and quality of the product.

The serum phase has a significant effect on the rheological and sensorial properties of the product. Decreased temperature and increased solids content increase serum viscosity by promoting freeze concentration. At high viscosity, ice crystal growth is inhibited and air cells are stabilized due to the limited molecular mobility of the unfrozen water (Sofjan and Hartel, 2004; Hagiwara *et al.*, 2006). Similarly, if serum phase viscosity is low due to low freeze concentration, less water is present as ice crystals, and the product has less solid-like rheological character. Thin serum lamellae created by air cells limit the mobility of solutes and unfrozen water, preventing crystallization of sugars and ice, while also producing a dry product that resists melting (Sofjan and Hartel, 2004). The properties of the serum phase are important to overall structure and quality of frozen desserts, but are also influenced by the ice, fat, and air phases in the product.

10.2.2 Ice Crystals

Ice crystals differentiate frozen dairy desserts from other dairy products. The size, number, and distribution of ice crystals influence the hardness and scoopability of the product, the rate of melting, and the smoothness during consumption. Thus, it is important to control the nucleation and growth of crystals during freezing, hardening, and storage. Nucleation of ice crystals only occurs during dynamic freezing in the scraped-surface freezer (SSF). Ice crystal growth occurs during hardening, as the phase volume increases with decreasing temperature. During storage and distribution, ice content fluctuates as temperatures change in an effort to maintain phase equilibrium.

In frozen dessert manufacturing, heterogeneous nucleation occurs at the walls of the SSF, where the temperature gradient between the refrigerant and the bulk solution is highest (Cook and Hartel, 2010). The refrigerant in the SSF must effectively remove heat from the friction of the dasher and the release of latent heat of fusion from ice crystallization while maintaining a temperature gradient sufficient to induce nucleation. The ice that forms at the wall is a concentrated slush containing many small dendritic-shaped crystals, which is then scraped off the walls by the continuous rotation of the scraper blades on the dasher and distributed into the interior bulk of the freezing barrel. During the initial stages of freezing, most of the crystal nuclei melt because of the relatively high bulk temperature. As freezing continues and the bulk temperature drops, the crystal nuclei ripen into disc-shaped crystals (Cook and Hartel, 2010). The temperature at the middle of the freezer barrel decreases along the length of the freezer as more ice is formed, and the ice phase is approximately at phase volume equilibrium at common draw temperatures of −6 to −8°C.

At draw from the SSF, approximately half of the initial water in the mix is frozen, and the average crystal diameter is 20–30 μm (Caldwell *et al.*, 1992; Donhowe and Hartel, 1996b; Muse and Hartel, 2004; Sofjan and Hartel, 2004; Cook and Hartel, 2010). Packaged products are then hardened in a forced-air convection process at −30°C, which reduces the core temperature of the product to near −18°C. Due to the increased ice phase volume as temperature is reduced, mean ice crystal size increases during hardening.

Ice crystal size is critical for product quality; an example of ice crystals in ice cream is shown in Figure 10.4. Large ice crystals greater than 50 μm in diameter are easily detected by the consumer, and cause a textural defect called coarseness or iciness (Arbuckle, 1966). According to a recent survey of U.S. commercial products, the mean size of ice crystals in hardened ice cream ranged from 26–67 μm (Warren and Hartel, 2014). This variation is caused by differences in formulation and process, but also through differences in storage and distribution.

At draw from the freezer, many small ice crystals and some larger crystals have formed as a result of ripening within the SSF. Crystals grow during hardening, and complex recrystallization phenomena alter the size and morphology of the crystals, resulting in a larger mean crystal size and a more normal distribution (Goff and Hartel, 2013). Typically, the distribution of ice crystals in hardened ice cream favors a large number of small ice crystals, as this promotes a smooth product texture.

Many studies have explored the effects of processing on ice crystal size. Both Russell *et al.* (1999) and Drewett and Hartel (2007) found a strong correlation between increased

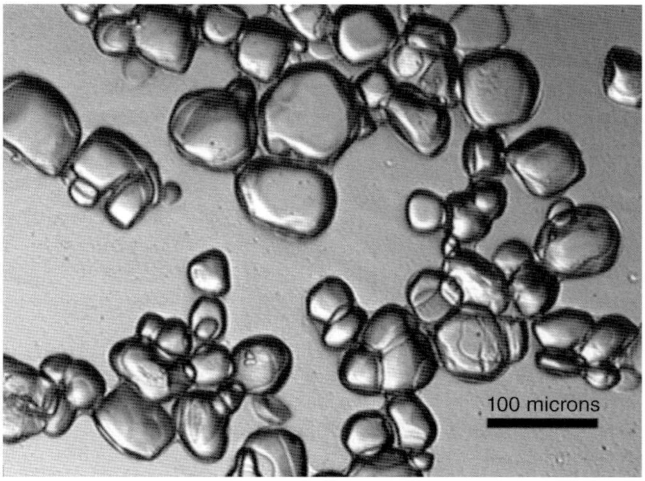

Figure 10.4 Ice crystals in hardened ice cream, as viewed by the optical microscopy method of Donhowe *et al.* (1991).

residence time in the SSF and increased ice crystal size due to the enhanced ripening of small crystals (Cook and Hartel, 2010). Flores and Goff (1999) and Sofjan and Hartel (2004) also showed that decreased overrun correlated to an increase in ice crystal size, though this is complicated by the additional implications of an increased air phase volume and decreased serum lamella thickness that would alter the development of ice crystals.

In many specialty frozen desserts, the ice crystal size distribution is controlled by using low-temperature extrusion (LTE) or liquid nitrogen (LN_2) instead of the conventional processes. These methods use rapid and efficient freezing to maximize the number of small crystals, and minimize ice crystal growth. LTE products are first frozen to −5°C in a conventional scraped-surface freezer, then continuously extruded to temperatures around −15°C before being packaged and hardened (Windhab and Bolliger, 1995). During the LTE process, temperatures are lower and more ice is present after extrusion; therefore, less liquid water is available for crystal growth during hardening. The extruder operates at low local shear conditions to maximize the ice phase volume during cooling, to minimize changes in ice crystal size and shape, while also affecting the development of air cells and partially-coalesced fat in the product (Bolliger *et al.*, 2000b; Wildmoser *et al.*, 2004). LTE ice creams are commonly found on the market today as novelty products or as multi-serve containers marketed as "slow-churned".

LN_2 can be used to rapidly harden containers, to aid in the hardening of dipped products, or to directly produce novelty ice cream drops (Clarke, 2004). In the case of rapid hardening, products are packaged and then immersed in liquid nitrogen to reduce the core temperature very quickly, minimizing the growth of ice crystals or air cells. As temperature of the product approaches the glass transition temperature, molecular mobility is inhibited by the glassy matrix and water will not crystallize as ice. Water remains unfrozen in the serum phase and ice crystal phase volume is lower than conventionally frozen and hardened products. However, the LN_2 process often results in a crumbly body, and the cold shock increases the risk of product shrinkage (Goff and Hartel, 2013).

Novelty ice cream "dots" are produced by controlling the drip rate of product from a nozzle into a vat of LN_2 at −196°C. The product freezes and solidifies on contact, and can be easily removed from the vat and packaged. When ice cream is frozen rapidly to below the glass transition temperature of the mix, only a portion of the ice phase is able to form and the product is not at phase equilibrium. As the dots are warmed above T_g' but below the freezing point of the mix, increased molecular mobility allows additional ice to form, and ice phase volume increases to maintain equilibrium. This process is known as irruptive recrystallization (Fennema, 1973). The release of latent heat caused by irruptive recrystallization may cause the product to melt if ambient storage temperatures are too high to counteract the local heating. It is recommended that these products be stored at or below −40°C for optimal product quality (Goff and Hartel, 2013).

Ice crystal size distribution and phase volume are responsible for many rheological and physical properties of frozen desserts. Oscillatory thermo-rheometry (OTR) measurements can be performed at a range of temperatures to describe properties of the frozen product and the product at normal room temperatures during consumption. Increased ice phase volume has been shown to correlate with an increased storage modulus, G' (Wildmoser *et al.*, 2004). This behavior is expected, as G' is used to describe the solid-like character of a material. The viscous character of the material is measured by the loss modulus, G″, which has been shown to decrease with increasing ice crystal size and phase volume (Bolliger *et al.*, 2000b; Wildmoser *et al.*, 2004). Increased ice crystal size correlates with increased product hardness (Sakurai *et al.*, 1996; Muse and Hartel, 2004). However, ice cream hardness is not exclusively a function of ice crystal size or volume, as the extent of fat destabilization, overrun, and solute concentration in the serum phase also influence hardness. Larger ice crystal size has also been found to correlate with decreased rate of melting (Muse and Hartel, 2004; Warren and Hartel, 2014). This may be because large crystals interrupt the microstructure, and when these crystals melt the serum phase is diluted and drips through other microstructural elements, causing the ice cream to melt more quickly.

Ice crystal size also contributes to many sensorial properties of the product. Increased size and number of ice crystals often correlates with a more pronounced coldness in the product because large crystals do not melt as quickly and therefore spend more time on the tongue (Roland *et al.*, 1999; Wildmoser *et al.*, 2004; Warren and Hartel, 2014; Amador *et al.*, 2017). Large ice crystal size has also been correlated to increased hardness of the frozen product (Muse and Hartel, 2004). The mean size of ice crystals is often reported in literature, but ice cream sensory quality is often more strictly related to the distribution of crystals. High quality products are smooth and creamy with a fine distribution of many small ice crystals, since ice crystals larger than 50 μm are detected by the consumer, resulting in a coarse or icy texture.

10.2.3 Fat Phase

The fat phase is one of the most important components in ice creams and frozen desserts. Most frozen dairy desserts primarily use milkfat in formulating, but non-dairy fats may be used in certain applications. In products that utilize non-dairy fats, the blend of fats and oils should be balanced to mimic the behavior of milkfat during processing and consumption. While the composition of the fat does not change during the processing of frozen desserts, the structure and functionality of the fat phase changes dramatically.

During freezing, some fat globules are destabilized and aggregate to form a large, complex, three-dimensional fat network within the frozen product. This behavior is called partial, or arrested, coalescence and describes how fat globules in the emulsion come together to share a membrane but do not fully coalesce. Shear caused by the dasher during the whipping and freezing step forces fat globules into contact, which initiates coalescence. Coalescence behavior of fat globules is a function of many properties in the globules, including solid fat content, interfacial energy, and properties of the membrane (Goff, 1997a; Thiel *et al.*, 2016). The free energy of two coalescing droplets is reduced as the droplets join; however, crystalline fat present in the globules prevents the droplets from fully coalescing (Figure 10.5). The solid fat content of the globules must be high enough to arrest coalescence, but small enough so that the droplet retains some liquid-like character, otherwise droplets will be fully stable to coalescence (Thiel *et al.*, 2016). If too little fat is crystalline, the droplets will coalesce completely to minimize surface free energy.

Interfacial tension of the fat globule is primarily determined by the composition of the fat globule membrane. Following homogenization, native milk proteins, predominantly caseins, immediately adsorb to the newly formed interface along with phospholipids

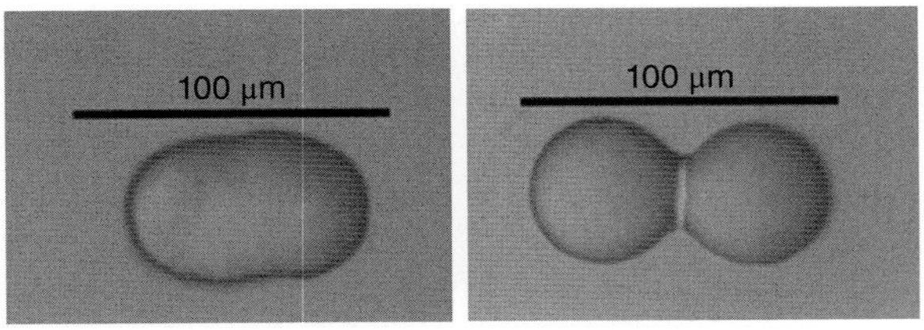

Figure 10.5 Different degrees of coalescence behavior of emulsions containing soybean and palm kernel oils, milk protein concentrate, mono- and diglycerides, and polysorbate 80, as shown by optical microscopy. The difference in degree of coalescence is caused by differences in solid fat content (higher on the right).

and serum proteins to stabilize the emulsion (Walstra *et al.*, 2006). During the aging process, caseins and serum proteins are displaced from the fat interface by small surfactant molecules such as mono- and diglycerides and polysorbate esters (Dickinson and Gelin, 1992). Polysorbate 80 (polyoxyethylene (20) sorbitan monooleate) is used in many commercial ice cream formulations because of its ability to promote partial coalescence (Goff *et al.*, 1987; Goff and Jordan, 1989; Muse and Hartel, 2004). These emulsifiers lower the interfacial tension of milkfat globules, destabilizing the mix emulsion, although they still retain enough stability to prevent creaming and flocculation in the mix prior to freezing.

The controlled destabilization of the emulsion is critical for creating the partially-coalesced fat network in frozen desserts. Bolliger *et al.* (2000b) found that a decrease in the amount of adsorbed protein, resulting from displacement by emulsifiers, results in a greater extent of partial coalescence. Compared to proteins, emulsifiers lower interfacial tension of the globules to a greater extent and promote coalescence, and the small molecular size provides for more rearrangement at the surface during coalescence (Goff, 1997b). The globules are then able to form clusters from 3–100 μm in size, provided the solid fat content provides adequate resistance to complete coalescence. In mixes made with higher amounts of whey or serum proteins, the extent of partial coalescence decreases because of the protein-stabilized membranes (Segall and Goff, 1999; Daw and Hartel, 2015). Frozen desserts produced by low temperature extrusion are subjected to extended shearing conditions at low temperatures, which promotes collisions of fat globules and results in a greater fraction of partially-coalesced fat (Eisner, 1998; Wildmoser *et al.*, 2004).

Partial coalescence of fat is irreversible and contributes to the final structure and melting properties of frozen desserts. In products made without stabilizers, air cells and partially-coalesced fat networks provide the majority of shape retention and structure once ice crystals have melted. As the extent of partial coalescence increases, the size and number of agglomerated globules increases. These provide barriers for the drainage of the serum during melting, reducing the meltdown rate (Muse and Hartel, 2004; Warren, 2015). While high fat destabilization often correlates with reduced meltdown rates, formulation, processing, and structural features in frozen desserts also affect the meltdown stability of the product (Warren and Hartel, 2014; Warren, 2015). Increased fat content and fat destabilization also correlate with increased elastic, solid-like character of the product, as networks of partially-coalesced fat promote structuring of the product (Adapa *et al.*, 2000; Muse and Hartel, 2004; Wildmoser *et al.*, 2004).

Fat enhances characteristic creaminess and smoothness in frozen dairy desserts by lubricating the palate and masking the coarse characteristics of ice crystals (Guinard *et al.*, 1997; Roland *et al.*, 1999; Warren and Hartel, 2014; Amador *et al.*, 2017). Fat also solubilizes many flavor compounds, improving product quality and flavor. However, if high-fat products are over-churned, large clusters of partially-coalesced fat can separate from the emulsion and form undesirable butter grains (Goff, 1997a).

10.2.4 Air Cells

Air is a primary component of frozen desserts, often comprising a majority of the product by volume. Air cells are incorporated during the dynamic freezing process and are stabilized by protein and emulsifier adsorption at the interface, networks of destabilized fat, and hydrocolloid stabilizers (Euston, 2008). In general, the initial size and distribution of air cells is highly dependent on the processing parameters of the whipping and freezing step and the final distribution is dependent on formulation and storage conditions. The air content of ice cream is measured by its percent overrun, which is expressed as the percent increase in volume of product greater than the amount of mix. For example, if 200 L of product is produced from 100 L of mix, the product has an overrun of 100%, and is 50% air by volume. Detailed overrun calculations for ice creams and frozen desserts have been previously described by Goff and Hartel (2013).

Strict control of overrun is important for achieving the desired physical properties of the product, but also for managing production costs and meeting regulatory requirements. Low-overrun products deliver more mix ingredients per unit volume, resulting in higher cost of production and generally a more premium quality product. Products with overrun greater than 100% for a 35–40% solids ice cream may not meet U.S. federal standards of identity for product container weight.

During freezing, the mix is aerated by the rotation of the dasher and inclusion of atmospheric or compressed air. In batch freezing operations, shear forces from the dasher allow air from the headspace to be incorporated. Overrun cannot be metered during batch freezing; instead, the relative proportions of air in the headspace and mix added to the freezer create overrun. Batch freezers allow small-scale manufacturers to produce a variety of products and flavors on a single model in short periods of time. Ice creams, sherbets, and other frozen desserts that will be hardened are easily frozen in batch operations. Soft-serve ice creams, frozen yogurt, frozen custard, gelato, and other products that are enjoyed immediately after freezing are almost exclusively produced on batch equipment. These products also generally have lower overrun (50–60%) than hardened products (80–100%).

Continuous freezers utilize an air injection system to meter the amount of air incorporated during freezing to achieve the desired overrun. In both batch and continuous freezers, shear forces created by the dasher blades break initial air bubbles into smaller cells. The forces needed to create and stabilize the smaller air cells must be greater than the increased Laplace pressure of smaller droplets, which promotes coalescence. Additional parameters in the freezer, including density, flow behavior, viscosity, and shear rate can be used to calculate an approximate maximum air cell size in a fluid system based on the Weber number (Chang and Hartel, 2002a).

At draw from the freezer, there are many small air cells and fewer large air cells present. The average size of air cells at draw is 20–25 μm in diameter (Caldwell *et al.*, 1992;

Goff *et al.*, 1999; Chang and Hartel, 2002b; Goff, 2002; Sofjan and Hartel, 2004). At this stage, the product exists in a semi-solid state that is often not representative of the final product. The distribution of air cells changes during hardening and storage by three main phenomena: drainage, coalescence, and disproportionation. Coalescence or disproportionation during hardening results in a reduction in the number of small air cells and a higher mean air cell size of 20–60 μm (Chang and Hartel, 2002b; Sofjan and Hartel, 2004; Warren and Hartel, 2014). Differences in formulation, processing, hardening, and other parameters influence the wide range of air cell sizes and distributions found in commercial products. Typical air bubbles in ice cream observed by optical microscopy (Chang and Hartel, 2002b) are shown in Figure 10.6. Products produced by LTE processes have been shown to have a fine distribution of small air cells due to the low temperatures achieved during extrusion (Wildmoser *et al.*, 2004).

During the dynamic freezing process, a new interface is introduced between air and serum. The surface of the air cells is hydrophobic, attracting fat globules and amphiphilic components to the initial interface. Air cells remain dispersed in the product and are stabilized by a highly viscous surrounding matrix. In the absence of emulsifiers, caseins predominate at the air interface; however, when emulsifiers are present in the mix, the air interface is occupied by emulsifiers and the local concentration of colloidal casein micelles dispersed in the serum phase increases (Goff *et al.*, 1999).

Milk proteins are known for their ability to form stable foams; however, increasing the protein content beyond what is commonly found in the MSNF fraction has little effect on the overrun or air cell size in ice creams (Daw and Hartel, 2015). This suggests that air cells may be stabilized by other microstructural components, such as hydrocolloids or fat. Studies have shown that whipping of mix above its freezing temperature does not sufficiently aerate the product; therefore, freezing is necessary to create and stabilize air cells (Rohenkohl and Kohlus, 1999; Chang and Hartel, 2002a; Windhab and Wildmoser, 2002). This is partially due to the increase in apparent viscosity caused by the ice phase,

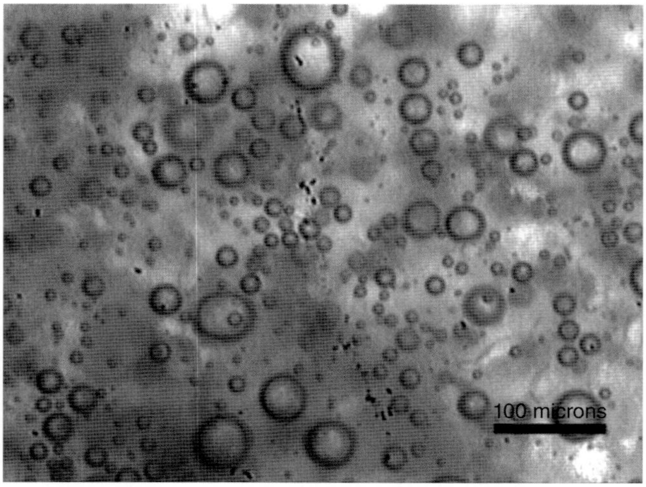

Figure 10.6 Air cells in commercial ice cream, as viewed by the optical microscopy method of Chang and Hartel (2002b).

which stabilizes the air cells and allows the action of the dasher blades to further reduce air cell size (Chang and Hartel, 2002a). Freezing also promotes the formation of large partially-coalesced fat networks, which provide stability to air cells in frozen and melted products. When melted ice cream is viewed under polarized light microscopy, these networks can be shown to associate with air cells and impart local stability (Figure 10.7).

Air cells act as an insulator, lowering the thermal diffusivity of the frozen dessert matrix and making it more difficult for heat to penetrate during melting. Sofjan and Hartel (2004) found that increased overrun in ice creams correlated with decreased meltdown rates, as expected; however, the development of overrun from high shear forces also results in increased fat destabilization and decreased meltdown rates (Warren, 2015; Amador *et al.*, 2017).

Other factors such as fat content, fat destabilization, and stabilizer addition can affect the stability of air cells. Sung and Goff (2010) found that increasing the extent of partial coalescence enhanced air cell stability but did not reduce air cell size. Caldwell *et al.* (1992) found that ice creams without added stabilizers had a higher mean air cell size than stabilized products, making them more prone to shrinkage during storage. Frozen desserts produced by low temperature extrusion have very fine dispersions of small air cells due to the high viscosity created during the low temperature process. Changes in air cell structure as a result of hardening are not observed since these products rarely undergo an additional hardening step, but air cell instability may still be observed during extended storage.

Warren and Hartel (2014) evaluated the sensory attributes of many commercial ice creams. In this study, high mean air cell size correlated with slow sensory melt rate, but melt rate did not correlate with overrun. In general, air cells disrupt the ice and fat phases and influence the heat transfer properties of the mix. Products with high overrun appear less greasy, creamy, and coarse because it is more difficult for panelists to perceive the fat or ice (Warren and Hartel, 2014). Large air cells and high overrun also contribute to a "fluffy" product texture.

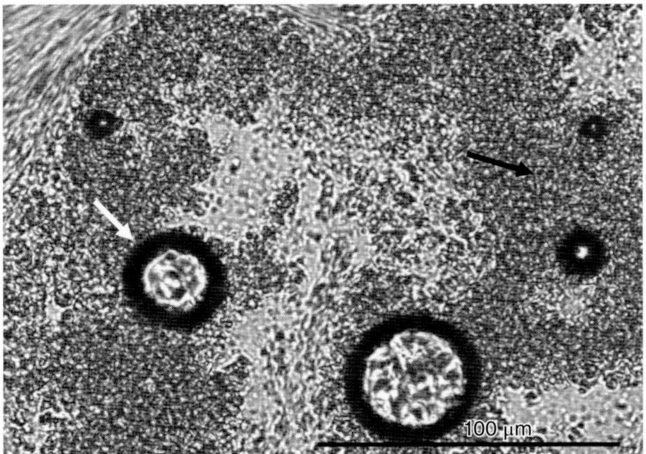

Figure 10.7 Air cells (white arrow) and destabilized fat (black arrow), as viewed under polarized light microscopy. Scale bar = 100 μm. (*See color plate section for the color representation of this figure.*)

The air phase also influences product rheology. Wildmoser *et al.* (2004) found that products with higher overrun values had lower loss moduli (G″) at temperatures below 0°C, but higher G″ at temperatures above 0°C, indicating that G″ is related to the ice phase and the air phase. High G″ values correspond to enhanced scoopability of the product, and more desirable sensory melt behavior. Small air cells, as produced by LTE, also increased the G″ of the product because of the ability of the air phase to disrupt the ice matrix.

10.2.5 Proteins and Hydrocolloids

Proteins and hydrocolloids provide additional stability to frozen desserts. Proteins adsorb to the fat interface and promote initial emulsion stability in the mix and contribute to air cell stability in the frozen product. Milk protein is present in varying amounts in frozen dairy desserts, depending on the source and amount of MSNF added to the product. Similarly, many hydrocolloid stabilizers are added to products to increase mix viscosity and to promote storage stability. Some nondairy frozen desserts use plant-based proteins or other added hydrocolloids for functional properties, while water ices rarely contain any protein in the formulation.

The protein content of ice creams prepared with nonfat dry milk powder and cream as sources of MSNF is generally 3–5%, depending on the actual amount of MSNF. Whey products can be used to supply additional protein and MSNF to frozen dessert mixes, provided the product still conforms to the legal Standard of Identity. High protein content may cause fouling on heat exchanger equipment during processing and may hinder fat destabilization, as the high concentration of adsorbed protein may not be fully displaced by emulsifiers (Daw and Hartel, 2015).

Hydrocolloid stabilizers are added to frozen desserts primarily to increase mix viscosity for desirable freezing, and to promote storage stability of the product by preventing ice recrystallization, lactose crystallization, and product shrinkage. There are many kinds of stabilizers used in frozen desserts, but the most common source of stabilizers is from polysaccharide gums. Many frozen desserts utilize a blend of polysaccharides to provide maximum stabilization in the product. Carrageenan, locust bean gum, and guar gum are added to many ice creams as a blend to maximize the various properties of each polysaccharide, whereas pectin is commonly used to stabilize more acidic or fruit-based products such as frozen yogurt and sherbet. Polysaccharides and proteins are thermodynamically incompatible in solution (Schorsch *et al.*, 1999); therefore, their behavior in frozen dairy desserts is governed by their phase behavior.

Certain polysaccharides, such as locust bean and xanthan gum, form gels within the serum phase, but the gelation behavior is also dependent on the presence of milk proteins (Regand and Goff, 2002), while guar gum and carboxymethyl cellulose do not form gels at frozen dessert temperatures. Recently, researchers have proposed novel plant-based stabilizers such as okra polysaccharides (Yuennan *et al.*, 2014) or basil seed gums (Bahram-Parvar and Goff, 2013) for use in frozen desserts. In soft-serve ice cream mixes, phases often separate due to prolonged quiescent storage before the product is frozen (Goff and Hartel, 2013). This separation can be detrimental to product quality if the mix is not sufficiently re-homogenized before freezing. When included in the formulation, κ-carrageenan complexes with κ-casein at the surface of casein micelles and decreases visible separation of phases (Thaiudom and Goff, 2003; Vega *et al.*, 2004; Vega and Goff, 2005).

Products made with stabilizers are perceived as smoother, possibly because stabilizers inhibit recrystallization and maintain small ice crystal sizes during storage (Caldwell *et al.*, 1992; Flores and Goff, 1999). However, Amador *et al.* (2017) found that increased product viscosity as a result of stabilizer addition masks sensory perception of iciness by providing lubricity in the mouth and reducing the sensation of large ice crystals. High levels of stabilizers are undesired since over-stabilized ice creams can become chewy and gummy (Goff and Hartel, 2013). In low-fat products, stabilizers are needed to impart a creamy mouthfeel to frozen desserts. Stabilizing fat replacers such as maltodextrin can be added to supplement solids and enhance the creaminess of low-fat and nonfat products (Roland *et al.*, 1999).

10.3 Storage

To maintain the carefully-balanced structure of the product, ideal storage and distribution temperatures would be similar to those in the hardening freezer, near −30°C, where molecular mobility is inhibited and the ice phase volume is high. However, since temperatures at the storage warehouse are generally −23 to −26°C and supermarket freezers can reach temperatures as high as −9°C (Ben-Yoseph and Hartel, 1998), products must be formulated to withstand these conditions. Fluctuations in temperature are often unavoidable during storage and transportation, and cause serious changes to product structure and quality. Most freezers are equipped with automatic freeze–thaw cycling to maintain refrigeration efficiency; however, this temperature cycling causes ice phase volume changes, recrystallization, and other undesirable changes to product microstructure.

10.3.1 Recrystallization

Ice recrystallization and growth of ice crystals result in coarsening of product structure, eventually causing a coarse or icy product texture. Many factors affect the rate and extent of recrystallization, including ice crystal size and ice phase volume at draw, solids content, stabilizer type and concentration, storage temperature, and temperature cycling.

Rates of recrystallization are accelerated by temperature fluctuations, but recrystallization will occur under isothermal conditions due to differences in ice crystal sizes (Donhowe *et al.*, 1991). Recrystallization phenomena such as Ostwald ripening, isomass rounding, and accretion occur at constant temperatures, while temperature cycling can cause melt-refreeze recrystallization (Fennema, 1973; Hartel, 2001). Ostwald ripening of crystals occurs when small ice crystals melt and nearby large crystals grow, but generally occurs at very slow rates in frozen desserts. Isomass rounding occurs when crystals with high radii of curvature (sharp protrusions or edges) reduce surface free energy by rounding to a lower radius of curvature (rounded or smooth surface). This phenomenon maintains a constant ice phase volume, but results in size and shape changes to ice crystals (Donhowe and Hartel, 1996b). Accretion of ice crystals occurs when nearby crystals join by the formation of a bridge or neck between the crystals. The crystal then undergoes isomass rounding, as the neck has a very large radius of curvature, resulting in the formation of a single larger crystal (Hartel, 1998).

The driving force for recrystallization is a reduction in surface free energy. Large crystals have smaller surface free energy compared to small crystals, thus, Ostwald ripening and accretion occur. Similarly, isomass rounding reduces the surface free energy of a protrusion. Ostwald ripening and isomass recrystallization of ice crystals are governed by the Kelvin equation (Hartel, 1998). For Ostwald ripening at a given bulk temperature, the Kelvin equation demonstrates that a critical radius exists where ice crystals smaller than this radius melt, and crystals larger will grow. However, true Ostwald ripening is generally rare in frozen desserts, as other mechanisms, particularly melt-refreeze, occur much faster (Donhowe and Hartel, 1996b).

Temperature fluctuations are unavoidable during storage, making melt-refreeze the main cause of undesirable ice crystal growth and coarsening in frozen desserts (Donhowe and Hartel, 1996b). When products are moved or when storage conditions change, due to automatic cycling or opening a freezer door, for example, ice in the system changes to maintain equilibrium. As temperatures increase, ice melts in accordance with the freezing curve, and re-freezes as temperatures decrease. Crystal morphology also changes, as block-shaped crystals become either more jagged and irregular or smoother and rounder, depending on temperature (Donhowe and Hartel, 1996b). Further nucleation of ice crystals does not occur after the dynamic freezing step, so small crystals that disappear as a result of Ostwald ripening or melt-refreeze recrystallization do not reappear. To maintain constant phase volume, all crystals grow larger to accommodate volume lost by the disappearance of the small crystals, resulting in product coarsening. Donhowe and Hartel (1996b) found that the mechanism for change in radius of a crystal during melt–refreeze recrystallization depends on non-convective diffusion processes, which depend on the formulation and microstructure of the product.

Stabilizers are added to products to increase the viscosity of the serum phase, limiting molecular mobility of water and thereby reducing migratory recrystallization during hardening and storage. The ability of stabilizers to retard ice recrystallization has been well studied by many researchers (Donhowe and Hartel, 1996a; Hagiwara and Hartel, 1996; Miller-Livney and Hartel, 1997; Flores and Goff, 1999; Bolliger *et al.*, 2000a; Regand and Goff, 2002, 2003; Bahram-Parvar and Goff, 2013; Yuennan *et al.*, 2014). Regand and Goff (2002) found that the formation of a gel-like network was not the only mechanism of stabilization, but that the thermodynamic incompatibility of proteins and polysaccharides was integral to preventing recrystallization. Proteins tend to localize at the interface with the ice crystal and impart water-holding capabilities directly at the interface to prevent crystal growth and regrowth, while polysaccharides primarily prevent the diffusion of water through the unfrozen serum (Regand and Goff, 2002). Hagiwara *et al.* (2006) and Herrera *et al.* (2007) showed that a decreased mobility of water as measured by pulsed NMR in different hydrocolloid systems correlated with reduced recrystallization rates. Some polysaccharides are blended to synergistically prevent recrystallization. Herrera *et al.* (2007) found that when xanthan and locust bean gums were combined, the recrystallization of ice in sugar solutions after storage was lower than when each stabilizer was used individually.

The best way to prevent extensive recrystallization during storage is to store the product at a temperature below –32°C. Below this temperature, molecular mobility is severely reduced and recrystallization occurs very slowly. Optimizing processing conditions can also prevent recrystallization from damaging product quality. While small ice crystals are necessary for product smoothness, small crystals are more prone to melting during Ostwald ripening or melt-refreeze recrystallization (Hartel, 1998). The melting of very small crystals (<5 μm) will cause the growth of large crystals, as ice maintains phase equilibrium.

Another way to reduce recrystallization is to minimize temperature fluctuations. Wide temperature fluctuations and higher storage temperatures increase the rate of recrystallization and reduce product quality (Donhowe and Hartel, 1996b; Ben-Yoseph and Hartel, 1998); however, since these fluctuations are unavoidable in almost all storage conditions, the best defense to recrystallization is preventing product instability through careful formulation and processing. Ingredients such as hydrocolloid stabilizers, proteins, or other molecules that increase the viscosity of the system are useful in preventing ice recrystallization in frozen desserts. Increasing the glass transition temperature of the mix through the use of sweeteners decreases the rate of recrystallization (Hagiwara and Hartel, 1996; Miller-Livney and Hartel, 1997); however, because storage temperatures are often still above the glass transition temperature, the moderate increase in glass transition temperature may be negligible in reducing recrystallization.

Novel ingredients such as propylene glycol monostearate (PGMS) or isolated ice structuring proteins (ISPs) can be added to frozen dessert mixes specifically to control ice recrystallization. PGMS directly affects the formation of ice at the surface of ice crystals, reducing ice crystal size before and after heat shock (Aleong *et al.*, 2008). The addition of PGMS was not found to affect viscosity of the serum phase, the mechanism by which hydrocolloid stabilizers commonly prevent recrystallization, suggesting that their action at the ice crystal surface was the primary mechanism of preventing recrystallization. ISPs are a subclass of so-called antifreeze proteins that prevent freezing of water to ice by interfering with the molecular structure of ice and can be found in cold-water fish, plants, and insects (Clarke *et al.*, 2002). Regand and Goff (2006) showed that ISP from acclimated winter wheat extract reduced recrystallization rates in ice cream by 46% at 0.004% ISP after heat shock.

10.3.2 Sugar Crystallization

Viscosity affects the crystallization of sugars in the product, though crystallization is also a function of solubility and supersaturation as well as molecular mobility. Crystallization requires supersaturation of the solute and adequate molecular mobility for crystals to grow. In frozen desserts stored at −15°C, the product is not in the glassy state and molecules are able to move freely; therefore, the primary factor in crystallization is supersaturation, controlled by the formulation (Hartel and Shastry, 1991). In products that contain high sugar content, such as sherbets and ices, the increased concentration of sucrose may result in crystallization. This can be prevented through the addition of corn syrups as humectants, though corn syrup imparts a heavy body to the product.

Lactose is present in the MSNF of the mix, and can crystallize at most storage temperatures. The α-monohydrate form of lactose crystallizes in a tomahawk shape, visible by light microscopy. Crystals in the α-monohydrate form as small as 12 μm are detectable in the mouth, causing a sandy mouthfeel during consumption (Fujimoto, 1926; Hunziker, 1926). This is much smaller than the 50 μm threshold of ice crystals because lactose crystals dissolve slowly on the tongue and therefore feel sandy during consumption. In frozen dessert mixes prepared with greater than 16% MSNF, lactose will readily crystallize at storage conditions from −10 to −15°C (Livney *et al.*, 1995).

Many factors influence the rate and extent of lactose crystallization, which has been well studied by Nickerson (1956). Lactose crystallization can be prevented by increasing the viscosity of the serum phase through the addition of solutes or hydrocolloids, reducing the molecular mobility of lactose.

10.3.3 Air Coarsening

Air cells incorporated during the freezing process are prone to instability during storage by coalescence, drainage, and disproportionation. Some changes to air cell structure occur during hardening (Chang and Hartel, 2002a); however, the majority of changes occur during extended storage and are accelerated by increased storage temperatures and temperature fluctuations.

The driving force for changes in air bubble size is the Laplace pressure, ΔP, which relates the interfacial tension and the droplet radius. Smaller air cells have larger Laplace pressure; therefore, there is a driving force to increase the size of the air bubbles through disproportionation. An increase in bubble size reduces the surface free energy, ΔG, where surface free energy is directly related to bubble surface area and interfacial tension. Air cell growth during storage is inhibited by low viscosity as a result of low storage temperatures or stabilizer addition (Caldwell *et al.*, 1992; Chang and Hartel, 2002a, 2002c). Low temperature extrusion promotes the formation of many small air cells that are well-stabilized by high ice phase volume at draw, resulting in fewer changes to air cells during storage (Euston, 2008).

Air cells in frozen desserts are stabilized by a highly viscous surrounding matrix. Coalescence, disproportionation, and distortion of air cells can be prevented by increasing the ice phase volume, adding stabilizers, increasing fat destabilization, or lowering the temperature to reduce flow (Goff *et al.*, 1999; Chang and Hartel, 2002a; Sung and Goff, 2010; Park *et al.*, 2015).

Coalescence of air cells occurs when two small cells fuse to form a single large air cell to reduce surface free energy. Over time, this results in coarsening of the system. Air cell coalescence leads to the formation of tunnels and channels throughout the product, which can adversely affect product quality and stability during storage and consumption (Chang and Hartel, 2002c). Drainage of the serum phase occurs when air cells rise to the top of the container, resulting in product shrinkage and foam collapse. Increasing the viscosity and structuring within the system through the inclusion of stabilizers, partially-coalesced fat networks, or ice crystals, maintains system integrity and decreases the rate of air cell drainage.

Disproportionation, also called Ostwald ripening, occurs when large air cells grow at the expense of small cells. Cells with a larger diameter have a smaller internal Laplace pressure compared to nearby smaller air cells, which creates a pressure gradient between nearby cells and results in the diffusion of gas from the small cell to the large cell (Chang and Hartel, 2002c). This creates a bimodal distribution of air cells within the product, with many small or large air cells present (Ronteltap and Prins, 1989). Disproportionation is inhibited when air cell size increases or matrix viscosity increases, and when air interfaces are stabilized against diffusion.

Channeling, or tunneling, of air cells is caused by partial coalescence of the air cells, which form channels through the frozen matrix. Initially, air cells exist as discrete entities and frozen desserts are considered a foam, but over time the interconnected channels cause the product to resemble a sponge (Chang and Hartel, 2002c). Turan *et al.* (1999) found that channeling was more apparent in frozen desserts stored at high temperatures and in products with high air phase volume and mean air cell size. Channeling of air cells was prevented by destabilized fat, addition of emulsifiers, and storage at low temperatures. The sponge-like structure created by channeled air cells causes undesirable changes to texture and mouthfeel, resulting in a crumbly body during scooping and eating, and can contribute to product shrinkage (Goff and Hartel, 2013).

10.3.4 Shrinkage

When air cells are not sufficiently stabilized, channels formed by air cells may cause shrinkage as product pulls away from the top and walls of containers. This can be a result of poor formulation or storage conditions, but is also commonly observed when frozen desserts are exposed to changes in external pressure. The volume of an air cell is governed by the ideal gas law; therefore, any changes to air phase volume as a result of temperature or pressure changes will affect the appearance of the product. Product shrinkage is common in products that are transported across areas of lower atmospheric pressure, such as mountains. As the pressure drops, air cell volume increases in accordance with the ideal gas laws, exerting a force on the surrounding matrix. When the pressure increases again as the product returns closer to sea level, air cell volume decreases, the surrounding matrix is no longer supported by forces exerted by the expanded air cell, and the product collapses to a level smaller than the original volume (Cole, 1940).

Although shrinkage is accelerated due to changes in external pressure or temperature cycling, it may still be observed in products stored at low temperatures due to an overall collapse in the foam matrix. Shrinkage is thought of as a two-step process, where the formation of channels and tunnels throughout the matrix cause the collapse of the foam and loss of product volume (Turan *et al.*, 1999). This foam collapse can be caused by the sinking of the product in the container, or the diffusion of air, which causes the product to pull away from the sides of the container (Dubey and White, 1997).

Shrinkage is somewhat understudied in modern literature. Dubey and White (1997) reviewed early literature on the subject and found that in most cases shrinkage was caused by small air cells, small ice crystals, high overrun, and temperature abuse. While some of these parameters would be likely to influence shrinkage in frozen desserts, many do not agree with current research. For example, Turan *et al.* (1999) found that small air cells formed fewer channels than large air cells, improving resistance to shrinkage. In a study of frozen and thawed dairy mousses, Duquenne *et al.* (2016) suggested that melting of large ice crystals may facilitate gas diffusion out of bubbles, which would result in more disproportionation and may cause more air to diffuse to the surface and out of the product matrix entirely. According to this theory, decreased ice crystal size and increased serum phase viscosity could prevent shrinkage in frozen desserts.

10.4 Conclusion

The microstructure of frozen dairy desserts is influenced by the formulation, processing, and storage parameters used in the manufacturing process. The complex interactions between the ice, water, fat, and air phases influence the physical, rheological and sensorial properties of the product. Frozen desserts are frozen quickly using a scraped-surface freezer to promote the formation of small ice crystals, even dispersion of small air cells, and the aggregation of fat globules to create a high-quality product. Ingredients in the mix including milkfat, milk solids not fat, sweeteners, emulsifiers, and stabilizers influence the behavior of the mix during freezing, and the storage stability of the product. The market category of frozen dairy desserts is incredibly diverse; therefore, it is only fitting that the product microstructure be equally complex.

References

Adapa, S., Dingeldein, H., Schmidt, K. A. and Herald, T. J. (2000) Rheological properties of ice cream mixes and frozen ice creams containing fat and fat replacers. *Journal of Dairy Science*, 83(10), 2224–2229. doi: 10.3168/jds.S0022-0302(00)75106-X.

Aleong, J. M., Frochot, S. and Goff, H. D. (2008) Ice recrystallization inhibition in ice cream by propylene glycol monostearate. *Journal of Food Science*, 73(9), E463–E468. doi: 10.1111/j.1750-3841.2008.00954.x.

Amador, J., Hartel, R. W. and Rankin, S. A. (2017) The effects of fat structures and mix viscosity on physical and sensory properties of ice cream. *Journal of Food Science*, 82(8), 1851–1860.

Arbuckle, W. S. (1966) Relation of freezing and hardening to the body and texture of ice cream. *Ice Cream Field and Trade Journal*, 48(6), 34–47.

Bahram-Parvar, M. and Goff, H. D. (2013) Basil seed gum as a novel stabilizer for structure formation and reduction of ice recrystallization in ice cream. *Dairy Science & Technology*, 93(3), 273–285. doi: 10.1007/s13594-013-0122-9.

Ben-Yoseph, E. and Hartel, R. W. (1998) Computer simulation of ice recrystallization in ice cream during storage. *Journal of Food Engineering*, 38(3), 309–329. doi: 10.1016/S0260-8774(98)00116-2.

Berger, K. G, Bullimore, B. K., White, G. W. and Wright, W. B. (1972) The structure of ice cream: Part 1. *Dairy Industries*, 37(8), 419–425.

Bolliger, S., Goff, H. D. and Tharp, B. W. (2000a) Correlation between colloidal properties of ice cream mix and ice cream. *International Dairy Journal*, 10(4), 303–309. doi: 10.1016/S0958-6946(00)00044-3.

Bolliger, S., Wildmoser, H., Goff, H. D. and Tharp, B. W. (2000b) Relationships between ice cream mix viscoelasticity and ice crystal growth in ice cream. *International Dairy Journal*, 10(11), 791–797. doi: 10.1016/S0958-6946(00)00108-4.

Caldwell, K. B., Goff, H. D. and Stanley, D. W. (1992) A low-temperature scanning electron microscopy study of ice cream. II. Influence of selected ingredients and processes. *Food Structure*, 11(1), 11–23.

Chang, Y. and Hartel, R. W. (2002a) Development of air cells in a batch ice cream freezer. *Journal of Food Engineering*, 55(1), 71–78. doi: 10.1016/S0260-8774(01)00243-6.

Chang, Y. and Hartel, R. W. (2002b) Measurement of air cell distributions in dairy foams. *International Dairy Journal*, 12(5), 463–472. doi: 10.1016/S0958-6946(01)00171-6.

Chang, Y. and Hartel, R. W. (2002c) Stability of air cells in ice cream during hardening and storage. *Journal of Food Engineering*, 55(1), 59–70. doi: 10.1016/S0260-8774(01)00242-4.

Clarke, C., Buckley, S. L. and Lindner, N. (2002) Ice structuring proteins – A new name for antifreeze proteins. *Cryoletters*, 23(2), 89–92.

Clarke, C. J. (2004) *The Science of Ice Cream*. London: Royal Society of Chemistry.

Cole, W. C. (1940) Factors affecting shrinkage. *Ice Cream Field National Trade Journal*, 36, 30–32.

Cook, K. L. and Hartel, R. W. (2010) Mechanisms of ice crystallization in ice cream production. *Comprehensive Reviews in Food Science and Food Safety*, 9(2), 213–222. doi: 10.1111/j.1541-4337.2009.00101.x.

Daw, E. and Hartel, R. W. (2015) Fat destabilization and melt-down of ice creams with increased protein content. *International Dairy Journal*, 43, 33–41. doi: 10.1016/j.idairyj.2014.12.001.

Dickinson, E. and Gelin, J. L. (1992) Influence of emulsifier on competitive adsorption of αs-casein + β-lactoglobulin in oil-in-water emulsions. *Colloids and Surfaces*, 63(3–4), 329–335. doi: 10.1016/0166-6622(92)80255-Z.

Donhowe, D. P. and Hartel, R. W. (1996a) Recrystallization of ice during bulk storage of ice cream. *International Dairy Journal*, 6(11–12), 1209–1221. doi: 10.1016/S0958-6946(96)00030-1.

Donhowe, D. P. and Hartel, R. W. (1996b) Recrystallization of ice in ice cream during controlled accelerated storage. *International Dairy Journal*, 6(11–12), 1191–1208. doi: 10.1016/S0958-6946(96)00029-5.

Donhowe, D. P., Hartel, R. W. and Bradley, R. L. (1991) Determination of ice crystal size distributions in frozen desserts. *Journal of Dairy Science*, 74(10), 3334–3344. doi: 10.3168/jds.S0022-0302(91)78521-4.

Drewett, E. M. and Hartel, R. W. (2007) Ice crystallization in a scraped surface freezer. *Journal of Food Engineering*, 78(3), 1060–1066. doi: 10.1016/j.jfoodeng.2005.12.018.

Dubey, U. K. and White, C. H. (1997) Ice cream shrinkage: A problem for the ice cream industry. *Journal of Dairy Science*, 80(12), 3439–3444.

Duquenne, B. Vergauwen, B., Capdepon, C., Boone, M. A., De Schryver, T., Van Hoorebeke, L., Van Weyenberg, S., Stevens P. and De Block, J. (2016) Stabilising frozen dairy mousses by low molecular weight gelatin peptides. *Food Hydrocolloids*, 60, 317–323. doi: 10.1016/j.foodhyd.2016.04.001.

Eisner, M. D. (1998) Low temperature ice cream extrusion technology and related ice cream properties. *European Dairy Magazine*, 24–29.

Euston, S. R. (2008) Emulsifiers in dairy products and dairy substitutes. In: *Food Emulsifiers and Their Applications*, Second Edition. New York, New York: Springer US, 195–232.

Fennema, O. R. (1973) *Low-Temperature Preservation of Foods and Living Matter*. New York, New York: Marcel Dekker, Inc.

Flores, A. A. and Goff, H. D. (1999) Ice crystal size distributions in dynamically frozen model solutions and ice cream as affected by stabilizers. *Journal of Dairy Science*, 82(7), 1399–1407. doi: 10.3168/jds.S0022-0302(99)75366-X.

Food and Drug Administration (2016) Frozen Desserts: Section 135. In: *Code of Federal Regulations, Title 21*.

Fujimoto, A. T. (1926) *Some Factors Influencing the Rate Of Crystallization of Lactose in Ice Cream*. Ohio State University.

Goff, H. D. (1992) Low-temperature stability and the glassy state in frozen foods. *Food Research International*, 25(4), 317–325. doi: 10.1016/0963-9969(92)90128-R.

Goff, H. D. (1997a) Colloidal aspects of ice cream – A review. *International Dairy Journal*, 7(6–7), 363–373. doi: 10.1016/S0958-6946(97)00040-X.

Goff, H. D. (1997b) Instability and partial coalescence in whippable dairy emulsions. *Journal of Dairy Science*, 80(10), 2620–2630. doi: 10.3168/jds.S0022-0302(97)76219-2.

Goff, H. D. (2002) Formation and stabilisation of structure in ice-cream and related products. *Current Opinion in Colloid & Interface Science*, 7(5–6), 432–437. doi: 10.1016/S1359-0294(02)00076-6.

Goff, H. D. and Hartel, R. W. (2013) *Ice Cream*, 7th Edition. New York, New York: Springer US.

Goff, H. D. and Jordan, W. K. (1989) Action of emulsifiers in promoting fat destabilization during the manufacture of ice cream. *Journal of Dairy Science*, 72(1), 18–29. doi: 10.3168/jds.S0022-0302(89)79075-5.

Goff, H. D., Liboff, M., Jordan, W. K. and Kinsella, J. E. (1987) The effects of polysorbate 80 on the fat emulsion in ice cream mix: Evidence from transmission electron microscopy studies. *Food Structure*, 6(2).

Goff, H. D. and Sahagian, M. E. (1996) Glass transitions in aqueous carbohydrate solutions and their relevance to frozen food stability. *Thermochimica Acta*, 280, 449–464. doi: 10.1016/0040-6031(95)02656-8.

Goff, H. D., Verespej, E. and Smith, A. K. (1999) A study of fat and air structures in ice cream. *International Dairy Journal*, 9(11), 817–829. doi: 10.1016/S0958-6946(99)00149-1.

Guinard, J. X., Zoumas-Morse, C., Mori, L., Uatoni, B., Panyam, D. and Kilara, A. (1997) Sugar and fat effects on sensory properties of ice cream. *Journal of Food Science*, 62(5), 1087–1094. doi: 10.1111/j.1365-2621.1997.tb15044.x.

Hagiwara, T. and Hartel, R. W. (1996) Effect of sweetener, stabilizer, and storage temperature on ice recrystallization in ice cream. *Journal of Dairy Science*, 79(5), 735–744. doi: 10.3168/jds.S0022-0302(96)76420-2.

Hagiwara, T., Hartel, R. W. and Matsukawa, S. (2006) Relationship between recrystallization rate of ice crystals in sugar solutions and water mobility in freeze-concentrated matrix. *Food Biophysics*, 1(2), 74–82. doi: 10.1007/s11483-006-9009-0.

Hartel, R. W. (1996) Ice crystallization during the manufacture of ice cream. *Trends in Food Science & Technology*, 7(10), 315–321. doi: 10.1016/0924-2244(96)10033-9.

Hartel, R. W. (1998) Mechanisms and kinetics of recrystalliztion in ice cream. In: *The Properties of Water in Foods: ISOPOW 6*, First edition. London: Thomson Science, 287–319.

Hartel, R. W. (2001) *Crystallization in Foods*. Gaithersburg, MD: Aspen Publishers.

Hartel, R. W. and Shastry, A. V. (1991) Sugar crystallization in food products. *Critical Reviews in Food Science and Nutrition*, 30(1), 49–112. doi: 10.1080/10408399109527541.

Herrera, M. L., M'Cann, J. I., Ferrero, C., Hagiwara, T. Zaritzky, N. E. and Hartel, R. W. (2007) Thermal, mechanical, and molecular relaxation properties of frozen sucrose and fructose solutions containing hydrocolloids. *Food Biophysics*, 2(1), 20–28. doi: 10.1007/s11483-007-9025-8.

Hunziker, O. F. (1926) Condensed milk defects, their causes and preventions. In: *Condensed Milk and Milk Powder*, 4th edition. La Grange, IL: Published by the author, 330–338.

Livney, Y. D., Donhowe, D. P. and Hartel, R. W. (1995) Influence of temperature on crystallization of lactose in ice-cream. *International Journal of Food Science & Technology*, 30(3), 311–320. doi: 10.1111/j.1365-2621.1995.tb01380.x.

Miller-Livney, T. and Hartel, R. W. (1997) Ice recrystallization in ice cream: Interactions between sweeteners and stabilizers. *Journal of Dairy Science*, 80(3), 447–456. doi: 10.3168/jds.S0022-0302(97)75956-3.

Muse, M. R. and Hartel, R. W. (2004) Ice cream structural elements that affect melting rate and hardness. *Journal of Dairy Science*, 87(1), 1–10.

Nickerson, T. A. (1956) Lactose crystallization in ice cream. II. Factors affecting rate and quantity. *Journal of Dairy Science*, 39(10), 1342–1350. doi: 10.3168/jds.S0022-0302(56)94858-5.

Park, S. H., Jo, J. Y., Chun, J. Y., Hong, G. P., Davaatseren, M. and Choi, M. J. (2015) Effect of frozen storage temperature on the quality of premium ice cream. *Korean Journal for Food Science of Animal Resources*, 35(6), 793–799. doi: 10.5851/kosfa.2015.35.6.793.

Regand, A. and Goff, H. D. (2002) Effect of biopolymers on structure and ice recrystallization in dynamically frozen ice cream model systems. *Journal of Dairy Science*, 85(11), 2722–2732. doi: 10.3168/jds.S0022-0302(02)74359-2.

Regand, A. and Goff, H. D. (2003) Structure and ice recrystallization in frozen stabilized ice cream model systems. *Food Hydrocolloids*, 17(1), 95–102. doi: 10.1016/S0268-005X(02)00042-5.

Regand, A. and Goff, H. D. (2006) Ice recrystallization inhibition in ice cream as affected by ice structuring proteins from winter wheat grass. *Journal of Dairy Science*, 89(1), 49–57. doi: 10.3168/jds.S0022-0302(06)72068-9.

Rohenkohl, H. and Kohlus, R. (1999) Foaming of ice cream and the time stability of its bubble size distribution. In: *Bubbles in Food*. St. Paul, MN: Eagan Press, 45–53.

Roland, A. M., Phillips, L. G. and Boor, K. J. (1999) Effects of fat replacers on the sensory properties, color, melting, and hardness of ice cream. *Journal of Dairy Science*, 82(10), 2094–2100. doi: 10.3168/jds.S0022-0302(99)75451-2.

Ronteltap, A. D. and Prins, A. (1989) Contribution of drainage, coalescence, and disproportionation to the stability of aerated foodstuffs and the consequences for the bubble size distribution as measured by a newly developed optical glass-fibre technique. In: Bee, R. D., Richmond, P. and Mingings, J. (Eds) *Food Colloids. The Proceedings of an International Symposium Organised by the Food Chemistry Group of the Royal Society of Chemistry*. Coloworth, UK.

Roos, Y. H. (2010) Glass transition temperature and its relevance in food processing. *Annual Review of Food Science and Technology*, 1, 469–496. doi: 10.1146/annurev.food.102308.124139.

Russell, A. B., Cheney, P. E. and Wantling, S. D. (1999) Influence of freezing conditions on ice crystallisation in ice cream. *Journal of Food Engineering*, 39(2), 179–191. doi: 10.1016/S0260-8774(98)00161-7.

Sahagian, M. E. and Goff, H. D. (1995) Influence of stabilizers and freezing rate on the stress relaxation behavior of freeze-concentrated sucrose solutions at different temperatures. *Food Hydrocolloids*, 9(3), 181–188. doi: 10.1016/S0268-005X(09)80214-2.

Sakurai, K., Kokubo, S., Hakamata, K. and Yoshida, S. (1996) Effect of production conditions on ice cream melting resistance and hardness. *Milchwissenschaft*, 51(8), 451–454.

Schorsch, C., Jones, M. G. and Norton, I. T. (1999) Thermodynamic incompatibility and microstructure of milk protein/locust bean gum/sucrose systems. *Food Hydrocolloids*, 13(2), 89–99. doi: 10.1016/S0268-005X(98)00074-5.

Segall, K. I. and Goff, H. D. (1999) Determination of protein surface concentration for emulsions containing a partially crystalline dispersed phase. *Food Hydrocolloids*, 13(4), 291–297. doi: 10.1016/S0268-005X(99)00011-9.

Smith, K. E. and Bradley, R. L. (1983) Effects on freezing point of carbohydrates commonly used in frozen desserts. *Journal of Dairy Science*, 66(12), 2464–2467. doi: 10.3168/jds.S0022-0302(83)82112-2.

Sofjan, R. P. and Hartel, R. W. (2004) Effects of overrun on structural and physical characteristics of ice cream. *International Dairy Journal*, 14(3), 255–262. doi: 10.1016/j.idairyj.2003.08.005.

Sung, K. K. and Goff, H. D. (2010) Effect of solid fat content on structure in ice creams containing palm kernel oil and high-oleic sunflower oil. *Journal of Food Science*, 75(3), C274–C279. doi: 10.1111/j.1750-3841.2010.01539.x.

Thaiudom, S. and Goff, H. D. (2003) Effect of κ-carrageenan on milk protein polysaccharide mixtures. *International Dairy Journal*, 13(9), 763–771. doi: 10.1016/S0958-6946(03)00097-9.

Thiel, A. E., Hartel, R. W., Spicer, P. T. and Hendrickson, K. J. (2016) Coalescence behavior of pure and natural fat droplets characterized via micromanipulation. *Journal of the American Oil Chemists' Society*, 93(11), 1467–1477. doi: 10.1007/s11746-016-2896-4.

Turan, S., Kirkland, M., Trusty, P. A. and Campbell, I. (1999) Interaction of fat and air in ice cream. *Dairy Industry International*, 64, 27–31.

Vega, C., Andrew, R. A. and Goff, H. D. (2004) Serum separation in soft-serve ice cream mixes. *Milchwissenschaft*, 59(5–6), 284–287.

Vega, C. and Goff, H. D. (2005) Phase separation in soft-serve ice cream mixes: rheology and microstructure. *International Dairy Journal*, 15(3), 249–254. doi: 10.1016/j.idairyj.2004.07.007.

Walstra, P., Wouters, J. T. M. and Geurts, T. J. (2006) *Dairy Science and Technology*, Second Edition. Boco Ratan, FL: CRC Press.

Warren, M. M. (2015) *Understanding melt behavior of ice cream: Influence of the microstructure and composition on drip-through rate of ice cream products*. University of Wisconsin-Madison.

Warren, M. M. and Hartel, R. W. (2014) Structural, compositional, and sensorial properties of United States commercial ice cream products. *Journal of Food Science*, 79(10), E2005–2013. doi: 10.1111/1750-3841.12592.

Wildmoser, H., Scheiwiller, J. and Windhab, E. J. (2004) Impact of disperse microstructure on rheology and quality aspects of ice cream. *LWT – Food Science and Technology*, 37(8), 881–891. doi: 10.1016/j.lwt.2004.04.006.

Windhab, E. J. and Bolliger, S. (1995) Combined aerator/freezer for ice cream manufacture. *European Dairy Magazine*, 7(1), 28–34.

Windhab, E. J. and Wildmoser, H. (2002) 'Extrusion: A model technology for the manufacture of ice cream', *Bulletin of the International Dairy Federation*, 374, 43–49.

Yuennan, P., Sajjaanantakul, T. and Goff, H. D. (2014) Effect of okra cell wall and polysaccharide on physical properties and stability of ice cream. *Journal of Food Science*, 79(8), E1522–E1527. doi: 10.1111/1750-3841.12539.

11

Whey Wastes and Powders

J. Chandrapala

School of Science, RMIT University, Bundoora VIC (3083), Australia

11.1 Whey

The dairy industry is a major contributor of liquid wastes to the environment. One of the dairy wastes that has received considerable attention as a source of value added products is "Whey" [1]. Whey is the watery and thin liquid residual fraction of milk after coagulation and separation of caseins during yoghurt and cheese manufacturing [2]. Whey is generally classified into two groups, "sweet" and "acid", based on the production process. Acid precipitation of caseins during manufacture of fresh cheeses such as Cottage and Ricotta or Greek style yoghurts yields "acid whey", whereas "sweet whey" originates from the production of rennet type hard cheeses. "Salty whey" is a secondary byproduct characterized by high salt concentrations applied during the salting step of some cheese varieties such as Cheddar or Colby [3]. In general 100 L of milk utilized during the manufacture of cheese/yoghurt, produces ~80–90 L of whey depending on the variety of the cheese or yoghurt being produced.

Whey in general contains about 50% of milk constituents, such as lactose (~85%), whey proteins (~10%), minerals and some fat. The exact chemical composition of whey varies in relation to the method used for its production, i.e. acid, sweet or salty whey. Generally, acid whey has a lower pH and lower protein content, but more calcium, phosphorus and lactic acid (LA) in comparison to sweet whey [4]. Acidification of milk during manufacturing of Greek yoghurt and soft cheeses via metabolic activity of lactic acid bacteria lowers the pH to the isoelectric point (pH ~4.6) of caseins, where decreasing pH results in re-equilibration of HPO_4^{2-} and PO_4^{3-} in the aqueous phase, leading to the dissolution of colloidal calcium phosphate to restore the balance and thus creating high amounts of calcium (Ca) and phosphates in the whey stream. In contrast, absence of the fermentation step and subsequent LA production in hard cheese manufacturing, allows the milk base to maintain the original pH of milk (~pH 6.7) and thereby minimal dissolution of colloidal calcium phosphate. Sweet whey is generated through the production of rennet-type hard cheeses, in which rennet/chymosin is used to initiate coagulation. The chymosin enzyme cleaves the Phe_{105}-Met_{106} bond in κ-casein, yielding para-κ-casein and glycomacropeptide (GMP). Most of the GMP ends up in the whey

Microstructure of Dairy Products, First Edition.
Edited by Mamdouh Mahmoud Abdel-Rahman El-Bakry, Antoni Sanchez, and Bhavbhuti M. Mehta.

resulting in a high protein content compared to acid whey. Salty whey contains significantly higher concentrations of sodium (Na) compared to other whey streams (Table 11.1) due to the addition of sodium chloride (NaCl) during the salting step of salted cheese varieties [5]. The application of salt to the cheese curd causes more moisture to be expelled due to an osmotic and salting effect on the proteins. Dry salting and brine salting are the two most commonly used salting modes within the dairy industry. The penetration of the salt into the cheese matrix depends on many factors such as the fat content, pH at the time of salting and temperature where increase in pH and fat content decreases the salt absorption, while increase in temperature increases the salt absorption.

Whey proteins are nutritionally one of the most valuable components in whey. They are composed mainly of β-lactoglobulin (β-Lg), α-lactalbumin (α-La), bovine serum albumin (BSA), immunoglobulin (Ig) and Lactoferrin (Lf) as well as thermostable proteose-peptones [2]. Whey proteins have a compact globular structure, with a relatively uniform distribution of non-polar, polar and charged amino acids. Whey protein has an exceptional biological value that exceeds that of egg protein by ~15% [6]. Whey proteins are rich in essential amino acids including many sulphur containing amino acids where they serve a critical role as anti-oxidants and as precursors to the potent intracellular anti-oxidant glutathione [7]. Whey proteins also act as good bioactive peptides which may affect the major body systems such as the cardiovascular, digestive, immune and nervous systems. Thus, the beneficial health effects of whey proteins can be classified as anti-microbial, anti-oxidative, anti-thrombotic, anti-hypertensive, or immunomodulatory [6]. Furthermore, whey proteins are easier to digest than caseins. Thus, they are used to manufacture infant formulas or to increase the nutritional value of dairy and

Table 11.1 Composition of sweet, salty and acid whey streams [3].

Component (%)	Sweet whey	Salty whey	Acid whey
Total solid	6.2–6.8	21.9–22.4	5.5–5.9
Ash	0.50–0.54	14.8–15.0	0.73–0.75
Total protein	1.1–1.2	0.80–0.85	0.24–0.27
Non protein nitrogen	0.01	0.02	0.01
Non casein protein	0.9–1.1	0.70–0.75	0.20–0.25
Lactose	3.1–3.2	2.8–3.0	3.20–3.25
pH	5.38–5.42	5.31–5.32	4.52–4.55
LA	0.25–0.28	0.28–0.31	0.55–0.57
Ca	0.059	0.072	0.132
K	0.12	0.13	0.133
Mg	0.017	0.016	0.017
Na	0.025	0.07–0.1	0.021
Citrate	0.064	0.072	0.071
Inorganic phosphate	0.057	0.1	0.1
Total phosphate	0.068	0.088	0.183

other food products. In addition to these beneficial health attributes, whey proteins have excellent functional properties, such as good solubility, viscosity, gelling and emulsifying properties [2, 3], which also attracted their inclusion into certain food products in attaining desirable functional properties of interest.

11.2 Current Whey Uses

Whey is being considered one of the most polluting food byproducts since it has a biochemical oxygen demand (BOD) of >435,000 ppm and a chemical oxygen demand (COD) of >460,000 ppm [2, 6]. For many years, whey produced as a byproduct was considered a waste material and was either dumped, sprayed on fields as a fertilizer or used as an animal feed. However, due to the increased production of sweet whey, utilization of whey was considered a priority by the industry. Thus, the sweet whey waste stream was successfully utilized by concentrating liquid sweet whey through evaporation to ~45–60% solids followed by crystallization of lactose and spray drying to produce a non-sticky free flowing whey powder, which can be used as a food ingredient in many food applications such as baby foods, soups, bakery products, beverages, salad dressings, confectionary, sausages and diet foods [8].

In contrast to sweet whey, an industrial solution for acid and salty whey processing has not been found, due to their unique compositional differences which limit certain steps in performance. For instance, failure of lactose to crystallize and the highly hygroscopic nature of LA present in acid whey limit the production of whey powders from the acid whey waste stream. Lactose remains in its amorphous form due to the presence of both LA and Ca [9]. In addition, lumping and caking of acid whey powder particles produced were observed [10]. So far, acid whey has mostly been utilized as an organic fertilizer. Supplementation of cattle feed was also tested, but caution was required due to energy overload and the suppression of metabolic activity of cows. Additionally, converting whey into biogas requires expensive anaerobic digesters to be installed underground, which has been deemed unfeasible at present [11]. High salt concentrations present in salty whey limits the concentration step which was performed through ultrafiltration (UF) during the manufacture of powders. High salt concentrations of salty whey foul the UF membrane quite easily, leading to longer processing times [12].

11.3 Processing of Liquid Whey

Figure 11.1 summarizes various treatments applied on the liquid whey streams converting it to whey-based ingredients. Production of whey powders were among the most popular conversions of liquid whey, although new technologies and scientific research has led to exploring new opportunities which transform an unwanted byproduct into an important raw material. Major processes in all of these treatments include:

1) Casein fines and cream recovery through separation.
2) Concentration of total solids.
3) Fractionation.

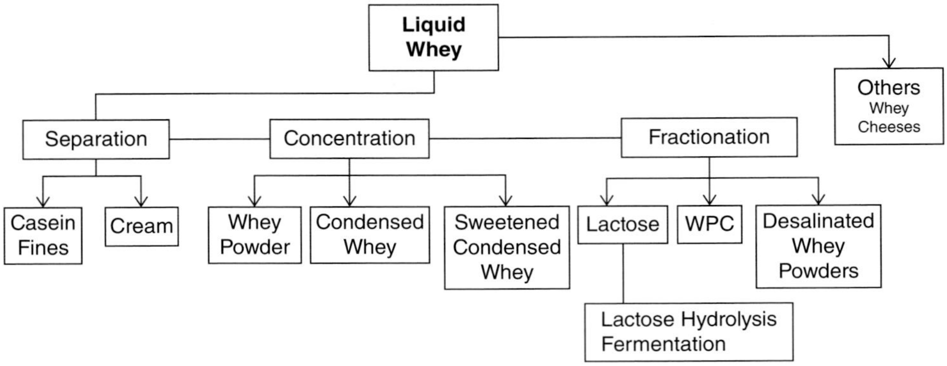

Figure 11.1 Whey based ingredients.

11.3.1 Recovery of Casein Fines and the Separation of Fat

Excessive casein fines are created due to over cutting or breaking of uncut curd particles, and will be lost with the whey. The separation of fines is important to increase the efficiency in the subsequent fat separation process. The best way is to use a centrifugal clarifier. The clarifier has one entry point and an exit point for the clarified material (Figure 11.2A). The sludge gets collected in the inner wall of the bowl with discs. The discs are conical in shape and are mounted on a spindle that rotates at high speed.

The centrifugal separator is used widely for the separation of fat. Whey enters through the bottom of the separator bowl holding a stack of conical discs, and rises up through

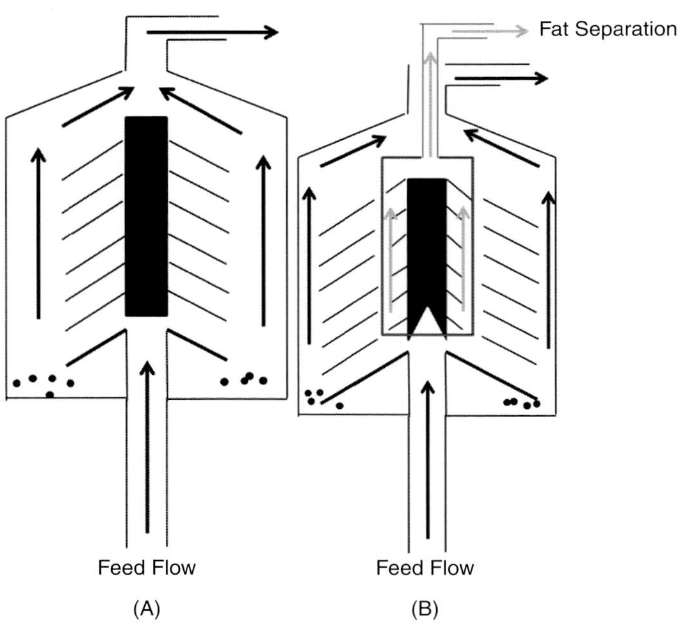

Figure 11.2 Schematic representation of (A) a clarifier and (B) a separator.

holes located in the middle of inner and outer edges of the discs. Whey between the discs is subjected to a centrifugal force in the rotating bowl and thereby tends to flow out from the center. The other fraction is heavy and moves away, forming a layer on the outer edge of the discs, whereas fat gathers on the inside edge (Figure 11.2B). The rate of separation is dependent on the speed of the separator bowl and the disc diameter. The resultant cream is sometimes used in the production of whey butter and in some instances cream is pasteurized and added to cheese milk [13].

11.3.2 Concentration of Total Solids

Concentration is essential to increase the total solid content prior to spray drying. Traditionally evaporators with two or more stages are used for the purpose. However, evaporators have different designs depending on the tubes that form the partitions between the steam and the product. They can be horizontal/vertical, or steam can be circulated either inside/outside the tubes [14]. Falling film evaporators are the most widely used. Milk is introduced at the top of a vertically arranged heated surface as a thin film. The residual film liquid and vapor are removed (Figure 11.3A). Usually multiple units are used to decrease the energy usage. In addition, vapor evolved can be used as a heating medium with the use of a thermocompressor.

In recent times, reverse osmosis (RO) is used to concentrate total solids. RO uses pressure in excess of osmotic pressure to force the solvent through a semipermeable membrane which allows water to separate [15]. The separation mechanisms are the size exclusion and diffusion. Thus, the applied pressure, flow rate and other external conditions such as temperature, nature of the feed solution, etc. can influence the efficiency of the separation. The membranes used for RO are very dense and are designed only to allow water to pass through and prevent the passage of solutes such as salts. This process uses a very high pressure of ~30–40 bar. In whey applications, the whey is concentrated with RO from 6% total solids to 10–20% total solids before ultrafiltration and/or

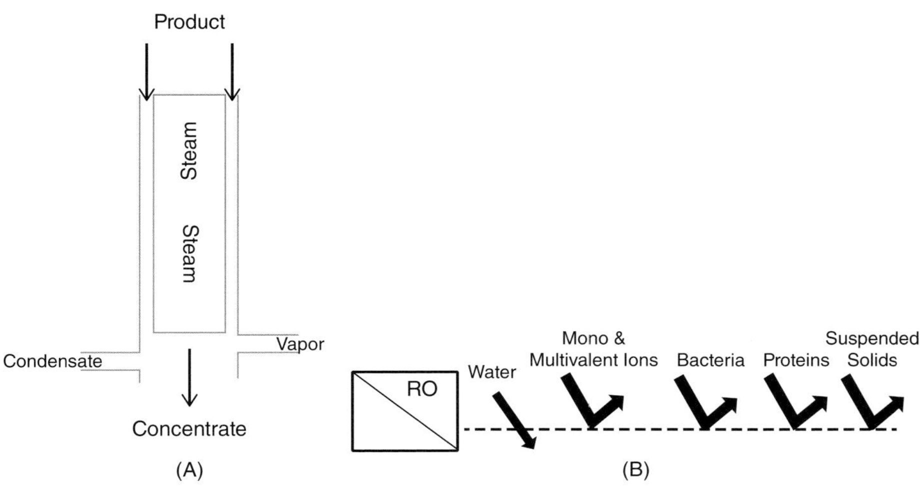

Figure 11.3 Schematic presentation of a falling film evaporator and RO separation.

concentration of the ultrafiltration permeate, which contains lactose from 5% total solids to 18–22% total solids to reduce crystallization and drying costs of the lactose powders being produced. The concentration factor attained by RO is much lower than that obtained by evaporation, due to its limitations by the osmotic pressure of the feed solution and the membrane pore blockage with minerals. For instance, the permeate flux obtained through RO of sweet whey was much greater than acid whey due to the high content of Ca in acid whey which acts as a foulant.

11.3.3 Drying

Whey is dried using single/double/twin drum dryers or spray dryers. During drum drying, steam heats the inner surface of the drum. The feed solution is uniformly applied in a thin layer onto the outer drum surface (Figure 11.4). Water evaporates when it gets in contact with the hot drum and a thin sheet of dried milk is scraped by a knife. Then the dried sheet moves away from the drums where a screw auger and a hammer break the large powder particles into small particles. The energy efficiency of drum dryers is great, although low output, high cost of replacing drums and sticking of high hygroscopic materials to the walls are some of the disadvantages [16].

Spray drying is carried out in two phases [17]: concentration to 45–55% total solids followed by drying in the drying chamber. A conventional spray dryer consists of a drying chamber, feed system, hot air system, atomizing device, powder separation system, pneumatic conveying and a cooling system. Feed is introduced to the atomizing device which is located on the top of the drying chamber (Figure 11.5). The drying air is drawn from the atmosphere via a fan. The atomized droplet meets with air and evaporation takes place where simultaneously air is cooled. The dried product falls to the bottom of the chamber. Fines will be entrained in the air and possess through the cyclone where they are collected (Figure 11.5). In the dairy industry, pressure nozzles and rotary atomizers are widely used. The former has advantages such as the ability to obtain powders with a high bulk density, improved flowability and fewer deposits in the chamber, while the latter has advantages such as the ability to use large quantities, use of high viscous samples and easy handling of powders with crystals [18]. Drying can be done in one stage or with two/three stages to achieve greater savings on the plant operational costs. Single stage drying is defined where the product is dried to the final moisture content in the spray drying chamber. During the two-stage drying, a pneumatic conveying system

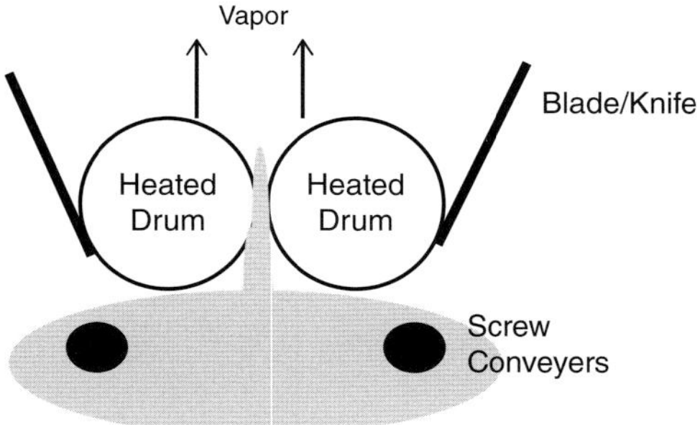

Figure 11.4 Schematic diagram of a drum dryer.

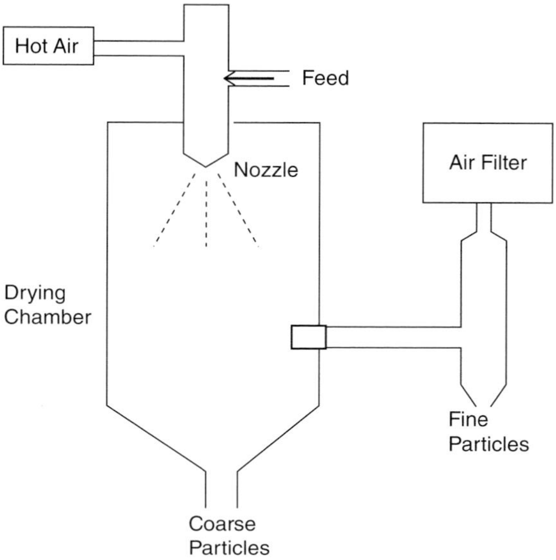

Figure 11.5 Schematic representation of a spray dryer.

is used to complete the evaporation of water [19]. This way enables achievement of good solubility, high bulk density, less powder emission, high capacity and a better economic value.

During drying, liquid milk droplets are transformed into solid particles with vacuoles of occluded air surrounded by a powder surface. Vacuoles in spray-dried milk particles originate during atomization where air is introduced to the concentrated milk. Spray drying of fluid milk into small droplets causes the spray-dried powder particles to be spherical with diameters in the range of 10 to 250 μm. The particles within the same powder may have different structures [20]. Figure 11.6(A) clearly highlights the presence of particles with smooth or wrinkled surfaces with/without indentations and fractures within the same skim milk powder (SMP) sample. The dimensions of vacuoles with occluded air vary within a powder sample. Figure 11.6(B) shows powder particles consisting of large central vacuoles or with small vacuoles.

The shape of a powder particle depends greatly on the processing conditions, the type of raw material, and other compositional parameters. For example, SMP samples processed on a spray dryer with nozzle atomization configuration showed indentations and wrinkles as mentioned before, while hand roller dried powder particles consisted of compact irregular shapes with edges (Figure 11.6C) [22]. The nozzle atomization system produces the smallest powder particles. However, the use of centrifugal and nozzle atomization configurations resulted in powder particles with similar morphology [23], although high and medium capacity nozzle atomization configurations resulted in different structures as indicated in Figure 11.7. In addition, the increase of the inlet air temperature of the spray dryer results in more wrinkles, cracks and vacuoles within the powder particles [24]. Increasing temperature overheats the milk and thereby expands the trapped air bubbles forming more cracks in the particle surface.

Figure 11.6 SEM image of a spray dried skim milk powder sample (A–B) using nozzle atomization and (C) hand roller configuration [21].

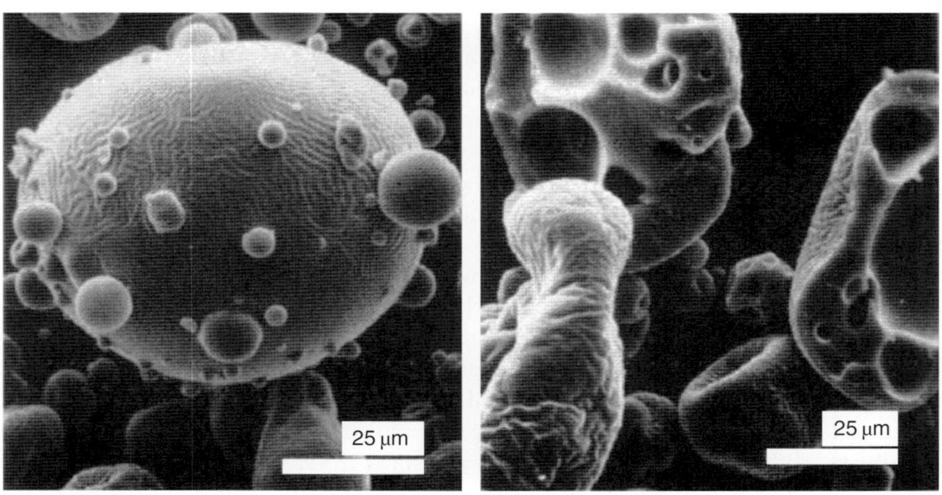

Figure 11.7 SEM images of a skim milk powder sample spray dried using a high and medium capacity nozzle atomization configuration respectively [25].

In addition to the influence of processing conditions towards the microstructure of powder particles, the composition of the feed solution does play a major role. As an example, SMP has a solid interior with many surface folds (Figure 11.8A), whereas a porous structure was observed for WMP (Figure 11.9). The formation of deep surface folds in skim milk is caused by the presence of caseins [26]. Schuck [27] studied the effect of lactose crystallization on the structure of powder particles. The study found that the presence of pre-crystallized or post-crystallized lactose crystals having varying shapes reflected the origin. If the lactose crystals were formed prior to spray drying, tomahawk shaped structures were observed, while needle-like structures were formed during post-crystallization. In another example, whole milk powder (WMP) showed a smooth surface, while non-fat dry milk showed deep wrinkles (Figure 11.8B) as was observed for skim milk.

The presence of varying amounts of a certain component within the feed solution also plays a role towards the microstructure of spray dried powder particles. Whole milk powder particles are porous due to the presence of small vacuoles and fat globules (Figure 11.9a). The confocal laser microscopy observations of spray dried WMP indicated fat droplets, occluded air and proteins entrapped within a continuous amorphous lactose phase [28]. Smaller globular fat droplets are present in the inner of the powder particles in low fat powders. On the contrary, a surface free fat layer can be observed for high fat powders. Surface free fat was specifically located at the joining points of agglomerated powder particles. Increase in surface free fat results in a smooth surface.

Figure 11.8 SEM images of spray dried (A) SMP and (B) non-fat dry milk powder samples.

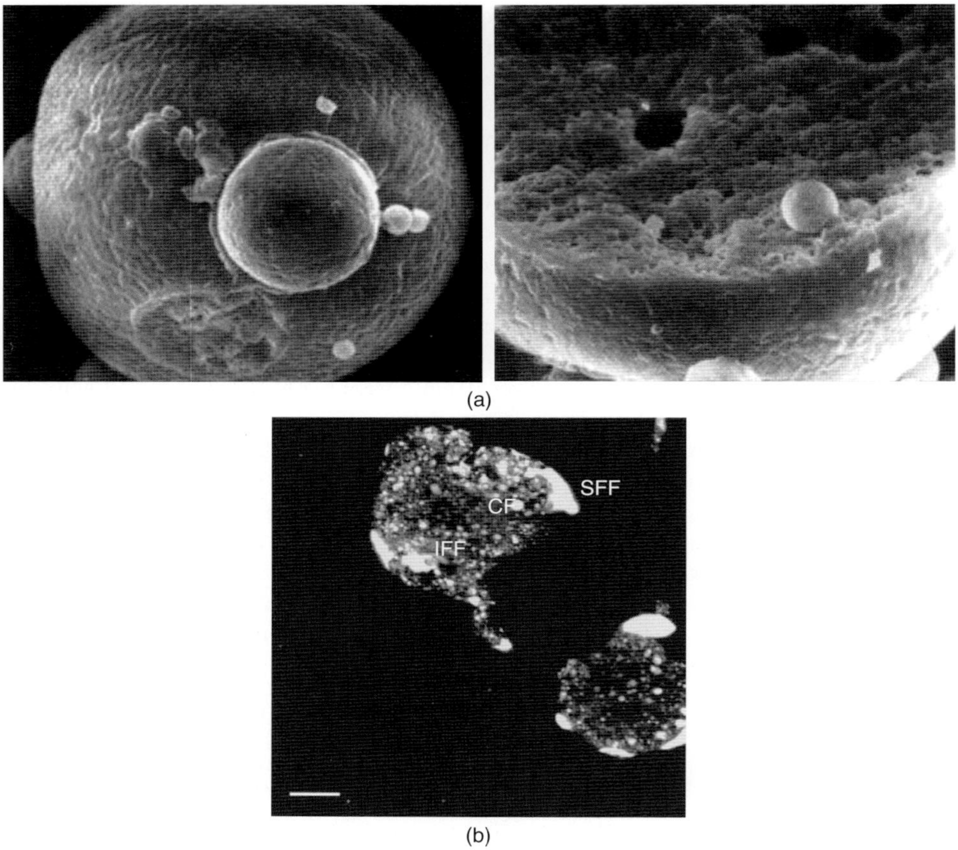

Figure 11.9 (a) SEM and (b) Confocal scanning laser microscopy images of whole milk powder (WMP). CF: Coalesced fat; IFF: Inner free fat; SFF: Surface free fat. (*See color plate section for the color representation of this figure.*)

11.3.4 Fractionation of Total Solids

Membrane technology has played an important role within the fractionation of total solids [29]. Membrane processing has been successful, because they can be effectively and economically implemented on the large scale levels required for most dairy applications. This technique is basically a filtration process in which porous membranes are used as the filtration media in order to separate solid components from the liquid phase. The theoretical principle of cross-flow filtration is derived from Fick's law of diffusion, where migration of solids flows towards the filtration surface, and the potential back-diffusion into the bulk system. The efficiency lies on appropriate use of a polymer membrane, module configuration, the system design and the operating conditions. Membrane processing has advantages over other processes, such as no phase changes, low temperature performance and good levels of separation, and can be used within a wide range of sizes. However, the degree of fractionation is somewhat limited due to the osmotic pressure changes, high viscosities and low mass transfer rates. A membrane material is a permeable or semi-permeable thin layer that restricts the motion of certain components [30]. Affectivity of a membrane is determined by the selectivity which is expressed as the permeation or retention and by the productivity which is expressed from the flux. These processes have been used for whey treatment using cellulose acetate membranes in earlier days, which were later replaced by more resistant and durable membranes made of poly-sulphones or poly-ether-sulphones [31]. Cellulose acetate membranes are hydrophilic and cheap. However, they can only be used in a narrow pH range and under low temperature conditions. Poly-sulphone membranes on the other hand can be used under high temperature conditions and within a wide pH range. However, they have major drawbacks of low pressure tolerance, and being hydrophobic resulting in extensive fouling. In recent years much attention for ceramic membranes was also given due to their increased thermal stability and chemical resistivity [32]. In addition, ceramic membranes are used in instances where pH, temperature and the cleaning chemistry affects the polymer membranes. Pressure driven membrane techniques used commonly within the dairy industry are microfiltration (MF), ultrafiltration (UF), nanofiltration (NF) and reverse osmosis (RO). However, in recent years, electrically driven membrane processes such as electro-dialysis and electro-electro-deionization methods have been investigated. MF separates particles between 0.1 to 10 μm, while UF separates particles in the range of 10–100 nm, and has a molecular weight cut-off (MWCO) range of 1000–300,000 Da. The pore size of NF membranes is between 1–10 nm with a MWCO of 200–1000 Da (Figure 11.10). MF and UF separation

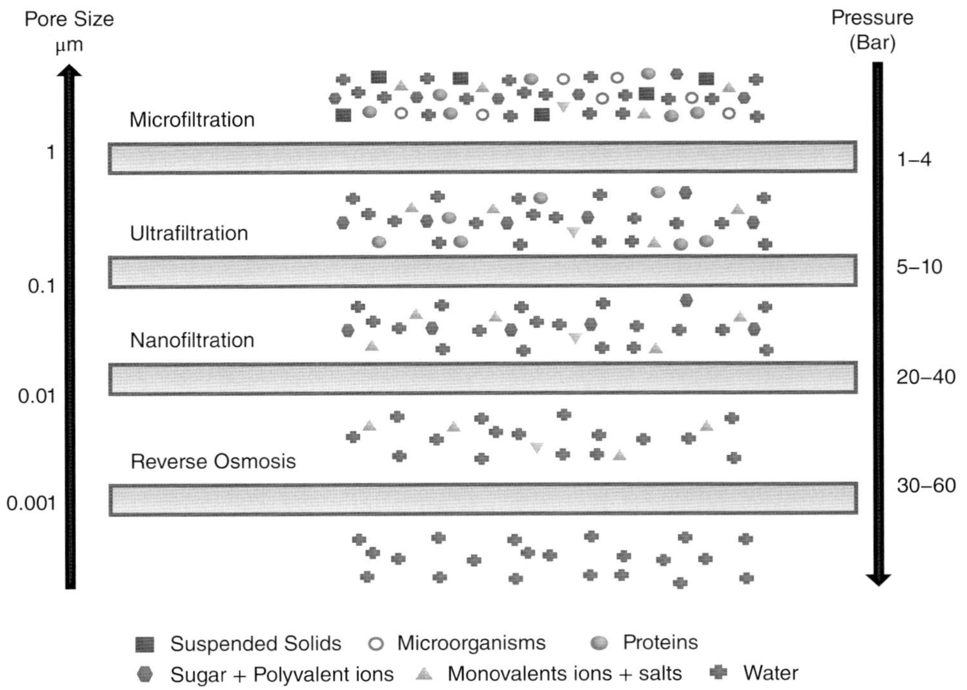

Figure 11.10 Membrane Filtration (adapted from www.nottingham.ac.uk/etc/sol_m_membranefiltration.php). (*See color plate section for the color representation of this figure.*)

is mainly on the basis of size, while the separations through NF and RO processes depend on the ionic charge, ionic composition of the feed solution and the affinity of the solutes towards the membranes.

Membranes can take different configurations such as tubular, flat, spiral wound and hollow fiber (Figure 11.11). Flat sheet is the simplest, consisting of two phases, although it has a low practical value in commercial applications. In tubular modules, the membrane is inside a tube and the feed solution is pumped through the tube. Spiral wound has a flat sheet membrane wrapped around a perforated permeate collection tube. Feed flow is from one side and the permeate is collected from the other side. The advantages of this configuration is that it can accept high viscosity feeds, is economical and allows scale up by adding more units in parallel. Hollow fiber consists of hollow fibers in a pressure vessel. The feed passes along the outside of the fibers and exits from the fiber ends [33]. The use of the hollow fiber configuration increases the surface area housed in a small volume.

Ultrafiltration (UF) has been used in the dairy industry to produce whey-protein concentrates. The lactose and mineral content in whey can be further reduced using a subsequent diafiltration (DF) step in which deionized water is continually added to the retentate while lactose and minerals are simultaneously removed from the filtrate [33]. This combined UF-DF yields a high value retentate of about 85% protein. DF can be performed in continuous or discontinuous modes. The continuous mode involves the

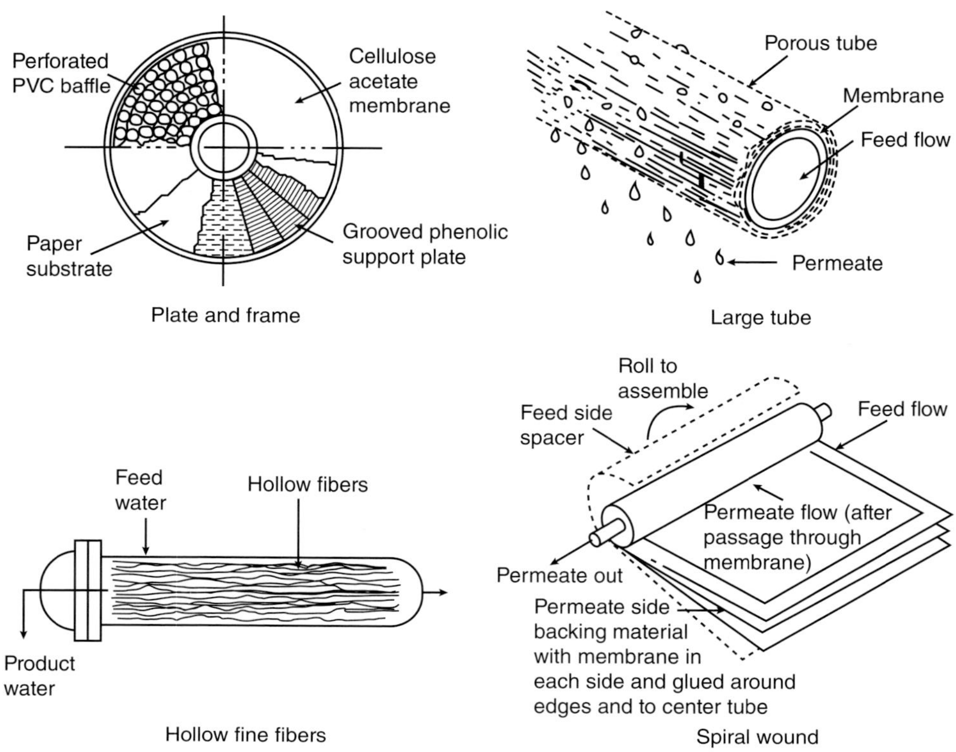

Figure 11.11 Membrane filtration modules (Source: SINCERO & SINCERO, 2003).

continuous addition of water at a rate equal to the rate of removal of the permeate. Continuous DF takes up less volume, although the discontinuous mode is easier to perform and has a less effect on the proteins. Concentration prior to DF reduces the required filtrate volume and time, although high viscosity can decrease the flux and subsequently increase the processing time. Further, in conjunction with ion exchange, these concentrates can be purified into whey protein isolates [34]. The state of the whey proteins in the form of native, denatured or aggregated proteins in different whey streams plays a significant role governing the membrane process and influences the resultant physico-chemical and functional properties of the concentrates based on the type of feed whey streams [12]. Numerous attempts to concentrate acid or salty whey through UF have encountered a number of difficulties leading to extended processing times, mainly due to membrane fouling and blockage [35]. A recent study by Nishanthi *et al.* [11] investigated the protein characteristics present in sweet whey concentrate (SWC), acid whey concentrate (AWC) and salty whey concentrate (StWC) with reference to native whey concentrate (NWC) which has not undergone any processing. Typical concentrations and the whey protein characterization of the concentrates are presented in Table 11.2. The study showed that the changes in the properties of whey

Table 11.2 Composition of sweet, salty and acid concentrates and their protein characteristics [12].

	Native whey	Sweet whey	Acid whey	Salty whey
Total solid (%)	2.5–2.6	16.8–17.0	2.9–3.0	8.9–9.0
Ash (%)	0.2	0.7–0.8	0.21	1.7–1.8
Total protein (%)	2.2	10.8–11.2	1.7	1.0–1.1
Non-protein nitrogen (%)	0.005	0.01	0.006	0.01
pH	6.50–6.53	6.40–6.42	4.20–4.22	5.50–5.52
Lactose (%)	0.1	2.9	0.7	2.4–2.7
Lactic acid (%)	0.001	0.1	0.2	0.07
Ca (%)	40–42	20–23	140–150	80–82
K (%)	30–32	100–120	10	50–56
Mg (%)	20–22	10–11	10–11	10–16
Na (%)	10–14	400–420	30–40	1100–1150
Inorganic phosphate (%)	10–16	10–13	3–4	10–13
Total phosphate (%)	20–22	30–36	10–13	20–21
Free thiol (μmol/g of protein)	39–44	72–76	6.6	7.1
Total thiol (μmol/g of protein)	380–385	781–786	409–411	354–370
Particle size (nm)	540–550	1170–1195	3327–3350	271–275
Zeta potential (mV)	−10.7	−7.0	−3.5	−7.2
Surface hydrophobicity	160–170	155–163	418–421	135–138

proteins during processing are clearly governed by the type of the whey stream. AWC and StWC were characterized with aggregated proteins through covalent cross linking among whey proteins (β-Lg, α-La, BSA and Lf) and/or caseins. These effects were mainly in response to the high acidity and salinity of acid and salty whey respectively.

NF allows separation of organics and monovalent salts in the molecular weight range ~200–1000 Da [36]. NF membranes have a high permeability for monovalent salts (such as NaCl, KCl) as compared to divalent salts. However, NF has a very low permeability for organic compounds of molecular weight greater than 300 Da such as lactose and proteins [33]. NF membranes are successfully applied in the separation of lactose and LA in cheese whey fermentation broths, while demineralization is also a useful effect [37, 38]. Furthermore, employing diafiltration (DF) increased the polyvalent ion retention of acid casein whey [39] due to the reduction of the membrane loading by the addition of the solvent [40]. The performance and selectivity of the separations was clearly dependent on pH [41, 42] and temperature [39], which affected mineral solubility and electrostatic protein interactions in the feed [42]. Valorization of acid whey (AW) is a challenge due to the presence of lactic acid (LA), which should be removed prior further processing of the waste stream. The presence of various levels of LA present in the original feed solution influenced the separation [37]. Chandrapala *et al.* [44] compared the performance of three commercially available NF membranes in the processing of AW as a function of pH (3, 4.5 or 7.3) and temperature (25 or 40°C). Additionally, NF under DF conditions was employed at 25°C. Lactose retention was dependant on the temperature only, while the transmission of LA depended on pH, membrane type and temperature. Both pH and temperature influenced the divalent and phosphate ion transmissions. Formation of calcium complexes with phosphates and lactates, which may consequently induce attraction with proteins, was responsible for the pH and temperature effects on the transmission of components [43]. Removal of almost half of the LA content of AW was possible in the best case within the study. Employing DF increased the transmission rates of Ca and monovalent ions, although the purification gained is connected with lower product concentrations in the permeate stream, which would require a concentration step after NF, e.g. a reverse osmosis plant.

The composition of the feed is important towards the microstructure as was highlighted earlier. Another example includes whey protein concentrate powders containing lower lactose content obtained through filtration, showing less wrinkling while high lactose content showed an increase in the amount of wrinkles within the powder particles. Similarly, ultrafiltered/DF skim milk powder particles with a low lactose content showed a smooth surface. However, a wrinkly surface could be obtained by the addition of lactose (Figure 11.12A). In contrast, the addition of proteins resulted in the disappearance of the wrinkles [44]. The microstructure of spray dried retentates obtained through membrane filtration showed similar external and internal structural characteristics as for SMP [22] (Figure 11.12B).

11.4 Whey Powders

The main industrial processing of liquid whey is through drying to manufacture whey powders. A number of improvements and innovations exist currently due to technological and research advancement. However, major developments are awaited on the

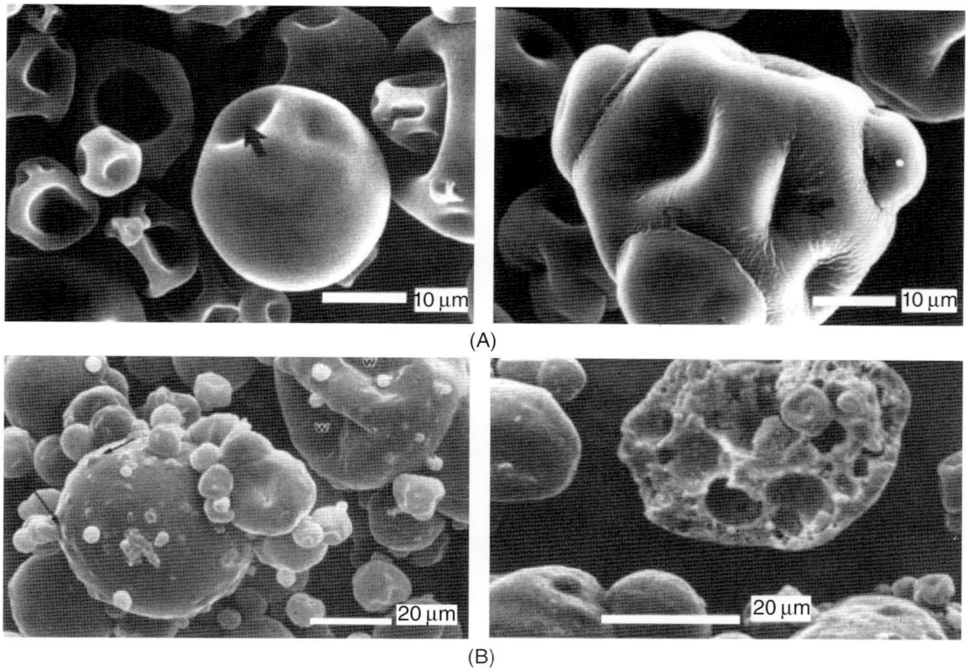

Figure 11.12 SEM images of spray dried (A) UF/DF SMP and (B) retentate powder.

sweet whey waste stream where the industry is still struggling, to dry acid and salty whey waste streams as mentioned before.

11.4.1 Whey Protein Concentrates

WPC powders are spray dried ultrafiltered whey retentates where the protein in dry matter ranged from 35–80% (w/w). Lactose and minerals are removed until the desired protein concentration is achieved. The use of diafiltration is necessary for producing higher protein powders (Figure 11.13). The lower retention of minerals, lactose and NPN depends on the type of the membrane being used, flux and the nature of the feed solution. Typical concentrations of WPC powders ranging from 35–80% are presented in Table 11.3. WPC 35 is commonly used as a replacement for skimmed milk and as a stabilizer. Water binding capacity of these powders increases with increase in total solids. Thus, WPC 50 and 80 are mostly used for applications such as soups, drinks, bakery goods, etc.

The powder quality depends on the spray dryer operating conditions. Some important parameters are inlet/outlet temperatures, atomizer speed, the feed flow rate, type of atomizer and the scale of drying. Increase in inlet and outlet temperatures leads to increase in particle size of the powders. Rotary atomizers create a high degree of air rotation promoting uniform temperatures throughout the drying process, although nozzle atomizers do not. Further, increases in the feed flow increases the particle size and bulk density of the powders and decrease the total solid content [46]. For example, SEM

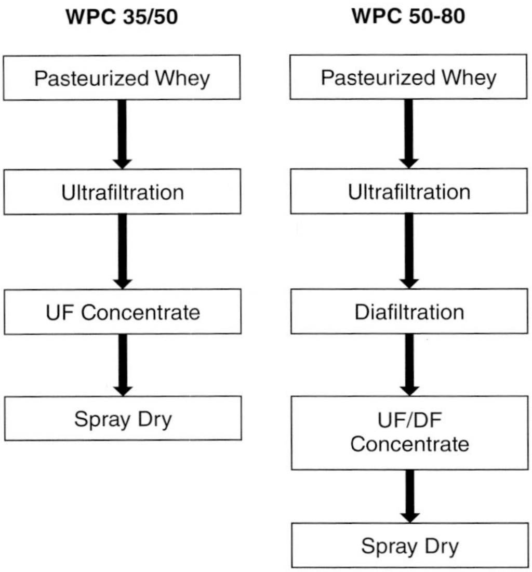

Figure 11.13 Production process of WPC 35–80 powders.

Table 11.3 Typical composition of WPC 35–80 powders [45].

Component (%)	WPC 35	WPC 50	WPC 65	WPC 80
Moisture	4.6	4.3	4.2	4.0
Protein	29.7	40.9	59.4	75
Lactose	46.5	30.9	21.1	3.5
Fat	2.1	3.7	5.6	7.2
Ash	7.8	6.4	3.9	3.1
LA	2.8	2.6	2.2	1.2

images (Figure 11.14) show that except for one powder sample, all others contain particles with a spherical shape which is characteristics to WPC powders [47]. However, the sample which was spray dried with a high capacity nozzle and followed by milling revealed fine particles with internal voids and crevices (Figure 11.14a). In addition, the microstructure of WPC is influenced by the chemical composition of the feed. WPC with no fat showed small smooth particles, while WPC containing fat have more fused particles fused and they are more aerated. Moreover, the storage conditions of these powders are of importance for them to retain their functionality characteristics. A recent study by Tunick *et al.* [53] showed that WPC 35 and WPC 80 powder samples became yellower and after 12 months of storage led to an unsatisfactory appearance (Figure 11.14b). Decreases in lysine residues indicative of protein aggregation, formation of volatile compounds and caking of the powders were observed. When stored in sealed bags, the shelf life of the powders is extended to 18 months at low temperature conditions.

Figure 11.14 (a) SEM images of WPC powders obtained under different spray drying conditions [53]; (b) Brown chip formation of WPC 35 powders stored at 90% RH conditions [54]. (*See color plate section for the color representation of this figure.*)

11.4.2 Whey Protein Isolates

WPI contain the higher percentage of proteins (>90%) and almost lactose free. The typical composition of WPI contains ~93% proteins, 3% lactose, 5% fat and 1% ash, although the protein ratios can differ with the production methods. The production of WPI can be manufactured through two processes (Figure 11.15).

1) Ion exchange followed by concentration and spray drying.
2) Microfiltration followed by ultrafiltration and spray drying.

During ion exchange, Ca and Mg are removed from water by adsorption onto a static resin bed replacing sodium ions at neutral pH. The pH of whey is adjusted to 3–3.5 using phosphoric acid and is introduced into a stirred resin tank and some proteins are adsorbed. After removal of the partially deproteinized whey from the tank, the adsorbed proteins are eluted, raising the pH with a sodium hydroxide solution. The pH of the

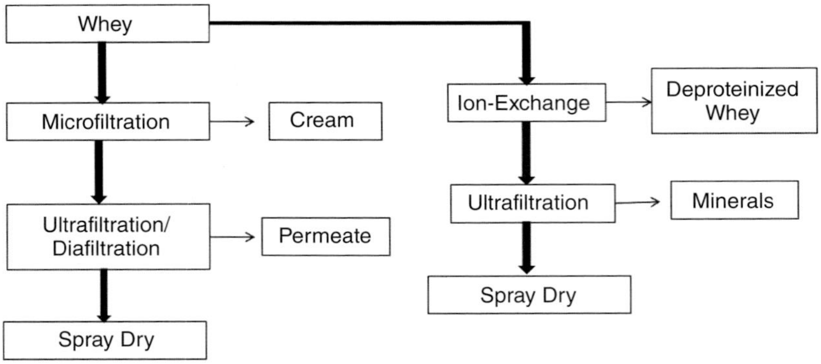

Figure 11.15 The two manufacturing processes of WPI.

resultant protein solution is adjusted as desired afterwards. One of the downsides of this method is that lactoferrin can be bound by the resin and there is a need to increase the pH to at least to 9–10 in order to recover lactoferrin, which is not economically viable. In addition, amino acids and nutrients can be lost due to the use of chemicals, although no denaturation of protein happens which is an added advantage. Regenerated cellulose ion exchangers are widely used for protein recovery. Jones [54] used CMC (carbonyl methyl cellulose) ion exchangers, but later Palmer [55] used a CMC derivative in a stirred batch system which was called "Vistec Protein Recovery". However, there is a low protein binding capacity and the recovery limits its wide spread application. In 1985, Ayers *et al.* [56] used sulphopryl spread cellulose to increase the protein binding capacity. However, the protein recoveries were found to be similar to CMC derivative ion exchangers, although the amount of acid/base used for the regeneration was 10% of what is used with the CMC. Spherosil-S which is a microporous silica based cation exchanger ($-SO_3H$) used to increase the protein recovery to 80% was trialled [52]. In this instance, whey is acidified to pH 4.5 to facilitate the adsorption of the proteins. However, substantial protein denaturation was observed. Spherosil-QMA which consists of $-N(CH_3)_3$ anion reactive groups was used by running pH 6.3 adjusted whey with HCl as the eluent.

MF is a pressure driven membrane filtration as discussed before. The MF defatted whey is then further concentrated by UF to >90%. However, in this case the stream has significant levels of GMP. No denaturation of proteins was experienced in this process, but more fat and lactose may be contained than the former method. Thus, the ratio of specific whey proteins is different depending on the method being used. For example, increased β-Lg and almost no GMP can be obtained from the use of a cation exchanger, while decreased lactoferrin can be expected with the use of anion exchangers (Table 11.4).

Inappropriate storage conditions can lead to decreases in shelf life of these powders. Thus, the control of storage is essential to preserve the protein functionality. Increase in surface hydrophobicity was observed when stored at 60°C and at the water activity of 0.2. WPI powders have a smooth surface and large dents. However, with storage, the powder surface has shown a crackled surface and a decrease in the presence of amino acids at the surface (Figure 11.16). The influence of the occurrence of the Maillard reaction has been attributed for these alterations [53]. Norwoord *et al.* [60] also showed an increase in protein lactosylation with storage at 20°C for 15 months, and an effect on the

Table 11.4 Whey protein rations obtained through different WPI processing methods.

Whey protein (%)	Cation exchange WPI	Anion exchange WPI	Microfiltered WPI
α-Lactoalbumin	18	5–10	19
β-Lactoglobulin	69	40–50	46
GMP	0	15–25	17
Immunoglobulin G	5	0.2–2	4
BSA	2	0.5–1	8
Lactoferrin	1–3	0.2	1–3
Proteose peptone	1–1.5	4.5	n/a

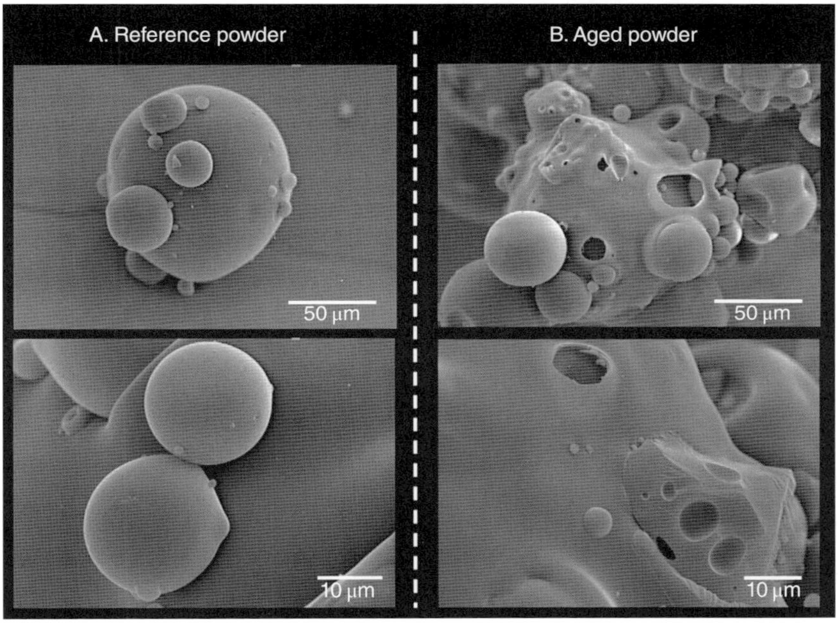

Figure 11.16 WPI powder particle surface as a function of aging [59].

denaturation and aggregation of proteins at 40°C in a dry state followed after lactosylation within the 1st two weeks.

11.4.3 Whey Protein Hydroxylates

Whey protein hydroxylates (WPH) are the products of hydrolysis of whey proteins by proteolytic enzymes. The manufacturing process of WPH powders is illustrated in Figure 11.17A. Amino acid bonds are broken by the enzyme and lead to improved and faster absorption by the body. However, they are considered bitter in taste which limits its application with certain food systems. In addition, WPH provide heat stability to the

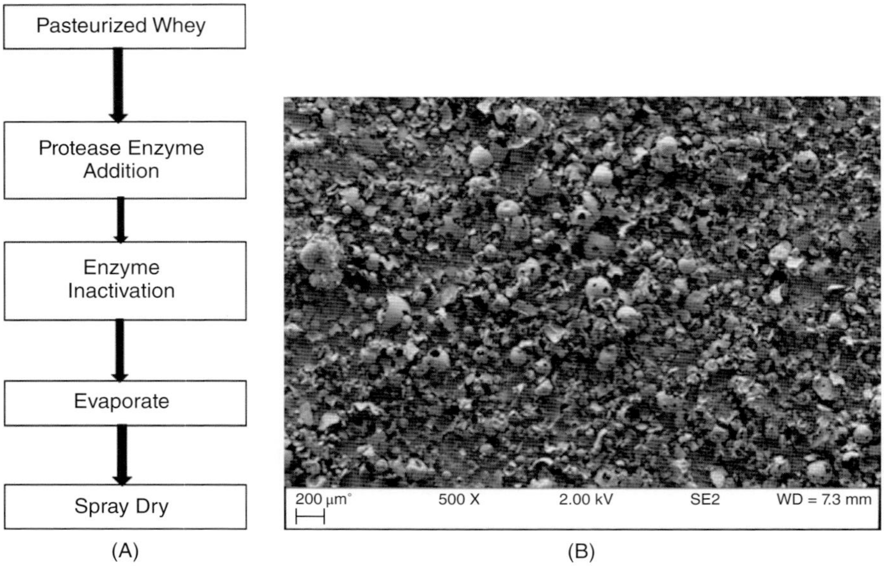

Figure 11.17 The manufacturing process of WPH powders and the SEM image of WPH powders.

denaturation of β-LG [55]. WPH powder particles are generally smaller than other whey powders (Figure 11.17B) and it was concluded that this was due to the lower viscosity of the hydrolyzed feed solution prior spray drying [56]. They can be considerably different depending on the specific enzymes used, time, temperature, etc. WPH is widely used in beverages and functional foods due to the increased health benefits related to bioactive peptides and decrease protein allerginicity. However, they have a high hygroscopicity and lower T_g values and were susceptible to drastic physical and chemical changes. Thus more restricted storage conditions should be implemented to ensure their stability [57].

11.4.4 Other Whey Powders

11.4.4.1 Defatted Whey Protein Concentrates

The presence of fat decreases the functional properties of WPC and reduces the storage time [58]. In some instances thermocalcic precipitation was used, which involved the precipitation of phospholipids under the influence of the presence of Ca and heating [59]. This mixture was then microfiltered to remove the precipitate followed by further concentration by UF to produce WPC. However, due to the increased demand for chemical free products, industry used only membrane filtration for formation of defatted WPC powders. The method comprises the separation of the cream, followed by UF to concentrate by 3-fold, and microfiltration of the retentate. The defatted MF permeate is then again subjected to UF under DF conditions and is spray dried. This method removes fat from 7.2% to <0.4%. Concurrently bacteria are removed.

11.4.4.2 Demineralized Whey Protein Concentrates

Utilization of whey is restricted by the presence of high salt concentrations. Varying degrees of demineralization can be achieved with 25, 50 and 90% removal of minerals through ion exchange, NF or electrodialysis methods.

Electrodialysis (ED) is a membrane process, during which ions are transported through semipermeable membranes under the influence of an electric field. Anion and cation exchange membranes are arranged alternatively in parallel between an anode and a cathode. Anion membranes carry cation groups which repel cations and are permeable to anions and vice versa (Figure 11.18). Water is circulated between alternate pairs of membranes, while whey is recirculated between the other pairs. When a DC current is applied, cations are drawn to the negative electrode and anions to the positive electrode. As the ions pass through the ion selective membranes they become trapped and demineralized; whey is then collected [60]. Usually, ED is applied after a pre-concentration step (total solids to ~20–30%) in order to increase the energy efficiency. The degree of demineralization through ED depends on the ash content of the whey, density, viscosity and the residence time. ED can be performed under batch and continuous conditions. In a batch ED, whey is re-circulated over the membrane stacks until a desired mineral removal is achieved. This process uses a temperature for around 30–40°C for 5–6 hours. These processing conditions led to questions on microbial safety. Therefore, addition of hydrogen peroxides or a cooling step to remove the heat arising from current were trialled. The continuous ED only runs for 10–40 min, although only 60–70% of mineral removal is achieved. Some of the disadvantages are the precipitation of calcium phosphate onto the cation membrane surfaces, and proteins on the anion membrane surfaces where replacement of the membrane surfaces is associated with a high cost.

Ion exchange is carried out in columns with ion exchange resins. Resins are plastic beads with diameters of 0.5–1 mm made from polystyrene and acrylic. These exchange resins can be of natural or synthetic origin composed of a rigid matrix and functional groups firmed through chemical bonds. During whey processing, whey is passed through a cation exchange resin followed by an anion exchange resin. Demineralization occurs when mineral ions are replaced with H^+ ions in the cation column and then by OH^- ions in the anionic column. The resins are then back flushed with water to obtain

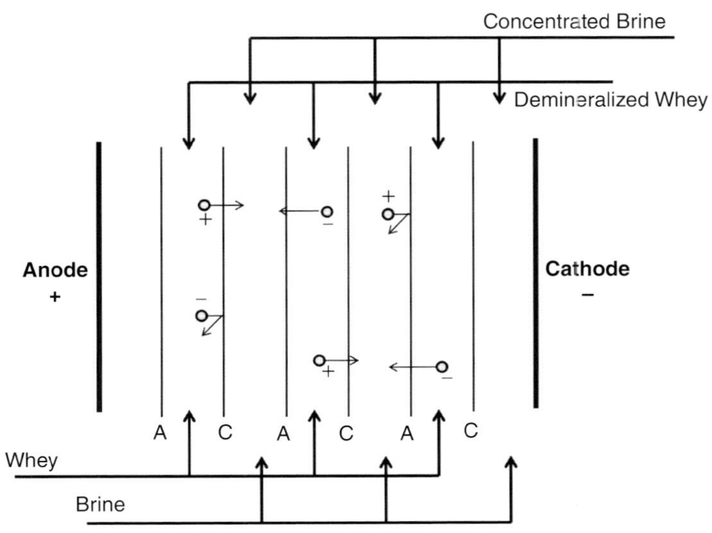

Figure 11.18 The working principle of electrodialysis.

the demineralized whey and then the cationic and anionic resins are regenerated with HCl and NaOH respectively [61]. A study by Greiter *et al.* [68] compared ED and ion-exchange towards demineralization of whey. The following results were found (Table 11.5) and concluded ED as the most preferred method. However, the Swedish Dairies Association has developed an alternative ion exchange process to reduce the consumption of the regeneration chemicals and thus a better waste management [46]. This process involves whey first entering to an anion column containing a weak base resin regenerated in the bicarbonate form followed by a cation column containing a weak acid cation exchange resin regenerated in the ammonium form.

Demineralized whey powders have also shown a spherical nature as WPC powders, indicating no effect from demineralization on the structure of the whey powders (Figure 11.19) [63]. Lower levels of demineralization such as 26–65% is acceptable for use in animal feeds, drinks, salad dressings, ice-cream and baking goods, although at

Table 11.5 Comparison of ED and ion exchange towards demineralization of whey

	Electrodialysis	Ion exchange
Demineralization (%)	90	99
Waste water (m^3)	1.25	3.7
Ash content in waste water (kg)	8.1	36.3
COD (kg/m^3 feed whey)	8.4	26
Energy Demand (kWh)	Pumping = 4.2 Electric current through ED cells = 5.4 Reduction of organic charge/m^3 whey = 3.2	Pumping = 0.15 Production of regenerates = 25.3 Reduction of organic charge/m^3 whey = 9.75

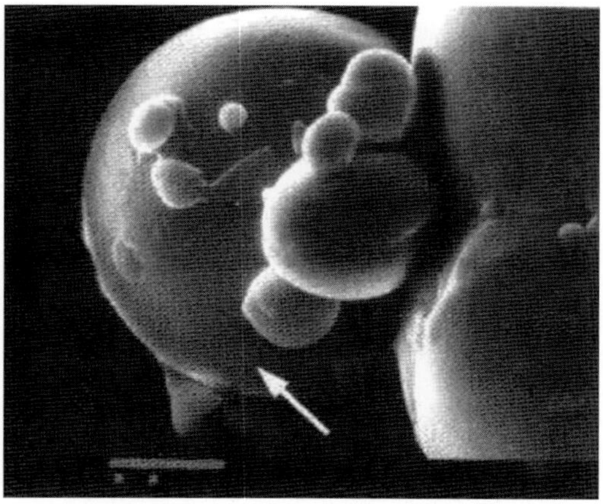

Figure 11.19 SEM image of a demineralized whey powder sample [63].

least >65% demineralization is needed to be used in infant formulas. Demineralized whey is either used as a cheap alternative to skim milk powder or to increase a specific desirable functionality.

The demineralized permeate powders consisted of globular and partly shaped particles. The proportion of globular to sharp-edged particles was influenced by the degree of demineralization; highest demineralization showed less presence of sharp edged particles, while lower degree of mineralization showed more sharp edged particles [64].

11.4.4.3 Delactosed Whey Powders

Delactosed whey powders are produced by concentrating the whey, crystallizing and separating the lactose followed by spray drying. Concentrated whey is immediately cooled and lactose crystals were seeded. In addition, lactose can be hydrolyzed by β-galactosidase followed by a UF step. The hydrolysis depends greatly on the time. The typical composition of delactosed whey powders is 18–24% protein, 52–58% lactose, 1–4% fat and 11–22% ash. During the process, the functional and nutritional properties of the proteins in delactosed powders are mostly retained. However, due to the low protein content, the uses are limited. Thus, amplifying the functionality of the protein fraction was attained through enzymatic proteolysis to increase the functionality [64].

11.4.4.4 Acid Whey Powders

Drying is the simplest operation used during whey utilization. Pre-crystallization of lactose prior drying is necessary to overcome the problems associated with spray drying and storage. However, the presence of lactic acid (LA) in acid whey leads to a more hygroscopic nature of the powders and results in lumping and caking of the powders with storage. In addition, sticking of the material to the spray drying walls also leads to limitations. As the demands for acid coagulated dairy products are still on the rise, it appears a quick solution is needed to fully utilize produced volumes.

Acid whey powders are similar to sweet whey powders except for the low lactose and high titrable acidity which makes itheir flavor slightly acidic. They can be mainly used in dry mixes, dry blends, salad dressings and frozen desserts. A recent study by Nishanthi *et al.* [11] found that AW powders contained a high protein surface coverage, as most dairy powders. Protein aggregation occurs, and created through covalent non-disulphide bonds. In addition, the protein secondary structural analysis revealed that less β-sheet, more α-helix exist along with intermolecular β-sheet crosslinking. An earlier study reported that manipulation of the concentration of LA and Ca in lactose containing model systems improved crystallization properties of lactose leading to enhanced yields of spray dried powders [64]. Thus, a recent study by Chandrapala and Vasiljevic [71] examined the effects of removing LA and Ca by membrane processing to varying extents in order to improve the feasibility of the spray drying of the acid whey stream. Removal of LA and Ca by 30% and 40%, respectively, by nanofiltration significantly increased the powder recovery from 18% to ~58% in the collection vessel. The powder particles decreased in size with well separated spherical powder particles (Figure 11.20). Thus, it was concluded that manipulating LA and Ca concentrations in a particular acid whey stream can improve the spray drying process and production of a non-sticky acid whey powder.

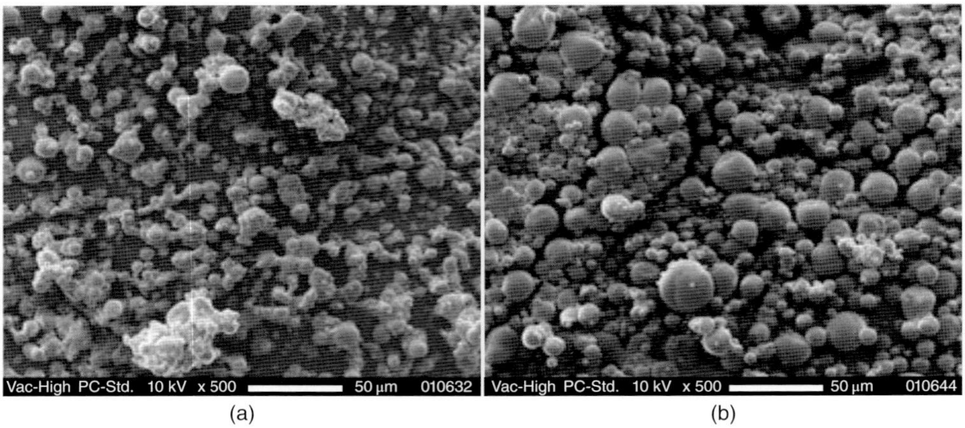

Figure 11.20 SEM image of (a) Acid whey powder and (b) LA and Ca reduced acid whey powder sample [72].

11.4.4.5 Salty Whey Powders

The high salt content in salty whey limits the concentration step through UF due to the increased fouling capacity of the membranes with salts. Thus, spray drying of the salty waste stream is hardly ever conducted by the industry prior to a demineralizing step. A recent study by Nishanthi *et al.* [11] investigated the salty whey powder properties. They showed the presence of protein aggregates via covalent disulphide interactions and a dense, compact powder structure (Figure 11.21). The powder surface was covered only ~40% by proteins. In contrast, ~37% of the surface was covered by fat which also increased the surface hydrophobicity. The protein secondary structural changes showed the presence of α-helix, β-sheet and intermolecular β-sheet aggregation. Further, the study showed these intrinsic powder characteristics are a cumulative effect of the properties of the original feed stream, UF concentrates and changes induced by the spray drying.

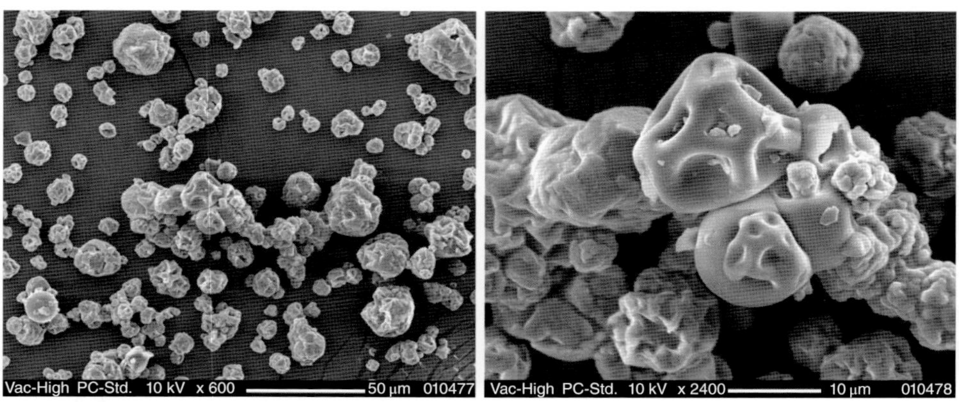

Figure 11.21 SEM images of salty whey powder [12].

11.5 Utilization and Applications of Whey Powders

Whey protein possesses excellent functional properties as stated in Table 11.6 and is widely used as a food ingredient. For example, WPC 35 is mostly used as a replacer for skimmed milk as well as a stabilizer and a fat mimetic due to its excellent water binding, fat-like mouthfeel and gelation properties. The ability of WPC 50–80 powders to be dissolved in a wide range of pH is being utilized for nutritional drinks, soups, bakery products, meat and protein fortified drinks. Defatted WPC is used as the egg white replacer in whipped products. Some of the functional properties utilized in certain products are highlighted in Table 11.6.

Functionality of whey proteins depends on the physico-chemical properties of their 3-dimensional protein structure including size, shape, amino acid composition, charge, hydrophobicity and hydrophillicity, molecular flexibility molecular underpinned by external factors such as pH, temperature, ionic strength and the presence of other components [66]. Microparticulation of whey proteins has received considerable attention in recent years in improving the functionalities of whey proteins. Microparticulation is a thermal and mechanical treatment to denature whey proteins and to create ideal particles with the influence of shear [66]. Microparticulation adds new dimensions to excellent functional properties such as creamy mouthfeel, increased water binding capacity, improved texture, excellent emulsifying properties, excellent foaming stabilities, superior gelation capabilities, high nutritional values and increased shelf life. During microparticulation, ~70% of whey proteins will form micro-particles following the order of β-Lg, α-La, Ig and BSA. Thus, different whey powders will perform to varying degrees depending on the ratio of whey proteins present. One of the popular methods used is extrusion where proteins are thermally denatured at high pressures [66]. The denatured proteins become fiber form, and sudden release of pressure evaporates water and expands. Modern extruders will exert a structural collapse although the nutritional value is increased. Fat reduction in certain dairy products leads to a decrease in its sensory rating, although there is an increase in demand by consumers for fat-free products. Thus microparticulated whey proteins are used as a replacement and have shown to be ideal due to the low calorie intake. They are also used as a starch substitution. A high native to denatured protein ratio provides yoghurts with a high creaminess,

Table 11.6 Utilization of a certain functional property of whey proteins in certain food systems.

Functional property	Mode of action	Food systems
Gelation/viscosity changes	Matrix formation	Salad dressings, cheeses, baked goods, meats
Emulsification	Ability to stabilize fat particles	Sausages, soups, cake, salad dressings
Water binding capacity	Increased retention of water	Meat, beverages, cakes, sausages
Foaming	Forms stable films	Whipped toppings, cakes, desserts
Browning/Flavor	Maillard reaction No off flavors	Confectionaries, meat, sauces, soup

increased viscosity, slow melt down in the mouth and a lower syneresis [67]. WPC 80 powder particles showed hollow microspheres with a smooth surface, while microparticulated whey powder samples exhibited smaller sized particles with a high degree of cross linking (Figure 11.22) [68]. Microparticulated whey proteins can be manipulated by pH, temperature and shear conditions [69] to obtain various functionalities of interest. Incorporation of microparticulated whey proteins to cheese has decreased the firmness in some instances and increased moisture retention depending on the type of cheese being produced [70]. Hays *et al.* [79] showed that the sensory profile for reduced fat cheddar cheese containing microparticulated whey proteins was similar to full fat cheddar cheese. Simpless R 100 is an example of a commercially available fat replacer made from microparticulated whey proteins in the size range of 0.1 to 2 μm [72].

In addition to their excellent functional performances, whey powders are vastly used as a nutritional ingredient in sports drinks, bars and formulas. Whey powders contain higher amounts of essential amino acids which can be absorbed by the body easily and large amounts of branched chain amino acids such as leucine, which is considered the most growth promoting amino acid [73]. They have shown benefits for depression, blood

Figure 11.22 SEM images of WPC80 and microparticulated WPC80 powder samples respectively.

sugar and pressure, and even help to treat symptoms of HIV and cancer. In addition, high amounts of cysteine amino acids are present, which will also help to boost levels of the cellular antioxidant glutathione. It was found that whey protein fortified sports beverages perform better rehydration due to the naturally occurring electrolytes and increased nutritional density [74]. In addition, the increased solubility, clear visibility and increased shelf life also encourage whey powders to be included in sports beverages.

Whey powders are successfully used in the production of edible films and coatings giving improved shelf life for certain food systems [75]. The on-going interest towards sustainable packaging materials throughout the use of biodegradable and edible biopolymers led to using whey powders as an excellent coating material. These coatings can protect the food systems from chemical and microbial degradation, and thereby increase the shelf life and maintain a high quality. Whey powders as coatings also show excellent mechanical and barrier properties. In addition, coatings can add a gloss to the surface which in turn improves the appearance. Whey powders have the ability to carry flavor compounds and the ability to trigger a slow release through food preparation or in the mouth by mastication.

In addition, the food and pharmaceutical industries are constantly developing novel products with a strong nutraceutical function using whey powder as a valuable source to either prevent or for a treatment of various diseases [76]. Whey powders also have the ability to serve as vesicles for bioactive compounds. Whey proteins are used as hydrogels and nanoparticle systems for encapsulation and the controlled delivery of bioactives [77]. Whey hydrogels not only function as a delivery systems, but also as texture and mouthfeel improvers. The pH sensitive swelling ability plays an important role in the kinetic release of the encapsulant material [77]. The ability of whey proteins in forming gels under cold conditions is being utilized in delivering heat-sensitive nutraceuticals [76, 78]. Furthermore, whey protein concentrates serve as an excellent protection against lipid oxidation [79, 80].

11.6 Conclusion

Environmental regulatory issues triggered industry to find pathways of utilizing the whey waste streams. Whey is an excellent source of functional proteins and peptides, lipids, vitamins, minerals and lactose. The increased production of different varieties of cheese and yoghurt products led industry to find solutions for drying other whey waste streams into powders apart from sweet whey, to be used as food ingredients. However, the microstructure of these powders differed with compositional changes of the feed solution and the processing conditions. Thus, novel approaches will be developed, refined and introduced to whey processing for the cost-effective manufacture of these whey powders, with excellent physical and functional properties which are preserved during storage.

References

1 Mirabella, N., Castellani, V., Sala, S. (2014). Current opinions for the valorization of food manufacturing waste: a review. *Journal of Cleaner Production*, **65**, 28–41.

2 Bozanic, R., Barukcic, I., Lisak, K., Tratnik, L. (2014). Possibilities of whey utilization. *Austin Journal of nutrition and Food Sciences*, **2**, 1036–1043.

3 Nishanthi, M., Vasiljevic, T., Chandrapala, J. (2017). Properties of whey proteins obtained from different whey steams. *International Dairy Journal*, **66**, 76–83.
4 Schmidt, R., Packard, V., Morris H. (1984). Effect of processing on whey protein functionality. *Journal of Dairy Science*, **67**, 2723–2733.
5 Lu, Y., McMahon, D. (2015). Effects of sodium chloride salting and substitution with potassium chloride on whey expulsion of Cheddar cheese. *Journal of Dairy Science*, **98**, 78–88.
6 Smithers, G. (2008). Whey and whey proteins – From gutter-to-gold. *International Dairy Journal*, **18**, 695–704.
7 Kinsella, J., Whithead, D. (1989). Proteins in whey: Chemical, physical and functional properties. *Advances in Food and Nutrition Research*, **33**, 343–438.
8 Chandrapala, J., Wijayasinghe, R., Vsiljevic, T. (2016). Lactose crystallization as affected by presence of lactic acid and Ca in model lactose systems. *Journal of Food Engineering*, **178**, 181–189.
9 Dec, B., Chojnowski, W. (2006). Characteristics of acid whey powder partially demineralised by Nano filtration. *Polish Journal of Food Nutritional Sciences*, **15**, 87–90.
10 Uduwerella, G., Chandrapala, J., Vasiljevic, T. (2017). *Pre concentration of yoghurt base by UF for reduction of acid whey generation during Greek yoghurt manufacturing. International Journal of Dairy Technology*, **84**(3), *346–354.*
11 Nishanthi, M., Vasiljevic, T., Chandrapala, J. (2017). *Properties of* whey proteins obtained from different whey steams. *International Dairy Journal*, **66**, *76–83.*
12 Shoveeler, A., Stoll B., Ball, R., Burrin, D. (2005). Nutritional and functional importance of intestinal sulphur amino acids metabolism. *Journal of Nutrition*, **135**, 1609–1612.
13 Fox, P., McSweeney, P. (1998). Dairy chemistry and biochemistry. *Lactose* (1st edn., Chapter 2), London: Blackie Academic and Professional.
14 Disanayake, M., Kasapis, S., George, P., Adhikari, B., Palmer, M., Meurer, B. (2013). Hydrostatic pressure effects on the structural properties of condensed whey/lactose systems. *Food Hydrocolloids*, **30**, 632–640.
15 Abrams, S., Griffin, I., Davila, P. (2002). Calcium and zinc absorption from lactose containing and lactose free infant formulas. *The American Journal of Clinical Nutrition*, **76**, 442–446.
16 Vince, A., Burrdge, S. (1980). Ammonia production by intestinal bacteria: the effects of lactose, lactulose and glucose. *Journal of Medicinal Microbiology*, **13**, 177–191.
17 Goodman, B. (2010). Insights into digestion and absorption of major nutrients in humans. *Advances in Physiology Education*, **34**, 44–53.
18 Birkhed, D., Imfeld, T., Edwardsson, S. (1993). pH changes in human dental plaque from lactose and milk before and after adaptation. *Caries Research*, **27**, 43–50.
19 Zhao, P., Li, H., Mohammadi, M., Kim, I. (2016). Effects of dietary lactulose supplementation on growth performance, nutrient digestibility. *Poultry Science*, **95**, 84–89.
20 Marwaha, S., Kennedy, J. (1988). Review: Whey pollution problem and potential utilization. *International Journal of Food Science and Technology*, **23**, 323–336.
21 Glanakis, C. (2012). Recovery of high added-value components from food wastes: Conventional, emerging technologies and commercialised applications. *Trends in Food Science and Technology*, **26**, 68–87.

22 Jelen, P. (1980). Industrial whey processing technology: An overview. *Journal of Agricultural and Food Chemistry*, **27**, 658–661.

23 Lucey, J., Kelly, J. (1994). Cheese yield. *Journal of Society of Dairy Technology*, **47**, 1–14.

24 Jenson, R., Iyer, M. (1991). Performances of falling film evaporators. *Journal of Dairy Research*, **58**, 29–38.

25 Kumar, P., Sharma, N., Ranjan, R., Kumar, S., Bhat, F, Jeong, D. (2013). Perspective of membrane technology in dairy industry: A review. *Asian-Australasia Journal of Animal Science*, **26**, 1347–1358.

26 Mustafa, L., Alsaed, A., Al-Domi, H. (2014). Drying of sweet whey using drum dryer technique and utilization of the produced powder in French type bread and butter cookies. *Pakistan Journal of Biological Science*, **17**, 812–820.

27 Schuck, P. (2002). Spray drying of dairy products: State of the art. *Lait*, **82**, 375–382.

28 Patel, R. Pate, M., Suthar, A. (2009). Spray drying technology: An overview. *Indian Journal of Science and Technology*, **2**, 44–47.

29 Mujumdar, A., Huang, L., Chen, X. (2010). An overview of the recent advances in spray drying. *Dairy Science and Technology*, **90**, 211–224.

30 Brans, G., Schroen, C., Van der Sman, R., Boom, R. (2004). Membrane fractionation of milk: State of the art and challenges. *Journal of Membrane Science*, **243**, 263–292.

31 Hausmann, A., Sanciolo, P., Vasiljevic, T., Ponnampalam, E., Quispe-Chavez, N., Weeks, M., Duke, M. (2011). Direct contact membrane distillation of dairy process streams. *Membranes*, **1**, 48–58.

32 Rezaei, H., Ashtiani, A., Fouladitajar, A. (2014). Fouling behaviour and performance of microfiltration membranes for whey treatment in steady and unsteady state conditions. *Brazilian Journal of Chemical Engineering*, **31**, 24–27.

33 Erdem, I., Ciftcioglu, M., Harsa, S. (2006). Separation of whey proteins by using ceramic composite membranes. *Desalination*, **189**, 87–91.

34 Duke, M., Vasiljevic, T. (2016). Whey Processing: Overview and role of membranes. *Encyclopedia of Membranes*, **4**, 2021–2024.

35 Duke, M., Vasiljevic, T. (2015). Whey protein isolation overview and membrane operations. *Encyclopedia of Membranes*, **2**, 1–3.

36 Kuo, K, Cheryn, M. (1983). Ultrafiltration of acid whey in a spiral wound unit: Effect of operating parameters on membrane fouling. *Journal of Food Science*, **48**, 1113–1118.

37 Pan, K., Song, Q, Wang, L., Cao, B. (2011). A study of demineralization of whey by nanofiltration membranes. *Desalination*, **267**, 217–221.

38 Li, Y., Shahbazi, A. (2006). Lactic acid recovery from cheese whey fermentation broth using combined UF and NF membranes, *Applied Biochemistry and Biotechnology*, **132**, 985–996.

39 Li, Y., Shahbazi, A., William, K., Wan, C. (2008). Separate and concentrate LA using combination of NF and RO membranes. *Applied Biochemistry and Biotechnology*, **147**, 1–9.

40 Alkhatim, H.,Alcaina, M., Soriano, E.,Iborra, M., Lora, J., Arnal, J. (1998). Treatment of whey effluents from dairy industries by NF membranes. *Desalination*, **119**, 177–184.

41 Luo, J., Ding, L., Qi, B., Jaffrin, M., Wan, A. (2011). A two stage ultrafiltration and nanofiltration process for recycling dairy waste water. *Bioresource Technology*, **102**, 7437–7442.

42 Kelly, J., Kelly, P. (1995). Desalination of acid casein whey by NF. *International Dairy Journal*, **5**, 291–303.

43 Ecker, J., Raab, T., Harasek, M. (2012). NF as a key technology for the separation of LA and AA. *Journal of Membrane Science*, **389**, 389–398.

44 Chandrapala, J., Duke, M., Gray, S., Palmer, M., Weeks, M., Vasiljevic, T. (2016). Nanofiltration and nano-diafiltration of acid whey as a function of pH and temperature. *Separation and Purification Technology*, **160**, 18–27.

45 Nguyen, M., Reynolds, N., Vigeneswaran, S. (2003). By product recovery from cottage cheese production by nanofiltration. *Journal of Cleaner Production*, **11**, 803–807.

46 Jelen, P., Buchheim, W. (1976). Norwegian whey "cheese." *Food Technology*, **30**, 62–74.

47 Jelen, P. (1992). Whey: Composition, properties, processing and uses in *Encyclopedia of Food Science and Technology*. Hui, Y. H., Ed., John Wiley & Sons, New York, 2835–2845.

48 Martin-Diana, A., Rico, D., Frias, J., Mulcahy, J., Henehan, G., Barry-Ryan, C. (2006). Whey permeate as a bio-preservative for shelf life maintenance 490 of fresh-cut vegetables. *Innovative Food Science and Emerging Technology*, **7**, 491, 112–123.

49 Maragkoudakis, P., Nardi, T., Bovo, B., Giacomini, A. (2010). Valorization of a milk industry by-product as substrate for microbial growth. *Journal of Biotechnology*, **150**, 340–343.

50 Koutinas, A., Papapostolou, H., Dimitrellou, D., Kopsahelis, N., Katechaki, E., Bekatorou, A., Bosnea, L. (2009). Whey valorization: A complete and novel technology development for dairy industry starter culture production. *Bioresource Technology*, **100**, 3734–3739.

51 Bylund, G. (1995). *Dairy Processing Handbook*. Tennotext, A.B. (Ed.) Tetra Pak Processing Systems AB, Lund, Sweden.

52 Chegini, G., Taheri, M. (2013). Whey Powder: Process technology and physical properties: A review. *Middle-East Journal of Scientific Research*, **13**, 1377–1387.

53 Onwulata, C., Konstance, R., Tomasula, P. (2004). Minimizing variation in functionality of whey protein concentrates from different sources. *Journal of Dairy Sciences*, **87**, 749–756.

54 Tunick, M., Thomas-Gahring, A., Van Hekken, L., Iandola, S, Singh, M., Qi, P., Ukkuku, D. *et al.* (2016). Physical and chemical changes in whey protein concentrate stored at elevated temperature and humidity. *Journal of Dairy Science*, **99**, 2372–2383.

55 Jones, H. (1976). A versatile continuous ion-exchange process: In: *Food Waste Applied Science*. Birch, G., Parker, K., Worgan, J. (Eds). Barking, pp. 242–255.

56 Palmer, D. (1977). High purity protein recovery. *Process Biochemistry*, **12**, 24–28.

57 Ayers, J., Peterson, M. (1985). Whey protein recovery using a range of novel ion exchangers. *New Zealand Journal of Dairy Science and Technology*, **20**, 129–142.

58 Nichols, J., Morr, C. (1985). Spherosil-S ion exchange process for preparing whey protein concentrate. *Journal of Food Science*, **50**, 610–614.

59 Burgain, J., Zein, R., Scher, J., Petit, J., Norwood, E., Francius, G., Gaiani, C. (2016). Local modification of whey protein isolate powder surface during high temperature storage. *Journal of Food Engineering*, **178**, 39–46.

60 Norwood, E., Chevallier, M., Floch-Flouere, C., Shuck, P., Jeantet, R., Croguennec, T. (2016). Heat induced aggregation properties of whey proteins as affected by storage conditions of whey protein isolate powders. *Food Bioprocess Technology*, **9**, 993–1001.

61 Wijayanti, H., Bansal, N., Sharma, R., Deeth, H. (2014). Effect of sulphydryl reagents on the heat stability of whey protein isolate. *Food Chemistry*, **163**, 129–135.

62 Hogan, S., O'Loughlin, I., Kelly, P. (2016). Soft matter characterization of whey protein powder systems. *International Dairy Journal*, **52**, 1–9.

63 Zhou, P., Liu, D., Chen, X., Labuza, T. (2014). Stability of WPH powders: Effects of relative humidity and temperature. *Food Chemistry*, **150**, 457–462.

64 Farbod, F., Kalbasi, A., Moini, S, Emam-Djomeh, Z., Razavi, H., Mortazavi, A., Beheshti, H. (2013). The effects of storage time on physico-chemical, rheological, micro-structural and sensory properties of feta cheese fortified with fish and olive oils. *Journal of Nutrition Food Science*, **3**, 230–234.

65 Boyd, L., Drye, N., Hansen, A. (1999). Isolation and characterization of whey phospholipids. *Journal of Dairy Science*, **82**, 2550–2557.

66 Valero, F., Barcelo, A., Arbos, R. (2011). *Electrodialysis Technology. Theory and Applications*. www.intechopen.com

67 Acikara, O. (2012). Ion-Exchange Chromatography and Its Applications. *Intech*, 31–58.

68 Greiter, M., Novalin, S., Wendland, M., Kulbe, K., Fishcher, J. (2002). Desalination of whey by electrodialysis and ion exchange resins: analysis of other processes with regard to sustainability by calculation their cumulative energy demands. *Journal of Membrane Science*, **210**, 91–102.

69 Saito, Z. (1988). Lactose crystallization in commercial whey powders and in spray dried lactose. *Food Structure*, **7**, 75–81.

70 Ismail, B. (2012). *Enhancing the functionality of delactosed whey by enzymatic hydrolysis using surface methodology approach*. Agricultural Utilization Research Institute, Minnesota, USA.

71 Chandrapala, J., Vasiljevic, T. (2017). Properties of spray dried lactose powders influenced by presence of lactic acid and Ca. *Journal of Food Engineering*, **198**, 63–71.

72 Chandrapala, J., Vasiljevic, T. (2017). *Properties of spray dried powders obtained from acid whey and nanofiltered acid whey retentates. Food and Bioprocess Technology*. Accepted.

73 Dissanayake, M. (2009). *Modulation of functional properties of whey proteins by microparticulation*. PhD thesis, Victoria University.

74 Torres, I., Jonhoj, T, Mikkelsen, B., Ipsen, R. (2011). Effect of microparticulated whey proteins with varying content of denatured protein on the rheological sensory characteristics of low fat yoghurt. *International Dairy Journal*, **21**, 645–655.

75 Sun, C., Liu, R., Wu, T., Zhang, M. (2015). Effect of superfine grinding on the structural and physico-chemical properties of whey protein and applications for microparticulated proteins. *Food Science and Biotechnology*, **24**, 1637–1643.

76 Dissanayake, M., Kelly, A., Vasiljevic, T. (2010). Gelling properties of microparticulated whey proteins. *Journal of Agriculture and Food Chemistry*, **58**, 6825–6832.

77 Tobin, J., Fitzsimons, S., Kelly, A., Kelly, P., Auty, M., Feneton, M. (2010). Microparticulation of mixtures of whey protein and inulin. *International Journal of Dairy Technology*, **63**, 32–40.

78 Gunasekaran, S., Xiao, L., Ould Eleya, M. (2006). Whey protein concentrate hydrogels as bioactive carriers. *Applied Polymer Science*, **99**, 2470–2476.

79 Hays, D., Meyer, G., Snook, R., Borr, T. (1990). Microparticulated proteins in food products: Nutritional and sensory attributes. Institute of Food Technologists, Dallas, June 1-5, Abstract 355.

80 Khan, M., Adrees, M., Arshad, M., Anjum, F., Jo, C., Sameen, A. (2015). Oxidative stability and quality characteristics of whey protein coated rohu fillets. *Lipids Health Discussion*, **14**, 58–62.

12

Microstructure of Selected Traditional Indian Dairy Products

Bhavbhuti M. Mehta

Dairy Chemistry Department, SMC College of Dairy Science, Anand Agricultural University, Anand, Gujarat, India

12.1 Introduction

Traditional Indian dairy products are much diversified and have varieties throughout the country. They are offered on religious occasions, festivals and social events. These products reflect the socio-cultural life in the Indian sub-continent. The wide range of indigenous dairy products are categorized into heat desiccated (e.g. khoa and khoa-based sweets); heat-desiccated milk-cereal based desserts/puddings (e.g. kheer); heat-acid coagulated (paneer, chhana and their based sweets); fermented (e.g. dahi, shrikhand); and fat rich (e.g. ghee) products. Khoa is a versatile intermediate base for a wide variety of sweets. The addition of sugar and other ingredients are required to prepare khoa-based sweets like burfi, peda, gulabjamun, kalakand, and other products. Gulabjamun has a round or cylindrical shape and smooth texture. It is soaked in thick sugar syrup. Rabri and basundi are concentrated liquid forms of milk solids with added sugar to sweeten the product for direct consumption (Shivashraya, 2014b). Kheer is a heat-desiccated, cereal-based sweetened and concentrated milk confection. Generally, rice is cooked in milk during kheer preparation. The paneer and chhana are the two most important heat-acid coagulated products and they seem to be a kind of unripened cheese. The most famous sweet prepared from chhana is rasogolla. Rasogolla is a delicacy, stored and served in sugar syrup (Aneja *et al.*, 2002). Shrikhand is a dessert prepared by the fermentation of milk with semi-soft consistency, sweet and sour taste.

The traditional dairy products are obtained by a wide variety of methods involving a range of unit operations that lead to a wide range of product structures and textures. The various processing treatments during manufacturing of milk products, and the microparticulate constituents undergo physical changes and interactions. This gives rise to the characteristic microscopic structure and physical properties of the products. Microstructure controls physical properties like viscosity, firmness, susceptibility and elasticity of the food products (Prasad, 1998). The microstructure of some Indian dairy products have been studied so far, but information is scattered and not well documented.

Microstructure of Dairy Products, First Edition.
Edited by Mamdouh Mahmoud Abdel-Rahman El-Bakry, Antoni Sanchez, and Bhavbhuti M. Mehta.

12.2 Heat Desiccated Dairy Products

The heat desiccated traditional Indian milk product is prepared by heat concentration of milk in an open pan with continuous stirring and scraping until a desirable consistency is obtained. It is a major intermediate base for a variety of sweets like gulabjamun, burfi, peda, kalakand, etc. The khoa prepared from cow milk has a sticky, moist surface with a sandy texture compared to that prepared from buffalo milk. Buffalo milk is generally preferable for preparation of khoa-based sweets. (Mann *et al.*, 2001; Choudhary *et al.*, 2015).

12.2.1 Khoa and Khoa-Based Sweets

Khoa is prepared by partial dehydration of milk. It is conventionally prepared by continuous boiling of milk in an open kettle until desired desiccation (normally 72–75% total solids) and texture is achieved. The heat treatment with continuous stirring/agitation applied to milk during khoa making causes various physical changes. These changes include coagulation of the milk proteins, coalescence of fat globules that leads to large lumps of fat, and supersaturation of lactose in the milk serum. The fat is solidified, and lactose crystallized during storage affects the structural properties of khoa. Khoa contains 18–42% moisture, 21–40% fat, 15–27% protein, 16–34% lactose and 3–5% ash (Mehta, 2015).

Khoa is used mainly as a base in the manufacture of a variety of sweets (De, 1980). There are three different types of khoa. These are known as dhap, danedar, and pindi. These types of khoa are used in preparation of various types of khoa-based sweet products. The moisture content of dhap khoa is 45%. Dhap khoa has a smooth and soft body and is suitable in preparing gulabjamun. Danedar khoa has 40% moisture and it has soft and granular texture. Kalakand can be made from danedar khoa. Pindi khoa has 35% moisture with a firm body and smooth grained texture. The peda and burfi can be prepared from pindi khoa. The pindi khoa undergoes textural changes during the first 2–3 days of storage at ambient temperature after its production. These changes are responsible for some structural rearrangements in the product (Garg *et al.*, 1989; Patil *et al.*, 1992).

12.2.1.1 Microstructure of Khoa

Patil *et al.* (1992) studied the microstructure of pindi khoa using light microscopy and electron microscopy. The granular structure with course particles was observed by the naked eye in khoa. The authors have mentioned that interstitial spaces between the granules were filled with viscous liquid representing a supersaturated solution of lactose, uncoagulated protein, and part of the fat, and this was shown by transmission electron microscopy (TEM). The free as well as globular fat was observed in khoa. Cryo-SEM (scanning electron microscopy) was also used by these authors who observed that the compact mass consisted of lactose crystals in the liquid phase and protein granules in stored khoa. The lactose crystals were only developed in the khoa that was cooled to 20 °C in about 3 hours.

Scanning electron microscopy (SEM) and transmission electron microscopy (TEM) were also carried out to study heat-induced structural changes that occur during the processing of khoa and gulabjamun by Adhikari (1993). During khoa making, a constant boiling of milk led to formation of casein–whey protein complexes. These complexes coalesced gradually on boiling, forming a fuzzy, agglomerated mass and finally precipitated as heat-induced milk gels, joined together by thick protein bridges. Further heat desiccation of this gel resulted in the compaction of the protein agglomerates with reduction in void spaces and fat globules interspersed in between khoa. Adhikari (1993) observed smooth and uninterrupted contours of the casein micelles in raw cow milk under TEM. Upon prolonged heating of the milk, several structural changes in both casein and whey proteins occurred. When milk was boiled for 5 min during khoa making, whey protein denatured and combined with casein micelles resulting in filamentous appendages all along the micelles' surface. Most of the casein micelles were ornamented with filamentous appendages projecting irregularly from their surfaces, while a few of them retained a smooth form. When a further 5 min of boiling continued, the casein micelles became ragged and fuzzy. The boiling continued up to 20 min, and the micelles gradually came closer to each other via protein bridging. On further boiling of the milk (total 30 min) just prior to milk coagulation, most of the micelles were fused together into clusters with large agglomerates and fat globules embedded in it with little fuzzy appearance. The colloidal state of the casein micelles were precipitated as heat-coagulated milk gels. There were greater numbers of small voids found in the matrix, giving a honeycomb-like structure. The coagulated milk gels when further heat desiccated for about 10–12 min, produced a semi-solid consistency of khoa. The conglomerated dense mass of protein with an uneven surface and voids in between were noticed. The fat globules cemented loosely in the protein matrix (Adhikari, 1993).

Recently, Choudhary *et al.* (2017) studied the effect of developed acidity and neutralization of milk on the microstructure of khoa prepared from cow and buffalo milk. They have examined the khoa using SEM (Figure 12.1). Khoa that can be prepared from acidic milk (>0.17% lactic acid) produces a grainy texture. With increase in acidity, the hardness and size of the grains increased.

Under SEM, a slight granular and smooth surface was observed in buffalo milk khoa as compared to cow milk khoa. The cow milk khoa has a rough and dry natural structure that can be due to the low fat content in cow milk. Khoa prepared from acidic milk (up to 0.18% lactic acid) had the shrinkage of natural microstructure. It also had a compact agglomerated irregular protein complex with larger granules. Khoa from neutralized milk (the acidity adjusted to 0.14% lactic acid using sodium bicarbonate) had a smooth surface, pasty consistency and slightly elongated loose matrix that suppressed grain formation. Khoa from neutralized milk possessed a smooth and plain surface as compared to fresh khoa.

12.2.1.2 Microstructure of Gulabjamun

Gulabjamun is prepared by deep-fat frying (in vegetable oil or ghee) balls of dough made of khoa, wheat flour and baking powder, and subsequent dipping in sugar syrup. Gulabjamun contains 22–34% moisture, 8–11% fat, 6–8% protein, 8–12% lactose, 1.7–3.4% ash and 17–42% sucrose (Mehta, 2015). Frying is the key operation in the manufacture of gulabjamun that induces the brown color and texture to the product. The uniform brown color with a smooth surface, special shapes and a uniform granular and fine-pored texture are considered as good quality gulabjamun. When balls of dough are fried in oil, moisture evaporates with uptake of oil in the product. The expansion in the

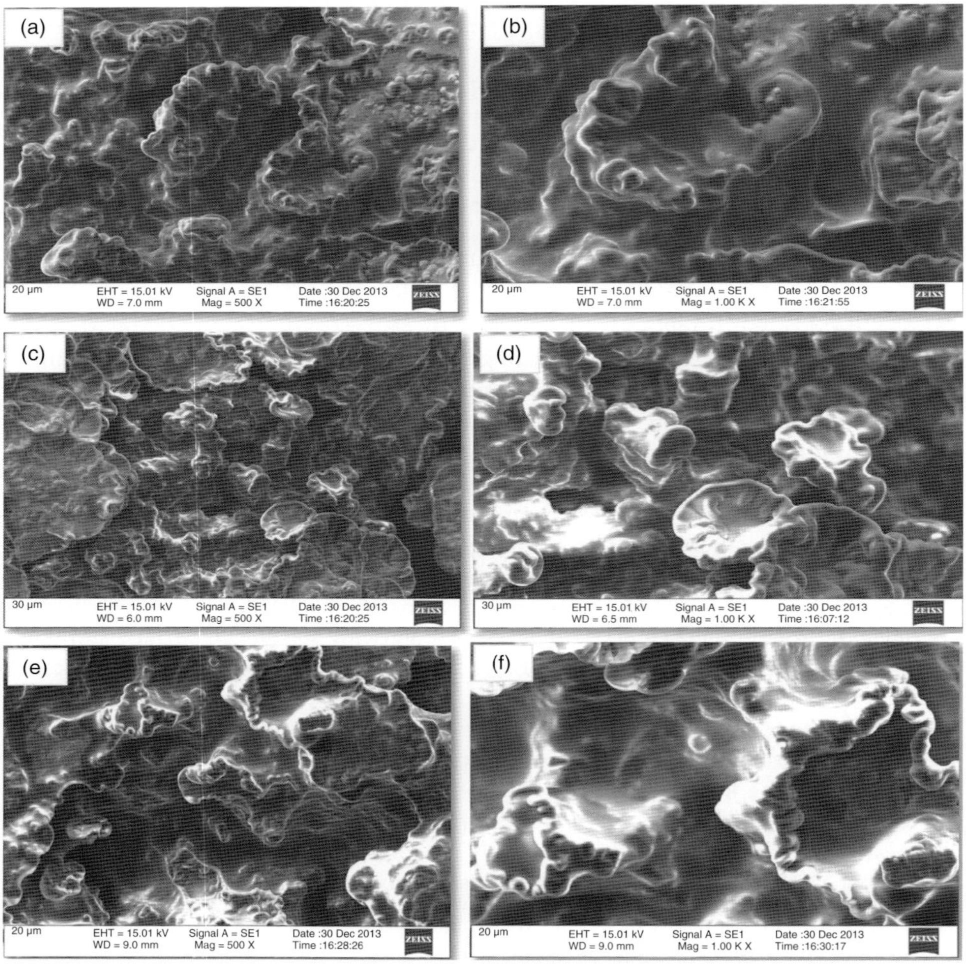

Figure 12.1 Khoa prepared from fresh, acidic and neutralized buffalo milk. (a) Acidic buffalo milk khoa at 500×. (b) Acidic buffalo milk khoa at 1.00 K×. (c) Fresh buffalo milk khoa at 500×. (d) Fresh buffalo milk khoa at 1.00K×. (e) Neutralized buffalo milk khoa at 500×. (f) Neutralized buffalo milk khoa at 1.00K×. From Choudhary *et al.*, 2017: Copyright © 2017 by Springer. Reprinted/Reproduced with permission.

volume of water causes a porous structure with interior voids and cracking, crevices and fine pores in the crust of the gulabjamun balls. During frying of the gulabjamun balls, starch gelatinization and denaturation of protein take place. The resultant protein–starch network produces the typical structure of the product (Jayendra Kumar *et al.*, 2006).

Adhikari (1993) studied the effect of frying of khoa in gulabjamun making. He prepared khoa using cow milk, the SEM of cow milk khoa having 68.5% total solid as shown in Figure 12.2. Frying of khoa in ghee alters the structure of gulabjamun. He found that frying of khoa in anhydrous milk fat (i.e. ghee) resulted in the enlargement of the voids,

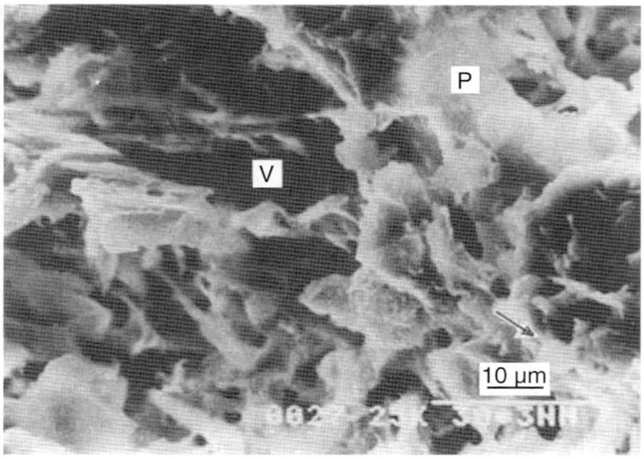

Figure 12.2 SEM of khoa (68.5% TS). P, protein matrix; arrow, fat globule, and V, void space. From Adhikari (1993): Copyright © 1993 by John Wiley and Sons. Reprinted/Reproduced with permission.

producing a loose matrix having starch particles interlinked loosely with the agglomerated protein bodies and the clumped fat globules cemented in gulabjamun. Frying leads to shattering of the interlinkages of the protein agglomerates, which produces a weak matrix in gulabjamun. The agglomerated proteins and starch particles were adhered loosely together in a defatted matrix of gulabjamun as the consequence of the removal of fat (evident as void spaces) from the matrix during the sample preparation for SEM (Figure 12.3).

Protein moieties looked like a carded cotton structure. The starch particles retained their separate identities. They showed threaded, flower petals (arrow) or a round-shaped (S) loosely bound mass. The fat globules and starch acted as weak plasticizers among the protein agglomerates in gulabjamun (Adhikari, 1993). Renuka *et al.* (2010)

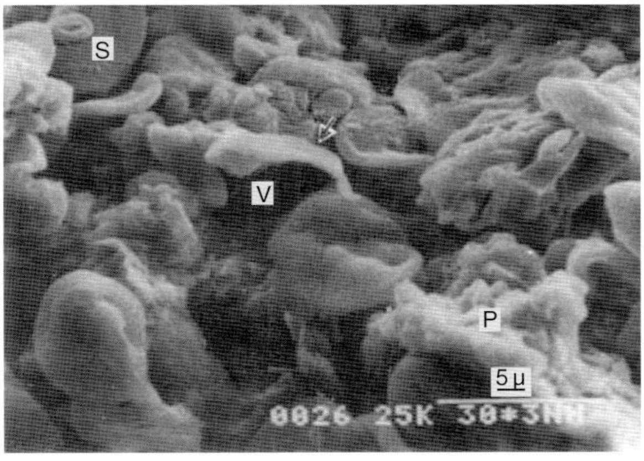

Figure 12.3 SEM of defatted gulabjamun matrix. P, protein matrix; S and arrow, starch particle; and V, void space. From Adhikari (1993): Copyright © 1993 by John Wiley and Sons. Reprinted/Reproduced with permission.

studied the microstructure of gulabjamun sweetened with fructooligosaccharides (FOS) and a FOS-sucrose blend over sucrose using SEM. A well-defined compact network was found in gulabjamun sweetened with FOS and a blend of FOS and sucrose, whereas loosely arranged structures with air pockets were observed in the case of sucrose sweetened gulabjamun. The slightly hard gulabjamun was obtained with FOS and blended indicating that FOS has the ability to bind water and maintain the structure of gulabjamun.

12.2.1.3 Microstructure of Burfi and Kalakand

Burfi is one of the most popular indigenous dairy products, where sugar (either in crystalline form or as syrup) is added into khoa for its preparation. The admixture of khoa and sugar are desiccated to obtain the desired body and soft, semi-hard or hard texture. The hot product with a semi-solid consistency is poured into molds and cooled to obtain the desired body as coarse or fine grained. Burfi has a mildly caramelized and pleasant flavor. Its color may range from off-white to creamy or light caramel. Burfi contains approximately 12–20% moisture, 10–20% fat, 13–21% protein, 12–18% lactose, 1.5–3.1% ash and 17–42% sucrose. Kalakand is a sweet heat-desiccated indigenous dairy product. It is generally made by evaporating acidified milk. Milk is desiccated with small quantities of acid (e.g. citric acid) and sugar to prepare kalakand. It has a granular texture, paste-like consistency and pleasant caramel flavor with off-white to light caramel color. The granular mass is fused and held together in a loosely compacted body. Kalakand contains approximately 20–35% moisture, 10–27% fat, 9–18% protein, 10–22% lactose, 1.7–3.4% ash and 16–43% sucrose.

Arora *et al.* (2007) studied the ultrastructure of burfi using SEM. The burfi prepared using sucrose showed a spongy 3-D structure as a result of the networking of casein micelles alternating with very small pockets (Figure 12.4). However, burfi prepared from saccharin, acesulfame-K, aspartame and sucralose (as low-calorie sweeteners) lacked a well-defined 3-D microstructure. The compactness of the network decreased with the use of sweeteners in burfi. Sucrose has the ability to bind water and it is responsible for maintaining the structure of burfi as compared to low-calorie sweeteners. The replacement of sucrose with low-calorie sweeteners resulted in the development of a

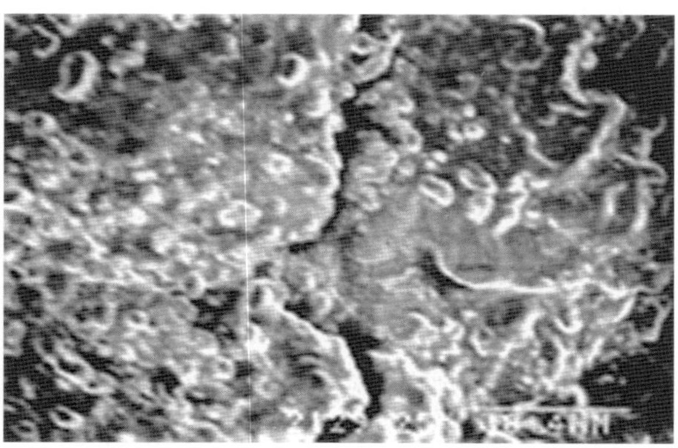

Figure 12.4 Scanning electron micrograph of burfi prepared from sucrose (100×). From Arora *et al.* (2007): Copyright © 2007 by John Wiley and Sons. Reprinted/Reproduced with permission.

microstructure which had lost most of its globular nature and showed an open or loose network. Similarly Arora *et al.* (2008) observed the lack of a well-defined 3D microstructure in kalakand prepared using low calorie sweeteners against that of sucrose.

12.3 Heat-Desiccated Milk Cereal Based Desserts

Kheer is an Indian traditional dessert prepared by partial dehydration of whole milk, with sugar and rice added to it. Kheer is similar to khoa, but the ratio of concentration is 3 to 4 as compared to 6 for khoa; it is usually sweetened before consumption.

12.3.1 Microstructure of Kheer

Kheer is a semi-solid-to-fluid dairy product with partially disintegrated cooked rice grains dispersed in viscous liquid comprising soluble starch from rice grains (Rangappa and Achaya, 1974). Kheer is made by concentrating milk with simultaneous cooking of rice grains and addition of sugar during the process (De *et al.*, 1976). Kheer has a dark creamish color as a result of prolonged cooking of rice grains in milk (Jha *et al.*, 2011).

When kheer is stored at refrigerated temperatures, the rice grains in the kheer tend to become tough and render the product coarse due retrogradation of the starch. Borad *et al.* (2017) studied the microstructure of cooked rice and kheer serum using SEM. Kheer were prepared from Basmati or Parimal rice. The scanning electron micrographs indicating the presence of starch crystals in rice grains and kheer serum are shown in Figure 12.5.

SEM revealed the presence of crystals of retrograded amylose, in rice as well as in serum. During the cooking and cooling process of kheer making, a translucent sheet-like ordered structure was developed in cooked rice grains. The open and solid three-dimensional as well as two-dimensional sheet-like structures were attributed to gelation of amylopectin and amylose crystals respectively. The amylase crystals diffuse out of amylopectin gel during gelatinization. The similar sheet-like particles were also found

(a) (b)

Figure 12.5 Scanning electron micrographs indicating the presence of starch crystals in (a) retrograded Basmati rice grains; (b) stored Basmati kheer serum; (c) retrograded Parimal rice grains; (d) stored Parimal kheer serum. From Borad *et al.*, 2017: Copyright © 2017 by Elsevier. Reprinted/Reproduced with permission.

(c) (d)

Figure 12.5 (Continued)

in the serum. These particles might be responsible for the coarseness of kheer serum and reduced the acceptability of the kheer during storage.

12.4 Heat-Acid Coagulated Dairy Products

Paneer and chhana are the two most important heat-acid coagulated Indian dairy products. The main steps in manufacture of paneer and chhana are: heating, coagulation of milk with an acid, draining of whey, consolidation of solid mass, its chilling and packaging. The heating and acidification of milk are the main factors that affect the final composition of products. Both paneer and chhana consist of entire milk casein, part of denatured whey proteins, almost all fat, and colloidal milk solids in proportion to the moisture content retained (Aneja *et al.*, 2002; Shivashraya, 2014b). Generally buffalo milk is preferred for paneer making, whereas cow milk is used for chhana making. Paneer is generally hard while chhana is relatively soft.

12.4.1 Microstructure of Paneer

Paneer is a heat-acid coagulated Indian dairy product. It is a vegetarian high protein-rich product in Indian food. Paneer is white in appearance with a spongy body, close-knit texture possessing a sweetish-acidic nutty flavor. It is slightly firm in body with good sliceability. Paneer retains its shape when cooked, yet being smooth and semi-soft (Umer Khan and Asraf Pal, 2011; Kumar *et al.*, 2014, Fox *et al.*, 2017). Paneer contains approximately 53–55% moisture, 23–26% fat, 17–18% protein, 2–2.5% carbohydrate and 1.5–2.0% minerals (Kanawjia and Singh, 2000). Coagulation of cow milk at 90°C and pH 5.5 lead to the formation of a core-and-lining ultrastructure of protein particles (Harwalkar and Kalab, 1981). Queso Blanco, the Latin-American white cheese, has a similar structure and it is made by a procedure related to that used to produce paneer (Kalab *et al.*, 1988).

12.4.1.1 Fried Paneer

Raw paneer is used as an ingredient in various vegetable preparations in vegetarian dishes. Paneer is also consumed after frying in edible vegetable oil. The fried paneer may also be further cooked by boiling in salt water. It is also considered as an ingredient in many other vegetable and food preparations. The additional heat treatment for paneer (milk curd) during frying in oil leads to dehydration to some extent. Moreover, the additional structural changes may develop while fried paneer is cooked in salt water (Kalab *et al.*, 1988). Kalab *et al.* (1988) studied development of microstructure in raw, fried, and fried and cooked paneer made from buffalo, cow, and mixed milks. In this study, buffalo, cow, and mixed (equal proportions (1:1) both buffalo and cow milks) milks were heated to 90°C. Cow milk was coagulated at 90°C with 1% (w/v) citric acid. Buffalo and mixed milks were first cooled from 90°C to 70°C and then coagulated with 1% (w/v) citric acid. The coagulated mass of curd was pressed to remove whey and a compact mass of paneer was thus obtained. The paneer was cut into cubes and deep fried in hydrogenated vegetable oil at 175°C for 4–5 min. The fried paneer was boiled in a 1.5% sodium chloride (w/v) solution for 5 min to simulate the usual cooking procedure (Kalab *et al.*, 1988).

The SEM and TEM were employed to study microstructure of the paneer samples by Kalab *et al.* (1988). The differences in the composition of cow and buffalo milks and the temperature of coagulation affected the structure of the paneer. The pH of paneer made from cow and buffalo milk were almost similar, i.e. 5.5 and 5.4 respectively. Both the paneers in the raw state consisted of aggregated protein particles. Under SEM, fat globules were evenly distributed in the protein network with protein particles varying in dimensions. Intact fat globules with casein particles attached to the fat globule membranes were frequently noticed in the raw paneer. While under TEM, the granular structure and internal ultrastructure of the protein particles were noticed. Raw paneer prepared from using cow milk has small protein particles that were uniformly in density and resembled those found in cottage cheese. The core-and-lining structure was found in larger particles. A similar core-and-lining structure was found in the buffalo milk paneer. In contrast to the cow milk paneer, protein particles were more densely packed and fused in the case of buffalo milk paneer. The pH of 5.5 is essential for development of core-and-lining structure in paneer. The local differences in the pH value during coagulation of milk are responsible for heterogeneity in the structure of the paneer. The areas of localized over-acidulations do not allow development of the core-and-lining structure of the casein particles. Moreover, core-and-lining ultrastructure in cow protein particles was also dependent on temperature of coagulation (90°C). When temperature of coagulation was kept at 90°C, the κ-casein and β-lactoglobulin complex was formed as compared with the lower coagulation temperature (70°C) in the case of buffalo and mixed milk. Hence, raw paneer prepared from buffalo and mixed milk did not show a clear lining around the casein particles (Kalab *et al.*, 1988).

When paneer is fried in oil, the severe compaction suppressed the fine granularity of the protein matrix as observed by both in SEM and TEM micrographs, but still a core-and-lining structure was noticed by Kalab *et al.* (1988) in cow milk paneer. Buffalo milk paneer did not show any granularity in its structure. The fat particles acquired sharp and pointed outlines with vanished or broken fat globule membranes in fried paneer as compared to their near-globular shapes in the raw paneer Kalab *et al.* (1988). Cooking of the fried paneer in salt water increased the moisture content and swelling of the protein matrix. The granular and core-and-lining structures were restored in fried paneer by cooking Kalab *et al.* (1988).

Rao and Mathur (1990) reported the texture and microstructure of traditional and sterilized paneer. The comparison of sterilized paneer prepared from milk concentrated by ultrafiltration and traditional paneer indicated that the sterilized paneer was softer and less chewy. SEM showed that protein particles were more densely packed and fused, with fewer void spaces, in the sterilized paneer, this being attributed to the fact that it was less cohesive and had a lower fat content.

12.4.2 Microstructure of Chhana and Chhana Based Sweets

Heat-acid coagulated milk protein mass is known as chhana. Chhana is considered as analogous to cottage cheese. Chhana consists of 57–58% moisture, 22–24% fat, 15–16% protein, 2.5–3.1% lactose and 1.5–2.1% minerals (Mehta, 2015).

SEM showed that cow milk chhana had a structure like cottage cheese or cream cheese. It was a conglomerated, compact protein matrix with numerous small, uniformly distributed pores. The coalesced, smooth protein bodies were joined with thick bridges. The fat globules embedded in casein matrix with some whey-filled spaces at the edge (Adhikari *et al.*, 1992b). SEM of buffalo milk chhana also shows a similar compact, coalesced protein network with numerous globular and irregular voids throughout the matrix, but slightly more uneven as compared to cow milk chhana.

12.4.2.1 Microstructure of Rasogolla

Rasogolla is manufactured by cooking chhana in concentrated sugar syrup to a soft, spongy consistency. The rasogolla consists of 44–46% moisture, 4–6% fat, 4–5% protein and 44–47% sucrose. Verma (1989) observed thread-like folded structures in both cow and buffalo milk rasogolla. Adhikari *et al* (1992a) studied the micro- and macrostructure of cow, buffalo and mixed milk rasogolla using the scanning electron and light microscope (Figures 12.6, 12.7, 12.8 and 12.9).

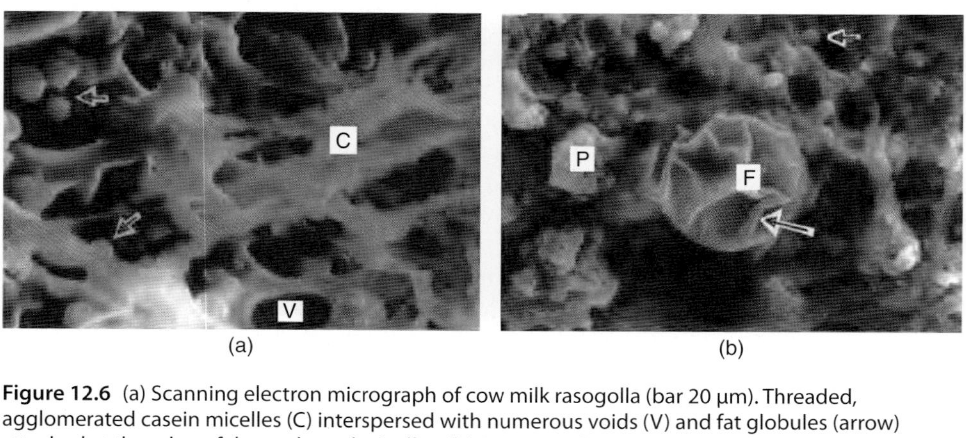

Figure 12.6 (a) Scanning electron micrograph of cow milk rasogolla (bar 20 μm). Threaded, agglomerated casein micelles (C) interspersed with numerous voids (V) and fat globules (arrow) attached at the edge of the coalesced micelles. (b) Scanning electron micrograph of a cow milk rasogolla fat globule (bar 5 μm). Collapsed fat globule (F) and shrunken membrane (large arrow), attached at the edge of the protein matrix (P). Small fat globules (small arrow) were least affected. From Adhikari, *et al.*, 1992a: Copyright © 1992 by Elsevier/INRA.

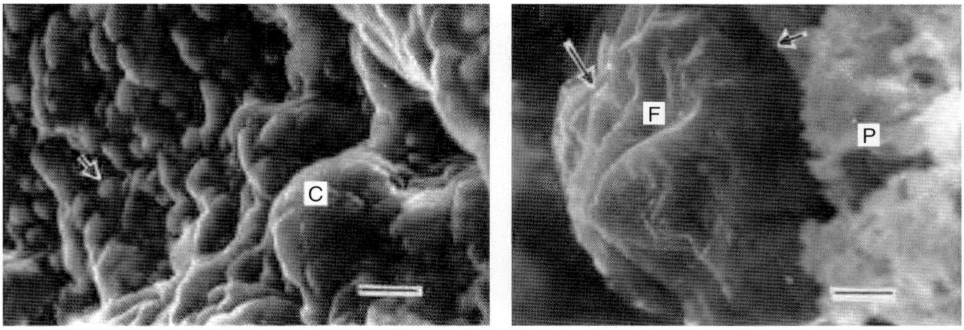

(a) (b)

Figure 12.7 (a) Scanning electron micrograph of buffalo milk rasogolla (bar 20 μm) showing dense, scale-type agglomerated casein micelles (C) interspersed with fat globules (arrow). The obscured surface was mainly due to fat and protein complexes produced during cooking. (b) Scanning electron micrograph of buffalo milk rasogolla fat globule (bar 5 μm). Fat globule (F) was embedded about half of its original size at the edge of the protein matrix (P) (obscured with fat and sugar syrup) and membrane wrinkled severely (large arrow). Fat globule partially detached from the matrix (small arrow). From Adhikari *et al.*, 1992a: Copyright © 1992 by Elsevier/INRA.

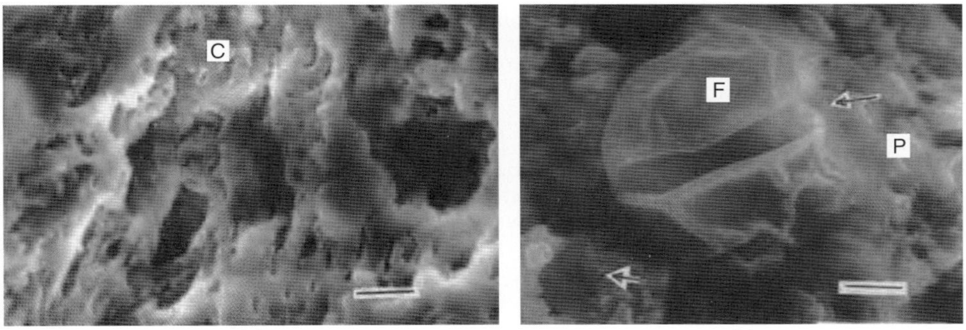

(a) (b)

Figure 12.8 (a) Scanning electron micrograph of a mixed milk rasogolla (bar 20 μm) showing threaded and scale-type coalesced casein micelles (C) admixed together. (b) Scanning electron micrograph of mixed milk rasogolla fat globule (bar 5 μm). Large collapsed fat globule (F), partially ruptured (large arrow), and was embedded in the protein matrix (P). No effect was noticed in the small fat globule (small arrow). From Adhikari, *et al.*, 1992a: Copyright © 1992 by Elsevier/INRA.

The protein matrix was obscured with fat and sugar complexes when rasogolla was prepared from cow milk. The structure of rasogolla prepared from cow milk showed the thick threaded protein bodies interspersed with numerous voids. The dense and scale type structure of buffalo milk rasogolla with fewer voids were noticed. A ragged protein matrix with threaded and scale type agglomerated casein micelles fusing densely together with numerous small voids was observed in mixed milk rasogolla. The fat globules were attached to the coalesced-threaded casein micelles. The SEM of rasogolla prepared from cow and buffalo milk with smaller fat globule having ≤10 μm diameter remained unaffected. The larger fat globules of both the rasogolla were collapsed and

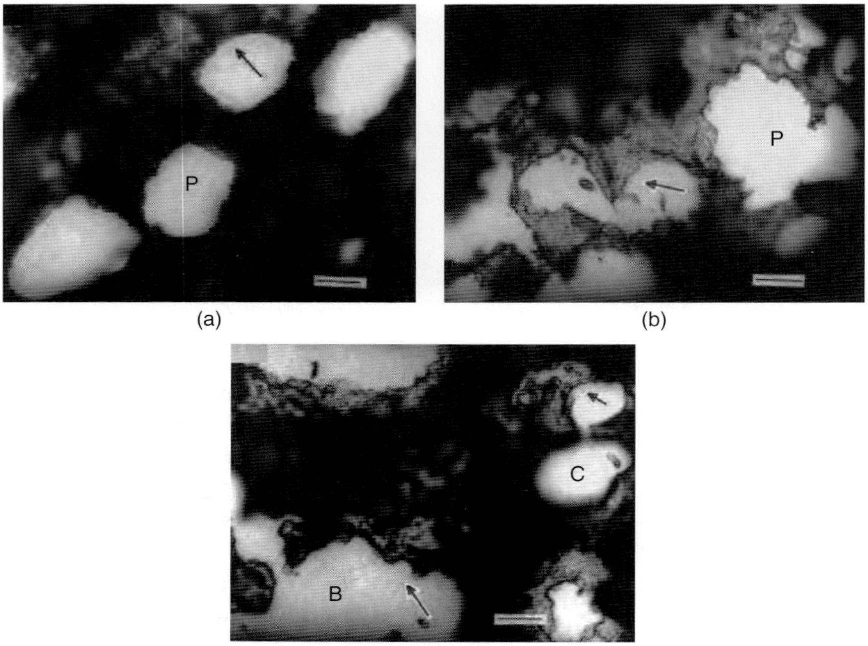

Figure 12.9 Light microscopic micrograph (50×) of various types of rasogolla matrix (bar 350 μm). (a) Cow milk rasogolla matrix. Pores are small (P), oval-shaped and smooth-edged (arrow) and are uniformly distributed throughout the matrix. (b) Buffalo milk rasogolla matrix. Pores are large (P), irregularly shaped, corrugated at the edges (arrow), and are unevenly distributed throughout the matrix. (c) Mixed milk rasogolla matrix. The structure shows both large, rough-edged (B) and small, smooth-edged (C) pores interspersed throughout. Adhikari, *et al.*, 1992a: Copyright © 1992 by Elsevier/INRA.

looked like empty balloons. The buffalo milk rasogolla showed a shrunk membrane, wrinkled to a greater extent than the cow milk rasogolla fat globule. The fat globules were ruptured and collapsed to a greater extent in mixed milk rasogolla. Under the light microscope, smooth, small oval-shaped pores were observed in cow milk rasogolla, whereas rough-edged, irregularly-shaped large pores were noticed in buffalo milk rasogolla (Figure 12.9). The heterogeneous structure was obtained in mixed milk rasogolla (Adhikari *et al.*, 1992a).

The pore size of rasogolla prepared by cow milk varied from 0.8–1 mm, buffalo milk from 0.9–1.9 mm and mixed milk from 0.8–1.8 mm. Chhana cooks in 60% sugar syrup to produce a soft, spongy consistency of rasogolla. During cooking, the compact, granular casein matrix of chhana becomes a fluffy, porous, thread- or scale-like folded structure with the development of numerous large and small voids throughout the matrix in rasogolla. Upon longer and rigorous heating in concentrated sugar syrup, a high osmotic pressure difference is produced surrounding the fat globule membrane, which results in the shrinkage and rupture of the membrane with release of free fat

(Adhikari *et al.*, 1992a). Adhikari *et al.* (1993) also found coalesced, compact casein micelles with associated fat globules in buffalo milk chhana. The numerous small voids interspersed throughout the matrix were noticed. When chhana was cooked, the fat globules were ruptured and finally coalesced to large masses. The void spaces increased distinctly producing a highly ragged and uneven matrix with cotton-like structure in rasogolla (Adhikari *et al.*, 1993). Rasogolla prepared from cow milk had a conglomerated, ragged and porous casein matrix with uneven surface and numerous large void spaces. The collapsed and ruptured fat globules were found embedded in the agglomerated protein particles of rasogolla (Adhikari *et al.*, 1992b). The denser protein network present in chhana reduced the mean free path of the coalesced casein micelles. Rasogolla has large voids between the coalesced proteins that give free access of the protein bodies to move freely during instron testing. This results in higher springiness. The higher springiness in rasogolla may be attributed to its porous, loose and ragged protein matrix.

12.4.2.2 Microstructure of Chhana Podo

Chhana podo is the only indigenous dairy product that is prepared by baking. Baking is the key step in the preparation of Chhana podo. Chhana podo has a brown crust with a white or light brown inner body. It has a typical cooked flavor with a rich taste. Chhana podo has a moist crumb, soft body and somewhat spongy cake-like texture. It is prepared by mixing chhana, semolina/refined wheat flour and sugar. Then the mixture of the dough is slowly baked for about 2–4 hours (Ghosh *et al.*, 1998, 2007; Kumari *et al.*, 2015).

Kumari *et al.* (2015) studied the effect of baking (120, 135 and 150°C) on characteristics of chhana podo using image analysis. The binary images of chhana podo at different baking times revealed that the pore structure of the crumb changed continuously during baking. The mean cell area of the fresh dough was 3.0 mm^2 which changed to 2.6–8.8 mm^2 in the final product depending on the baking temperature. The initial air cell density was 4.4 cells/cm^2 in dough and a higher baking temperature produced more vacuoles or air cells/unit area in the chhana podo. The cell-to-total area ratio, i.e. porosity (void fraction) was increased during baking due to evaporation of moisture and formation of new air cells. The authors have also reported the fractal dimension and shape factor for crumb cells of the product. The fractal dimension is a very useful parameter to characterize the product morphology. The lower fractal dimension is associated with a smoother product. The fractal dimension decreased initially and then increased further. The decrease in fractal dimension was due to melting of fat, which filled up the pores to produce a more uniform and less jagged structure. The data of the shape factor revealed that air cells in chhana podo were not of spherical shape. Recently, Franklin *et al.* (2016) observed that the specific volume increased from 0.94 cm^3/g in fresh dough to 1.09, 1.24 and 1.35 cm^3/g, after baking for 120 min at 120, 135, and 150°C, respectively. The higher baking temperatures imparted a puffing effect due to high pressure of water vapor within the crumb that increased the product volume. The denatured milk proteins interact with starch during baking in chhana podo preparation. The starch–protein matrix contributed to volumetric expansion and structure development in the dough (Franklin *et al.*, 2016). A similar phenomenon was observed during the baking of bread by Thorvaldsson and Skjöldebrand (1998).

12.5 Fermented Dairy Products

Fermented Indian dairy products like dahi, misti dahi, lassi and shrikhand are very popular in the country. Dahi is considered the oldest Indian fermented milk product. Shrikhand (sweetened concentrated curd) and lassi (stirred curd) are also equally important in the Indian culture.

12.5.1 Microstructure of Dahi

Dahi is a typical fermented Indian dairy product with more nutritional and therapeutic values. Dahi resembles plain yoghurt in appearance and consistency and differs in having lesser acidity and use of different starter organisms in its manufacture. Dahi is generally prepared using mixed cultures. Dahi contains generally 85–88% water, 5–8% fat, 3–3.4% protein, 4.2–5.1% lactose, 0.7–0.73% ash, 0.5–1.1% lactic acid (Shivashraya, 2014a).

Prasad and Prasad (1986) examined the interaction of milk casein and lactic microorganisms in dahi prepared from buffalo milk using SEM. The microstructure of curd coagulum showed a compact matrix with small cavities and a uniform aggregation of casein micelles. A number of well-preserved spherical fat globules were seen embedded in the casein background. Presence of *Streptococci* in short chains or singly were observed evenly throughout the matrix. Normally these starter microorganisms were localized in or around deep pockets.

12.5.2 Microstructure of Shrikhand

Shrikhand is an Indian dessert prepared by the fermentation of milk. It has semi-soft consistency with a sweet and sour taste. It resembles sweetened Quarg or Quark produced in Germany (Patel and Abd-El-Salam, 1986). Shrikhand is prepared using chakka (strained dahi/curd). The curd is partially strained through a cloth to remove the whey and thus produce a solid mass called chakka. Chakka has milky white color, smooth texture and mild acidic flavor. Chakka is finely mixed with sugar and flavoring agents and the finished product is called shrikhand. It has a pH of about 4.2–4.4. The proximate composition of shrikhand is: moisture: 34–40%, sugar (sucrose): 43–45%, fat: 4–6% and milk solid not fat: 10–12%, protein: 3.6–3.9%; ash: 0.72%; titratable acidity: 0.9–0.95% lactic acid and pH: 4.4–4.6 (Boghra and Mathur, 2000; Shivashraya, 2014a).

Kulkarni *et al.* (2006) studied the microstructure of shrikhand using a cross-polarized optical microscope. Shrikhand consists of two different microstructures: one composed of crystallites of milk fats having a length scale of ~50–100 μm, and the other composed of aggregates of colloidal casein micelles of ~0.5–10 μm in size as reported by Kulkarni *et al.* (2006). At room temperature fine birefringent structures were uniformly distributed in the shrikhand. More prominent birefringent structures were observed around an entrapped air bubble. The birefringence is caused by crystallized fat globules that are uniformly dispersed in the shrikhand during preparation. The air bubbles provide a hydrophobic surface for the fat to accumulate and that produces the strong birefringent structures. The lipid structure was found to melt in the range of 10–40°C. The protein microstructure was believed to be the cause of high shear sensitivity of shrikhand. Thus, the temperature sensitivity of the visco-elastic parameters was dominated by the semi-crystalline milk fat microstructure, while the shear sensitivity was largely governed by the protein network (Kulkarni *et al.*, 2006).

12.6 Conclusion

India with its divergent food habits has a number of traditional foods. The chemical composition, food additives and specific processing conditions of traditional Indian dairy products determine the microstructure of the product. Knowledge of this microstructure helps in process development and monitoring of textural quality. Khoa, paneer, chhana, chakka, etc. are the main Indian dairy products and these provide a base for the preparation of a variety of other dairy products. The traditional Indian dairy products have wide compositional differences which produce variation in microstructure, and some of the products have not yet been studied for these aspects. There is ample scope for future research in the area of microstructure of traditional Indian dairy products, and systematic works and documentation are needed in this area.

References

Adhikari, A.K. (1993). Microstructure and texture of khoa and gulabjamun made from cows' milk: heat-induced changes during processing and frying. *Journal of Food Science and Agriculture*, **61**: 7–15.

Adhikari, A.K., Mathur, O.N. and Patil, G.R. (1992a). Micro- and macro-structure of cow, buffalo and mixed milk rasogolla: a comparative scanning electron and light microscope study. *Le Lait*, **72**(5): 475–489.

Adhikari, A.K., Mathur, O.N. and Patil, G.R. (1992b). Texture and microstructure of chhana and rasogolla made from cows' milk. *Journal of Dairy Research*, **59**: 413–424.

Adhikari, A.K., Mathur, O.N. and Patil, G.R. (1993). Texture and microstructure of chhana and rasogolla made from buffalo milk. *Australian Journal of Dairy Technology*, **48** (2): 52–58.

Aneja, R.P., Mathur, B.N., Chandan, R.C. and Banergee, A.K. (2002). *Technology of Indian Milk Products*. A Dairy India Publication, New Delhi, India.

Arora, S., Gawande, H., Narendra, K., Yarrakula, S., Singh, V.P., Sharma, V., Wadhwa, B.K., Tomer, S.K. and Sharma, G.S. (2008). Textural and structural properties of *kalakand* made with artificial sweeteners – saccharin, acesulfame-k, sucralose and aspartame. *Journal of Food Science and Technology*, **45**(3): 263–266.

Arora, S., Singh, V.P., Yarrakula, S., Gawande, H., Narendra, K., Sharma, V., Wadhwa, B.K., Tomer, S.K. and Sharma, G.S. (2007). Textural and microstructural properties of *burfi* made with various sweeteners. *Journal of Texture Studies*, **38**: 684–697.

Boghra, V.R. and Mathur, O.N. (2000). Physico-chemical status of major milk constituents and minerals at various stages of Shrikhand preparation. *Journal of Food Science and Technology*, **37**(2): 111–115.

Borad, S.G., Patel, A.A., Singh, A.K., S.K. and Singh, R.R.B. (2017). Effect of storage and reheating on textural properties of rice in dairy dessert as related to its pasting properties and microstructure. *LWT – Food Science and Technology*, **80**: 485–491.

Choudhary, S., Kumari, A. and Arora, S. (2015). Heat induced changes in *khoa:* A Review. *Indian Journal of Dairy Science*, **68**(5): 415–424.

Choudhary S., Arora, S., Kumari A., Narwal V., Tomar S.K. and Singh A.K. (2017). Effect of developed acidity and neutralization of milk on sensory, microstructural and textural changes in khoa prepared from cow and buffalo milk. *Journal of Food Science and Technology* **54**(2): 349–358.

De, S., Thompkinson, D.K., Gahlot, D.P. and Mathur, O.N. (1976). Studies on methods of preparation and preservation of kheer. *Indian Journal of Dairy Science*, **29**: 316–318.

De S. (1980). *Outlines of Dairy Technology*, Oxford Univ. Press, Delhi, India, pp. 385–393.

Fox, P.F., Guinee, T.P., Cogan T.M. and McSweeney, P.L.H. (2017). Fresh cheese products: principals of manufacture and overview of different varieties. In: *Fundamentals of Cheese Science*, 2nd Edition. Fox, P.F., Guinee, T.P., Cogan, T.M. and McSweeney, P.L.H. (Eds). Springer, New York, pp. 579–580.

Franklin, M.E.E., Pushpadass, H.A., Ghose, B.C., Nath, B.S., Kumari, A. and Simha V. (2016). Evaluation of process conditions and their optimization for baking of an Indian dairy dessert-chhana podo. *Journal of Culinary Science and Technology*, **14**(2): 136–152.

Garg F.C., Patel A.A., Patil, G.R., Rajorhia, G.S. and Gupta, S.K. (1989). Textural changes in Khoa during post-manufacture holding. *Indian Journal of Dairy Science*, **42**(4): 804–809.

Ghosh, B.C., Rao, K.J. and Kulkarni, S. (1998). Chhana Podo – Baked indigenous delicacy. *Indian Dairyman*, **50**: 13–14.

Ghosh, B.C., Rao, K.J., Balasubramanyam, B.V. and Kulkarni, S. (2007). Process standardization and shelf life evaluation of Chhana Podo. *Indian Journal of Dairy Science*, **60**: 322–326.

Harwalkar, V.R. and Kalab, M. (1981). Effect of acidulants and temperature on microstructure, firmness and susceptibility to syneresis of skim milk gels. *Scanning Electron Microscopy*, **3**: 503–513.

Jayendra Kumar, A., Singh, R.R.B., Patel, A.A. and Patil, G.R. (2006). Kinetics of color and texture changes in Gulabjamun balls during deep-fat frying *LWT – Food Science and Technology*, **39**: 827–833.

Jha, A., Patel, A.A., Gopal, T.K.S. and Ravishankar, C.N. (2011). Development of a process for manufacture of long-life dairy dessert kheer and its physicochemical properties. *International Journal of Dairy Technology*, **64**(4): 591–597.

Kalab, M., Gupta, S.K., Desai, H.K. and Patil, G.R. (1988). Development of microstructure in raw, fried, and fried and cooked paneer made from buffalo, cow, and mixed milks. *Food Structure*, **7**(1): 83–91.

Kanawjia, S.K. and Singh, S. (2000). Technological advances in paneer making. *Indian Dairyman*, **52**(10): 45–50.

Kulkarni,C., Belsare, N. and Lele, A. (2006). Studies on shrikhand rheology. *Journal of Food Engineering*, **74**:169–177.

Kumar, S., Rai, D.C., Niranjan, K. and Bhat, Z.F. (2014). Paneer – an Indian soft cheese variant: a review. *Journal of Food Science and Technology* **51**: 821–831.

Kumari, A., Franklin, M.E.E., Simha, V. and Pushpadass, H.A. (2015). Effects of baking conditions on color, texture and crumb grain characteristics of Chhana Podo. *International Journal of Dairy Technology*, **68**(2): 270–280.

Mann, B., Kumar, R., Sangwan, R.B. and Vij, S. (2008). *Chemistry of milk in relation to manufacture of Indian dairy products*. Dairy Year Book, Sadana Publishers, New Delhi, pp. 73–75.

Mehta B.M. (2015). Chemical composition of milk and milk products. In: *Handbook of Food Chemistry*. Cheung, P.C.K. and Mehta, B.M. (Eds). Springer-Verlag, Berlin Heidelberg, pp. 511–554.

Patel, R.S. and Abd-El-Salam, M.H. (1986). Shrikhand – an Indian analogue of western quarg. *Cultured Dairy Produce Journal*, **21**(1): 6–7.

Patil, G.R., Patel, A., Paula A.-W. and Kalab, M. (1992). Microstructure and texture of khoa. *Food Structure*, **11**(2): 155–163.

Prasad, D.N. and Prasad, N. (1986). Scanning electron microscopy of Indian dairy product – Dahi. In: *MILK the vital force*. International Dairy Federation. D. Reidel Publishing Company, Dordrecht, Holland. p. 206.

Prasad, D.N. (1998). Microstructure of traditional dairy products. *In: Lecture compendium on Advances in Traditional Dairy Products*. Fourth Short Course, Centre of Advanced Studies, NDRI, Karnal. pp: 134–139.

Rangappa, K.S. and Achaya, K.T. (1974). *Indian Dairy Products*. Asia Publishing House, Bombay, India, pp. 104–115.

Rao, K.V.S.S. and Mathur, B.N. (1990). *Texture and microstructure of traditional and sterilized paneer*. Brief Communications of the XXIII International Dairy Congress, Montreal, October 8–12, 1990, Vol. II. p. 459.

Renuka, B., Prakash, M. and Prapulla, S.G. (2010). Fructooligosaccharides based low calorie gulabjamun: studies on the texture, microstructure and sensory attributes. *Journal of Texture Studies*, **41**: 594–610.

Shivashraya, S. (2014a). *Dairy Technology: Milk and Milk Processing* (vol. 1). New India Publishing Agency, New Delhi.

Shivashraya, S. (2014b). *Dairy Technology: Dairy Products and Quality Assurance* (vol. 2). New India Publishing Agency, New Delhi.

Thorvaldsson, K. and Skjöldebrand, C. (1998). Water diffusion in bread during baking. *LWT – Food Science and Technology*, **31**(7–8): 658–663.

Umer Khan, S. and Ahraf Pal, M. (2011). Paneer production: a review. *Journal of Food Science and Technology*, **48**: 645–660.

Verma, B.B. (1989). *Technological studies on production of rasogolla*. PhD Thesis, Kurukshetra University, Haryana, Kurukshetra, India, pp. 66–104.

13

Using Microscopy for Microorganism Localization within Dairy Products

I.T. Smykov

All-Russian Scientific Research Institute for Butter and Cheese making, Uglich, Russia

13.1 Introduction

The quality of many dairy products, their taste, flavor and storage ability largely depend on microorganisms found inside them. Microorganisms are an integral part of cheese, sour cream, yoghurt and other fermented products. Not only taste but also the safety of use by the consumer depends on how active the growth of various microorganisms in the product will be.

A study of the role of microorganisms in the formation properties of fermented dairy products shows that the issue continues to be of interest to researchers and manufacturers. However, despite numerous studies, there are still many unexplored phenomena associated with the interaction of microorganisms with each other, with proteins and milk fat.

One of the main methods in these investigations is microscopy. Its use enables the development of microorganisms to be observed, and their impact on product microstructure to be assessed. Methods of microscopic visualization and new and existing methods of sample preparation, allowing better understanding of the processes, are being developed and improved. However, interesting and new information, even for well-known microscopy methods, is becoming available.

This chapter is divided into two major parts. The first part is dedicated specifically to the description of the role of microorganisms in the formation of taste, aroma and consistency of the fermented dairy products, to a short review of methods of visualization of microstructure and microorganisms localization in it, and also an account of the interaction of microorganisms and milk components. The second part concentrates on some results of research of microorganism localization in cheese microstructure and yoghurt, and the interactions observed by the author using a transmission electron microscope.

The main objective of this chapter is to introduce to the reader some results of research and problems related to the study of the role of microorganisms in dairy products, which may be resolved using various microscopy methods, and to illustrate them with separate results of characteristic research.

Microstructure of Dairy Products, First Edition.
Edited by Mamdouh Mahmoud Abdel-Rahman El-Bakry, Antoni Sanchez, and Bhavbhuti M. Mehta.

PART 1

13.1.1 Microorganisms and Starters

All microorganisms found in fermented dairy products can be divided into several groups and subgroups. The first group includes the microorganisms required for production of a fermented dairy product with desirable properties (texture, taste, flavor, etc.). The second group includes foreign organisms, which appear in milk products as a result of contamination.

Microorganisms included in the first group belong to ferments (starter cultures); in their turn, they can be divided into two subgroups: defined starter cultures and undefined complex starter cultures (Smid *et al.*, 2014). Defined starter cultures consist of one or more thoroughly studied microbial strains with well-known properties and are produced by specialized enterprises for a wide range of customers. Undefined complex starter cultures have an experimental origin and were originally extracted at small cheese-making factories from dairy products popular among consumers. These undefined ferments contain a unique combination of various microbial strains forming a specific product taste and flavor, and are generally owned by one company. Undefined starter cultures are usually stored frozen to keep their original composition for a long period of time and are used as rarely as possible.

A key difference between defined starter cultures and undefined complex starters is their sensitivity to bacteriophage attack. Defined starters are more vulnerable to bacteriophage attack than undefined complex starters during dairy production (Stadhouders, 1986). This feature as well as their function as a source for isolation of new dairy strains with interesting functional characteristics, explains the renewed interest in undefined complex starter cultures (de Vos, 1989).

Foreign microorganisms may be found in raw milk, air, on the surface of production equipment in contact with milk, as well as on factory workers. Foreign microorganisms according to their influence on milk product properties may be divided into the following subgroups: technically harmful, causing flaws in dairy products and reducing their quality; pathogenic, capable of causing food-borne diseases; potentially pathogenic, causing disease among people with low body resistance and under specified conditions.

Each of these groups consists of tens and hundreds of microorganisms and microbial strains; each of them in the process of evolution has developed specific environmental requirements which provide for the vital functions and successful reproduction. Milk starter microorganisms are originally distributed at random and exist in different living conditions, which leads to the accelerated growth of one type and the extinction of the others. Starter microorganisms are not capable of their own movements; in contrast, many pathogenic and potentially pathogenic microorganisms are capable of active movements in search of better conditions for development, and this complicates the struggle against them.

Spatial location and distribution of bacterial colonies and their growth rate define ripening and development of flavor and aroma of many dairy products, including cheese, yoghurts and sour cream. Initial distribution of colony-forming microorganisms in dairy products at the macroscopic level is defined by their initial concentration and distribution technology. Initial distribution of microorganisms in dairy products is not homogeneous (Fitzsimons *et al.*, 2001), but in the process of protein matrix formation and product ripening, distribution tends to equalize by diffusion (Floury *et al.*, 2010). Thereafter, the location of the microorganisms changes.

Growth rate, size and location of microbial colonies in the ultimate milk product depend on the initial location of colony-forming microorganisms at the microscopic level. Growth rate, size and degree of impact on the dairy product properties depend on the initial conditions of the development of a colony. In order to have the ability to control dairy product ripening and formation of required flavor and aroma, one must know for sure the location of microorganisms at the microscopic level, the conditions of development of a colony at this point, the interaction of microorganisms with the milk components and the degree of impact of microorganisms on dairy products.

13.1.2 Techniques Used in the Microstructure Analyses

Some microorganisms are very small in size and are invisible to the naked eye; their colonies in dairy products may be large and they can be defined visually. Initially, research on the localization of dairy product microorganisms was conducted visually, with the use of specific coloring of colonies that allowed observation of quite big colonies. Later on for research on the localization of dairy products microorganisms, the usual light microscopy was used. However, it soon became clear that the resolution of light microscopy was not enough for accurate definition of the location of microorganisms and their interaction with milk components. At the same time it became clear that the task of visualization of microorganisms in dairy products is technically very complex, and its solution is very important for the production process and consumers of the ultimate product. Therefore, for visualization of microorganisms and their localization in microstructure of the product, researchers started to use more and more sophisticated instruments, working on the basis of different physical principles, and to develop new methods and protocols for sample preparation.

Nowadays, popular high resolution devices used for visualization of microscopic objects can be divided into three main groups according to the imaging principle.

In the first group the received image is the result of interaction of light with the surface or internal structure of the object. This group includes the light microscope (LM), fluorescence microscope (FM), confocal laser scanning microscope (CLSM), and two-photon excitation microscope (TPEM).

In the second group, the received image is the result of the interaction of electrons with the surface or internal structure of the object. This group includes transmission electron microscope (TEM), scanning electron microscope (SEM), scanning transmission electron microscope (STEM), energy filtered transmission electron microscope (EFTEM), as well as cryo-scanning electron microscope (cryo-SEM).

In the third group image is the result of interaction of a micromechanical probe with the surface of the object. This group includes scanning probe microscope (SPM) and its modifications, scanning tunneling microscope (STM) and atomic force microscope (AFM).

Each of these groups provides only the basic types of microscopic devices, in reality their amount is much higher and every year new and more sophisticated types appear. There is no need to describe their operating principles and application, for more information consult books on light and electron microscopy: (Murphy, 2001a; Slayter and

Slayter, 1992 and Amelinckx *et al.*, 2008); on transmission electron microscopy: (Deepak *et al.*, 2015 and Reimer, 1995); on confocal and fluorescence microscopy: (Herman and Herman, 1998; Cornea and Conn, 2014; Pawley, 2006 and Kubitscheck, 2013), on two-photon excitation microscopy: (Diaspro, 2001) and on scanning probe microscopy (Kalinin and Gruverman, 2007). Furthermore, there are other groups of devices for visualization of microstructure and localization of microorganisms of objects using other physical operating principles, but they are not mentioned because they are rarely used for research on localization of microorganisms in dairy products.

In order to obtain a good image of the microstructure of a dairy product, to define the location of microorganisms therein and to identify the nature of interaction of microorganisms with milk components, a good microscope is not enough, it is necessary to select and perform a thorough sample preparation method for microscopic examination. For each type of microscopy, there are a number of standard methods, which have been developed by researchers over many years and provide good results. These methods, as applied to the preparation of dairy product samples, are widely represented in early articles (Kaláb, 1981; Buchheim, 1982; Schmidt, 1982) and in more modern books (Glauert and Lewis, 1998; Bozzola and Russell, 1998; Hayat, 2000a,b,c and Kuo, 2007). However, it should be noted that there is no perfect method for sample preparation, but you can select or modify those, which are the least likely to create artifacts in the object and provide the most reliable information.

Microscopy is the main method for defining the location of microorganisms in fermented dairy products. It is used by many researchers, but with different types of microscopy and different methods of sample preparation. As a result, we can obtain significantly different results studying the same object. This may happen due to a difference in physical and chemical effects on the object in the process of sample preparation. Various modes and conditions of visualization, for example, different electron energy in the microscope, its settings etc. also influence the received results. It does not mean that you need to conduct research using only one type of microscopy on the samples prepared by one method. On the contrary, the use of different types of microscopy and various sample preparation methods may provide additional information about the object and eliminate possible artifacts of preparation and visualization (Ong *et al.*, 2011).

No less important step in microscopic study is interpretation of the results. Thus, first of all, it is necessary to study isolated components of the object, for example, separately casein micelles and separately bacteria, using different methods of sample preparation; identify artifacts and choose a method which provides a satisfactory combined visualization. This allows identifying or eliminating possible artifacts and correctly describes elements of the received image.

A large number of research results on microstructure of dairy products and localization of microorganisms therein using different types of microscopy have been published. A review of current information on the use of microscopic methods on research of growth and localization of bacteria colonies in fermented dairy products (Hickey *et al.*, 2015a) has shown that such microscopic methods as CLSM and cryo-SEM allow rapid sample analysis, positioning and distribution of lactic-acid bacteria, foreign microflora and pathogenic bacteria. Detection of foreign microflora and pathogenic bacteria, using CLSM, is carried out using specific fluorescent paint. The authors of the review point out that the possibility of allocating technically harmful, pathogenic bacteria which create flaws in dairy products and constitute a minority of all starters and foreign microflora, is the basis for an operational quality analysis of dairy products in the process of production and has a great perspective for the entire food industry.

When conducting microscopic examination of dairy products, one of the most important tasks is to assess the viability of microorganisms during sample preparation. It is known, for example, that during cheese ripening, some kinds of bacteria develop intensively, and then die, but at that time other kinds begin to develop. The ability to allocate living and dead microorganisms in the received image makes it possible to quickly develop recommendations for manufacturers on how to control production of dairy products. Research results, conducted with the help of CSLM (Auty *et al.*, 2001) have shown that the method of coloring a sample before preparation, suggested by the authors, makes it possible to assess the viability of bacteria and reduce the time of milk products analysis.

The ability of bacterial growth in fermented dairy products requires that vital substances as well as metabolites could freely transfer to the location of bacteria colonies and pass through the protein matrix. Vital functions of a colony lead to the appearance of pH microgradients and concentrations of different substances in and around bacterial colonies (Floury *et al.*, 2010), which in its turn enhance the diffusion processes and lead to the appearance of capillary-concentration flows. Early studies have shown the existence of interactions among milk fat globules and bacteria colonies because of generally observed location of bacterial colonies in close proximity or direct contact with milk fat globules and their membranes (Laloy *et al.*, 1996). Apart from localization of bacteria on the surface of milk fat globules, they often appear in residual droplets of whey in dairy products and at the junction of cheese grain borders, i.e. predominantly in hydrophilic regions.

Confocal microscopy is widely used for studying microstructure of dairy products and localization of microorganisms; however advanced light microscopy methods are not used very often, but in future they may be widely used in food research. The potential of advanced microscopic methods, namely, fluorescence-lifetime imaging microscopy(FLIM);confocalRamanmicroscopy(CRM);andnonlinearmicroscopy–two-photon excitation microscopy (TPEM), and second harmonic generation microscopy (SHGM), was demonstrated for studying microstructure of cheese and localization of its individual components and microorganisms (Burdikova *et al.*, 2015). These new methods of light microscopy allow defining localization of components and their spatial orientation in three dimensions more accurately, and a better understanding of the impact of production processes on the quality of cheese. This study concluded that the second harmonic generation microscopy method (SHGM) facilitates visualization of calcium lactate crystals and their interaction with fat and protein components of cheese. Furthermore, the fluorescence-lifetime imaging microscopy method (FLIM) produces more accurate definition and location of fats and phospholipids in the matrix of cheese, and the confocal Raman microscopy method offers a new and very powerful methodology for analyzing the localization of various chemical components of the cheese matrix.

13.1.3 Interactions Occurring in the Microstructure

Interactions between microbial cells and milk proteins are important for localization of a cell in the microstructure of the dairy product matrix. In a study by Burgain *et al.* (2013), such interactions between two probiotic strains: *Lactobacillus rhamnosus GG* and *Lactobacillus rhamnosus GR*-1 and milk proteins (micellar casein, intact and denatured whey proteins) were analyzed. During the research, X-ray photoelectron spectroscopy (XPS) and atomic force microscope (AFM) were used. These studies defined specific interactions between whey proteins and bacteria, in contrast to observed indefinite interactions with micellar casein. This study for the first time defined and compared interactions between related probiotic strains and milk proteins with the help of AFM.

Better understanding of interactions occurring in the microstructure of dairy products and lactic acid bacteria and various milk components in a changing environment can successfully improve their transportation to the target site of action. This may contribute to the design of new and more efficient dairy products. The review by Burgain *et al.* (2014) provides a framework for better understanding and interpreting the interaction of bacteria with milk components. It has considered in detail research methods (CLSM, EM, AFM) for the bacterial surface, and focused on deciphering interactions between bacteria and milk proteins. The review has noted that full understanding of the interaction of bacteria with milk components requires combining several research methods. It has also presented and considered appropriate results for studying and understanding these interactions for industry.

Research results of interactions that occur between bacteria and isolated whey proteins performed using atomic force microscopy (AFM) are represented in work by Guerin *et al.* (2016). This work has clearly demonstrated that *Lactobacillus rhamnosus GG* bacteria can interact with β-lactoglobulin, but do not interact with α-lactalbumin and bovine serum albumin. This research result can be used to create new dairy products. Since β-lactoglobulin is able to establish strong connections with bacteria, it can be assumed that adding β-lactoglobulin to the product with *Lactobacillus rhamnosus* will isolate bacteria from environmental influence, and their targeted delivery to a human body in a live condition.

Clostridium tyrobutyricum is the main cause of late blowing defect (LBD) in such cheeses as Grana Padano and Parmigiano-Reggiano. Early studies have shown that the interaction between milk fat globules and these bacteria are the cause of such a defect, but there was no direct evidence to prove that. According to the studies, (D'Incecco, *et al.*, 2015) cream, infected with *Clostridium tyrobutyricum* spores, was analyzed by light, luminescent, and electron microscopy. The results have shown that most bacterial spores were close to the surface of the milk fat globules. The most likely candidates responsible for the adhesion of spores are immunoglobulins, but other systems (e.g., bacterial extracellular polysaccharides) may also provide such adhesion.

Taste, aroma and rheology of fermented dairy products predominantly depend on the microstructure, localization and intensity of microorganism growth. The review by El-Bakry and Sheehan (2014) considers the results of some popular studies on cheese microstructure, distribution and localization of bacteria therein, conducted using light and electron microscopy. In the above-mentioned studies, bright-field, polarization, fluorescence and confocal laser scanning microscopes were used. These studies also used electron microscopy of two types – transmission and scanning. The review has considered the problems of definition and localization of bacterial cells in cheese matrix. It was recommended that these studies be improved, including the development of standard methods, coloring protocols and sample preparation that could be used to study cheese in the process of ripening.

Intoxication or infection due to bacterial contamination of dairy products are serious economic and health problems. Receiving the information on the presence of pathogens and their distribution in the product is a complex technical problem of signal detection from a small number of pathogens in such complex microbial, chemical and physical ecosystems to which dairy products belong. In the study by Fleurot *et al.* (2014) contamination of real cheese with *Staphylococcus aureus*, producing enterotoxin which causes food poisoning was analyzed. Immuno-chemical sample preparation methods and laser scanning confocal microscopy allowed the authors to define that in the process of cheese ripening *Staphylococcus aureus* is preferably localized on its surface. In the conclusion of this study, it was suggested that the developed procedure for the preparation of samples for visualization of lactic acid bacteria and pathogenic flora in other products be used.

The growth of lactic acid bacteria colonies, which causes cheese ripening, occurs due to product assimilation of enzyme reactions taking place nearby and the intensity of its growth will depend on how accessible these products are for the growing colony. The purpose of the study by Floury *et al.* (2013), conducted with the use of confocal laser scanning microscopy, was to define whether fluorescent dextrans of different sizes could penetrate through *Lactococcus lactis* subsp. *lactis* LD61, immobilized in a solid medium, agar and cheese. The studies have shown that *Lactococcus lactis* colonies are of high porosity and high molecular dextrans and enzyme permeable. However, for better understanding of the ripening process, it is necessary to conduct additional studies with other bacteria and other sorts of cheese.

In a review by Hickey *et al.* (2015b) the authors examined the impact of production process parameters on cheese microstructure, localization of microorganisms and interaction between cheese matrix and bacteria during cheese ripening. The methods of defining the location of bacteria in cheese and their interaction with membranes of fat globules received specific attention. It was noted that the existing knowledge about the role of fat globules membranes in the development of microorganisms and cheese ripening is not sufficient and needs more in-depth research in this direction. The most promising method for such studies is considered to be fluorescence microscopy and flavocytochrome.

One of the problems for the dairy industry is regulation of salt content in cheese. Sodium chloride plays a significant role in the development of microorganisms in cheese during ripening and at the same time it defines taste of cheese. The purpose of the work by Søndergaard *et al.* (2015) was to define the effect of sodium chloride concentration in semi-hard cheese on distribution, localization, growth and autolysis of lactic acid bacteria. In these studies, confocal laser scanning microscopy was used, which showed that lactic acid bacteria were distributed in cheese as single cells, groups of two or three cells, and rarely more than four cells. During cheese ripening the number of viable bacteria was mainly decreased in colonies of more than four cells. The studies have shown that during the ripening of unsalted cheese a much higher amount of bacteria is retained, compared to cheese with normal and small salt concentration. However, the intensity of lactic acid bacteria autolysis during the ripening of cheese with different concentrations of sodium chloride heavily depended on the type of the ferment used. Since cheese aroma and structure directly depend on the viability of the ferment in the process of ripening, this study highlights the need for careful selection of starter cultures in the production of salt reduced cheese.

This brief review of research results on the role of microorganisms in the formation properties of fermented dairy products shows that the issue continues to be relevant. And, despite numerous studies, there are still many unexplored phenomena connected with interaction of microorganisms with each other, proteins and milk fat. One of the main methods in these investigations is microscopy. Its use allows watching the development of microorganisms and assessing their impact on product microstructure. Methods of microscopic visualization as well as new and already existing methods of sample preparation, which allow better understanding of the processes, are being developed and improved. However, even well-known, widely used microscopy methods result in a lot of interesting, new information.

A substantial bank of knowledge has been accumulated on cheese microstructure and microorganism localization. It is inevitable that this bank of knowledge will be revised and expanded, providing a basis for further research.

PART 2

13.2 Materials and Methods

13.2.1 Bacterial Strains and Dairy Products

Starter cultures of bacterial strains *Lactobacillus plantarum* RCMB 28, *Lactococcus lactis* subsp. *lactis* RCMB 722_3, *Lactococcus lactis* subsp. *lactis diacetylactis* RCMB 985_2, *Lactococcus lactis* subsp. *cremoris* RCMB 318_2 and *Propionibacterium freudenreichii* subsp. *shermanii* RCMB 56_4 were obtained from the Russian Collection of lactic acid bacteria from "Experimental Biofactory" of Uglich, (RCMB, 2017), under a limited use agreement.

Bacterial strains in the form of powder received at low-temperature lyophilization were used to prepare a liquid leaven and subsequent production of cheese. Uglichsky and Kostromskoy semihard Russian cheeses with a fat content in the dry matter of 45% min, moisture of 45% max, and 1.5–2.5% of salt (Gudkov, 1999) were produced in an experimental workshop of the All-Russian Scientific Research Institute of Butter and Cheese making, Uglich (ARSRIBC, 2017).

Starter cultures of bacterial strains *Streptococcus salivarius* subsp. *thermophilus*, *Bifidobacterium bifidum* and *Lactobacillus delbrueckii* subsp. *bulgaricus* were obtained from the lactic acid bacteria collection of the All-Russian Research Institute of the Dairy Industry, Moscow (ARRIDI, 2017). These cultures are used for the production of low fat probiotic yoghurts.

13.2.2 Electron Microscopy

Electron microscopic study of the microstructure of milk, cheese, and yoghurt components and localization of microorganisms was performed using the EM-410 transmission electron microscope ("Philips", the Netherlands).

Samples for electron microscopy were prepared using three different methods:

- shadowing of surface topography by vacuum deposition of heavy metals;
- negative staining of sample with heavy metal solution;

- replication of swift-frozen sample fracture.

Use of three sample preparation methods for electron microscopy studies is motivated by a desire to evaluate the artifacts of preparation and obtain more information on the structural elements of the study sample.

To reduce the distortions arising from the impact of an electron microscope beam on the sample, the study was carried out at a minimum current of microscope beam and a voltage of 40 kV.

13.2.2.1 Surface Topography Heavy Metal Shadowing

Samples of skim milk were selected from the general tank at specified time intervals after adding rennet and bacterial culture. Samples were fixed at room temperature with 2% solution of glutaraldehyde in 0.1 M Sorenson phosphate buffer, pH 7.2, for 20 min (0.1 ml of milk to 1 ml of fixative). After this, fixed samples were diluted with distilled water in a ratio of 1:100.

A drop (5 μl) of fixed sample was transferred using a pipette to a copper 400-mesh grid coated with parlodion film and held for 2 minutes. Subsequently, excess solution was removed from the surface of the grid with a filter paper and the grid was placed in a vacuum chamber for final drying at room temperature.

To enhance the sample's contrast, a 3 to 5 nm thick layer of platinum (Pt) was applied to the surface of the sample at a 24° angle in the vacuum chamber using the vacuum metal deposition method (Matsko *et al.*, 2013). Thickness of the deposited metal was controlled with quartz sensor. Shades formed at metal deposition displayed the topography of the sample surface. The obtained sample was transferred to an electron microscope for studies.

13.2.2.2 Negative Staining Transmission Electron Microscopy

Samples of skim milk were selected from the general tank at specified time intervals after adding rennet and bacterial culture; samples of low fat yoghurt were taken from the final product provided by the customer. All collected samples were fixed at room temperature with 2% solution of glutaraldehyde in 0.1 M Sorenson phosphate buffer, pH 7.2, for 20 min (0.1 ml of milk to 1 ml of fixative). After this, fixed samples were diluted with distilled water in a ratio of 1:100.

A drop (5 μl) of fixed sample was transferred using a pipette to a copper grid coated with parlodion film and held for 2 minutes for settling of dispersed particles on the surface of the nitrocellulose film. Contrast of the sample was increased through its negative staining (Hayat, 2000b; Kavanagh, 2000; Matsko *et al.*, 2013). For this purpose, a drop (2 μl) of 4% uranyl acetate solution was transferred using a pipette to the drop of the fixed sample at the grid and held for 4 minutes. Subsequently, excess solution was removed from the surface of the grid with a filter paper and the grid was placed in a vacuum chamber for final drying at room temperature. The obtained sample was transferred to an electron microscope for studies.

13.2.3 Freeze-Fracture Replication

This method (Buchheim, 1982; Bozzola and Russell, 1998) is used to prepare cheese samples for studies. Cheese samples were collected at different stages of cheese maturing. Samples were taken from different cheese places, including small eyes in the form of non-deformed parallelepipeds of about 2×2×8 mm in size.

The samples were frozen using a massive copper block cooled by liquid nitrogen to a temperature of −180°C. Freezing rate was not less than $-100°C \cdot s^{-1}$, which prevented the formation of ice crystals in the sample. The frozen sample was placed in a cold vacuum chamber where it was fractured at the temperature of −100°C and held for one minute to remove surface ice.

To obtain a contrast replica of the prepared surface of the cheese sample fracture, a 3 to 5 nm thick layer of platinum (Pt) was applied to the surface of the sample at a 24° angle in the vacuum chamber using vacuum metal deposition method. Subsequently, to obtain a solid replica, a 20 to 30 nm thick layer of carbon (C) was applied at a 90° angle. Thickness of the deposited material was controlled with a quartz sensor.

A cold fractured sample with the obtained replica was immediately removed from the vacuum chamber and the replica was separated from the sample using heated sodium hypochlorite (NaOCl) solution, it was washed with toluene, acetone, and ethanol, and eventually rinsed with distilled water. The obtained replica of the sample surface fracture was placed on the copper grid and transferred to an electron microscope for studies.

13.3 Results and Discussion

In order to better understand the mechanisms of dairy product structuring, to determine the role of microorganisms in the formation of their rheological properties, taste and flavor, first of all it is necessary to examine and study the morphological and structural features of the major components of milk and their transformation during the production processes. This knowledge will allow correct interpretation of the results of dairy product structure visualization and localization of microorganisms within this structure. Next will be shown some results of private research related to localization of microorganisms in protein structure of dairy products, conducted with the use of a transmission electron microscope.

13.3.1 Casein Micelle

Casein micelles are the main component which determines the aggregate and thermodynamic stability of the native milk as a dispersed system. Casein micelles determine the structure, organoleptic and physical and chemical properties of most dairy products obtained on the processing of milk.

The most part of casein in milk, 95%, is organized into micelles, spherical colloidal particles with a diameter ranging from 50 to 500 nm, and the average value of 120 nm (Fox and McSweeney, 1998). Casein micelles of different sizes are composed of hundreds or thousands of individual molecules of different types of casein, and have different molecular composition and different physical and chemical properties, but in general they represent dense gel microparticles with an individual nanostructure.

Figure 13.1 shows a photograph of casein micelles of raw cow's milk, which is similar to other images obtained by other authors. The photo was obtained by a transmission electron microscope and the sample for microscopy was prepared using a heavy metal

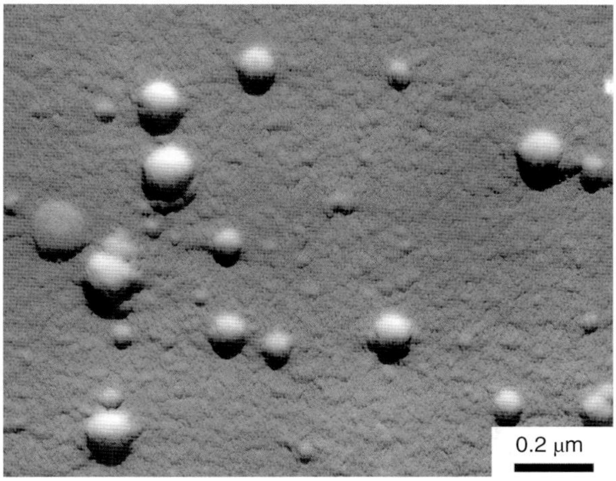

Figure 13.1 Casein micelles in raw milk. TEM-image, heavy metal shadowing.

shadowing method. This image of casein micelles is presented in this study due to the need of further comparative identification of morphological changes in micelles in the milk production processes.

During the microscopic examination, it is necessary to keep in mind that the results of measurements of diameters of the micelles from the same sample which were obtained from different microscopy methods will be significantly different. For example, vacuum drying of sample prior to electron microscopy reduces the apparent size of micelles, and measurement of diameter of micelles that are in the liquid medium gives a value of the hydrodynamic diameter with a different physical meaning (Gebhardt *et al.*, 2006). In addition, frequently used term "average size of the micelles" also requires clarification: average by the number or distribution in the sample, average by reactivity of the surface, average by weight, or other options.

Different transformation may occur during the production of fermented dairy products in the casein micelles under the influence of the environment – temperature, pH, enzymes, etc. Structure, size and shape of the micelle change in this case. Micelles can break, join, settle out or form a complex three-dimensional structures, so it is not easy to determine microstructures formed by micelles, in particular micelles which have undergone significant external changes from the image.

Production of milk gel, the structure of which is formed by the casein micelles is the intermediate stage of cheese manufacturing. Milk gel can be obtained in various ways, but the most common way in cheese production is the enzymatic method in which the spatial structure is formed through destabilization of casein micelles with the use of rennet.

It is known (Fox *et al.*, 2004, 2015), that there are a large number of κ-casein molecules on the surface of native casein micelles. Hydrophilic glycomacropeptides of κ-casein which are represented by threads approx ~10 nm long and ~2 nm in diameter cover almost the entire surface of micelles making it generally hydrophilic. Such micelle structure allows them to associate with each other, making the milk, i.e. dispersion system, thermodynamically stable.

When adding rennet into the milk, it removes extrusive threads of glycomacropeptides from the surface of micelles and releases hydrophobic parts of para-κ-casein on the surface of micelles, which is energetically unfavorable for the dispersion system. The system aims to reduce the internal free energy by forcing the micelles to contact each other with released hydrophobic parts and form a loosely joined cluster of micelles. Initially, connection between the micelles in the cluster is hydrophobic, weak and depends on the properties of the water molecules, so it has little influence on the viscosity of the milk. Figure 13.2 shows a photograph of a cluster of casein micelles at this moment of time.

Here it is seen that the micelles, as structural units, are clearly expressed, but if compared with the micelles in Figure 13.1, it becomes obvious that their shape is distorted and deformed. Some micelles are located at a very short distance from the cluster, which can confirm the weakness of crossmicellar connections in the cluster at this stage of gelation.

Electron microscopy studies of cluster forms of casein micelles at the enzymatic gelation show that the micelles in clusters also change shape at the final stage of the enzymatic gelation phase, immediately before the gel point, i.e., the time when the viscosity starts to increase sharply. Figure 13.3 shows a photograph of a cluster of micelles at this moment of time. It is possible to see here that there was a destruction of the micelles; the nature of relationships in the cluster was possibly changed and this is no longer a cluster of casein micelles but the aggregate of protein particles.

Destruction of casein micelles at the enzymatic gelation process is a very complex and important process for the production of dairy products, which largely determines the interaction of microorganisms with protein and, subsequently, the rheological and organoleptic properties of the finished product. The full explanation and description of the casein micelle destruction mechanism represents a major problem and starts with a description of the general mechanism of protein molecule denaturation.

It is possible (Smykov, 2015a,b) that at the final stage of the enzymatic gelation phase a cooperative phase of first-order transition of "all-or-nothing" type occurs in the casein

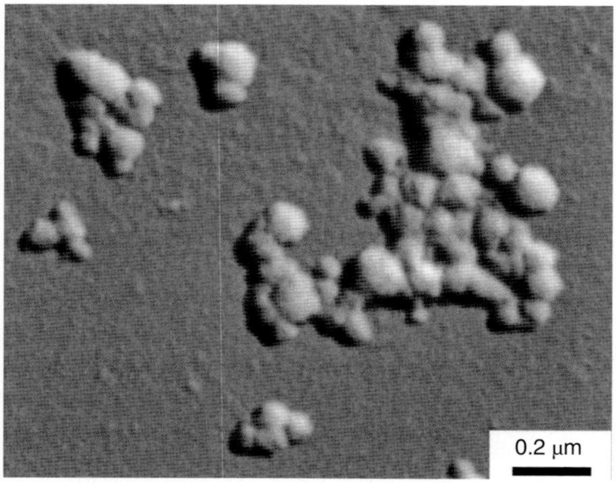

Figure 13.2 Casein micelles cluster before gel point. TEM-image, heavy metal shadowing.

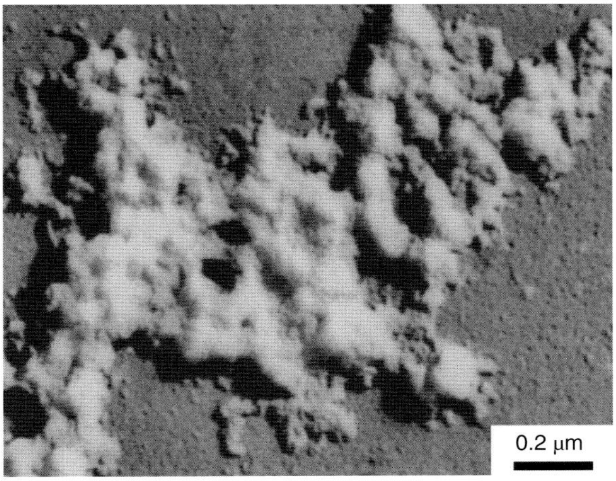

Figure 13.3 Casein micelles aggregate after gel point. TEM-image, heavy metal shadowing.

molecules that form micelles, as well as in the micelles cluster (Privalov, 1989). Initially, denaturation of protein molecules in the micelles can occur as "globule–globule"; such a denatured state is usually called "molten" (Ptitsyn, 1995; Arai and Kuwajima, 2000; Farrell *et al.*, 2002).

The beginning of this process is caused by a point of well-determined (for these conditions) release by the enzyme of the most surface of the micelles from κ-casein glycomacropeptides. It is widely believed that the active flocculation of micelles begins when the degree of their proteolytic hydrolysis reaches approximately 80%. Physically, in our opinion, this value characterizes a part of the micelle surface free from glycomacropeptides, which is critical for commencement of the cooperative phase transition in the protein molecules of this micelle. Therefore, the main part of the micelles turns out to be in a critical metastable state almost at the same time when a cooperative phase transition starts in the protein molecules of the micelle, and cooperative phase transition begins for the majority of the micelles in the cluster.

The beginning of a cooperative phase transition of a "globule–molten globule" type is perhaps observed at the final stage of flocculation (Figure 13.2) and is determined by deformation and change of the micelles size. After transition of micelles to the "molten globule" state, water molecules and contained ions penetrate into the gaps formed in the secondary structure of casein molecules. It is enough to have six to eight water molecules penetrating inside in order to initiate the process of deployment of the secondary structure of casein molecules (Murphy, 2001b; Gavina and Britz-McKibbin, 2007) and implementation of the subsequent phase transition "molten globule–loose tangle". These transitions are separated between themselves by a certain time interval. Actually, the phase transition "molten globule–loose tangle" is a basic transition and it determines further development of the process of coagulation and formation of the dairy gel structure. By itself, the co-operative phase transition is accompanied by minor changes in entropy and enthalpy of the system, but it is crucial for the development of subsequent gelation processes; moreover, from a physical point of view this transition has the right to be called a gel point. After the gel point, weakly joined micelle clusters are converted into aggregates of protein particles (Figure 13.3), which have a large excess of free energy after the phase transition. This forces the dispersion system to minimize free energy due to the formation of new chemical and hydrogen bonds and van der Waals and hydrophobic interactions which appear in the subsequent combination of the aggregates of protein particles into a single three-dimensional structure and its contraction (syneresis).

13.3.2 Bacteria

Lactic acid bacteria are the most common and important type of bacteria for the production of fermented milk products. The first and primary objective of their use is to ferment natural lactose found in the milk into the lactic acid, i.e. to perform its glycolysis. This ensures a rapid decrease of pH, which inhibits the growth of technologically harmful and pathogenic organisms, while leaving open the possibility of growth for acid-resistant bacteria (Johnson and Steele, 2013).

Another objective of lactic acid bacteria use is to provide basic background taste and flavor of the product. This is due to the fact that during the growth of bacterial colonies and their death, various intracellular enzymes are spread into the environment, which first induce lipolysis and proteolysis, and subsequently promote the formation of various aldehydes, esters, alcohols and acids.

In the production process, lactic acid bacteria are usually introduced into the milk as starter cultures, i.e. a combination of several types of bacteria, selected by type and quantity necessary to obtain the final high quality product. The following bacteria are usually included in the cheese starter: *Lactococcus lactis* subsp. *lactis, Lactococcus lactis* subsp. *lactis diacetylactis, Lactococcus lactis* subsp. *cremoris, Lactobacillus plantarum* and *Propionibacterium freudenreichii* subsp. *shermanii.* Yoghurt starter usually includes the following bacteria: *Streptococcus salivarius* subsp. *thermophilus,* and *Lactobacillus delbrueckii* subsp. *bulgaricus.* Figure 13.4 shows, for example, a TEM photograph of the sample of the culture microflora used in the production of cheese. The sample was

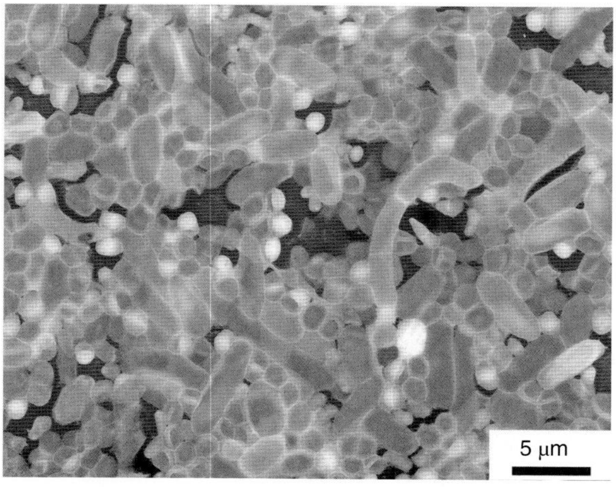

Figure 13.4 Starter cultures for cheese making. TEM-image, UA negative staining.

prepared for electron microscopy studies using negative staining with uranyl acetate. This picture illustrates the diversity of lactic acid bacteria found in the starter. It is clearly visible that the starter contains various sizes of coccus that are differently stained with uranyl acetate and bacillary forms of bacteria that have different morphology.

First, added starters dominate over their own cheese microflora, suppressing the development of technically harmful and pathogenic microorganisms. Secondly, starter bacteria create conditions for active growth of subsequent species of bacteria, which will create the desired consistency, taste and flavor of the product. When getting into the milk and later cheese, starter bacteria begin to grow and form colonies, which alter the microscopic environment. This changes many factors, including pH, oxidation-reduction potential, levels of organic acids, etc.

The growth of bacterial colonies in the cheese depends on the primary conditions of localization at the time when the starter is added to the milk. The colonies will develop faster under certain conditions, while in others they will develop slower. However, the final localization of bacteria after addition of starter to the milk remains uncertain until formation of a sufficiently dense clot.

Figure 13.5 shows a photograph of the bacteria *Lactococcus lactis* subsp. *lactis* in milk at its enzymatic coagulation 15 minutes after addition of rennet. This photo was obtained with TEM and preparation of the sample using a heavy metal (Pt) shadowing method. It is clear from the picture that at this point of time the bacteria had not yet come into full contact with the casein micelles; they had not formed colonies and their subsequent localization in the product is uncertain.

L. lactis is perhaps the most prevalent microorganism used in the production of dairy products. *L. lactis* decomposes lactose in the milk without formation of gases into two molecules of lactic acid which ensures a pH decrease. The most favorable temperature for their development is +30 to + 35°C. *L. lactis* produces different bacteriocins and antibiotic substances of a protein nature. Bacteriocins play a major role in the antagonism of lactic acid bacteria and have an inhibitory effect on the growth of a number of

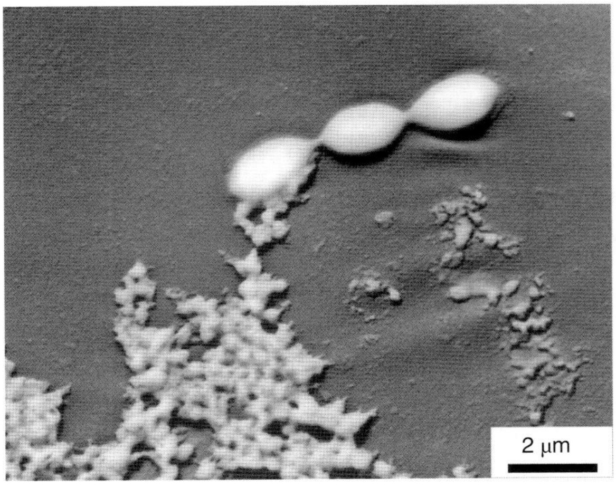

Figure 13.5 *Lactococcus lactis* subsp. *lactis* and casein micelles cluster. TEM-image, heavy metal shadowing.

microorganisms including pathogenic forms, i.e. *Listeria, Bacillus, Staphylococcus* etc., which develop in the dairy products during their storage. Nisin is the most famous of the bacteriocins.

Other bacteria – *Propionibacterium freudenreichii* play an important role in the production of hard and semi-hard cheeses such as Emmental and Maasdam. Their concentration in these cheeses is higher than in any other type of cheese. During maturation of Emmental type cheeses, *P. freudenreichii* produce a number of enzymes that contribute to the formation of acetate, propionate and carbon dioxide. Unlike most other lactobacilli, *P. freudenreichii* mainly perform lipolysis of milk fat, forming fatty acids in the free state. This provides the desired taste and flavor of cheese. The produced carbon dioxide is responsible for the formation of characteristic "eyes" in the cheese. The size and number of "eyes" in the cheese depend on the intensity of the growth of such bacteria; it is possible to control the growth of bacteria and, accordingly, the formation of "eyes" by changing the acidity and temperature.

P. shermanii often take the form of short sticks, and they are easy to confuse with coccus forms of lactic acid bacteria. Figure 13.6 shows a TEM photograph of the bacterium *Propionibacterium freudenreichii* subsp. *shermanii* in milk 15 minutes after addition of rennet. Here *P. shermanii* also have the form of a short rod and are similar to coccuses, while bacteria are not localized in a protein matrix, they are in the milk.

Pathogenic and extraneous microflora can also be found in milk together with starter microflora. Figure 13.7 shows a TEM photograph of the bacteria *Escherichia coli. E. coli* can produce lactate, succinate, ethanol, acetate and carbon dioxide as waste products. Often, the product can be defected or damaged in such a case. *E. coli*, unlike lactic acid bacteria, can have peritrichous flagellation and is able to move purposefully. The high mobility of these bacteria, possible for their individual movement in the milk, give them an advantage as compared with immobile lactobacilli.

They are located in the product matrix in a non-random manner and only in the spots where environmental conditions are optimal for their livelihoods. According to our

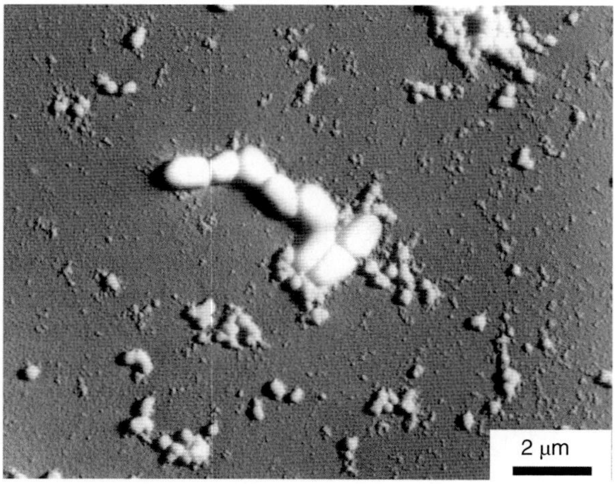

Figure 13.6 *Propionibacterium freudenreichii* subsp. *shermanii* and casein micelles. TEM-image, heavy metal shadowing

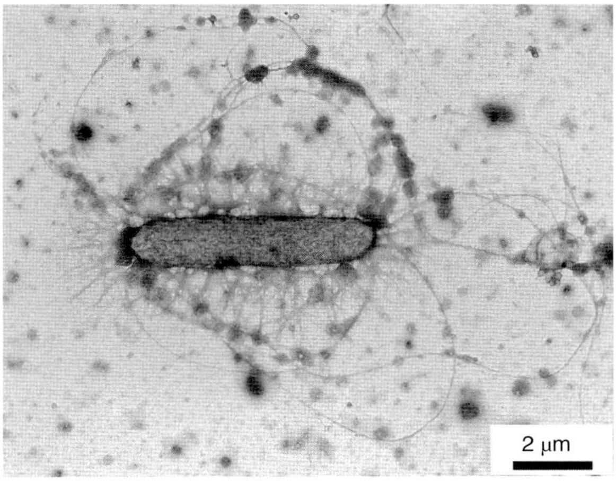

Figure 13.7 *Escherichia coli* in milk. TEM-image, UA negative staining.

observations, the speed of their movement in the milk may reach 20 μm per second, and the preferable movement of *E. coli* in the milk is from one fat globule to the next.

13.3.3 Bacteria in a Protein Matrix

After the gel point of the gelation process, a single dimensional structure, which limits the possibility of the microorganism movement throughout the product, starts to form. They are located in the dimensional structure randomly and are forced to exist in these conditions. If the environmental conditions are favorable, the bacteria grow, if the conditions are bad, their growth slows down or stops. Figure 13.8 illustrates the localization of *Lactococcus lactis* in the aggregate of protein particles in the dairy gel formed after the conformational phase transition in the molecules of casein micelles after the rennet action.

A large colony of bacteria or a random set of closely spaced and possibly related bacteria is localized on the surface of the aggregates of the protein particles. At this stage of milk gelation the large number of the bacteria localized in one place can be a casual event and cannot reliably confirm the presence of favorable conditions for their development in this place. It is possible that this colony existed in the starter and appeared to be in this place as micelles aggregate randomly and directly on the addition of the starter.

However, not all bacteria in the milk gel have an equal position. Figure 13.9 shows another image of *Lactococcus lactis* localization in the same milk gel and at the same time.

The bacteria are isolated from each other and located at a greater distance. Moreover, unlike the previously discussed case, being located in the aggregate of protein particles, bacteria remain isolated from them. It is possible that there is a specific capsule around bacteria or they are surrounded by the proteolysis products of surrounding proteins or other metabolic products. In any case, when comparing Figures 13.8 and 13.9, it can be stated that the primary localization of bacteria can be very important for the possibility of their further development.

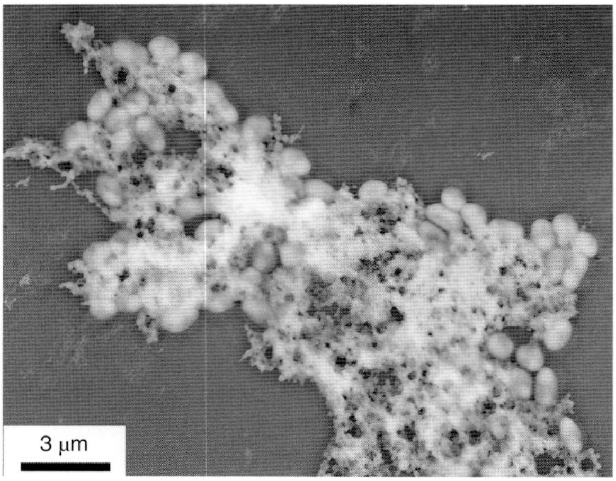

Figure 13.8 *Lactococcus lactis* on protein particles aggregate in milk gel. TEM-image, UA negative staining.

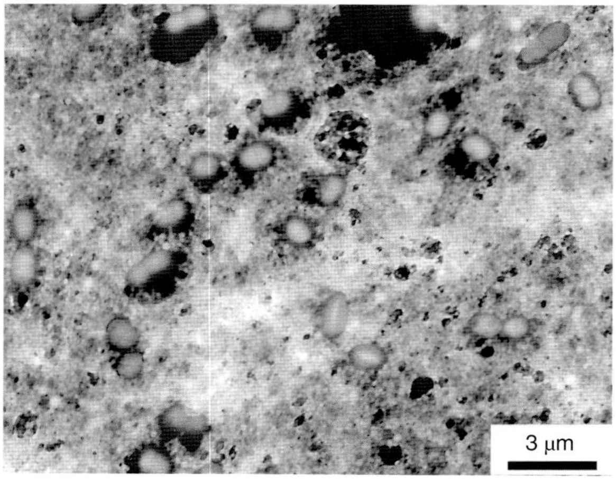

Figure 13.9 *Lactococcus lactis* in protein particles aggregate in milk gel. TEM-image, UA negative staining.

After cutting the formed milk gel, its stirring, and the cessation of thrusting and pressing, the bacterial localization changes dramatically. Before the bacteria were located mostly in a liquid medium, and some were removed from the clot by thrusting; the environment for the remaining bacteria was then mostly solid after pressing.

Figure 13.10 shows a TEM image of the microstructure of Uglichsky cheese, a cheese made of cow's milk, after whey removal; this image was obtained using the rapid freeze-fracture-etching method. Milk fat globules (F) were included in the cheese protein matrix (P) and bacteria (B) were localized. There are no large colonies of bacteria here; isolated bacteria of up to 2 to 4 in the group can be found. Perhaps this is due to mechanical separation of large colonies (found in the milk gel after addition of starter)

Figure 13.10 Bacteria and fat globule in Uglichsky cheese microstructure after whey removal. TEM-image, freeze-fracture replication.

for small groups or isolated cells that were separated during mechanical processing of gel.

Most often, it is possible to find information in the literature that the bacteria are localized near the fat globules; however there is less information about their location in the residual thrusting located in the interstices of the protein matrix. Our studies have shown preferential localization of bacteria in the cavities of the matrix filled with thrusting.

Figure 13.11 shows a similar image of the microstructure of low-fat cheese after whey removal. Here, small groups of bacteria are predominantly localized in the protein matrix voids, where there is a residual thrusting. These photographs were obtained after removal of thrusting and cheese pressing, i.e. almost at the beginning of the

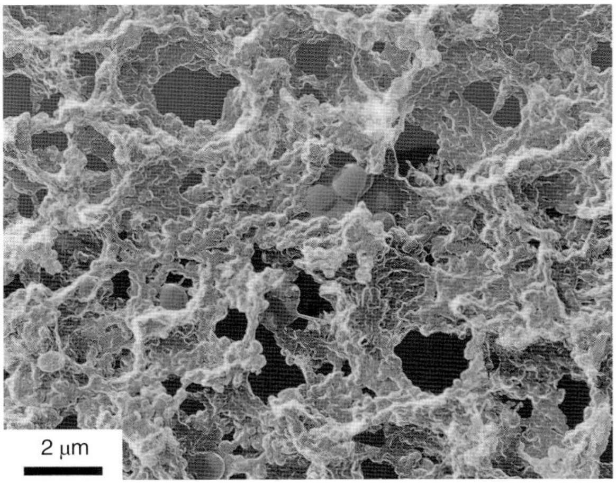

Figure 13.11 Bacteria in low-fat cheese microstructure after whey removal. TEM-image, freeze-fracture replication.

process of bacteria development, and therefore their further growth will depend on the conditions of the environment in the places of their localization.

Analysis of a series of similar images of cheese microstructure and localization of bacteria shows that there is from 0.5 to 2 billion cells in one cubic centimeter of cheese. In contrast to the estimates of the number of cells obtained by microbiological methods, these are not necessarily viable cells and this refers to the number of cells, not to the number of colony forming units.

The maturing of cheese and the activity of microorganisms in it are two inseparable and complementary processes. Developing in the cheese, starter microflora initiates complex, multi-step biochemical reactions in it, and changes and converts almost all the original components of the cheese. This initial maturation of cheese is necessary to create conditions for the development of other microorganisms that cause other biochemical processes which create taste and flavor of the cheese required by humans. This is the final maturation of the cheese and the way it occurs will largely depend on the quality of the finished product.

Electron microscopic studies of the microstructure of the maturing cheese showed that the most intensive development of bacteria at this stage takes place at the junction of large blocks of protein ("grains") which have undergone proteolysis. The TEM image in Figure 13.12 shows the development of bacteria in the maturing cheese based on the same scenario. Bacteria (B), which occupy all the available space, are located at the junction of protein blocks (P). It is possible that at the junction of these blocks there is sufficiently large amount of moisture or, specifically, solution of proteolysis products and other nutrients essential for vital functions of bacteria that were formed from the residual thrusting. Water may also be required for removal of metabolic products. Such conditions of existence were suitable for active growth of bacteria colonies, which, therefore, strongly influenced the maturation of cheese.

The opposite occurs when bacteria are localized in dense protein blocks. Transfer of necessary substances for the bacteria and removal of the products of their metabolism are obstructed here, i.e. living conditions are not favorable. Figure 13.13 shows a TEM

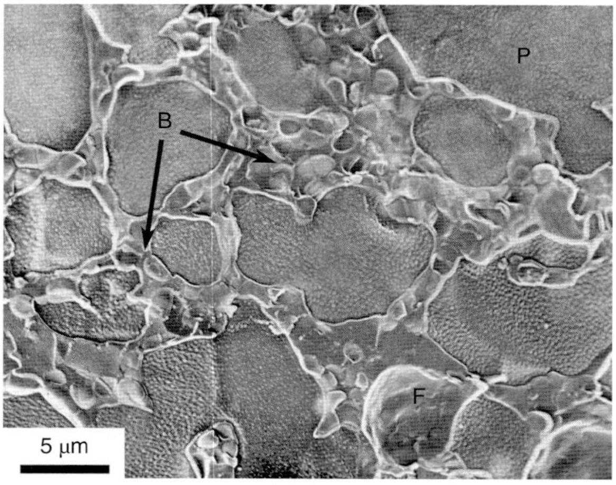

Figure 13.12 Bacteria in ripened cheese microstructure. TEM-image, freeze-fracture replication.

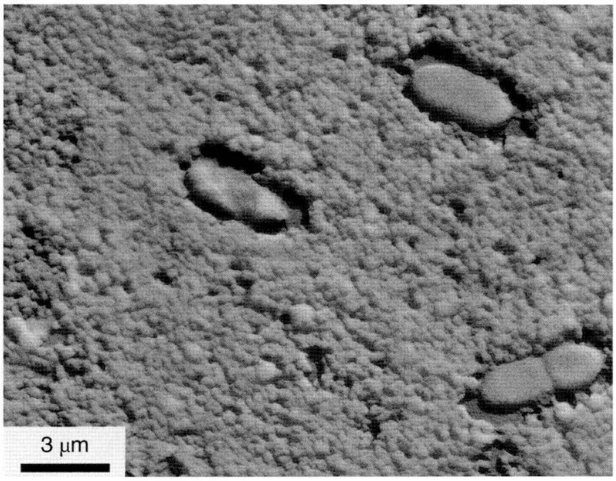

Figure 13.13 Bacteria in ripened cheese microstructure. TEM-image, freeze-fracture replication.

image of this possible bacteria localization. It can be seen that bacteria in the dense protein medium are isolated from each other; they do not develop well, and it is even unclear how they are viable.

A method of rapid freeze-fracture and replication was used when preparing samples of maturing cheese for microscopic examination, the results of which are shown in Figures 13.12 and 13.13. A method of staining with heavy metal salt (Figures 13.8 and Figure 13.9) was used in preparation of cheese samples after pressing for microscopic examination.

Both of these methods have their own shortcomings and are accompanied by occurrence of artifacts. However, despite this and the fact that in one case the cheese was examined after pressing, and in another case the cheese was examined at its active maturation, the resulting images are respectively similar, and findings of the analysis are also similar.

13.3.4 Bacteria in Cheese Eyes

Gas bubbles or as they are called "eyes" of different sizes and shapes may be present in the structure of semi-hard cheeses. Apparently, the first work which considered this question, was the work of W.M. Clark (Clark, 1912; 1918) in studies of the gas composition and formation of eyes in Emmental cheese. At the present time a lot of attention has been given to the research of eye formation and growth mechanisms. This is due to the practical observation that the number, shape and size of eyes are directly connected with the taste, flavor and texture of the cheese.

The results of the study on growth kinetics and microstructure of cheese near these eyes was conducted using three types of microscopy – light, confocal laser scanning and electronic (Huc *et al.*, 2014). Magnetic resonance imaging has provided visualization of the distribution of eyes in the cheese and the eye growth kinetics during the ripening of cheese. Light microscopy showed that the initial microscopic eyes of 50 μm to 200 μm in diameter are always located at the contacting surfaces of the cheese grains and they appear in the initial stage of ripening. Most of the bacterial colonies were localized in the center of the cheese; a smaller number were on the sides. The eyes in cheese were distributed accordingly. However, the mechanism of formation and growth of cheese eyes, as indicated by a number of researchers, remains to be fully understood (Daly *et al.*, 2010; Guggisberg *et al.*, 2015). Bacteria localization and their development directly in the cheese eye have not been fully studied yet.

We have carried out a study on localization of bacteria in cheese eyes using a transmission electron microscope. The sample was prepared for the research in the following sequence. An approximately two millimeters thick plate is cut from the central part of the cheese. A place with a completely closed eye of about one millimeter in diameter was selected visually, against the light, on such a plate of average maturity cheese. A rectangular block 2×2×10 mm in size was cut out from this plate so that the eye was inside the block, avoiding the eye opening. The obtained sample of cheese with an eye was quickly frozen to a temperature of −180°C and placed in a vacuum chamber where it was fractured along the eye diameter, and the replica of the inner surface of the eye was formed. Next, the replica was removed, washed and placed in a TEM. Two types of semi-hard cheeses were used for this study: Uglichsky and Kostromskoy made in the laboratory of experimental production. Figure 13.14 shows a TEM image of the inner surface of the eye in Uglichsky cheese.

A bacterial cell colony of coccus forms which covers the surface of an eye is visible in the picture, and it can be assumed that these bacteria release gas that causes the formation of eyes. The cells are located at some distance from each other, which is not typical for the colony of immovable bacteria. Such an arrangement can only be explained in the way that when the gas pressure is increased, the size of eye is increased and its surface is stretched. Surface stretching improves the process of colony growth due to the release of additional space and better access to nutrients. This leads to the conclusion that the release of gas by bacteria is not a secondary process and it is not a byproduct of metabolism, as it is usually presented (McSweeney and Fox, 2004), rather it is a strategic process that ensures improvement of the conditions of the colony. Strategically important

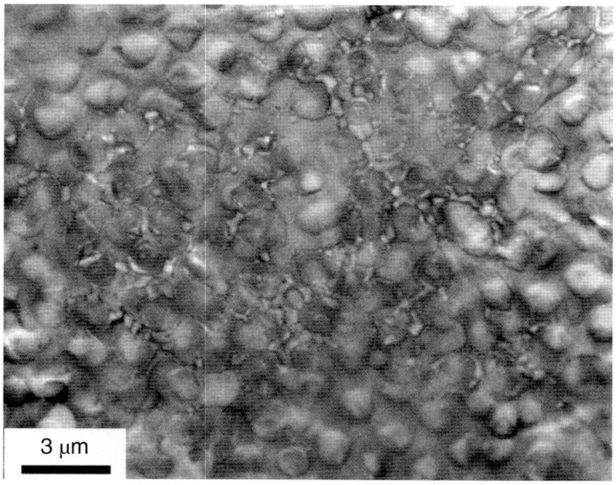

Figure 13.14 Bacteria in Uglichsky cheese eye. TEM-image, freeze-fracture replication.

for bacterial colonies, the task of expansion of their habitat cannot be solved without effective eye sealing. In other words, in the process of evolution, bacteria must develop or use this path of metabolism, which will cause release into the environment of products that transform the underlying surface into a structure which ensures the required impermeability and elasticity.

Figure 13.15 shows a TEM image of the inner surface of the eye in Kostromskoy cheese. This image of the cheese eye surface is fundamentally different from the image on the Figure 13.14. There are no compact surface colonies of bacteria on the surface of the eye, but there is are separate, small groups of cells, which are located around the fibers of unknown origin. It is only clear that the living conditions for bacteria around fibers are better than in the distance, and that these fibers come directly from the surface of the eye. It can be stated that in this case the strategy of survival and development of colonies of bacteria in cheese is completely different, even though the external result – formation of eyes – for us is same.

13.3.5 Bacteria in Yoghurt

Many countries for thousands of years produced and consumed milk fermented by wild microorganisms. However, it was not until the beginning of the 20th century, when Russian scientist I.I. Metchnikoff published his research results (Metchnikoff, 1905; 1907; 1908) on the positive effect of yoghurt on health and life expectancy and was awarded the Nobel Prize, that mass production and consumption of yoghurt by people at large began (Gordon, 2008).

It took more than a hundred years to study this product; however, many physical and biochemical mechanisms that occur during its manufacture are still unknown. The most interesting mechanism to draw attention of many researchers is the mechanism for increasing milk viscosity and changing its microstructure, according to the development of various bacteria in it. Exopolysaccharides (EPS), separated from lactic acid bacteria and their role in the formation of yoghurt attracted particular interest among

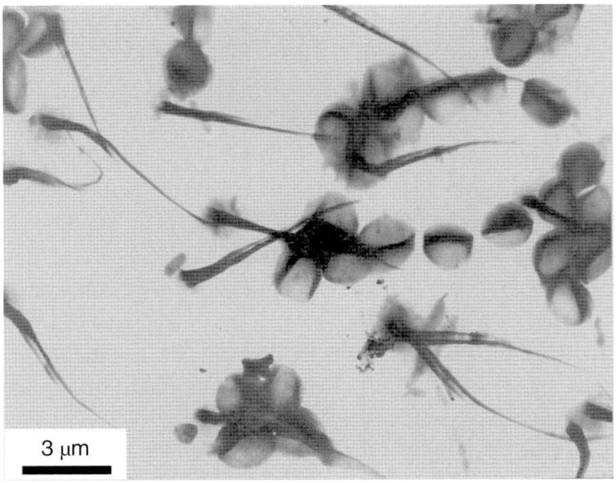

Figure 13.15 Bacteria in Kostromskoy cheese eye. TEM-image, freeze-fracture replication.

researchers. Due to the variety of composition, properties and structure of exopolysaccharides and their location towards bacteria, visualization of bacterial exopolysaccharides in dairy products is not an easy task to do. As far as EPS localization is concerned, we distinguish capsular and free EPS. Capsule EPS forms a dense capsule around the bacteria, the size of which depends on the type and strain of bacteria. Free bacteria EPS may bear the form of filaments or colloidal particles and be removed at a considerable distance from them. Many physical and chemical properties of EPS are closely linked to three-dimensional biopolymer structure (Dogsa *et al.*, 2005). For example, different *Streptococcus thermophilus* and *Lactobacillus delbrueckii* subsp. *bulgaricus* bacteria strains may bear a neutral, hard and slightly branched, anionic, stiff and linear, neutral, flexible and highly branched EPS form (Harding *et al.*, 2005; Sanchez-Medina *et al.*, 2007; Sawen *et al.*, 2010; Martinovic et al., 2013). Furthermore, the EPS state heavily depends on environment pH.

In the process of preparation of yoghurt samples for different microscopy methods, various kinds of artefact related to three-dimensional EPS structure may arise. Therefore, the interpretation of visualization results will be ambiguous. This was highlighted in one of the first works (Kaláb, 1993; Kaláb et al., 1995), when with the help of scanning electron microscopy the appearance of fibrous EPS structures in yoghurt was interpreted as a preparation artifact related to the need for sample dehydration. However, scanning electron microscopy is often used to study bacteria in yoghurt using new methods of sample preparation. SEM-image of filaments of *Lactococcus lactis* ssp. *cremoris* EPS bacteria, interconnecting milk protein elements, are introduced in the work of Ayala-Hernandez *et al.* (2008). This study obviously demonstrates the existence of interactions between polysaccharides and proteins, but the nature of this interaction remains ambiguous.

A radical way to eliminate sample preparation artifacts for electron microscopy by dehydration is quick-freeze sample preparation microscopy, in which water is in a vitrified, amorphous state. For this research, cryo-scanning electron microscopy (cryo-SEM) is used. A study conducted by Hassan *et al.* (2003) with the use of cryo-SEM, reported the first observation of lactic acid bacteria EPS microstructure in the form of a three-dimensional network. However, the results of these studies have not proved direct interactions among EPS proteins.

Confocal laser scanning microscopy is widely used to study the localization of bacterial EPS in yoghurt; however, it also has a number of drawbacks. The first one is low resolution and difficulties in selecting reagents for fluorescence visualization of the objects. Hassan *et al.* (1995; 1996; 2002) have developed sample preparation for CSLM in the study of microstructure of fermented dairy products in their natural hydratable state, and localization of bacterial exopolysaccharides in yoghurts. Zhang *et al.* (2016) with the help of CSLM, studied the effect of capsular EPS of various bacterial strains on the microstructure of low-fat yoghurt. Fluorescent staining of bacterial EPS of starters of one type revealed their localization in yoghurt and their impact on microstructure and rheological properties.

For yoghurt microstructure visualization and bacteria and bacterial EPS localization, researchers use a variety of microscopic methods, which allow more accurate interpretation of the received images. Conventional light microscopy combined with transmission electron microscopy used in the study of Yang *et al.* (2010) made it possible to confirm the formation of capsules by *Lactobacillus rhamnosus* JAAS8. In the work of Costa *et al.* (2012) on the study of the role of bacterial EPS in gel formation and its effect on the viscosity of fermented milk, scanning electron and atomic force microscopy were used. This study has shown that EPS contributes to unification of bacterial cells in large branching clusters. Combination of atomic force and transmission electron microscopy in the studies of bacteria capsules (Stukalov *et al.*, 2008) revealed sample preparation artifacts and artifacts of their visualization. Suggestions on how to eliminate artifacts and the results of direct observation of capsules on the bacteria are provided here.

Transmission electron microscopy, in spite of all its drawbacks and difficulties in sample preparation, is still used for localization of microorganisms in fermented dairy products and gives good results. Figure 13.16 shows the image of a starter culture bacteria obtained in the process of TEM studies on localization of bacteria and bacterial EPS in low-fat probiotic yoghurt. Sample preparation for TEM studies was performed by negative staining with uranyl acetate solution. *Lactobacillus bulgaricus* bacteria in Figure 13.16a, and *Bifidobacterium bifidum* in Figure 13.16b have no visible capsules or any other metabolites. *Streptococcus thermophilus* in Figure 13.16c and d have clearly visible "clouds" of products of metabolism around them. In our opinion these are bacterial-free polysaccharides which further on interact with milk proteins and, eventually, affect yoghurt microstructure.

Figure 13.17 shows a TEM image of a *Streptococcus thermophilus* bacteria colony on the aggregate of casein micelles in yoghurt.

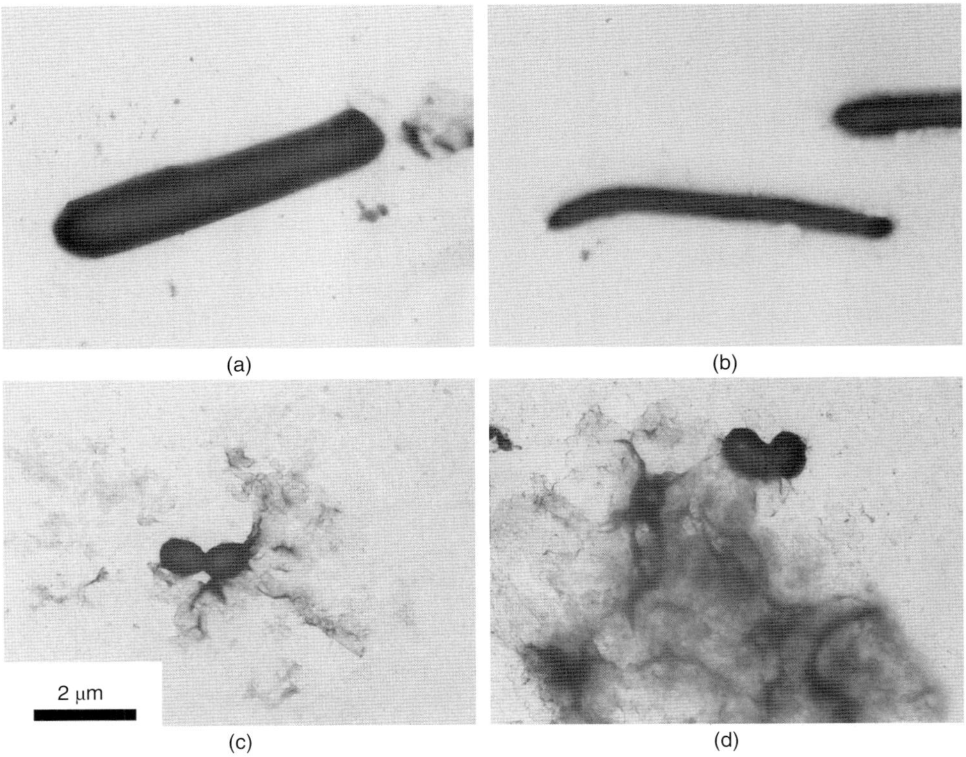

Figure 13.16 *Lactobacillus bulgaricus* (a), *Bifidobacterium bifidum* (b), and *Streptococcus thermophilus* (c, d) in yoghurt. TEM-image, UA negative staining.

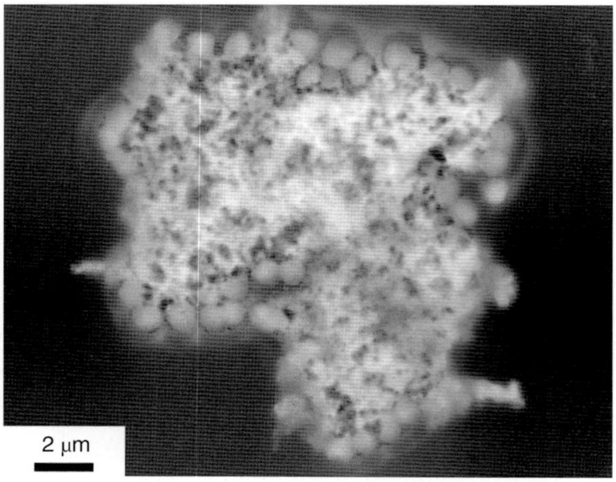

Figure 13.17 *Streptococcus thermophilus* and EPS on casein micelles aggregate in yoghurt. TEM-image, UA negative staining.

Here the entire bacteria colony and casein micelles are surrounded by metabolites separated from bacteria. If we compare this image with the image of micelles aggregate in Figure 13.8, the difference in interactions between *Lactococcus lactis* and *Streptococcus thermophilus* with casein micelles in milk becomes obvious.

13.3.6 Bacteriophages

Among microorganisms playing an important role in the production of fermented dairy products, bacteriophages, which can attack starter cultures, occupy a dangerous place. The consequences of bacteriophage attack on lactic acid bacteria can lead to milk ripening, quality change and, in the most serious cases, to a complete loss of product. Therefore, in the production of fermented products, it is necessary to pay much attention to protection from bacteriophages. One of the most effective ways to prevent phage attacks is to use phage-resistant strains of lactic acid bacteria. In order to obtain such strains, bacteriophages are separated and studied. The collections include thousands of species and varieties of bacteriophages, which differ from each other in morphological characteristics, in their specificity toward particular species and strains of bacteria, as well as towards the organization of genetic material. The vast majority of identified bacteriophages of lactic acid bacteria belong to three families: Myoviridae, Siphoviridae and Podoviridae. These families differ morphologically, for example, Myoviridae have a long contractile tail, Siphoviridae have a long non-contractile tail, while Podoviridae have a short non-contractile tail. Other morphological features are also used for classification, e.g. head shape, baseplate etc. *Lactococcus* phages mainly belong to the Siphoviridae family and only a small part belongs to Podoviridae family (Mc Grath and van Sinderen, 2007).

Morphological characteristics of bacteriophages are generally defined by transmission electron microscopy and sample staining with salt of heavy metals. Figure 13.18 shows TEM images of some species of lactic acid bacteria phages, obtained in the process of studying their morphology. The images are shown at the same magnification, and allow comparison of the size and the shape of the bacteriophages.

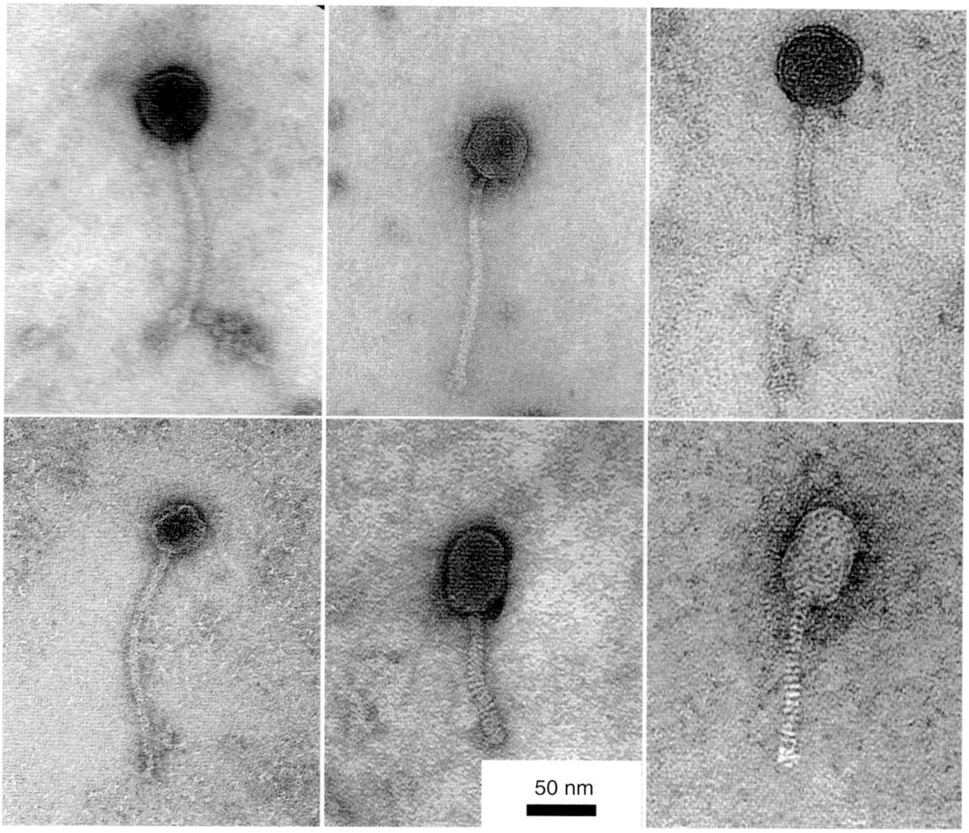

Figure 13.18 Lactic acid bacteria phages. TEM-image, UA negative staining.

Assessing the number of phage elements and their interactions with the phage host is usually performed by microbiological methods using special breeding grounds and bacterial testing culture sensitive to a particular bacteriophage species. Electron microscopic examination of samples, taken from a testing culture infected with bacteriophage, allows us to see the time of infection and lysis of bacteria. Figure 13.19 shows the time of bacteriophage absorption on the surface of *Lactobacillus plantarum*.

However, some bacteriophages cannot completely lyze testing cultures, and thus they significantly underestimate their quantity. In a study by Zago *et al.* (2012) for estimating the number of bacteriophages in infected *Lactobacillus helveticus* cultures, epifluorescence microscopy was used. For phages and bacteria visualization during various stages of the lytic cycle, the atomic force microscope (AFM) was used. The research results have shown that a combination of these two methods can be used to estimate the amount of phages in the culture and to study the effect of phage on the host bacterium during the infection. In particular, epifluorescence microscopy allows detection of the presence of bacteriophage in such complex bacteria populations as natural fermentation starters, in which phages usually develop. Using atomic force microscopy we can determine the shape and the size of bacterial cells, visualize changes in cell surface at

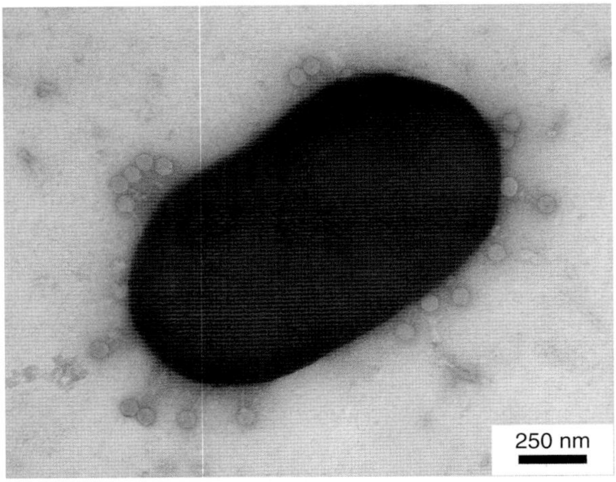

Figure 13.19 Bacteriophages on a *Lactobacillus plantarum* bacteria. TEM-image, UA negative staining.

different stages of infection, and easily distinguish an intact bacterial cell from an infected one.

13.4 Conclusions

Natural fermented dairy products have a very complex microstructure, which to a large extent determines their physical, structural and sensory properties. Microorganisms, their localization, growth and autolysis rate have a great impact on the formation of this microstructure. Many direct observation methods have been used by researchers to study microstructure of fermented products and localization of microorganisms in it, including the following: TEM, cryo-SEM, CLSM, AFM and others. However, sample preparation for any microscopy method is accompanied by the appearance of artifacts which must be considered when interpreting the received images. Nevertheless, the use of various microscopy methods and sample preparation techniques for studying microstructure of dairy products and localization of microorganisms within provided valuable information on some widely observed phenomena.

The research results of microorganism localization in the fermented dairy products given in this chapter, and also obtained by different researchers with the use of various microscopy methods and sample preparation techniques were individual and not complete. These works do not completely provide an understanding of the role of microorganisms in the serial formation of the microstructure of dairy products at various stages of manufacture.

For future research on the phenomena of microorganism localization on the microstructure of fermented dairy products, the following can be suggested:

- It would be interesting to develop in light microscopy a research method of growth kinetics of bacterial colonies and their interaction with milk components *in vivo*, for example, in thin transparent cuvettes with a thickness of 5–7 micrometers.
- Determination of the 3-D microstructure of cheese and localization of microorganisms within can be achieved using sample preparation techniques for confocal laser

scanning microscopy (multiphoton fluorescence microscopy) to ensure the transparency of the large three-dimensional object and selective coloring components and microorganisms, like a "transparent brain".
- Electron microscopy may be a possibility for developing a method of elemental analysis, and mapping of areas near bacterial colonies in cheese using microstructure energy-dispersive X-ray spectroscopy (EDXS) and/or a wavelength dispersive X-ray spectroscopy (WDXS).

The research results could provide the best comprehension of the different phenomena occurring at various stages of manufacture, relating to the biochemical and microstructural changes. A deep insight into the microstructure of the products, especially into spatial distribution and the role of microorganisms, and their interaction with milk components, is a key tool for the development of new high-quality dairy products with the required rheological and organoleptic properties.

Acknowledgment

The author is grateful to Dr. Marina B. Zakharova for the many stimulating discussions during this work.

References

Amelinckx, S, van Dyck, D, van Landuyt, J and van Tendeloo, G 2008, Electron Microscopy: Principles and Fundamentals, Wiley, Oxford.

Arai, M and Kuwajima, K 2000, 'Role of the molten globule state in protein folding', *Advances in Protein Chemistry*, vol. 53, pp. 209–82.

ARRIDI, 2017, All-Russian Research Institute of the Dairy Industry, Moscow, viewed 28 March 2018, http://www.vnimi.org

ARSRIBC, 2017, All-Russian Scientific Research Institute of Butter and Cheese making, Uglich, viewed 28 March 2018, http://vniims.info

Auty, MAE, Gardiner, GE, McBrearty, SJ, O'Sullivan, EO, Mulvhill, DM, Collins, JK, Fitzgerald, GF, Stanton, C and Ross, RP 2001, 'Direct In Situ Viability Assessment of Bacteria in Probiotic Dairy Products Using Viability Staining in Conjunction with Confocal Scanning Laser Microscopy', *Applied and Environmental Microbiology*, vol. 67 no. 1, pp. 420–5.

Ayala-Hernandez, I, Goff, HD and Corredig, M 2008, 'Interactions Between Milk Proteins and Exopolysaccharides Produced by *Lactococcus lactis* Observed by Scanning Electron Microscopy', *Journal of Dairy Science*, vol. 91, pp. 2583–90.

Bozzola, JJ and Russell, LD 1998, Electron Microscopy: Principles and Techniques for Biologists, 2nd edn, Jones and Barlett Publishers, Sudbury, MA.

Buchheim, W 1982, 'Aspects of sample preparation for freeze-fracture/freeze-etch studies of proteins and lipids in food systems: a review', *Food Microstructure*, vol. 1, pp. 189–208.

Burdikova, Z, Svindrych, Z, Hickey, C, Wilkinson, MG, Auty, MAE, Samek, O, Bernatova, S, Krzyzanek, V, Periasamy, A and Sheehan, JJ 2015, 'Application of advanced light microscopic techniques to gain deeper insights into cheese matrix physico-chemistry', *Dairy Science and Technology*, vol. 95, pp. 687–700.

Burgain, J, Gaiani, C, Francius, G, Revol-Junelles, AM, Cailliez-Grimal, C, Lebeer, S, Tytgat, HLP, Vanderleyden, J and Scher, J 2013, 'In vitro interactions between probiotic bacteria and milk proteins probed by atomic force microscopy', *Colloids and Surfaces B: Biointerfaces*, vol. 104, pp. 153–62.

Burgain, J, Scher, J. Francius, G, Borges, F, Corgneau, M, Revol-Junelles, AM, Cailliez-Grimal, C and Gaiani, C 2014, 'Lactic acid bacteria in dairy food: Surface characterization and interactions with food matrix components', *Advances in Colloid and Interface Science*, vol. 213, pp. 21–35.

Clark, WM 1912, 'A study of the gases of Emmental cheese', *United States Department of Agriculture, Bureau of Animal Industry, Bulletin*, 151.

Clark, WM 1918, 'On the formation of "eyes" in Emmental cheese' *Journal of Dairy Science*, vol. 1, pp. 91–113.

Cornea, A and Conn, PM (eds) 2014, Fluorescence Microscopy, Academic Press, London.

Costa, NE, Wang, L, Auty, ME, Hannon, JA, McSweeney, PLH and Beresford, TP 2012, 'Rheological, microscopic and primary chemical characterization of the exopolysaccharide produced by *Lactococcus lactis* subsp. *cremoris* DPC6532', *Dairy Science and Technology*, vol. 92, pp. 219–35.

D'Incecco, P, Faoro, F, Silvetti, T, Schrader, K and Pellegrino, L 2015, 'Mechanisms of *Clostridium tyrobutyricum* removal through natural creaming of milk: A microscopy study', *Journal of Dairy Science*, vol. 98, pp. 5164–72.

Daly, DFM, McSweeney, PLH and Sheehan, JJ 2010, 'Split defect and secondary fermentation in Swiss-type cheeses – A review', *Dairy Science and Technology*, vol. 90, pp. 3–26.

de Vos WM 1989, 'On the carrier state of bacteriophages in starter lactococci: an elementary explanation involving a bacteriophage resistance plasmid', *Netherlands Milk and Dairy Journal*, vol. 43, pp. 221–7.

Deepak, FL, Mayoral, A and Arenal, R (eds) 2015, Advanced Transmission Electron Microscopy, Springer International Publishing.

Diaspro, A (ed) 2001, Confocal and Two-Photon Microscopy: Foundations, Applications and Advances, Wiley-Liss, Wilmington, DE.

Dogsa, I, Kriechbaum, M, Stopar, D and Laggner, P 2005, 'Structure of Bacterial Extracellular Polymeric Substances at Different pH Values as Determined by SAXS', *Biophysical Journal*, vol. 89, pp. 2711–20.

El-Bakry M and Sheehan, J 2014, 'Analysing cheese microstructure: A review of recent developments', *Journal of Food Engineering*, vol. 125, pp. 84–96.

Farrell, HM Jr, Qi, PX, Brown, EM, Cooke, PH, Tunick, MH, Wickham, ED and Unruh JJ 2002, 'Molten Globule Structures in Milk Proteins: Implications for Potential New Structure-Function Relationships', *Journal of Dairy Science*, vol. 85, pp. 459–71.

Fitzsimons, N, Cogan, T, Condon, S and Beresford, T 2001, 'Spatial and temporal distribution of non- starter lactic acid bacteria in Cheddar cheese', *Journal of Applied Microbiology*, vol. 90, pp. 600–8.

Fleurot, I, Aigle, M, Fleurot, R, Darrigo, C, Hennekinne, J.-A, Gruss, A, Borezée-Durant, E and Delacroix-Buchet, A 2014, 'Following Pathogen Development and Gene Expression in a Food Ecosystem: the Case of a *Staphylococcus aureus* Isolate in Cheese', *Applied and Environmental Microbiology*, vol. 80, no. 16, pp. 5106–15.

Floury, J, Jeanson, S, Aly, S and Lortal, S 2010, 'Determination of the diffusion coefficients of small solutes in cheese: a review', *Dairy Science Technology,* vol. 90, pp. 477–08.

Floury, J, Jeanson, S, Madec, M-N and Lortal, S 2013, 'Porosity of *Lactococcus lactis* subsp. *lactis* LD61 colonies immobilised in model cheese', *International Journal of Food Microbiology,* vol. 163, pp. 64–70.

Fox, PF, McSweeney, PLH, Cogan, TM and Guinee, TP (eds) 2004, *Cheese: Chemistry, Physics and Microbiology. General Aspects,* V.1, Elsevier Academic Press, Amsterdam.

Fox, PF, Uniacke-Lowe, T, McSweeney PLH and O'Mahony, JA 2015, *Dairy Chemistry and Biochemistry*, Springer International Publishing, New York.

Gavina, JMA and Britz-McKibbin, P 2007, 'Protein Unfolding and Conformational Studies by Capillary Electrophoresis', *Current Analytical Chemistry,* vol. 3, pp. 17–31.

Gebhardt, R, Doster, W, Friedrich, J and Kulozik, U 2006, 'Size distribution of pressure-decomposed casein micelles studied by dynamic light scattering and AFM', *European Biophysics Journal*, vol. 35, pp. 503–9.

Glauert, AM and Lewis, PR 1998, *Biological Specimen Preparation for Transmission Electron Microscopy,* Portland Press, London.

Gordon, S 2008, 'Elie Metchnikoff: father of natural immunity', *European Journal Immunology,* vol. 38, pp. 3257–64.

Gudkov, AV 1999, 'Cheeses of the former USSR' in *Cheese: Chemistry, Physics and Microbiology, V. 2 Major Cheese Groups,* 2nd edn, Fox, PF (ed), Aspen Publishers Inc., Maryland.

Guerin, J, Bacharouche, J, Burgain, J, Lebeer, S, Francius, G, Borges, F, Scher, J and Gaiani, C 2016, 'Pili of *Lactobacillus rhamnosus* GG mediate interaction with β-lactoglobulin', *Food Hydrocolloids,* vol. 58, pp. 35–41.

Guggisberg, D, Schuetz, P, Winkler, H, Amrein, R, Jakob, E, Fröhlich-Wyder M-T, Irmler, S, Bisig, W, Jerjen, I, Plamondon, M, Hofmann, J, Flisch, A and Wechsler, D 2015, 'Mechanism and control of the eye formation in cheese', *International Dairy Journal,* vol. 47, pp. 118–27.

Harding, LP, Marshall, VM, Hernandez, Y, Gu, Y, Maqsood, M, McLay, N and Laws AP 2005, 'Structural characterization of a highly branched exopolysaccharide produced by *Lactobacillus delbrueckii* subsp. *bulgaricus* NCFB2074', *Carbohydrate Research,* vol. 340, pp. 1107–11.

Hassan, AN, Frank, JF, and Qvist KB. 2002, 'Direct Observation of Bacterial Exopolysaccharides in Dairy Products Using Confocal Scanning Laser Microscopy', *Journal of Dairy Science,* vol. 85, pp. 1705–08.

Hassan, AN, Frank, JF, Farmer, MA, Schmidt, KA, and Shalabi, SI 1995, 'Observation of encapsulated lactic acid bacteria using confocal scanning laser microscopy', *Journal of Dairy Science*, vol. 78, pp. 2624–28.

Hassan, AN, Frank, JF, Schmidt, KA and Shalabi, SI 1996, 'Textural properties of yogurt made with encapsulated nonropy lactic cultures', *Journal of Dairy Science*, vol. 79, pp. 2098–103.

Hassan, AN, Frank, JF and Elsoda M 2003, 'Observation of bacterial exopolysaccharide in dairy products using cryo-scanning electron microscopy', *International Dairy Journal,* vol. 13, pp. 755–62.

Hayat, MA 2000a, 'Positive staining', in *Principles and Techniques of Electron Microscopy: Biological Applications*, 4th edn, Cambridge University Press, Cambridge.

Hayat, MA 2000b, 'Negative staining', in *Principles and Techniques of Electron Microscopy: Biological Applications*, 4th edn, Cambridge University Press, Cambridge.

Hayat, MA 2000c, 'Applications of microwave heating to microscopy', *Principles and Techniques of Electron Microscopy: Biological Applications*, 4th edn, Cambridge University Press, Cambridge.

Herman, Br and Herman, B 1998, *Fluorescence Microscopy (Microscopy Handbooks)*, 2th edn, Springer-Verlag, New York.

Hickey, CD, Sheehan, JJ, Wilkinson MG and Auty, MAE 2015a, 'Growth and location of bacterial colonies within dairy foods using microscopy techniques: a review', Frontiers in Microbiology. Food Microbiology, vol. 18, no. 6, pp. 99–107.

Hickey, CD, Auty, MAE, Wilkinson, MG and Sheehan, JJ 2015b, 'The influence of cheese manufacture parameters on cheese microstructure, microbial localization and their interactions during ripening: A review', *Trends in Food Science and Technology*, vol. 41, no. 2, pp. 135–48.

Huc, D, Mariette, F, Challois, S, Barreau, J, Moulin, G and Michon, C 2014, 'Multi-scale investigation of eyes in semi-hard cheese', *Innovative Food Science and Emerging Technologies*, vol. 24, pp. 106–12.

Johnson M and Steele J 2013, 'Fermented dairy products', in Food Microbiology: Fundamentals and Frontiers, eds Doyle MP and Buchanan RL, ASM Press, Washington.

Kaláb, M 1981, 'Electron microscopy of milk products: a review of techniques', *Scanning Electron Microscopy*, vol. 3, pp. 453–72.

Kaláb, M 1993, 'Practical aspects of electron microscopy in dairy research', *Food Structure*, vol. 12, pp. 95–114.

Kaláb, M, Allan-Wojm, P and Miller SS 1995, 'Microscopy and other imaging techniques in food structure analysis', *Trends in Food Science and Technology*, vol. 6, pp. 177–86.

Kalinin, S and Gruverman A (eds) 2007, *Scanning Probe Microscopy*, Springer International Publishing, New York.

Kavanagh, GM, Clark, AH and Ross-Murphy, SB 2000, 'Heat-induced gelation of globular proteins: part 3. Molecular studies on low pH b-lactoglobulin gels', International Journal of Biological Macromolecules, vol. 28, pp. 41–50.

Kubitscheck, U (ed) 2013, *Fluorescence Microscopy:* From Principles to Biological Applications, Wiley-Blackwell, Weinheim.

Kuo, J (ed) 2007, *Electron Microscopy Methods and Protocols*, 2nd edn, Humana Press Inc. Totowa, NJ.

Martinovic, A, Moe, KM, Romeih, E, Aideh, B, Vogensen, FK, Østlie, H and Skeie, S 2013 'Growth of adjunct *Lactobacillus casei* in Cheddar cheese differing in milk fat globule membrane components', *International Dairy Journal*, vol. 31, pp. 70–82.

Matsko, NB, Letofsky-Papst, I, Albu, M and Mittal, V 2013, 'An analytical technique to extract surface information of negatively stained or heavy-metal shadowed organic materials within the TEM', *Microscopy and Microanalysis*, vol. 19, no. 3, pp. 642–51.

Mc Grath, S and van Sinderen, D (eds) 2007, Bacteriophage: Genetics and Molecular Biology, Caister Academic Press, Norfolk.

McSweeney, PLH and Fox, PF 2004, 'Metabolism of residual lactose and of lactate and citrate', in Cheese: Chemistry, Physics and Microbiology, V. 1, General Aspects, 3rd edn, Fox, PF, McSweeney, PLH, Cogan, TM and Guinee, TP (eds), Elsevier Applied Science, Amsterdam.

Metchnikoff, E 1905, Immunity in Infective Diseases, trans. FG Binnie, Cambridge University Press, London.

Metchnikoff, E 1907, Essais Optimistes, A Maloine, Paris.

Metchnikoff, E 1908, The Prolongation of Life; Optimistic Studies, trans. PC Mitchell, GP Putnam's Sons Publishing, New York and London.

Murphy, DB 2001a, Fundamentals of Light Microscopy and Electronic Imaging, Wiley-Liss, New York.

Murphy KP 2001b, Protein Structure, Stability and Folding, Humana Press, Totowa, New Jersey, U.S.

Ong, L, Dagastine, RR, Kentish, SE *and* Gras, SL 2011, 'Microstructure of milk gel and cheese curd observed using cryo scanning electron microscopy and confocal microscopy', *LWT – Food Science and Technology,* vol. 44, pp. 1291–302.

Pawley, J (ed) 2006, Handbook of Biological Confocal Microscopy, 3rd edn, Springer, New York.

Privalov, PL 1989, 'Thermodynamic problems of protein structure', *Annual Review of Biophysics and Biophysical Chemistry,* vol. 8, pp. 47–69.

Ptitsyn, OB 1995, 'Molten globule and protein folding', *Advances in Protein Chemistry,* 47, pp. 83–9.

RCMB, 2017, Russian Collection of lactic acid bacteria of "Experimental Biofactory", Uglich, viewed 12 February 2017, http://uglich-biofabrika.ru

Reimer, L (ed) 1995, Energy-Filtering Transmission Electron Microscopy, Springer International Publishing, Berlin.

Sanchez-Medina, I, Gerwig, GJ, Urshev, ZL and Kamerling, JP 2007, 'Structure of a neutral exopolysaccharide produced by *Lactobacillus delbrueckii* ssp. *bulgaricus LBB.B26', Carbohydrate Research,* vol. 342, pp. 2430–39.

Sawen, E, Zhang, X, Yang, Z and Widmalm, G 2010, 'Structural analysis of the exopolysaccharide produced by *Streptococcus thermophilus* ST1 solely by NMR spectroscopy', *Journal of Biomolecular NMR,* vol. 47, pp. 125–34.

Schmidt, DG 1982, 'Electron microscopy of milk and milk products: problems and possibilities', *Food Microstructure,* vol. 1, pp. 151–65.

Slayter, EM and Slayter, HS 1992, Light and Electron Microscopy, Cambridge University Press, Cambridge, UK.

Smid, EJ, Erkus, O, Spus, MJ, Wolkers-Rooijackers, CM, Alexeeva, S and Kleerebezem, M. 2014, 'Functional implications of the microbial community structure of undefined mesophilic starter cultures', *Microbial Cell Factories,* vol. 13, pp. 1–9.

Smykov, IT 2015a, 'Kinetics of Milk Gelation. Part I. Coagulation Mechanism', in *Rheology: principles, applications and environmental impacts,* Karpushkin, E (ed), Nova Science Publishing, New York.

Smykov, IT 2015b, 'Kinetics of Milk Gelation. Part II. Mathematical Modeling', in Rheology: principles, applications and environmental impacts. Karpushkin, E (ed), Nova Science Publishing, New York.

Søndergaard, L, Ryssel, M, Svendsen, C, Høier, E, Andersen, U, Hammershøj, M, Møller, JR, Arneborg, N and Jespersen, L 2015, 'Impact of NaCl reduction in Danish semi-hard Samsoe cheeses on proliferation and autolysis of DL-starter cultures', *International Journal of Food Microbiology,* vol. 213, pp. 59–70.

Stadhouders, J 1986, 'The control of cheese starter activity', *Netherlands milk and dairy journal,* vol. 40, pp. 155–73.

Stukalov, O, Korenevsky, A, Beveridge, TJ, and Dutcher, JR 2008, 'Use of atomic force microscopy and transmission electron microscopy for correlative studies of bacterial capsules', *Applied and Environmental Microbiology,* vol. 74, pp. 5457–65.

Yang, Z., Li, S, Zhang, X, Zeng, X, Li, D, Zhao, Y and Zhang, J 2010, 'Capsular and slime-polysaccharide production by *Lactobacillus rhamnosus* JAAS8 isolated Chinese sauerkraut: Potential application in fermented milk products', *Journal of Bioscience and Bioengineering*, vol. 110, pp. 53–7.

Zago, M, Scaltriti, E, Fornasari, ME, Rivetti, C, Grolli, S, Giraffa, G, Ramoni, R and Carminati, D 2012, 'Epifluorescence and atomic force microscopy: Two innovative applications for studying phage–host interactions in *Lactobacillus helveticus*', *Journal of Microbiological Methods*, vol. 88, pp. 41–6.

Zhang, L, Folkenberg, DM, Amigo, JM and Ipsen R 2016, 'Effect of exopolysaccharide-producing starter cultures and post-fermentation mechanical treatment on textural properties and microstructure of low fat yoghurt', *International Dairy Journal*, vol. 53, pp. 10–9.

14

Microstructure of Dairy Products: Challenges and Future Trends

Maricê Nogueira de Oliveira

Department of Biochemical and Pharmaceutical Technology, University of São Paulo, Brazil

14.1 Introducing Microstructure of Dairy Products

Dairy products are constituted by proteins, carbohydrates and lipids that are organized in a matrix at different spatial scales. The interaction of these constituents with structural elements forms its microstructure which is associated to its functional properties (Tamime *et al.*, 2007a,b).

Studying microstructure of dairy products – milk powder, fermented milks, and cheeses, provides information concerning the physico-chemical interactions taking place in the matrix. Moreover, understanding dairy product microstructure supports information concerning texture, solubility, flow, viscoelasticity and fracture features of the product that could affect its sensory, texture, nutrition characteristics, and functionality (Dewettinck, 2011). In addition, the obtained data could enable intervention in technological aspects of manufacture in order to improve safety and quality of the products.

The literature of microstructure on dairy foods is vast and continues to develop. The origin of the techniques employed may be found in Morris and Groves (2013). Studies on dairy product microstructure mostly employ scanning electron microscopy – SEM (Fallico *et al.*, 2006; Le Feunteun and Mariette, 2007; Luo *et al.*, 2014); transmission electron microscopy (TEM); confocal laser scanning microscopy – CLSM (Ercili-Cura *et al.*, 2013; Morand *et al.*, 2012); and automated image analysis (Silva *et al.*, 2015).

After micrographs are obtained, they must be analyzed and interpreted. Lately a wide range of numerical applications have been used to provide qualitative and quantitative data of dairy microstructure (Wium *et al.*, 2003; Aguilera *et al.*, 2000; Geng *et al.*, 2011; Ong *et al.*, 2011a,b; Rovira *et al.*, 2011).

These aspects will be demonstrated and discussed in this chapter concerning mainly dairy products in which yoghurt is the leading product, and cheese.

Microstructure of Dairy Products, First Edition.
Edited by Mamdouh Mahmoud Abdel-Rahman El-Bakry, Antoni Sanchez, and Bhavbhuti M. Mehta.

14.2 Microstructure of Fermented Milks

Fermentation is one of the oldest methods of long-term preservation of milk. According to the Codex standard for fermented milks (CODEX STAN 243-2003; (FAO/WHO, 2010):

> Fermented milk is a milk product obtained by fermentation of milk, which milk may have been manufactured from products obtained from milk with or without compositional modification by the action of suitable microorganisms and resulting in reduction of pH with or without coagulation (iso-electric precipitation). These starter microorganisms shall be viable, active and abundant in the product to the date of minimum durability. If the product is heat treated after fermentation the requirement for viable microorganisms does not apply.

This category of dairy products includes many popular dairy foods such as yoghurts – concentrated or drinking yoghurts, Scandinavian fermented milks, acidophilus milks, kefir, kumys and a variety of modern fermented dairy foods.

Fermented milks can be classified on the basis of the type of fermentation they undergo, as lactic (sour cream), yeast-lactic (Kefir, Koumiss, acidophilus-yeast milk) and mold-lactic (Villi). Depending on the characteristics of the employed bacteria, fermented milks could be also classified as mesophilic (Filmjolk, Nordic ropy milk), thermophilic (Yoghurt, Shrikhand, Skyr, Bulgarian buttermilk), and probiotic or therapeutic ("Bio"-fermented milks, Acidophilus milk) (Robinson, 2002; Walstra *et al.*, 2006; Tamime and Robinson, 2007; Oliveira, 2009; Oliveira, 2014). Each product has its particular structure that is related to sensory characteristics, texture and even functional properties.

The production of fermented milk products may vary slightly, but always includes the steps of preparation of milk base and standardization, homogenization, heat treatment of milk or milk-base, and lactic acid fermentation. During these steps, chemical changes and structure formation occur which could affect quality and nutritional properties of the products.

Many factors affect the structure of the gel as a composition of milk base, temperature and time of heat treatment, the type and amount of culture used, temperature and duration of fermentation, stirring the gel or not, and storage conditions until the end of shelf-life of the product. Many authors have studied microstructure of dairy products focusing the effects of composition and supplementation of milk base (Damin *et al.*, 2009; Marafon *et al.*, 2011; Morell *et al.*, 2015; Li and Nie, 2015), manufacture processing (Wium *et al.*, 2003; Penna *et al.*, 2007; Peng *et al.*, 2009; Ong *et al.*, 2011a,b; Devi *et al.*, 2013), and storage conditions until the end of shelf-life of the product (Oliveira *et al.*, 2002; Nguyen *et al.*, 2014).

The structure of fermented milks is complex, although it contains high moisture content (i.e. same amount as in the milk base), the gel formed behaves as solid-like fluid. This behavior is due to the aggregation of casein micelles in the milk and formation of a three-dimensional network, with the fat embedded in the matrix (Tamime *et al.*, 2007a,b).

Factors as enrichment of milk base in solids-not-fat, modifying fat content, and addition of other added ingredients (prebiotics, fruits pulp, sweetener) as well as processing conditions of the base (i.e. homogenization pressure or heat-treatment level) can affect the physical properties of the acidified gel of fermented milk such as: stability, viscosity, texture and microstructure (Sodini *et al.*, 2004; Sodini *et al.*, 2005; Tamime *et al.*, 2007a,b; Devi *et al.*, 2013).

14.3 Microstructure of Yogurt and Milk Drinks

14.3.1 Yogurt

Traditional yoghurt is a fermented product made from milk, and by using a mixed starter culture that consists of *Lactobacillus delbrueckii* subsp. *bulgaricus* and *Streptococcus thermophilus* (Tamime and Robinson, 2007; Oliveira, 2014). Initially, the manufacture of yoghurt involves mainly fortification of the milk base with dairy ingredients (i.e. concentrated milk or different powders) in order to increase the protein concentration to 40–50 g.kg^{-1} (Sodini *et al.*, 2005). Afterwards, the milk base is pre-heated to 60–70°C homogenized (17–20 MPa), heated (95°C for 5 min), cooled to fermentation temperature (42°C), and inoculated with starter cultures. In general, yogurt is presented to consumers in either a gel-like form (set yogurt) or as a viscous fluid (stirred yogurt), and the main difference between these two types of yoghurt is whether the gel is broken or not. The gel formation is one of the main properties during the manufacture of yoghurt (Lucey and Singh, 1997; Xu *et al.*, 2008). Stirred yoghurt could be added with fruit pulps and other ingredients.

Supplementation of milk employed in yogurt manufacture in order to increase the protein concentration affects its microstructure. It is well established that the fortification of the protein content in the milk-base also improves the texture (Séverin and Wenshui, 2005; Vasiljevic and Shah, 2008), and is related to stronger protein matrices which lead to stronger firmness and more compact microstructures (Marafon *et al.*, 2011).

Many articles showed that textural properties of the yogurt, are governed by its microstructure due to physico-chemical interactions between its different structural components, such as water, fat globules, protein resulting from milk base and formulation (Tamime, 2007; Damin *et al.*, 2009; Marafon *et al.*, 2011).

Tamime *et al.* (1984) reported that supplementing milk base of set-type yogurt with sodium caseinate caused a most open casein matrix, consisting of robust casein particle chains and larger clusters. Transmission electron microscopy of the same yogurt showed some fusion of the casein particles that were relatively "smooth" without any spikes or appendages.

Bhullar *et al.* (2002) studied the effect of supplementation with milk whey protein concentrate, whey powder and skimmed milk on the microstructure and texture of yogurt. Pores were easily observed in the samples with whey powder and skimmed milk powder, but were not visible when it was added to concentrated whey protein. They considered that this behavior was due to the heat treatment that formed a complex of whey protein with casein protein resulting in denser network.

Marafon *et al.* (2011) evaluated the quality of stirred type skim milk probiotic yogurt fortified by partially replacing skim milk powder (SMP) with whey protein concentrate (WPC) and sodium caseinate during cold storage for 28 days, and compared it with non-fortified yogurt. The observed differences in microstructure between the fortified yogurts compared to the non-fortified product were due to changes in chemical composition of the milk base, and to the colloidal characteristics of the product. Scanning electron microscopy of the yogurts during storage is shown in Figure 14.1, and the micrographs illustrate the differences in the microstructure of the products (i.e., compactness of the protein particles and pore sizes). The yogurts had different pore sizes that varied during the storage. Compared with those in the fortified yogurt, the pore sizes in the non-fortified yogurt in the beginning of storage (d 1) were larger and more evenly distributed, and the pore size of both yogurts became similar at the end of shelf-life (d 28). In general, the microstructure of all samples was similar, i.e. composed of casein micelle chains and clusters. However, the ways in which the casein micelles were linked to each other differed slightly. The resulting matrices appeared to be relatively more compact (denser) in the fortified yogurt which has more protein content than in the non-fortified yogurt. The more open matrices were due to larger pore spaces, and were evident in the microstructure of both yogurts at the end of the storage period.

Yogurt with a denser matrix should suffer less syneresis as observed by Trachoo and Mistry (1998). They observed a more open matrix in low fat yogurt than in those enriched with buttermilk ultra-filtrate powder, an effect attributed to the higher protein content of this product which also resulted in stronger texture. Smaller pores and more

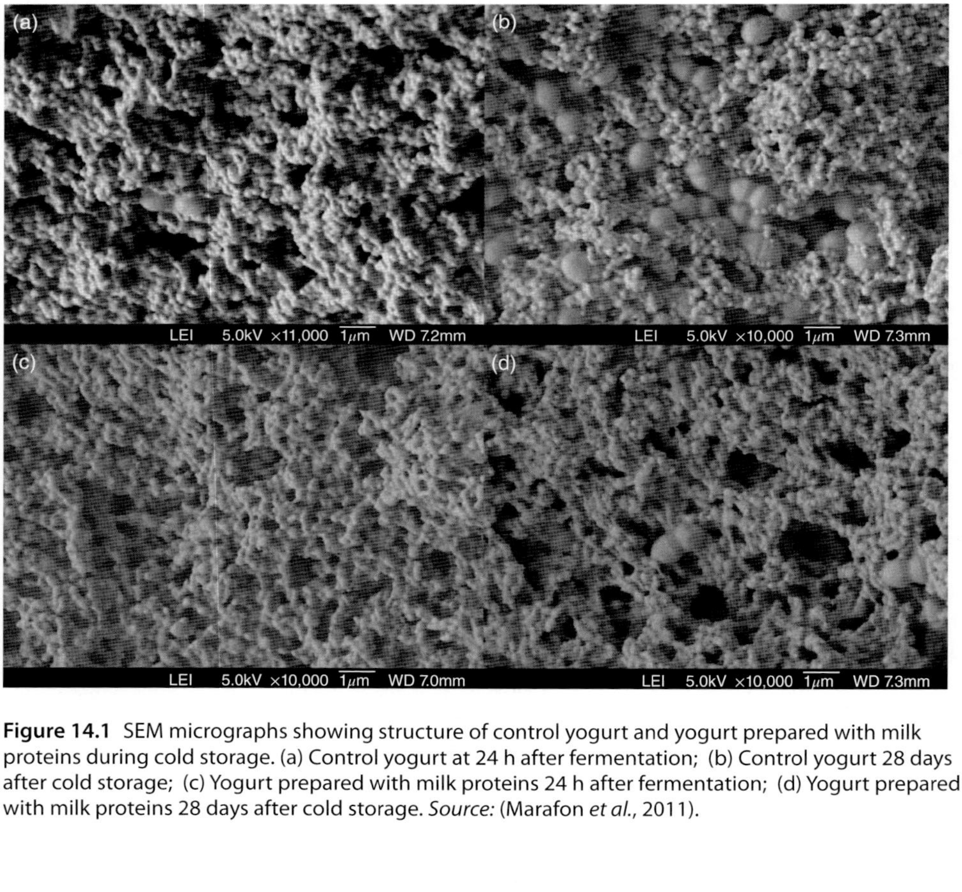

Figure 14.1 SEM micrographs showing structure of control yogurt and yogurt prepared with milk proteins during cold storage. (a) Control yogurt at 24 h after fermentation; (b) Control yogurt 28 days after cold storage; (c) Yogurt prepared with milk proteins 24 h after fermentation; (d) Yogurt prepared with milk proteins 28 days after cold storage. *Source:* (Marafon *et al.*, 2011).

compact structure were found in labneh with higher protein content when compared to products of lower values (Ozer *et al.*, 1999). These publications did not mentioned methods for micrographs analysis.

According to Lee and Lucey (2003), the presence of high numbers of pores indicates a weaker gel, suggesting that weak interactions were present between the casein particles. Sodini *et al.* (2005) studied the microstructure of yogurt and observed changes in the formation of the gel due to denaturation of proteins when the samples were enriched with milk proteins. The compactness of casein micelle matrix in liquid or yogurt has been studied by some researchers (Pang *et al.*, 2015).

Ramirez-Santiago *et al.* (2010) evaluated syneresis and microstructure properties of stirred yogurt prepared with milk base supplemented with yam soluble fiber extracted from *Pachyrhizus erosus* (1 g per 100 mL) comparing it with the product without fiber addition. The supplemented yogurt exhibited a more compact casein micelle aggregate network than that of the not-supplemented one that was more open, relaxed and covered with fibrous structures. Moreover, incorporation of the fiber significantly reduced the syneresis.

The effects of the addition of polysaccharides with a different ionic charge on rheology, microstructure, texture and water holding capacity of acid milk gels were studied and compared to that of gelatine addition. The gel microstructure was not changed greatly, but showed smaller particle size at a carrageenan concentration of 0.05% than the control sample (Pang *et al.*, 2015).

Morell *et al.* (2015) studied yogurts prepared by the addition of skimmed milk powder and whey protein, as well as a control without extra protein. Three more samples were prepared by adding 2% of a physically modified starch to each in order to increase the consistency and impart creaminess. Light microscopy images showed denser areas in samples containing extra milk protein than in control, and two protein networks could be distinguished in samples containing whey protein. In the samples with added starch, starch granules were embedded in the protein networks.

Damin *et al.* (2009) evaluated the simultaneous effect of milk supplementation with casein hydrolysate, whey protein concentrate and skim milk powder in the production of bio-yogurt. Milk was fermented by *Lactobacillus bulgaricus* (Lb), *Lactobacillus acidophilus* (La) and *Bifidobacterium lactis* (Bl) in co-cultures with *Streptococcus thermophiles* (St). Samples of StLa fermented milk without (control) and with protein supplementation (test) were analyzed by scanning electron microscopy to study the microstructure. Examination of the images shows dense structures with irregular aggregates which were interconnected with bacteria, and pore structures with different sizes. The augmented images, 5000×; 10,000× (Figures 14.2 and 14.3), were analyzed to determine the quantity of pores and their size to observe the differences between them. Using the Image Pro-Plus it was possible to analyze the diameter and amount of pores setting a diameter range. After many tests, the diameter range was chosen between 0 and 10 micrometers for analysis. The pore counts in the enriched milk micrographs showed the presence of more pores with diameter less than 1 microns for 5000 × and 10,000 × images. In enriched milk 54% and 67.2% of the pores were below 1 μm, while for control milk, 38.5% and 58.7% were in this range, for increases of 5000 × and 10,000 × respectively (Figure 14.4). The highest protein content of the enriched sample showed a greater number of pores of smaller diameter, which represents a denser matrix, while the control milk micrograph shows the highest percentage of pores of

Figure 14.2 SEM micrograph of control milk fermented by *S. thermophilus* and *L. acidophilus*. *Source:* (Damin *et al.*, 2009).

Figure 14.3 SEM microphotography of fermented milk enriched with casein hydrolysate, whey protein concentrate and skim milk powder and fermented by *S. thermophilus* and *L. acidophilus*. *Source:* (Damin *et al.*, 2009).

larger diameter. The presence of *S. thermophilus* linked to the protein matrix was observed in both milks, but it was not possible to identify *L. acidophilus* in these images.

The enrichment of probiotic yogurt matrix with passion fruit peel-powder was investigated initially by Espírito-Santo *et al.* (2013) and recently by Perina (2015). In the previous work, Espírito-Santo *et al.* (2013) showed throughout SEM micrographs that a normal structure of non-fat yogurt with a casein network surrounded by a high

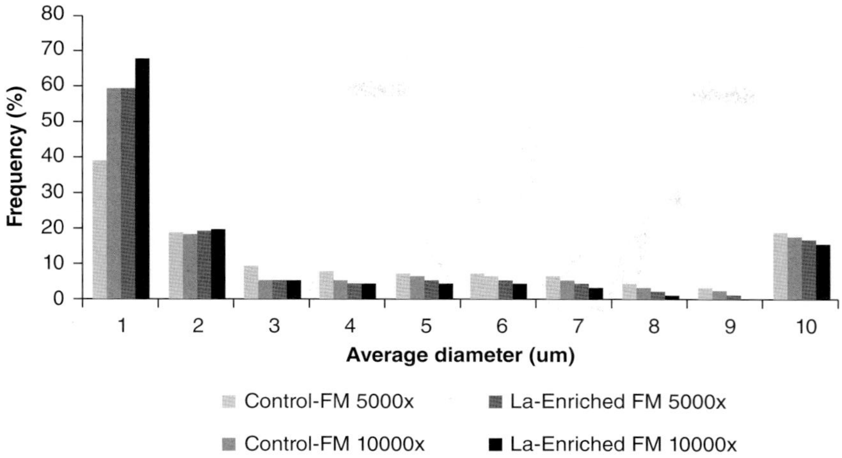

Figure 14.4 Distribution of average diameter pores in control fermented milk (Control-FM), and in enriched fermented milk (Enriched-FM) issued from photomicrography. *Source:* (Damin *et al.*, 2009).

number of pores of whey could be seen in Figure 14.5a. In yogurts enriched with passion fruit-fiber, the casein gel overlaid the fiber and sometimes was nestled in it (Figure 14.5c). In control yogurts, but not in enriched yogurts, exopolysaccharides formed several filaments and covered some areas of the protein gel (Figure 14.5b). Also this observation can be explained by the presence of pectin in fiber-enriched yogurts, which may have interacted with proteins and bacterial exopolysaccharides forming a more compact structure. The passion fruit fiber presented a needle-shaped blade and was clearly observed in the micrographs nesting the casein gel and, sometimes, the yogurt bacteria (Figure 14.5c).

Perina (2015) studied the simultaneous supplementation of yogurt with passion fruit powder and vegetal-oil emulsion was investigated, concerning viable cell-counts, physical properties (composition, fatty acids, profile and firmness) and microstructure after one day of cold storage at 4°C. In addition, scanning electron microscopy was used to evaluate the gel microstructure, and used to explain changes of the physical properties of yogurts, after the addition of fiber and substituting milk-fat. Figure 14.6 shows the microstructure of probiotic yogurts 24 h after fermentation. For Y1 and Y2, a normal structure of non-fat yogurt with a casein network, surrounded by a high number of pore-serum was seen compared to the other yogurts. According to Lee and Lucey (2003), the presence of high numbers of pores corresponded to a weaker gel, suggesting that there were weak interactions between the casein particles.

The structures of yogurt samples, enriched with passion fruit peel powder (Y3–Y6) were not much different from each other, except for the appearance of fat in the Y6 image. In fact, the process parameters used to homogenize cream milk and vegetal-oil emulsion have not broken or spread fat into particles/fragments. Therefore, fat recovered by casein network, as could be seen in Figure 14.6(d) and (f), mainly when using vegetal-oil emulsion (Y6). The findings of this study agree with Tamime (2007) who reported that the fortification method used to increase solids' non-fat, i.e., the addition

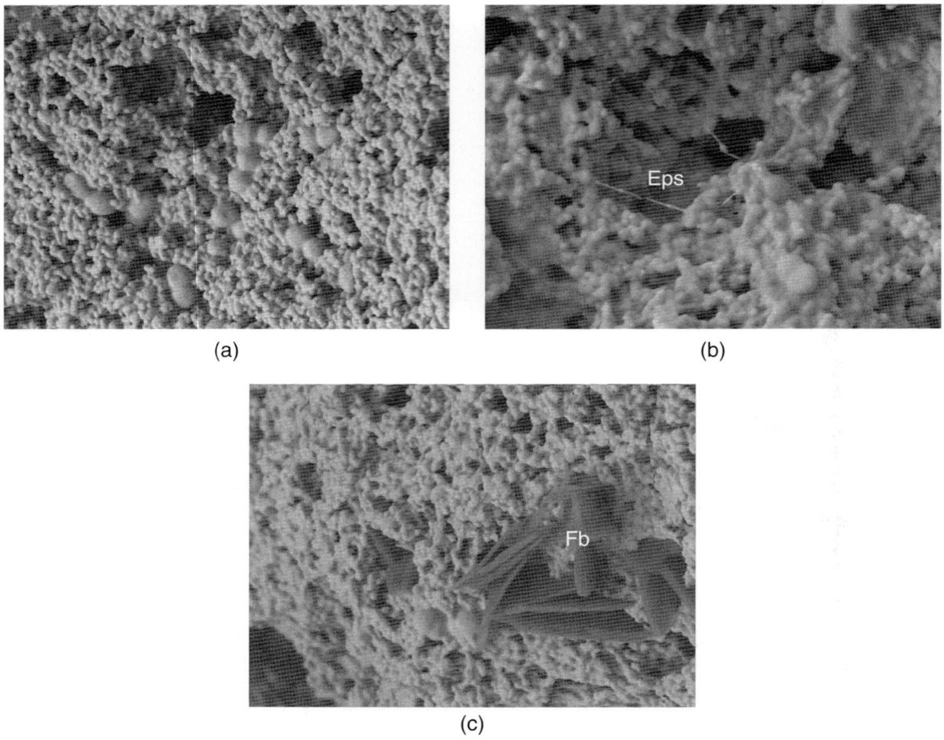

Figure 14.5 Microstructure of yogurts co-fermented by *Bifidobacterium lactis* subsp. *animalis* Bl04 at ×10,000 magnification. (a) passion fruit fiber yogurt; (b) passion fruit fiber yogurt shown with the highest contrast; (b) control yogurt; and (c) passion fruit fiber yogurt. Bar = 1 μm. Abbreviations: Eps = filaments of exopolysaccharides; Fb = passion fruit fiber. *Source:* Espírito-Santo *et al.* (2013).

of skimmed milk powder, or where protein concentrates can produce a yogurt with higher firmness, and a more regular, dense, protein-network microstructure. Finally, in Figure 14.7, it could be seen how the structures of passion fruit peel-powder interact with the casein networks matrices network.

The effect of milk processing on the microstructure of probiotic low-fat yogurt was studied by Penna *et al.* (2007). The microstructure of heat-treated milk yogurt had fewer interconnected chains of irregularly shaped casein micelles, forming a network that enclosed the void spaces. On the other hand, microstructure of high hydrostatic pressure yogurt had more interconnected clusters of densely aggregated protein of reduced particle size, with an appearance more spherical in shape, exhibiting a smoother more regular surface and presenting more uniform size distribution. The combined treatment of high hydrostatic pressure and heated milk led to compact yogurt gels with increasingly larger casein micelle clusters interspaced by pores, and exhibiting a high degree of cross-linking. The rounded micelles tended to fuse and form small irregular aggregates in association with clumps of dense amorphous material, which resulted in improved gel texture and viscosity.

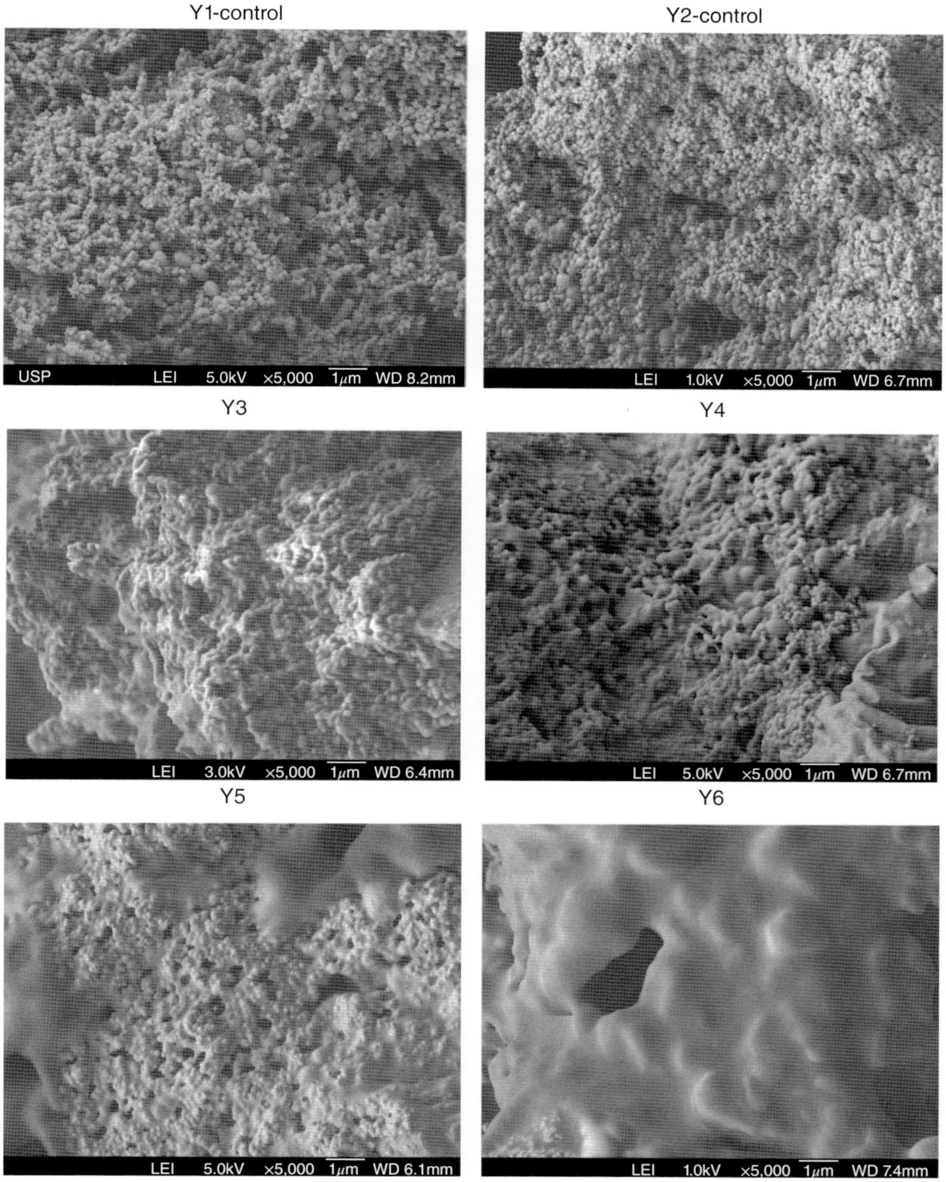

Figure 14.6 SEM of probiotic yogurt containing passion fruit fiber after 1 (d1) day of cold storage. Source: (Perina, 2015). Y1-control: skimmed milk, P1; Y2-control: skimmed milk, P2; Y3: fat from milk cream, P1; Y4: fat from milk cream, P2; Y5: fat from vegetal emulsion, P1; Y6: fat from vegetal emulsion, P2. P1: pre-heating to 50°C – 15 min/heating to 90°C – 5 min/cooled to 10°C; P2: pre-heating to 55°C/ homogenization at 15 MPa heating to 95°C – 5 min/ cooled to 10°C.

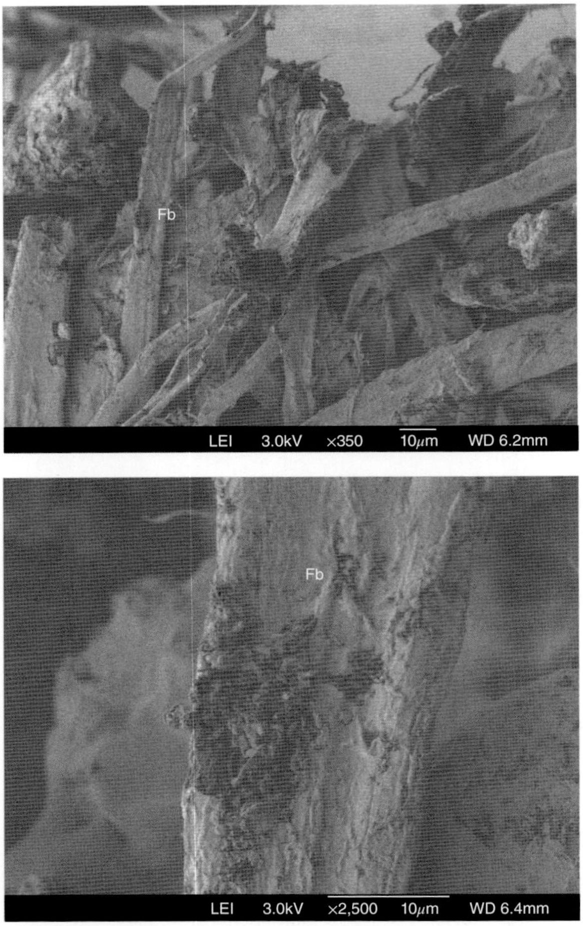

Figure 14.7 SEM of probiotic yogurt containing passion fruit peel powder fiber (Fb) showing interaction with casein network. *Source:* (Perina, 2015).

14.3.2 Milk Drinks or Lactic Beverages

Milk drinks or lactic beverages are part of the variety of the modern fermented dairy foods group. According to the Technical Regulation Identity and Quality of Milk Beverages in Brazil (ANVISA, 2005), milk drink is the resulting dairy product of milk mixture (fresh, pasteurized, sterilized, reconstituted concentrate, powdered, whole, semi-skimmed or partially skimmed and skim) and whey (liquid, concentrate and powder) or without the addition of product(s) or substance(s), food(s), vegetable fat, fermented milk(s), selected lactic ferments and other dairy products. Milk drinks could be fermented or not.

The process of manufacture of fermented milk drinks is based on the mixture of yogurt and whey, added or not with fruit pulp and other ingredients such as stabilizers, colors, flavors, and prebiotics, for example, occasioning a formulated dairy product. The former became very popular due the use of whey – a sub-product of the cheese industry, and its price, much cheaper than the yogurt itself. The addition of whey dilutes the yogurt affecting its structure and functionality.

Oliveira *et al.* (2002) studied the rheological characteristics, and microstructure in fermented milk drinks containing only yogurt bacteria (*Streptococcus thermophiles* and *Lactobacillus bulgaricus*), and in products containing yogurt and probiotic bacteria (*Streptococcus thermophilus* and *Lactobacillus acidophilus*, and *Streptococcus thermophilus* and *Lactobacillus rhamnosus*). Microscopy observations showed a dispersion of protein particles. Filamentous and spherical structures corresponding to guar and starch that were added as stabilizers are detected. Scanning electron microscopy (SEM) showed that lactic beverages consist of a protein matrix composed of a branched network of casein particles forming chains and clusters, immobilizing the continuous phase into pores of small sizes (Figure 14.8 a,b). These acid-induced microstructures have been observed many times for yogurt (Tamime *et al.*, 1984; Oliveira *et al.*, 2002) as well as the presence of chains that are common in yogurt made from heat treated milk at 85°C. Whey proteins are part of the matrix in yogurt prepared from heated milk. SEM showed that milk beverages containing probiotics had similar structures as compared with the ones containing yogurt bacteria including the porosity of protein matrix. At low magnification (Figure 14.8a), cavities were visible in all illustrations. A higher magnification (Figure 14.8b), revealed that the void spaces are characterized by interstitial spaces (pores), the dimensions of which depend on the protein content in that matrix, and the void spaces could be occupied with whey and/or gas bubbles (Tamime *et al.*, 1984).

Lactic acid and probiotic bacteria not only ferment milk, but their cells became part of the structural components of the lactic beverages (Figure 14.8c). It could be also observed that the micellar chains did not always form a continuous matrix. Casein

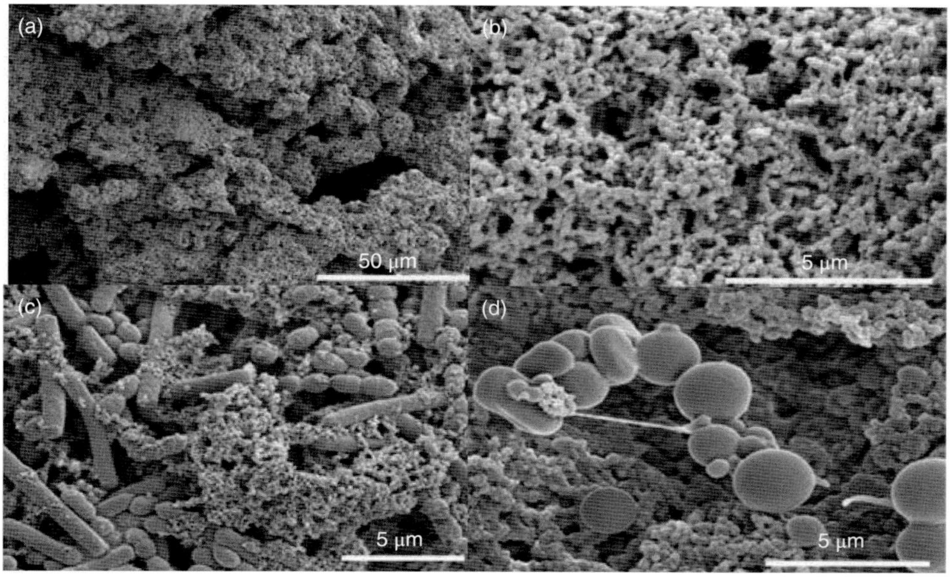

Figure 14.8 SEM micrographs of lactic beverages containing probiotics after 7 d of storage at 4°C. (a) general view of structure; (b) high magnification showing chains and clusters; (c) lactic acid bacteria being part of the structural components; and (d) presence of filamentous and spherical structures. *Source:* (Oliveira *et al.*, 2002).

micelles not linked to each other were deposed at the surface of the bacteria cells (Figure 14.8c). This is due to milk beverage manufacture, which included stirring the yogurt, dilution of the yogurt by adding liquid whey, and mixing. Thus, milk drink could be considered as a dispersion of protein particles, for stirred yogurt contrary of milk beverage that is a diluted dispersion due to the addition of whey (40.2% of fermented milk against 49.0% of whey). SEM revealed the presence of filamentous and spherical structures located either at the surface of casein particles or in the bulk (Figure 14.8d). In fact, such structures are due to the presence of guar gum and starch grains (modified starch of wheat) in the formulation of the lactic beverages.

Simões (2015) studied the microstructure of symbiotic yogurt and milk drinks added with red fruit pulps which SEM is presented in Figure 14.9. The images allow observing, comparing and evaluating the interaction of all components of the product´s matrix. Figures 14.9a to 14.9e show the structures of symbiotic yogurt whilst in Figures 14.9f to 14.9j the structures of symbiotic milk drinks are presented. When comparing photomicrographs Figures 14.9a and 14.9f corresponding respectively to symbiotic control yogurt and milk drink control both without addition of fruit pulp, a significant difference in porosity of the structure are observed. In the former, Figure 14.9a, a more compact structure is noted while in image 14.9f the presence of whey results in a less compact porosity. In addition to microbial cultures used for fermentation of milk base and is present in all the obtained images. Moreover, in photomicrographs from Figures 14.9g and Figure 14.9h corresponding to milk beverages with strawberry pulp and raspberry, respectively, while fruit structures are not detected, the production of extracellular structures of polysaccharide produced by the bacterium *Streptococcus thermophiles* are eminent. It is believed that the production of exopolysaccharides by cultures used in the production of fermented milks may directly affect the physical properties of the product, thus contributing to a higher consistency.

14.4 Microstructure of Cheeses

Man has used milk in the diet since the dawn of our civilization. The various types of fermented milk, resulting from natural fermentation which usually occurs in stored milk in environmental conditions, arose from perished milk. The syneresis of coagulated milk promoted by agitation or breaking the clot, gave rise to a protein curd mainly casein surrounded by fat, thus giving a fatty protein concentrate of high nutritional value called cheese. The low moisture content, acidity and relatively high salt usually added, made cheese one of the most traditional and most expressive forms of milk

Figure 14.9 SEM of yogurts and lactic beverages fermented by Streptococcus thermophilus (TA040), *Lactobacillus bulgaricus* (LB340) and *Lactobacillus acidophilus* (L104) 36 h after fermentation at 42°C until pH 4.7. Source: Simões (2015) (a) Control Symbiotic Yogurt: without fruit addition; (b) Symbiotic Yogurt 1 plus strawberry; (c) Symbiotic Yogurt 2 plus raspberry; (d) Symbiotic Yogurt plus "pitanga"; (e) Symbiotic Yogurt 4 plus same proportions of strawberry, raspberry and "pitanga" pulps; (f) Control Symbiotic Milk drink without fruit addition; (g) Symbiotic Milk drink 1 plus strawberry; (h) Symbiotic Milk drink 2 plus raspberry; (i) Symbiotic Milk drink plus "pitanga"; (j) Symbiotic Milk drink 4 plus Milk drink same proportions of strawberry, raspberry and "pitanga" pulps.

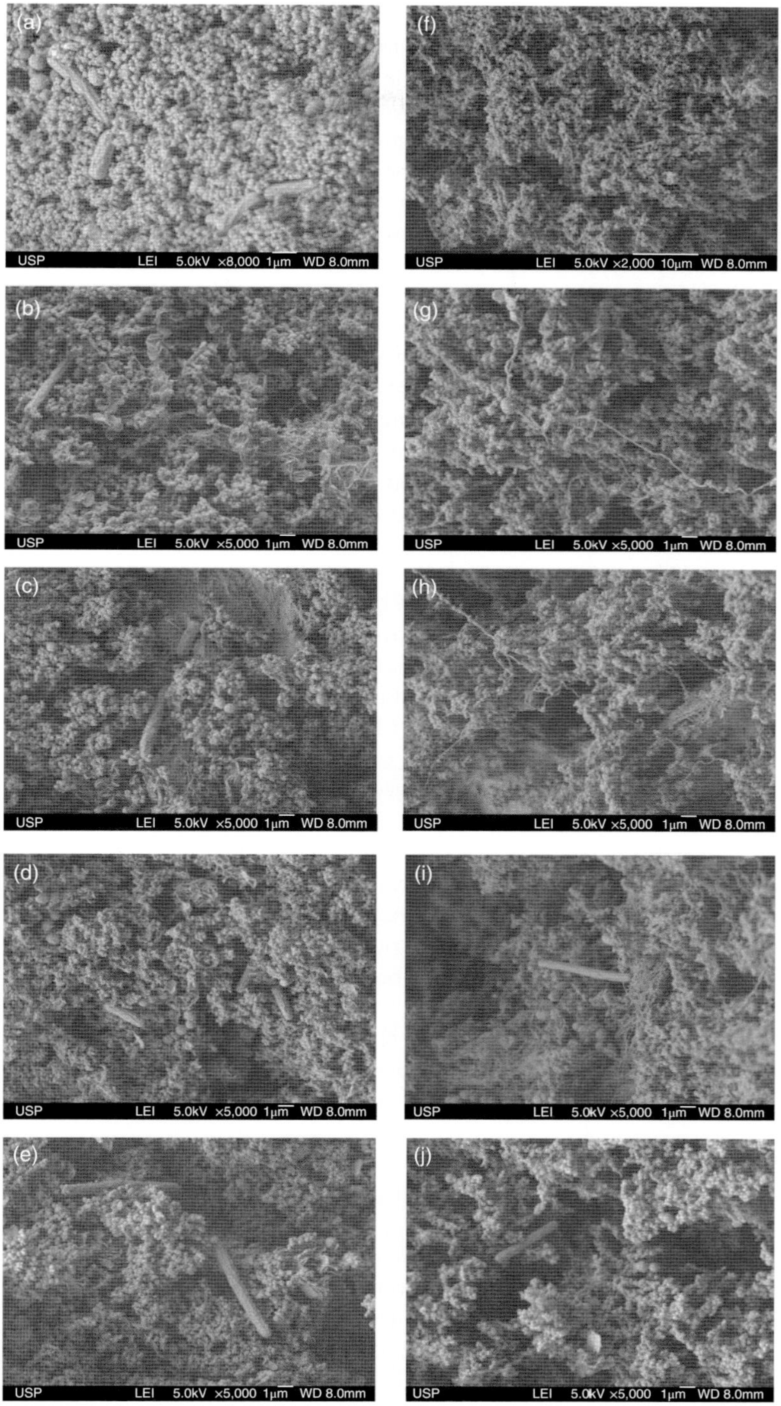
(a)
USP LEI 5.0kV ×8,000 1μm WD 8.0mm
(f)
USP LEI 5.0kV ×2,000 10μm WD 8.0mm
(b)
USP LEI 5.0kV ×5,000 1μm WD 8.0mm
(g)
USP LEI 5.0kV ×5,000 1μm WD 8.0mm
(c)
USP LEI 5.0kV ×5,000 1μm WD 8.0mm
(h)
USP LEI 5.0kV ×5,000 1μm WD 8.0mm
(d)
USP LEI 5.0kV ×5,000 1μm WD 8.0mm
(i)
USP LEI 5.0kV ×5,000 1μm WD 8.0mm
(e)
USP LEI 5.0kV ×5,000 1μm WD 8.0mm
(j)
USP LEI 5.0kV ×5,000 1μm WD 8.0mm

conservation (Fox, 1993; Hill and Prashanti, 2013). Additionally, cheeses are the most common of the dairy products, and the basic manufacture procedure is simple.

Cheese is a concentrated milk product and mainly consists of a casein-fat matrix being enclosed by it (Fox, 2001; Walstra *et al.*, 2006). The ability to convert the major components of milk into cheese, has many advantages as increased stability during storage, ease of transportation. It is also is a way to diversify the human diet.

The cheese-making technology is closely linked to tradition and to the type of product. Although specific variations in the manufacturing process of different kinds of cheeses occur, in general, the typical steps include the selection and pasteurization of milk, coagulation, cutting the curd, whey drainage or syneresis, molding, brining, maturation and storage.

Several studies have examined the microstructure of cheese (Everett and Olson, 2003; Ong *et al.*, 2010; 2011a,b; 2012; 2013). It's a consensus that the microstructure of cheese can determine its yield and consistency, influencing its functional properties such as texture, syneresis, shredability, melting, and stretching.

Protein and fat globules are the main components in influencing cheese structure formation. Their interactions during manufacture and shelf-life, define cheese structure.

The main step in cheese fabrication is coagulation in which proteases break down κ-casein present on the surfaces of casein micelles in milk. Deprived of its protective action, casein micelles coagulate and form a gel. When examined by electron microscopy, the coagulum consists of casein micelle clusters and short chains. They encapsulate fat globules – the natural large corpuscular particles present in milk. Void spaces in the casein matrix are filled with the liquid milk serum called whey which is a solution of lactose, minerals, and vitamins, and a suspension of whey proteins (Tamime, 2007).

Hickey *et al.* (2015) reviewed the relationship between cheese manufacture process parameters and protein standardization techniques on the resultant cheese microstructure, microbial activity within the cheese matrix and the interactions between the cheese matrix and the entrapped bacteria during ripening. In particular, the effects of processing parameters, both prior to and during manufacture, on the milk fat globule membrane structure and the phospholipids which occupy it merit investigation with a reference to achieving increased cheese quality at both localized level and throughout cheese blocks.

Changes in protein network structures affect cheese texture (Ong *et al.*, 2013). However, according to (Soodam *et al.*, 2014), milk protein concentration has little effect on the cheese microstructure and texture especially when an adequate ripening period is provided.

A compact cheese structure resulted from dialysis of rennet gels could be achieved according to Koutina *et al.* (2015). As demineralization decreased, many aggregates are formed, and small protein particles are able to increase the potential connection points among proteins, support particle fusion and producing a compact structure.

During cheese manufacture fat globules remain with the proteins in the curd. Simultaneously, a small number of fat globules that were exposed during curd cutting are washed away with the whey. In cheese in which curd was heat treated, microstructure techniques show how fat may be lost during processing.

In processed cheese analogues, milk fat can be partially replaced with inulin to improve the functional properties of the final product (Solowiej *et al.*, 2015). Confocal microscopy revealed that the highest hardness values could be attributed to more intense distribution of fat droplets in the sample with the highest fat content. Moreover, addition of inulin to different kinds of cheese can be beneficial in the manufacture of a low-fat, texturized, symbiotic product (Karimi *et al.*, 2015). The published information in this review leads some directions on uses of inulin in different types of cheese and its effect on texture and rheological properties. Otherwise, further studies should be conducted to elucidate the mechanisms of interaction between casein and inulin in forming cheese's structure.

14.5 Conclusion

This chapter has documented the current state of the art in relation to dairy microstructure, focusing on fermented milk, milk drinks and cheeses.

Many publications have shown a great interest in the microstructure of dairy products, which has a vital impact on texture, sensory attributes, shelf life, and functionality of dairy foods. One of the most important publications relating to dairy product microstructure is Tamime (2007), and is essential for understanding a basic knowledge about these products.

New techniques have been developed, and significant knowledge is available. However, further studies should be conducted concerning mainly the protein network formation, milk fat globule and bacterial interactions, and their role in dairy product manufacture. Finally, the appraisal of microstructure of dairy products is indispensable for knowledge about the products' functional properties.

References

Aguilera, J., Stanley, D., and Baker, K. (2000). New dimensions in microstructure of food products. [Review]. *Trends in Food Science and Technology, 11*(1), 3–9. doi: 10.1016/S0924-2244(00)00034-0.

ANVISA. Ministério da Agricultura e do Abastecimento, Secretaria de Defesa Agropecuária. Departamento de Inspeção de Produtos de Origem Animal. Regulamento Técnico de Identidade e Qualidade de Bebidas Lácteas. Aprovado pela Instrução Normativa n° de Brasília, 2005.

Bhullar, Y.S., Uddin, M.A., and Shah, N.P. (2002). Effects of ingredients supplementation on textural characteristics and microstructure of yoghurt. *Milchwissenschaft, 57*, 328–332.

Damin, M.R., Alcantara, M.R., Nunes, A.P., and Oliveira, M.N. (2009). Effects of milk supplementation with skim milk powder, whey protein concentrate and sodium caseinate on acidification kinetics, rheological properties and structure of nonfat stirred yogurt. *Lwt-Food Science and Technology, 42*(10), 1744–1750. doi: 10.1016/j.lwt.2009.03.019.

Devi, A., Buckow, R., Hemar, Y., and Kasapis, S. (2013). Structuring dairy systems through high pressure processing. *Journal of Food Engineering, 114*, 106–122.

Dewettinck, K. (2011). IDF Symposium on Microstructure of Dairy Products, 9–11 June 2010, Tromso, Norway. [Editorial Material]. *International Dairy Journal, 21*(9), 663-663. doi: 10.1016/j.idairyj.2011.06.001.

Ercili-Cura, D., Lille, M., Legland, D., Gaucel, S., Poutanen, K., Partanen, R., and Lantto, R. (2013). Structural mechanisms leading to improved water retention in acid milk gels by use of transglutaminase. [Article]. *Food Hydrocolloids, 30*(1), 419–427. doi: 10.1016/j.foodhyd.2012.07.008.

Espírito-Santo, A.P., Lagazzo, A., Sousa, A., Perego, P., Converti, A., and Oliveira, M. (2013). Rheology, spontaneous whey separation, microstructure and sensorial characteristics of probiotic yoghurts enriched with passion fruit fiber. [Article]. *Food Research International, 50*(1), 224–231. doi: 10.1016/j.foodres.2012.09.012.

Everett, D., and Olson, N. (2003). Free oil and rheology of cheddar cheese containing fat globules stabilized with different proteins. [Article]. *Journal of Dairy Science, 86*(3), 755–763. doi: 10.3168/jds.S0022-0302(03)73656-X.

Fallico, V., Tuminello, L., Pediliggieri, C., Horne, J., Carpino, S., and Licitra, G. (2006). Proteolysis and microstructure of Piacentinu Ennese cheese made using different farm technologies. [Article]. *Journal of Dairy Science, 89*(1), 37–48.

FAO/WHO. (2010). *CODEX Standard for fermented milks.*

Fox, P.F. (1993). Cheese: an overview. In Fox, P. (Ed.) *Cheese: chemistry, physics, and microbiology. General aspects.* (Vol. v.1, pp. 1–36). London: Chapman and Hall.

Fox, P.F. (2001). Milk proteins as food ingredients. *International Journal of Dairy Technology, 54,* 41–55.

Geng, X., van den Berg, F., Bager, A., and Ipsen, R. (2011). Dynamic visualization and microstructure of syneresis of cheese curd during mechanical treatment. [Article]. *International Dairy Journal, 21*(9), 711–717. doi: 10.1016/j.idairyj.2011.01.013.

Hickey, C., Auty, M., Wilkinson, M., and Sheehan, J. (2015). The influence of cheese manufacture parameters on cheese microstructure, microbial localisation and their interactions during ripening: A review. [Review]. *Trends in Food Science and Technology, 41*(2), 135–148. doi: 10.1016/j.tifs.2014.10.006.

Hill, A., and Prashanti, K. (2013). Dairy products: cheese and yogurt. In A.M. Eskin and F. Shahidi (Eds.) *Biochemistry of foods,* 3rd edn. London: Elsevier, pp. 319–362.

Karimi, R., Azizi, M., Ghasemlou, M., and Vaziri, M. (2015). Application of inulin in cheese as prebiotic, fat replacer and texturizer: A review. [Review]. *Carbohydrate Polymers, 119,* 85–100. doi: 10.1016/j.carbpol.2014.11.029.

Koutina, G., Knudsen, J., Andersen, U., and Skibsted, L. (2015). Influence of colloidal calcium phosphate level on the microstructure and rheologicalproperties of rennet-induced skim milk gels. *LWT – Food Science and Technology.* doi: 10.1016/j.lwt.2015.03.035.

Le Feunteun, S., and Mariette, F. (2007). Impact of casein gel microstructure on self-diffusion coefficient of molecular probes measured by H-1 PFG-NMR. [Article]. *Journal of Agricultural and Food Chemistry, 55*(26), 10764–10772. doi: 10.1021/jf071982v.

Li, J.-M., and Nie, S.-P. (2016). The functional and nutritional aspects of hydrocolloids in foods. *Food Hydrocolloids, 53,* 46–61.

Lucey, J.A., and Singh, H. (1997). Formation and physical properties of acid milk gels: a review. *Food Research International, 30*(7), 529–542.

Luo, J., Wang, Z., Wang, F., Zhang, H., Lu, J., Guo, H., and Ren, F. (2014). Cryo-SEM images of native milk fat globule indicate small casein micelles are constituents of the membrane. [Article]. *RSC Advances, 4*(90), 48963–48966. doi: 10.1039/c4ra06171c.

Marafon, A., Sumi, A., Granato, D., Alcantara, M., Tamime, A., and de Oliveira, M. (2011). Effects of partially replacing skimmed milk powder with dairy ingredients on rheology, sensory profiling, and microstructure of probiotic stirred-type yogurt during cold storage. [Article]. *Journal of Dairy Science, 94*(11), 5330–5340. doi: 10.3168/jds.2011-4366.

Morand, M., Dekkari, A., Guyomarc'h, F., and Famelart, M. (2012). Increasing the hydrophobicity of the heat-induced whey protein complexes improves the acid gelation of skim milk. [Article]. *International Dairy Journal, 25*(2), 103–111. doi: 10.1016/j.idairyj.2012.03.002.

Morell, P., Ramirez-Lopez, C., Velez-Ruiz, J., and Fiszman, S. (2015). Relating HPMC concentration to elicited expected satiation in milk-based desserts. [Article]. *Food Hydrocolloids, 45*, 158–167. doi: 10.1016/j.foodhyd.2014.11.011.

Morris, V., and Groves, K. (2013). *Food Microstructures:Measurement and Modelling.* Cambridge: Woodhead Publishing.

Nguyen, H., Ong, L., Lefevre, C., Kentish, S., and Gras, S. (2014). The microstructure and physicochemical properties of probiotic buffalo yoghurt during fermentation and storage: a comparison with bovine yoghurt. [Article]. *Food and Bioprocess Technology, 7*(4), 937–953. doi: 10.1007/s11947-013-1082-z.

Oliveira, M.N. (2009). *Tecnologia de Produtos Lácteos Funcionais.* São Paulo: Atheneu.

Oliveira, M.N. (2014). Fermented milks/Yoghurt. In: *Encyclopedia of Food Microbiology.* Elsevier: Oxford.

Oliveira, M.N., Sodini, I., Remeuf, F., Tissier, J.P., and Corrieu, G. (2002). Manufacture of fermented lactic beverages containing probiotic cultures. *Journal of Food Science, 67*(6), 2336–2341.

Ong, L., Dagastine, R., Kentish, S., and Gras, S. (2010). The effect of milk processing on the microstructure of the milk fat globule and rennet induced gel observed using confocal laser scanning microscopy. [Article]. *Journal of Food Science, 75*(3), E135–E145. doi: 10.1111/j.1750-3841.2010.01517.x.

Ong, L., Dagastine, R., Auty, M., Kentish, S., and Gras, S. (2011a). Coagulation temperature affects the microstructure and composition of full fat Cheddar cheese. [Article]. *Dairy Science and Technology, 91*(6), 739–758. doi: 10.1007/s13594-011-0033-6.

Ong, L., Dagastine, R., Kentish, S., and Gras, S. (2011b). Microstructure of milk gel and cheese curd observed using cryo scanning electron microscopy and confocal microscopy. [Article]. *Lwt-Food Science and Technology, 44*(5), 1291–1302. doi: 10.1016/j.lwt.2010.12.026.

Ong, L., Dagastine, R., Kentish, S., and Gras, S. (2012). The effect of pH at renneting on the microstructure, composition and texture of Cheddar cheese. [Article]. *Food Research International, 48*(1), 119–130. doi: 10.1016/j.foodres.2012.02.020.

Ong, L., Dagastine, R., Kentish, S., and Gras, S. (2013). The effect of calcium chloride addition on the microstructure and composition of Cheddar cheese. [Article|Proceedings Paper]. *International Dairy Journal, 33*(2), 135–141. doi: 10.1016/j.idairyj.2013.03.002.

Peng, Y., Horne, D., and Lucey, J.A. (2009). Impact of preacidification of milk and fermentation time on the properties of yogurt. *Journal of Dairy Science*, 2977–2990. doi: DOI 10.3168/jds.2008-1221.

Penna, A., Subbarao-Gurram, and Barbosa-Canovas, G. (2007). High hydrostatic pressure processing on microstructure of probiotic low-fat yogurt. [Article]. *Food Research International, 40*(4), 510–519. doi: 10.1016/j.foodres.2007.01.001.

Robinson, R.K. (2002). *Dairy Microbiology Handbook* (3rd edn). New York: Wiley-Interscience.

Rovira, S., Lopez, M., Ferrandini, E., and Laencina, J. (2011). Hot topic: Microstructure quantification by scanning electron microscopy and image analysis of goat cheese curd. [Article]. *Journal of Dairy Science, 94*(3), 1091–1097. doi: 10.3168/jds.2010-3996.

Silva, J., Legland, D., Cauty, C., Kolotuev, I., and Floury, J. (2015). Characterization of the microstructure of dairy systems using automated image analysis. [Article]. *Food Hydrocolloids, 44,* 360–371. doi: 10.1016/j.foodhyd.2014.09.028.

Simões, T. (2015). *Behavior of elagic acid in symbiotic milk drinks supplemented with L-tryptophan and red fruit pulp.* Master, São Paulo University.

Sodini, I., Montella, J., and Tong, P.S. (2005). Physical properties of yogurt fortified with various commercial whey protein concentrates. *Journal of the Science of Food and Agriculture, 85,* 853–859.

Sodini, I., Remeuf, F., Haddad, S., and Corrieu, G. (2004). The relative effect of milk base, starter, and process on yogurt texture: a review. *Critical Reviews in Food Science and Nutrition, 44,* 113–137.

Solowiej, B., Glibowski, P., Muszynski, S., Wydrych, J., Gawron, A., and Jelinski, T. (2015). The effect of fat replacement by inulin on the physicochemical properties and microstructure of acid casein processed cheese analogues with added whey protein polymers. [Article]. *Food Hydrocolloids, 44,* 1–11. doi: 10.1016/j.foodhyd.2014.08.022.

Soodam, K., Ong, L., Powell, I., Kentish, S., and Gras, S. (2014). The effect of milk protein concentration on the microstructure and textural properties of full fat cheddar cheese during ripening. [Article]. *Food and Bioprocess Technology, 7*(10), 2912–2922. doi: 10.1007/s11947-014-1342-6.

Séverin, S., and Wenshui, X. (2005). Milk biologically active components as nutraceuticals: review. *Critical Reviews in Food Science and Nutrition, 45,* 645–656.

Tamime, A.Y. (2007). *Structure of Dairy Products.* West Sussex: Wiley-Blackwell.

Tamime, A.Y., Kalab, M., and Davies, G. (1984). Microstructure of set-style yoghurt manufactured from cow's milk fortified by various methods. *Food Microstructure, 3,* 83–92.

Tamime, A.Y., Hassan, A., Farnworth, E., and Toba, T. (2007a). Structure of fermented milks. In A. Tamime (Ed.), *Structure of dairy products* (pp. 134–169). Oxford, UK, Iowa, USA. Victoria, AUS: Blackwell Publishing Ltd.

Tamime, A.Y., Hasan, A., Farnworth, E.R., and Toba, T. (2007b). *Structure of fermented milks* (Vol. Chapter 6): Wiley-Blackwell.

Tamime, A.Y., and Robinson, R.K. (2007). *Yoghurt: Science and Technology* (3rd edn). Boca Raton: CRC.

Vasiljevic, T., and Shah, N.P. (2008). Probiotics-From Metchnikoff to bioactives. *International Dairy Journal, 18,* 714–728.

Walstra, P., Wouters, J.T.M., and Geurts, T.J. (2006). *Dairy Science and Technology* (2nd edn).

Wium, H., Pedersen, P., and Qvist, K. (2003). Effect of coagulation conditions on the microstructure and the large deformation properties of fat-free Feta cheese made from ultrafiltered milk. [Article]. *Food Hydrocolloids, 17*(3), 287–296. doi: 10.1016/S0268-005X(02)00079-6

Xu, Z.M., Emmanouelidou, D.G., Raphaelides, S.N., and Antoniou, K.D. (2008). Effects of heating temperature and fat content on the structure development of set yogurt. *Journal of Food Engineering, 85,* 590–597.

Index

Microstructure of Dairy Products, First Edition.
Edited by Mamdouh Mahmoud Abdel-Rahman El-Bakry, Antoni Sanchez, and Bhavbhuti M. Mehta.

n

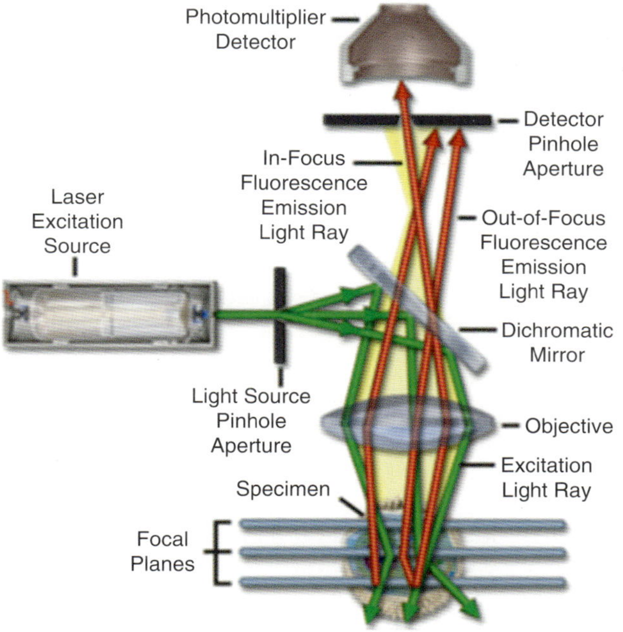

Figure 1.3 Schematic of confocal scanning laser microscope.

Microstructure of Dairy Products, First Edition.
Edited by Mamdouh Mahmoud Abdel-Rahman El-Bakry, Antoni Sanchez, and Bhavbhuti M. Mehta.

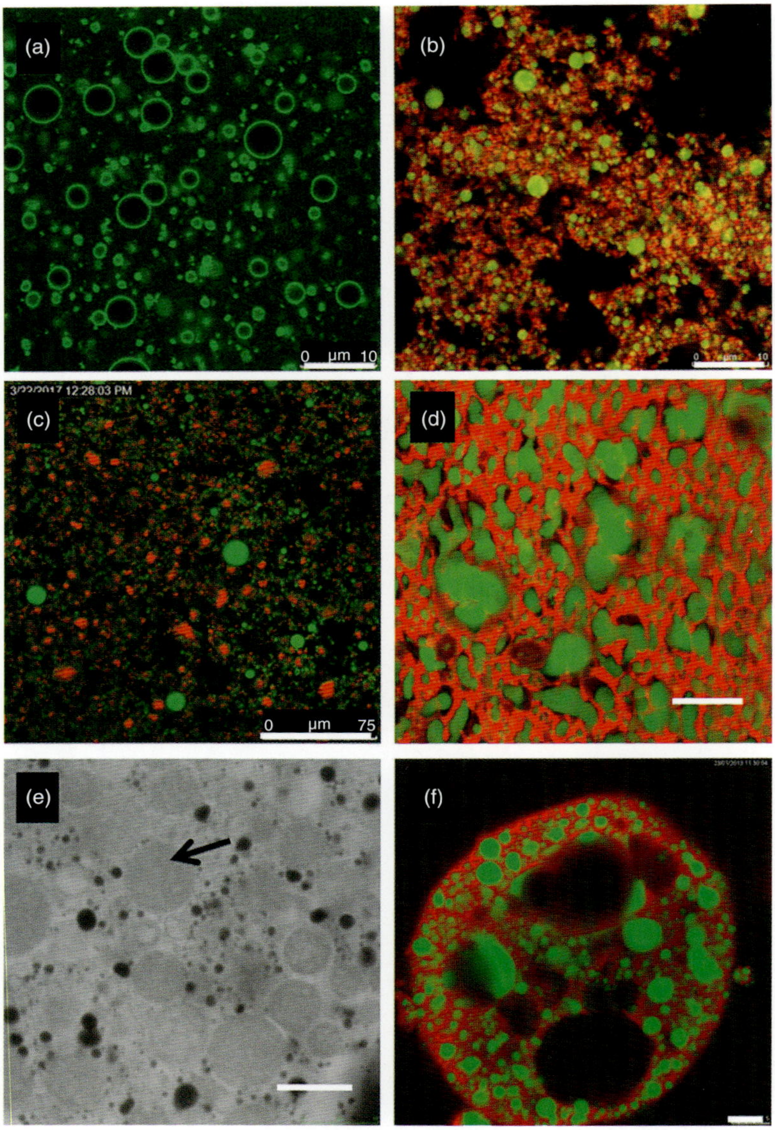

Figure 1.4 Confocal scanning laser micrographs of various dairy products labeled with fluorochromes. b, c, d and f are dual labeled with Nile Red and Fast Green FCF to show fat (green) and protein (red), respectively. (a) Whey protein-stabilized emulsion labeled with Rhodamine B (pseudo-colored green) to visualize protein at the droplet interface, scale bar = 10 µm. (b) Full-fat dairy yoghurt showing protein network and small fat droplets in addition to pores (dark regions), scale bar = 10 µm. (c) Mayonnaise showing fat droplets and discrete aggregated protein particles, scale bar = 75 µm. (d) Cheddar cheese showing irregular shaped fat pools and continuous protein matrix, scale bar = 25 µm. (e) Butter labeled with Nile Red (greyscale image) showing characteristic rounded butterfat crystals (dark grey, arrowed), water droplets (dark circles) and continuous fat phase (bright), scale bar = 5 µm. (f) Fat-filled whey protein enriched dairy powder showing fat droplets entrapped in continuous protein matrix, scale bar = 5 µm.

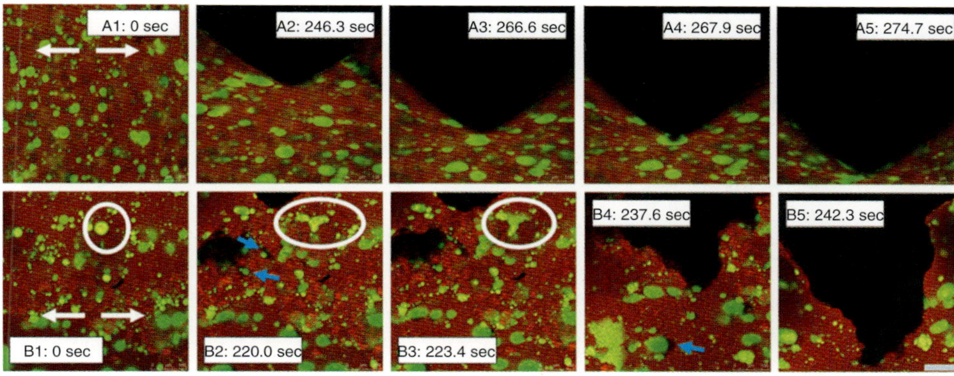

Figure 1.6 Confocal scanning laser micrographs of fat-filled whey protein gels monitored using a microtensile stage. Protein is pseudocolored red, fat pseudocolored yellow/green. A1–A5 gelled at pH 7.0 and is fine stranded, B1–B5 is gelled at pH 5.4 and is particulate. Notch propagation is vertical and fracture properties measured in extension. Note: deformation and movement of fat in the particulate gel (circle) while the fine stranded gel fractures through fat droplets. Adapted from Abhyankar (PhD thesis). Scale bar is 25 μm.

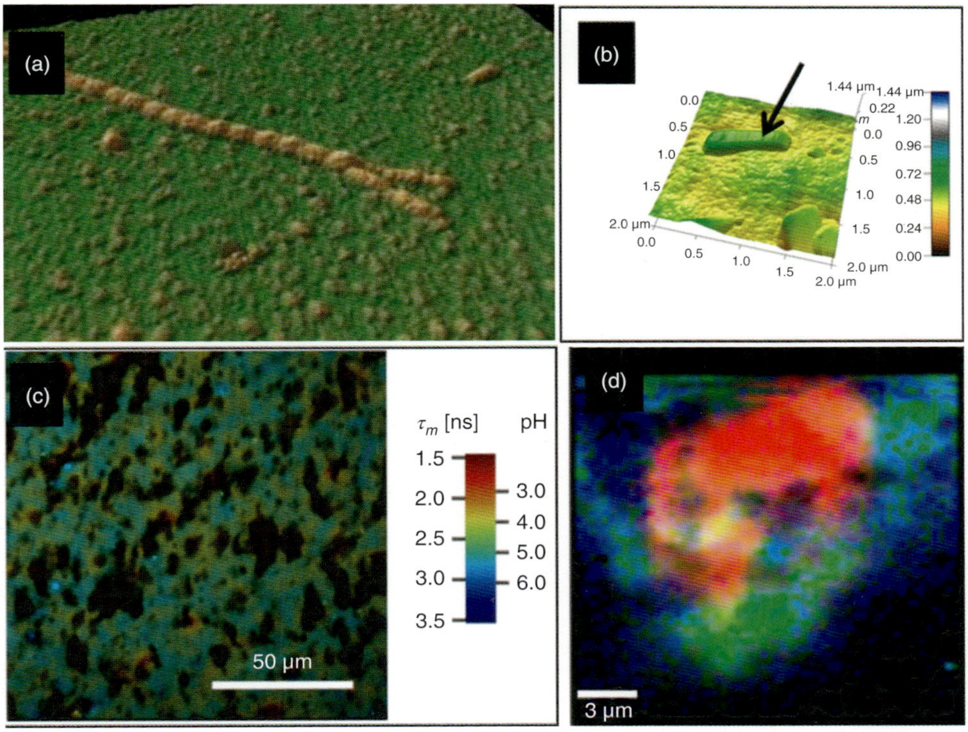

Figure 1.10 Some examples of advanced/emerging imaging techniques. (a) Atomic force 3D micrograph (height image) of β-lactoglobulin fibrils imaged in water; fibrils are ~3 nm thick. (b) Atomic force microscope image of surface of spray dried skim milk powder stored at 55% relative humidity, showing lactose crystallizing at the surface (arrow). (c) Fluorescence lifetime imaging (FLIM) of Cheddar cheese labeled with Oregon Green 488 nm, showing heterogeneous pH distribution (adapted from Burdikova *et al.*, 2015b). (d) Confocal Raman microscope z-depth image of Cheddar cheese from a discolored region showing presence of lycopene (red).

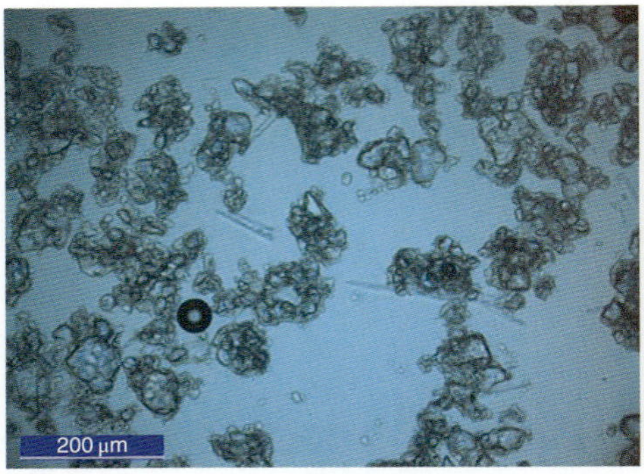

(a)

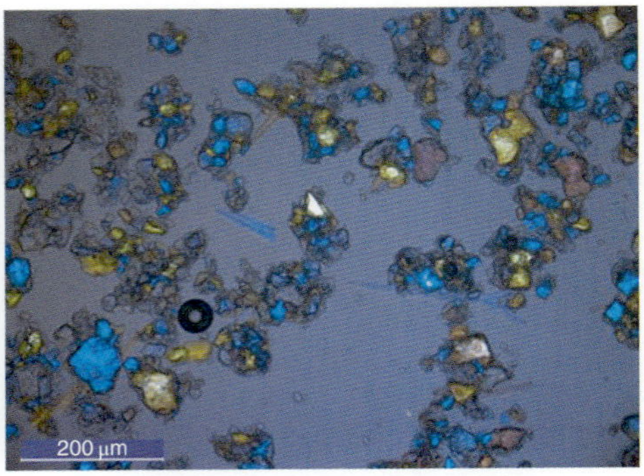

(b)

Figure 2.3 Images of ice crystals isolated from ice cream. The image depicted in (a) shows the presence of disintegrated ice cream consisting of clusters of various size. The polarized light image (b) identifies the presence, size and shape of the ice crystals.

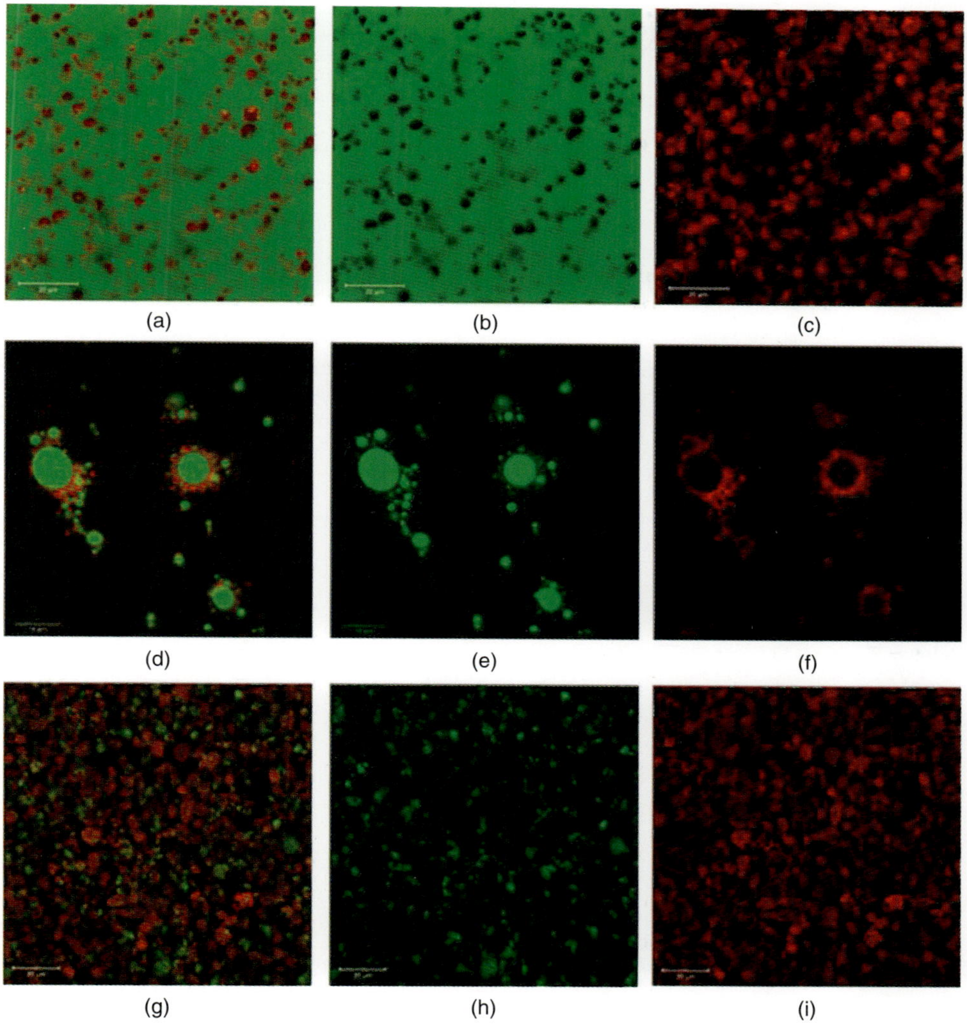

Figure 2.4 CSLM images of different emulsions stained with old Nile Blue, co-localizing protein and fat contents by assigning the protein contents of the emulsion to a red channel and the oil phase to a green channel. The images depicted in (a–c) show a liquid margarine as an example of a water in oil emulsion whereas the images shown in (d–f) show an ice cream emulsion at high magnification, as an example of an oil in water emulsion clearly showing the stabilization of the fat droplets by protein material. The fresh cheese sample as an example of a more complex system, depicted in the images (g–i) shows the co-localization, the homogenization and the distribution of the two phases.

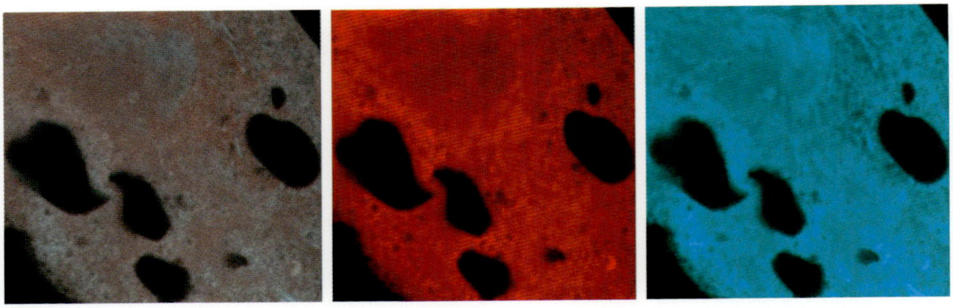

(a) (b) (c)

Figure 2.5 Fluorescence imaging of a frozen aerated dessert substituted at low temperature in the presence of fluorescent stains and imaged as a resin embedded block (a). The protein phase was stained using Rhodamine and assigned to a red color (b); Calcofluor White was used to stain carbohydrate stabilizers, depicted in cyan (c). The fat phase was extracted during the freeze substitution procedure; air and ice show as voids.

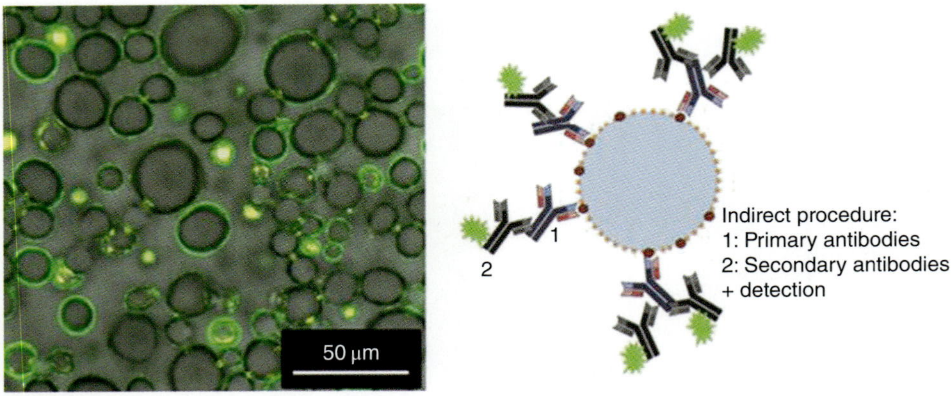

(a) (b)

Figure 2.6 Immunofluorescent staining of molecules present at the air–water interface of a foam using a combination of fluorescent imaging and transmitted light (a) and (b) a schematic representation of a two-step universal immunolabeling in which a highly specific primary antibody binds to its targeted epitope, itself targeted by a secondary antibody-marker complex that is visualized by the imaging system. Similar mechanisms apply for lectin and CBM labeling procedures.

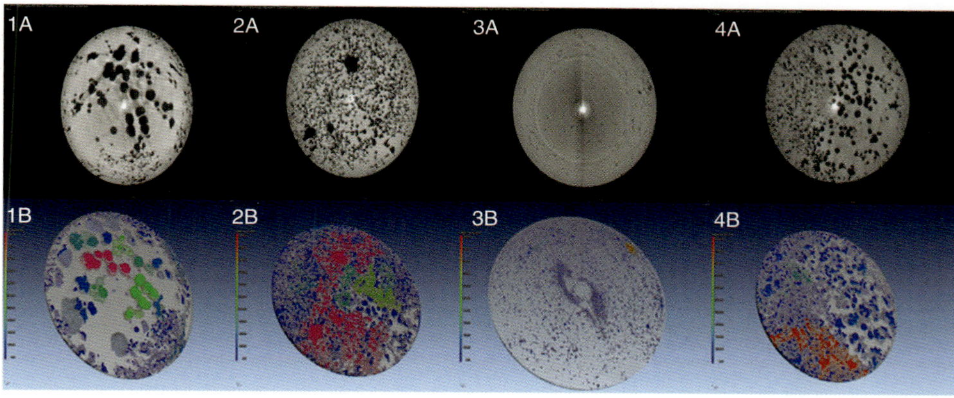

Figure 4.2 Eye formation in Swiss-type cheeses as determined by X-ray computed tomography. Colors in the blue spectrum represent voids of 0 to 6000 mm^3, those in green represent 9000 to 21,000 mm^3 and those in red represent 24,000 to 30,000 mm^3 (O'Sullivan *et al.*, 2016).

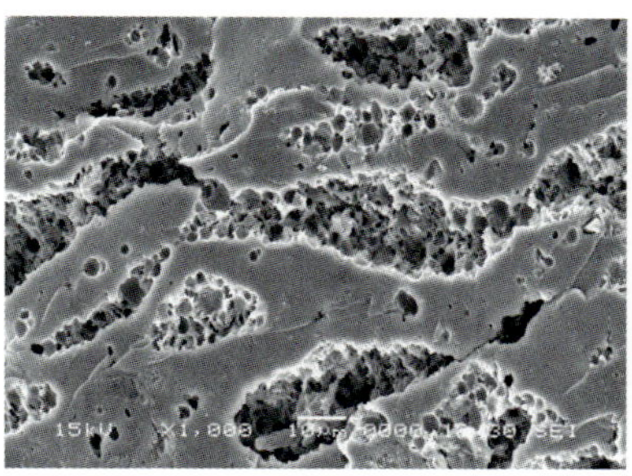

(a) SEM micrograph.

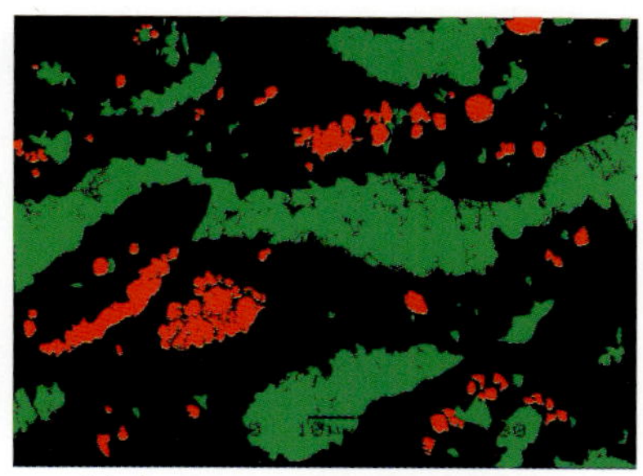

(b) Segmented image with three classes (two plus background).

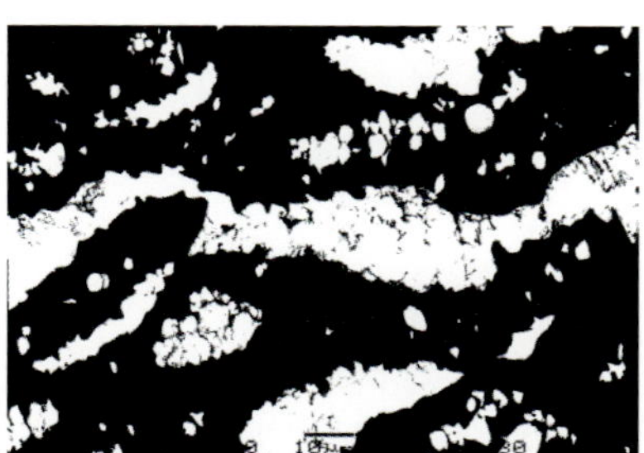

(c) Binarization (two classes: foreground and background).

Figure 5.1 An SEM micrograph of a cheese sample. Examples of segmentation are shown with three and two classes (binarized).

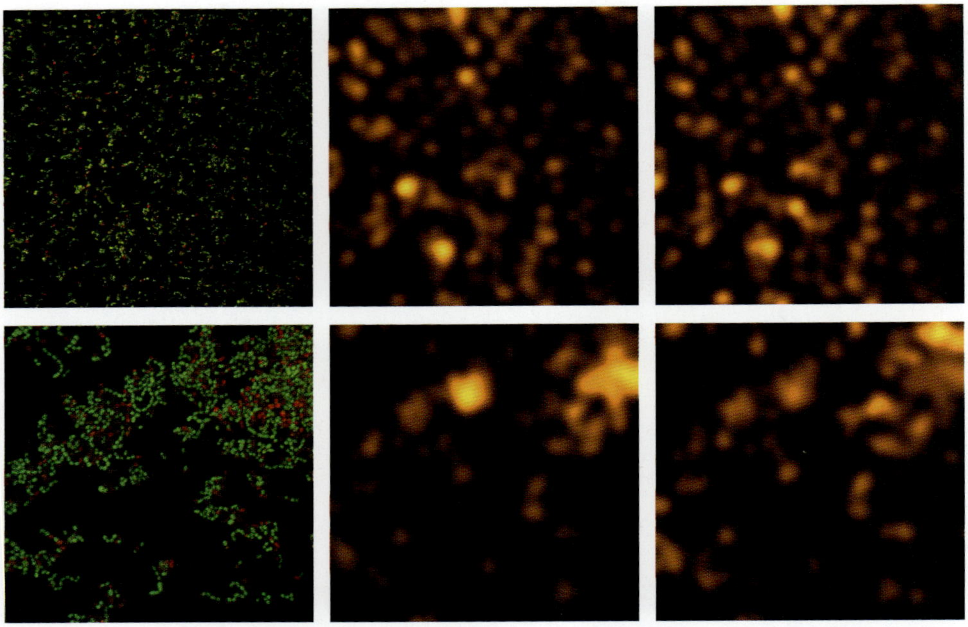

Figure 5.2 Confocal laser scanning micrographs of live (green) and dead (red) cells. Visual representations of two different measures for density are shown in the middle and right columns.

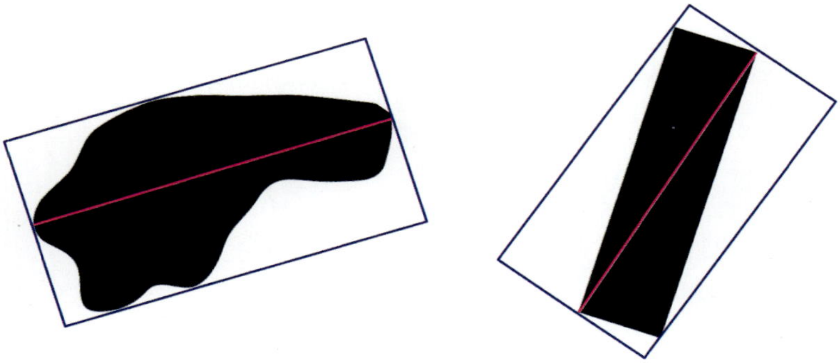

Figure 5.3 Two image features with *directionality* traced as the lines linking the farthest points in their boundaries. The example on the right shows a failure of this simple definition.

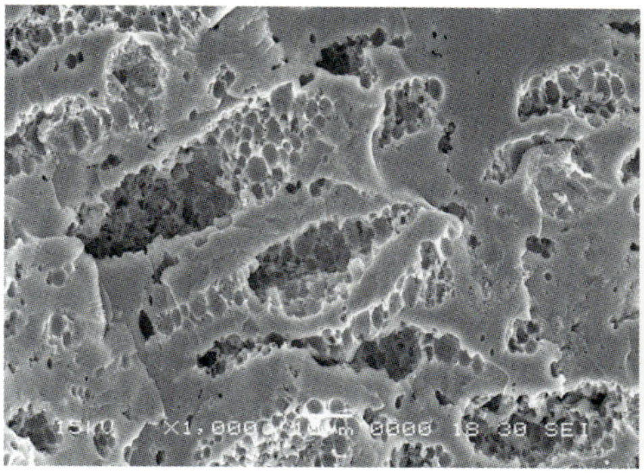

(a) SEM scan.

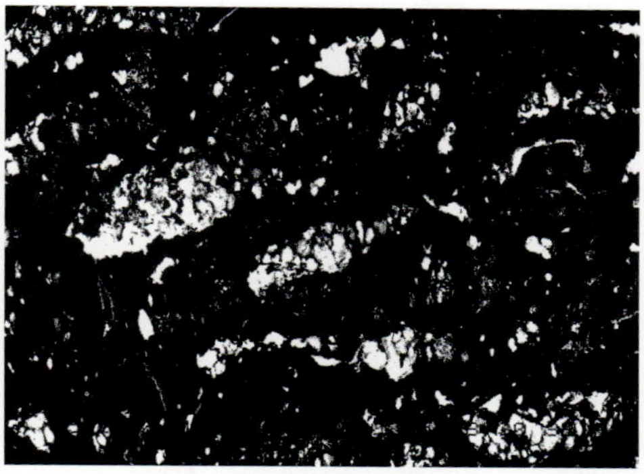

(b) Thresholding with high threshold.

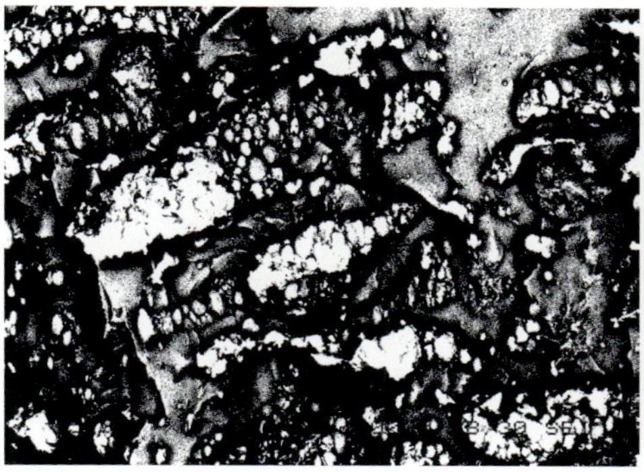

(c) Thresholding with low threshold.

(d) A more sophisticated thresholding algorithm tailored to SEM imagery (see [17]).

Figure 5.4 A scanning electron microscope (SEM) image of Ragusano cheese.

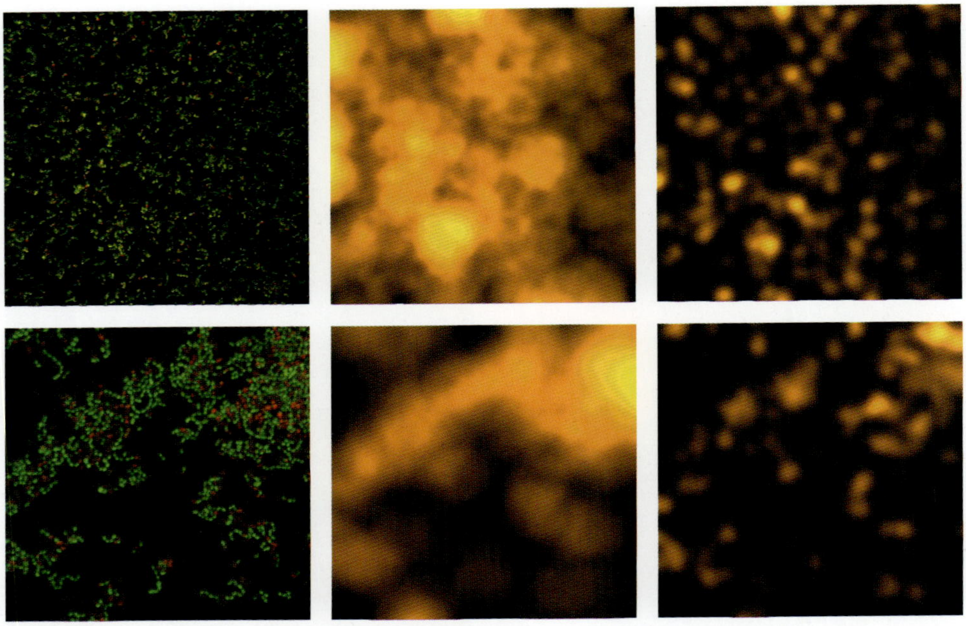

Figure 5.10 Confocal laser scanning micrographs of Figure 5.2. In the right and middle columns are shown, respectively, visual representations of a measure for density (right) and a different measure (middle) not trying to reproduce density (see text).

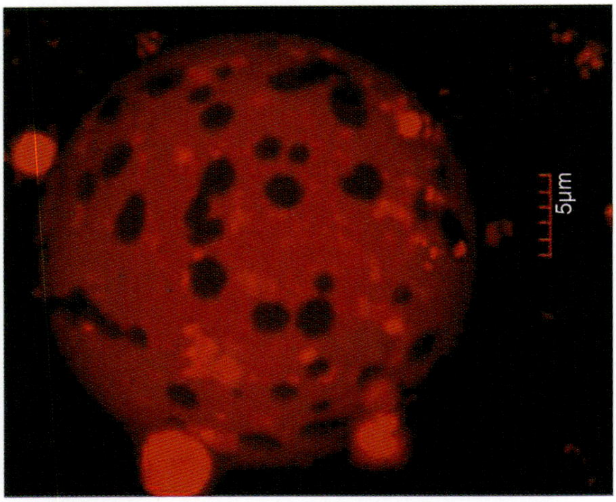

Figure 6.1 Confocal laser scanning micrographs of a milk fat globule in raw cream from bovine milk, showing red liquid-disordered and dark liquid-ordered phases. From Gallier *et al.* (2010), reprinted with permission from the *Journal of Agricultural and Food Chemistry*, copyright 2010 American Chemical Society.

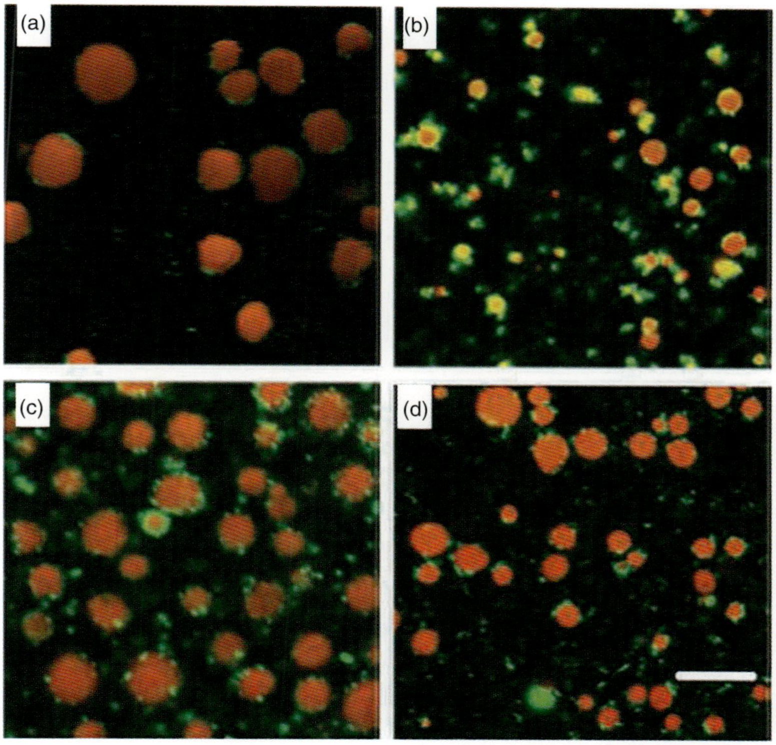

Figure 6.2 Confocal laser scanning micrographs of (a) raw bovine milk; (b) pasteurized and homogenized milk; (c) raw ultrafiltered retentate; and (d) pasteurized and standardized cheesemilk stained so that fat appears as red and protein as green. Samples were diluted 50 times in simulated milk ultrafiltrate solution. Bar at lower right, 10 μm. Adapted from Ong *et al*. (2010), reprinted with permission from the *Journal of Food Science*, copyright 2010 Wiley.

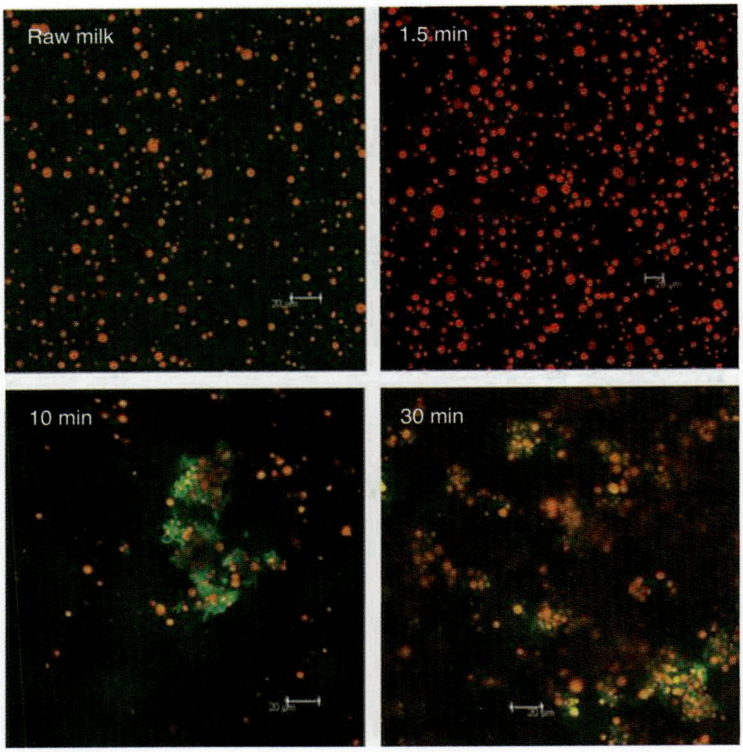

Figure 6.16 Confocal laser scanning micrographs of raw bovine milk 0, 1.5, 10, and 30 min after addition of simulated gastric fluid. Fat appears as red/orange and protein as green. Bars = 20 μm. From Ye *et al.* (2011). Reprinted from *Journal of Dairy Science*, copyright 2011, with permission from Elsevier.

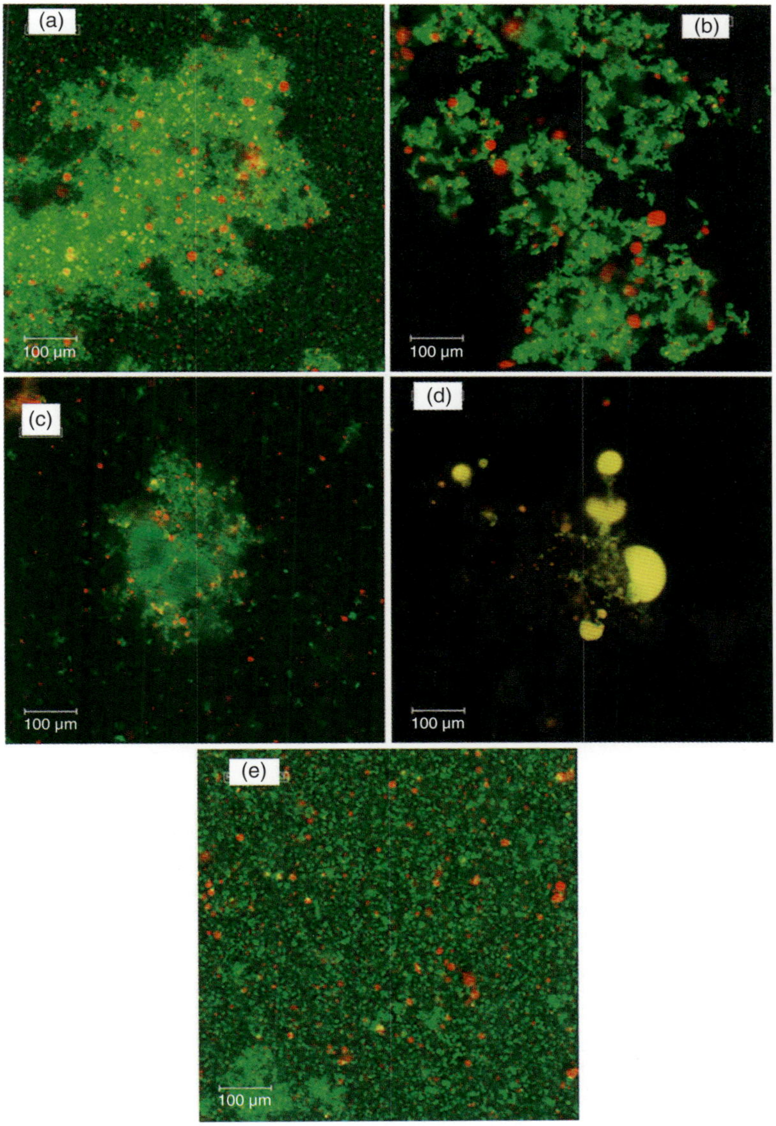

Figure 6.17 Confocal laser scanning micrographs of homogenized bovine milk during simulated digestion. (a) Raw milk before digestion; (b) 60 min after addition of simulated gastric fluid; (c) immediately after addition of simulated intestinal fluid; (d) 120 min later; (e) homogenized and ultra-high temperature pasteurized milk before digestion. Fat appears as red/orange/yellow and protein as green. Bars, 10 μm.

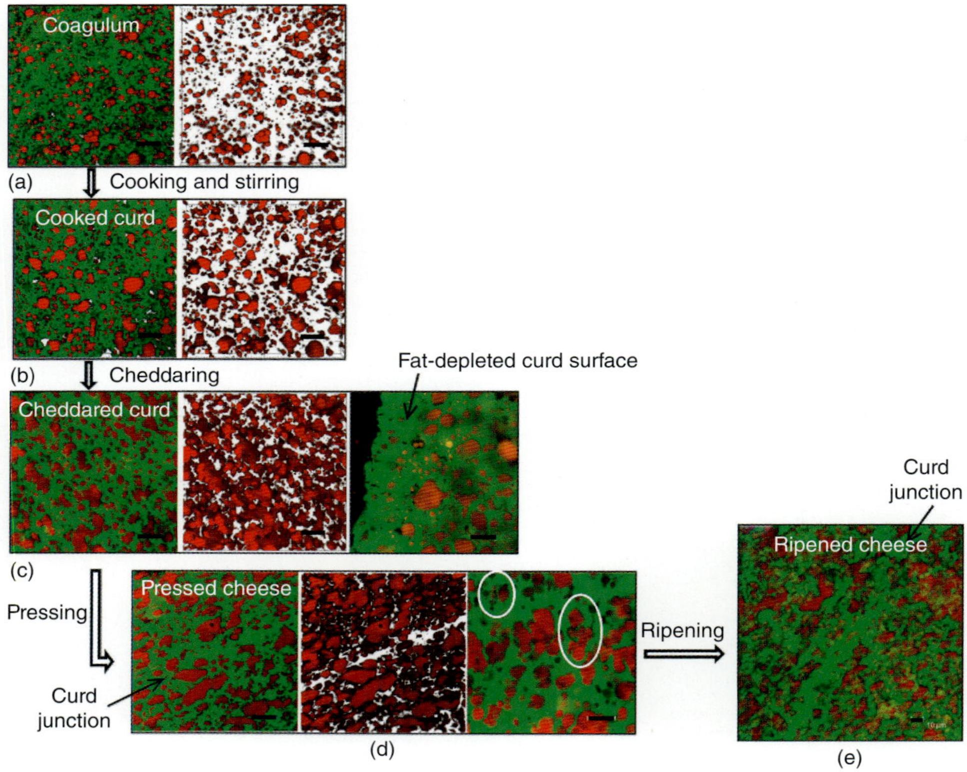

Figure 7.1 Development of microstructure in the initial rennet coagulum through to Cheddar cheese observed using confocal laser scanning microscopy (CSLM). (a) Rennet coagulum; (b) cooked curd; (c) cheddared curd; (d) pressed cheese; and (e) aged Cheddar cheese. Nile red-stained fat appears *red*, fast green FCF-stained protein appears *green*, and the round spots highlighted by the circle in the pressed cheese indicate bacteria. The scale bar is 10 μm in length. Ong *et al.* (2017): Copyright © 2017 by Elsevier Ltd. Reprinted/Reproduced with permission.

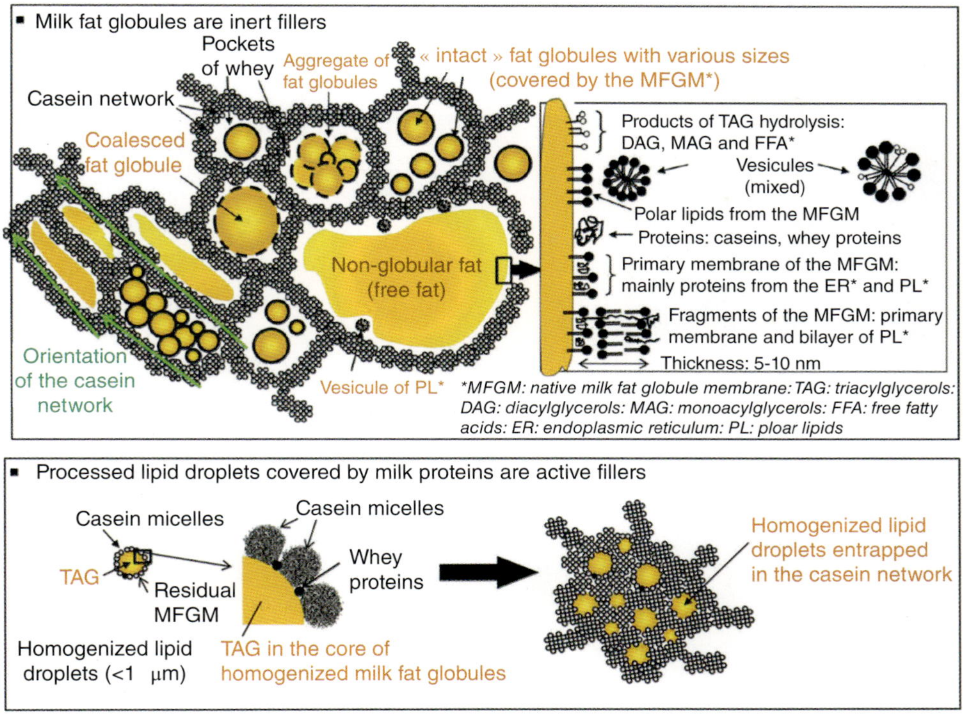

Figure 7.2 Schematic representation of the organization of milk fat in a casein matrix (e.g. cheeses). Lopez *et al.* (2015): Copyright © 2015 by Springer. Reprinted/Reproduced with permission.

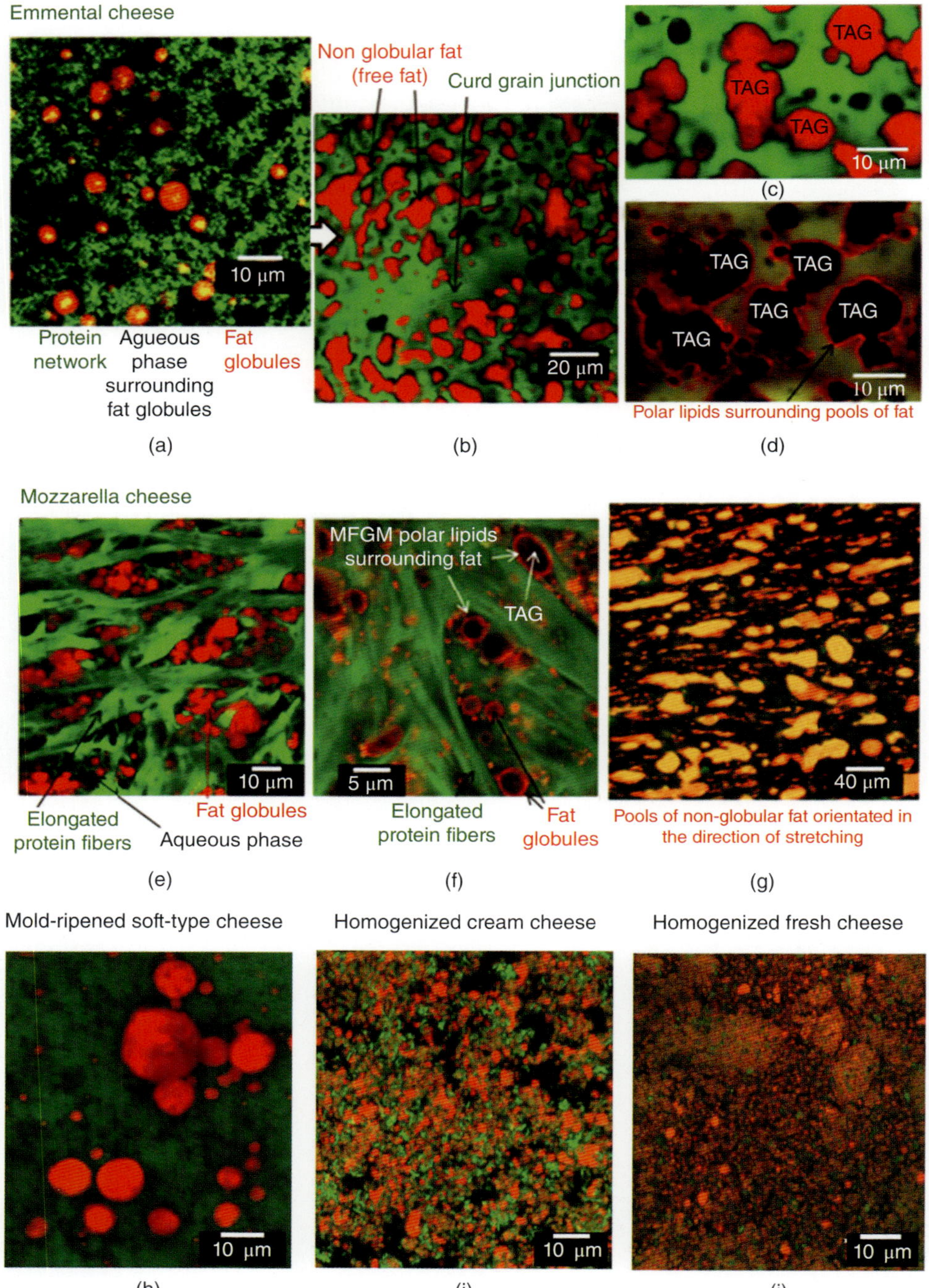
Emmental cheese
10 μm
Protein network
Agueous phase surrounding fat globules
Fat globules
(a)
Non globular fat (free fat)
Curd grain junction
20 μm
(b)
TAG
TAG
TAG
10 μm
(c)
TAG
TAG
TAG
TAG
TAG
10 μm
Polar lipids surrounding pools of fat
(d)
Mozzarella cheese
10 μm
Elongated protein fibers
Aqueous phase
Fat globules
(e)
MFGM polar lipids surrounding fat
TAG
5 μm
Elongated protein fibers
Fat globules
(f)
40 μm
Pools of non-globular fat orientated in the direction of stretching
(g)
Mold-ripened soft-type cheese
10 μm
(h)
Homogenized cream cheese
10 μm
(i)
Homogenized fresh cheese
10 μm
(j)

Figure 7.3 Microstructure of cheeses showing the effect of dairy technological processes on the organization of lipids. Confocal laser scanning microscopy (CLSM) images of (a–d) Emmental cheeses: (a) after coagulation, (b) after pressing of the curd grains; (c), (d) in the final product; (e), (f) high-moisture Mozzarella cheese; (g) low-moisture Mozzarella cheese; (h) mold-ripened soft-type cheese; (i) homogenized cream cheese; (j) homogenized fresh cheese. In all the images, proteins are labeled by using fast green FCF fluorescent dye (green color). Images a–c, e, g, h–j: Total fat is labeled using Nile red (red color). Images d and f: Phospholipids from the milk fat globule membrane are labeled using Rhodamine-DOPE fluorescent dye (red color). Scales bars are in the images. Lopez *et al.* (2015): Copyright © 2015 by Springer. Reprinted/Reproduced with permission.

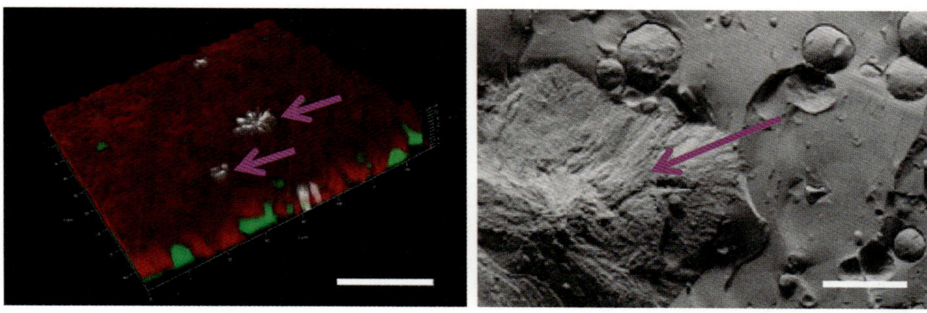

(a) (b)

Figure 7.4 Calcium lactate crystals inside the Cheddar cheese matrix. (a) (red) two-photon fluorescence of protein matrix stained with Nile blue; (green) two-photon fluorescence of fat stained with Nile blue; (grey) SHG signal from the calcium lactate crystals (purple arrows); scale bar 50 μm. (b) Freeze-fracture cryo-SEM image showing the details of fat droplets in protein matrix and a single calcium lactate crystal (purple arrow); scale bar 5 μm. Burdikova *et al.* (2015): Copyright © 2015 by Springer. Reprinted/Reproduced with permission.

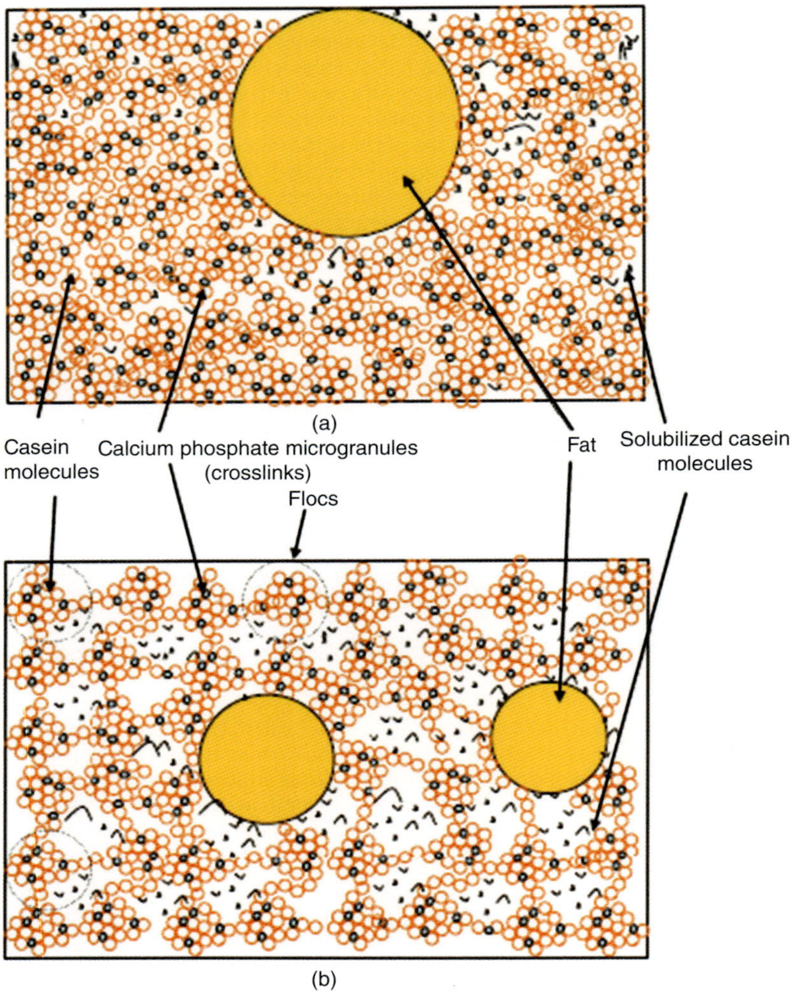

Figure 7.5 Schematic microstructure of (a) natural cheese and (b) processed cheese. Kapoor and Metzger (2008): Copyright © 2008 by Wiley. Reprinted/Reproduced with permission.

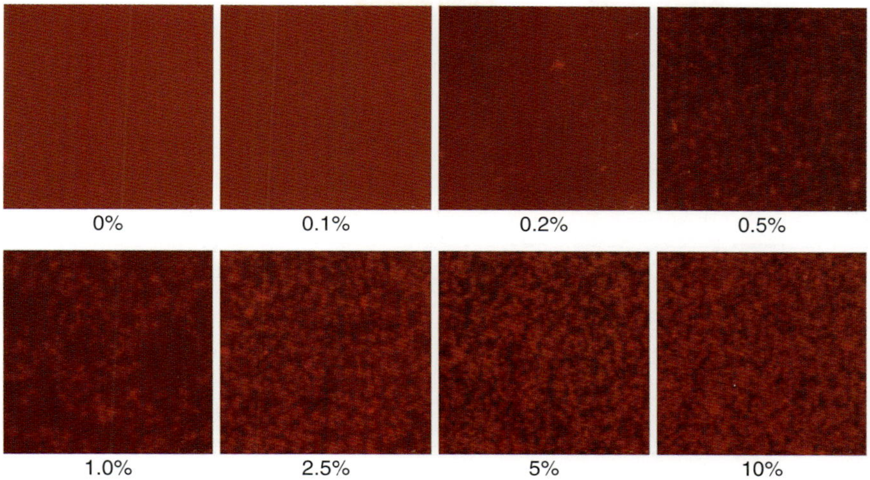

Figure 8.5 Confocal laser scanning microscopy images of 0–10% (w/w) amylomaltase-treated starch in yoghurt serum after 1 d of storage at 5°C. Image size is 62 μm x 62 μm. Structures in bright red are amylomaltase-treated starch domains. Adapted from Kanning *et al.* (2012).

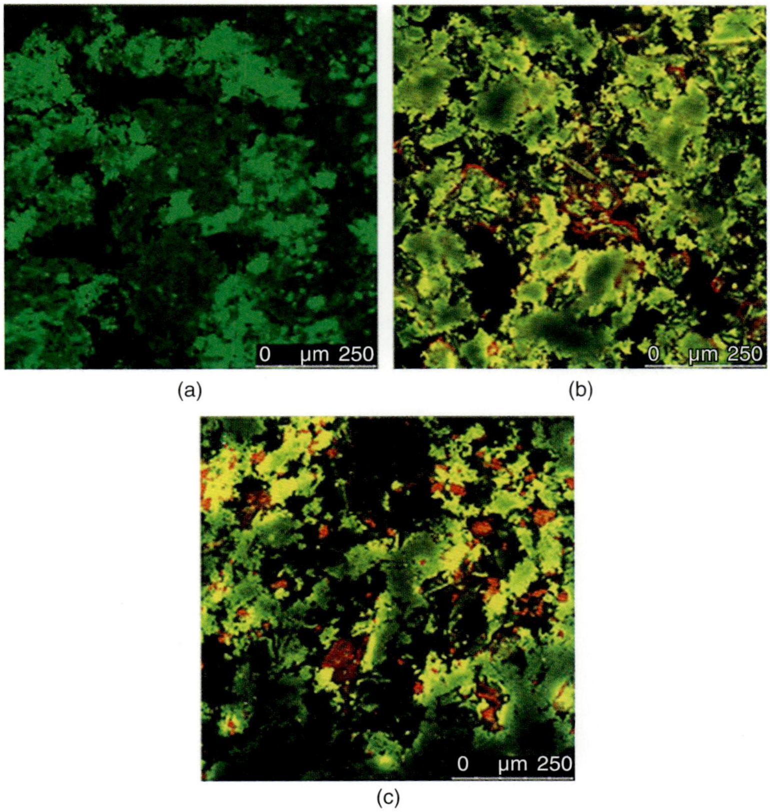

Figure 8.7 CLSM images of the structure of milk–CWP gels containing: (a) 0%; (b) 0.5% and (c) 1.5% rehydrated carrot cell wall fiber particles ($d_{0.5}$ = 71 μm, dry). Adapted from McCann *et al.* (2011).

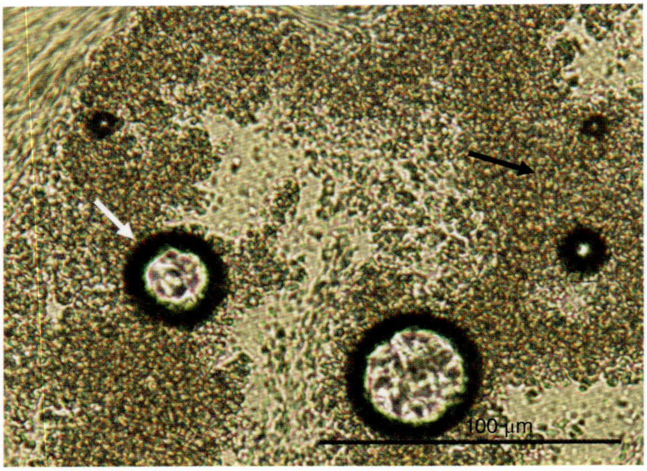

Figure 10.7 Air cells (white arrow) and destabilized fat (black arrow), as viewed under polarized light microscopy. Scale bar = 100 μm.

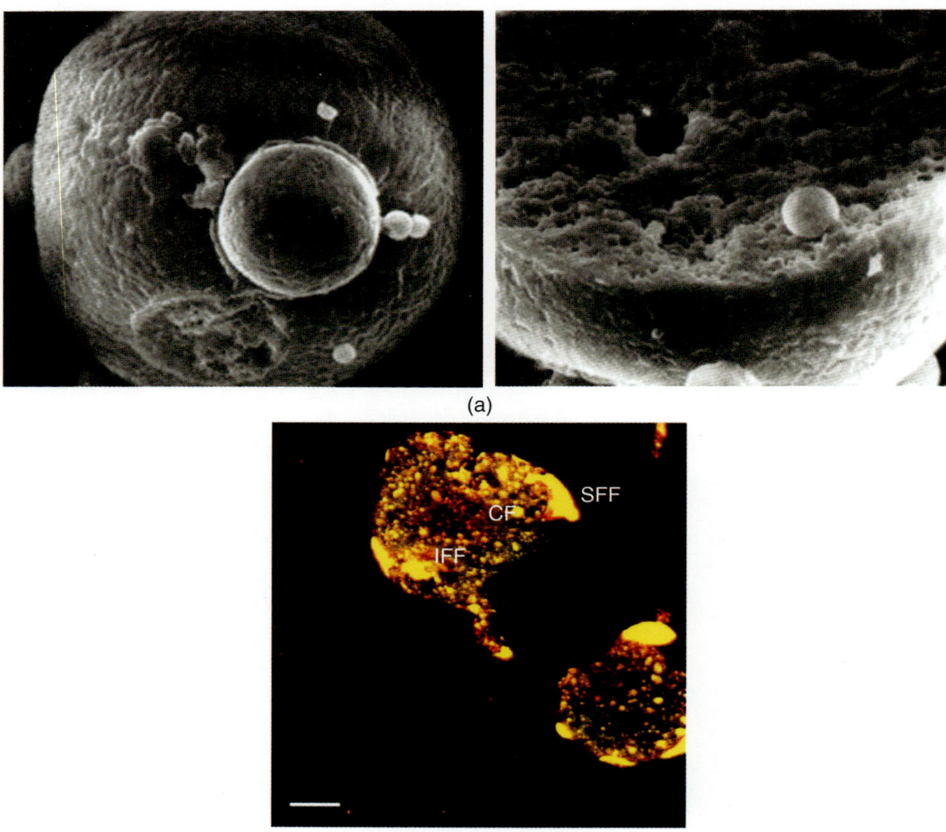

Figure 11.9 (a) SEM and (b) Confocal scanning laser microscopy images of whole milk powder (WMP). CF: Coalesced fat; IFF: Inner free fat; SFF: Surface free fat.

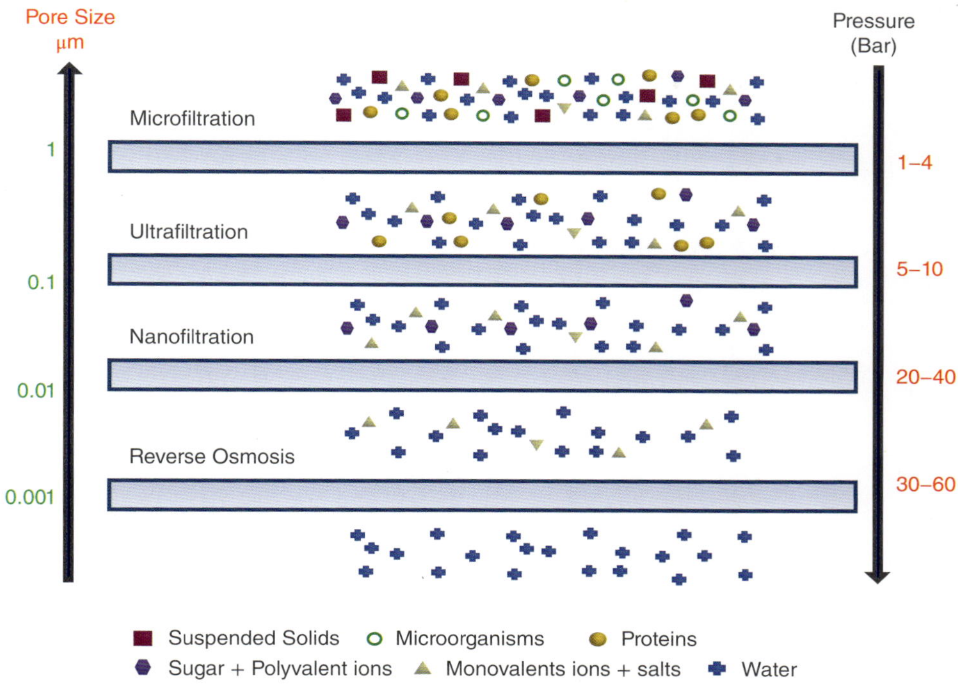

Figure 11.10 Membrane Filtration (adapted from www.nottingham.ac.uk/etc/sol_m_membranefiltration.php).

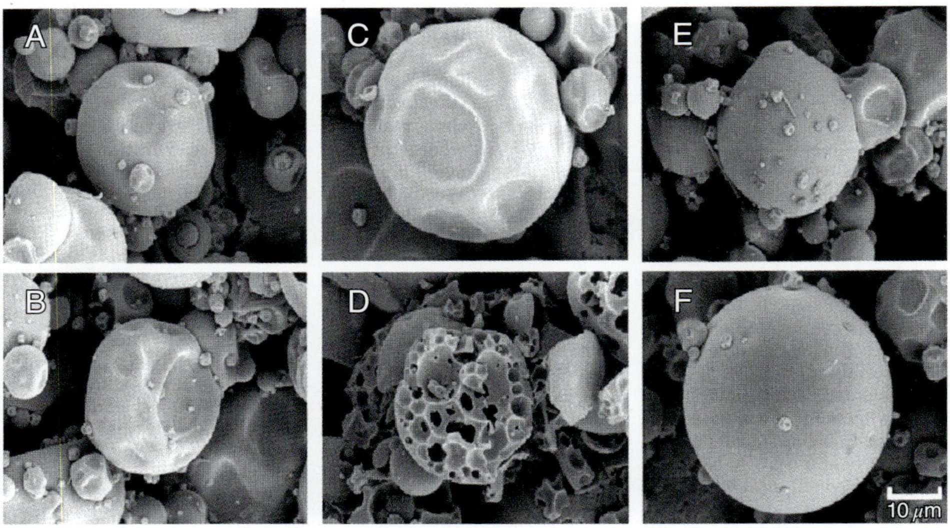

(a)

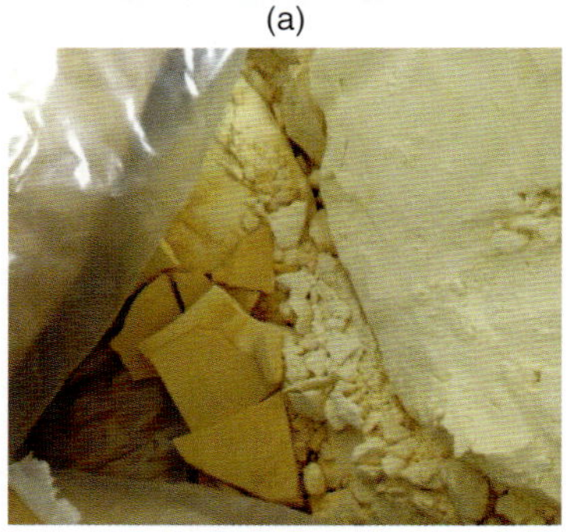

(b)

Figure 11.14 (a) SEM images of WPC powders obtained under different spray drying conditions [53]; (b) Brown chip formation of WPC 35 powders stored at 90% RH conditions [54].